Introduction to Mathcad® 15

Introduction to Mathcad® 15

RONALD W. LARSEN
Montana State University

Prentice Hall
Upper Saddle River Boston Columbus San Francisco New York Indianapolis London
Toronto Sydney Singapore Tokyo Montreal Dubai Madrid Hong Kong Mexico City
Munich Paris Amsterdam Cape Town

Vice President and Editorial Director, ECS: Marcia J. Horton
Executive Editor: Holly Stark
Editorial Assistant: Keri Rand
Vice-President, Production: Vince O'Brien
Director of Marketing: Margaret Waples
Executive Marketing Manager: Tim Galligan
Marketing Assistant: Mack Patterson
Senior Managing Editor: Scott Disanno
Production Manager: Wanda Rockwell
Composition: Aptara® Inc.
Full-Service Project Management: Sudip Sinha/Aptara® Inc.
Printer/Binder: RR Donnelley
Typeface: 10/12 Times Ten

Pearson Education Ltd., London
Pearson Education Singapore, Pte. Ltd
Pearson Education Canada, Inc.
Pearson Education—Japan
Pearson Education Australia PTY, Limited
Pearson Education North Asia, Ltd., Hong Kong
Pearson Educación de Mexico, S.A. de C.V.
Pearson Education Malaysia, Pte. Ltd.
Pearson Education, Inc., Upper Saddle River, New Jersey

CIP to come

10 9 8 7 6

Prentice Hall
is an imprint of

www.pearsonhighered.com

ISBN-10: 0-13-602513-7
ISBN-13: 978-0-13-602513-9

Contents

ESource Reviewers

We would like to thank everyone who helped us with or has reviewed texts in this series.

Naeem Abdurrahman, *University of Texas–Austin*
Sharon Ahlers, *Cornell University*
David G. Alciatore, *Colorado State University*
Stephen Allan, *Utah State University*
Anil Bajaj, *Purdue University*
Grant Baker, *University of Alaska–Anchorage*
William Bard, *University of Texas*
William Beckwith, *Clemson University*
Haym Benaroya, *Rutgers University*
John Biddle, *California State Polytechnic University*
Ray Biswajit, *Bloomsburg University of PA*
Donald Blackmon, *University of North Carolina–Charlotte*
Tom Bledsaw, *ITT Technical Institute*
Fred Boadu, *Duke University*
Gregory Boardman, *Virginia Tech*
Stuart Brand, *The Ohio State University*
Jerald Brevick, *The Ohio State University*
Tom Bryson, *University of Missouri–Rolla*
Ramzi Bualuan, *University of Notre Dame*
Dan Budny, *Purdue University*
Betty Burr, *University of Houston*
Fernando Cadena, *New Mexico State University*
Joel Cahoon, *Montana State University*
Dale Calkins, *University of Washington*
Monica Cardella, *Purdue University*
Linda Chattin, *Arizona State University*
Harish Cherukuri, *University of North Carolina–Charlotte*
Vanessa Clark, *Washington University in St. Louis*
Arthur Clausing, *University of Illinois*
Barry Crittendon, *Virginia Polytechnic and State University*
Donald Dabdub, *University of California–Irvine*
Richard Davis, *University of Minnesota–Duluth*
Kurt DeGoede, *Elizabethtown College*
John Demel, *Ohio State University*
James Devine, *University of South Florida*
Heidi A. Diefes-Dux, *Purdue University*
Jeffrey A. Doughty, *Northeastern University*
Jerry Dunn, *Texas Tech University*
Ron Eaglin, *University of Central Florida*
Dale Elifrits, *University of Missouri–Rolla*
Timothy Ellis, *Iowa State University*
Nurgun Erdol, *Florida Atlantic University*
Christopher Fields, *Drexel University*

Patrick Fitzhorn, *Colorado State University*
Julie Dyke Ford, *New Mexico Tech*
Susan Freeman, *Northeastern University*
Howard M. Fulmer, *Villanova University*
Frank Gerlitz, *Washtenaw Community College*
John Glover, *University of Houston*
Richard Gonzales, *Purdue University–Calumet*
John Graham, *University of North Carolina–Charlotte*
Hayden Griffin, *Virginia Tech*
Laura Grossenbacher, *University of Wisconsin–Madison*
Ashish Gupta, *SUNY at Buffalo*
Otto Gygax, *Oregon State University*
Malcom Heimer, *Florida International University*
Robin A. M. Hensel, *West Virginia University*
Donald Herling, *Oregon State University*
Orlando Hernandez, *The College of New Jersey*
David Herrin, *University of Kentucky*
Thomas Hill, *SUNY at Buffalo*
A. S. Hodel, *Auburn University*
Susan L. Holl, *California State University–Sacramento*
Kathryn Holliday-Darr, *Penn State, The Behrend College–Erie*
Tom Horton, *University of Virginia*
David Icove, *University of Tennessee*
James N. Jensen, *SUNY at Buffalo*
Mary Johnson, *Texas A&M Commerce*
Vern Johnson, *University of Arizona*
Jean C. Malzahn Kampe, *Virginia Polytechnic Institute and State University*
Moses Karakouzian, *University of Nevada Las Vegas*
Autar Kaw, *University of South Florida*
Kathleen Kitto, *Western Washington University*
Kenneth Klika, *University of Akron*
Harold Knickle, *University of Rhode Island*
Terry L. Kohutek, *Texas A&M University*
Thomas Koon, *Binghamton University*
Reza Langari, *Texas A&M University*
Bill Leahy, *Georgia Institute of Technology*
John Lumkes, *Purdue University*
Mary C. Lynch, *University of Florida*
Melvin J. Maron, *University of Louisville*
Christopher McDaniel, *University of North Carolina–Charlotte*
Khanjan Mehta, *Penn State–University Park*
F. Scott Miller, *University of Missouri–Rolla*
James Mitchell, *Drexel University*
Robert Montgomery, *Purdue University*
Naji Mounsef, *Arizona State University*
Nikos Mourtos, *San Jose State University*
Mark Nagurka, *Marquette University*
Romarathnam Narasimhan, *University of Miami*
Shahnam Navee, *Georgia Southern University*
James D. Nelson, *Louisiana Tech University*

Soronadi Nnaji, *Florida A&M University*
Sheila O'Connor, *Wichita State University*
Matt Ohland, *Clemson University*
Paily P. Paily, *Tennessee State University*
Kevin Passino, *Ohio State University*
Ted Pawlicki, *University of Rochester*
Ernesto Penado, *Northern Arizona University*
Michael Peshkin, *Northwestern University*
Ralph Pike, *Louisiana State University*
Andrew Randall, *University of Central Florida*
Dr. John Ray, *University of Memphis*
Marcella Reekie, *Kansas State University*
Stanley Reeves, *Auburn University*
Larry Richards, *University of Virginia*
Marc H. Richman, *Brown University*
Jeffrey Ringenberg, *University of Michigan*
Paul Ronney, *University of Southern California*
Christopher Rowe, *Vanderbilt University*
Blair Rowley, *Wright State University*
Liz Rozell, *Bakersfield College*
Mohammad Saed, *Texas Tech University*
Tabb Schreder, *University of Toledo*
Heshem Shaalem, *Georgia Southern University*
Randy Shih, *Oregon Institute of Technology*
Howard Silver, *Fairleigh Dickenson University*
Avi Singhal, *Arizona State University*
Greg Sun, *University of Massachusetts–Boston*
John Sustersic, *Penn State–University Park*
Tim Sykes, *Houston Community College*
Murat Tanyel, *Geneva College*
Toby Teorey, *University of Michigan*
Scott Thomas, *Wright State University*
Virgil A. Thomason, *University of TN at Chattanooga*
Neil R. Thompson, *University of Waterloo*
Dennis Truax, *Mississippi State University*
Raman Menon Unnikrishnan, *Rochester Institute of Technology*
Thomas Walker, *Virginia Tech*
Michael S. Wells, *Tennessee Tech University*
Ed Wheeler, *University of Tennessee at Martin*
Joseph Wujek, *University of California–Berkeley*
Edward Young, *University of South Carolina*
Garry Young, *Oklahoma State University*
Steve Yurgartis, *Clarkson University*
Mandochehr Zoghi, *University of Dayton*

Introduction to Mathcad® 15

CHAPTER 1

Mathcad: The Engineer's Scratch Pad

Objectives

After reading this chapter, you will

- begin to see how Mathcad solves problems
- understand how Mathcad can assist the engineering design process
- see how Mathcad can be used to solve math problems
- learn how Mathcad handles unit conversions
- see how Mathcad can help you present your results to others
- know what to expect from the text
- become familiar with some nomenclature conventions that are used throughout the text

1.1 INTRODUCTION TO MATHCAD

Mathcad[1] is an equation-solving software package that has proven to have a wide range of applicability to engineering problems. Mathcad's ability to display ***equations*** the same way you would write them on paper makes a Mathcad ***worksheet*** easy to read. For example, if you wanted to calculate the mass of water in a storage tank, you might solve the problem on paper as shown in Figure 1.1.

The same calculation in Mathcad is shown in Figure 1.2.

One of the nicest features of Mathcad is its ability to solve problems much the same way people do, rather than making your solution process fit the program's way of doing things. In contrast, you could also solve this problem in a spreadsheet by entering the constants and the equations for volume and mass into various cells (Figure 1.3).

In the C programming language, the problem might be solved with the program shown in Figure 1.4.

While spreadsheets and programming languages can produce the solution, Mathcad's presentation is much more like the way people solve equations on paper. This makes Mathcad easier for you to use. It also makes it easier for others to read and understand your results.

Mathcad's user interface is an important feature, but Mathcad has other features that make it excel as a ***design*** tool, a mathematical problem solver, a unit converter, and a communicator of results.

1.2 MATHCAD AS A DESIGN TOOL

A Mathcad ***worksheet*** is a collection of ***variable definitions***, equations, ***text regions***, and ***graphs*** displayed on the screen in pretty much the same fashion

[1]Mathcad is a registered trademark of the Parametric Technology Corporation of Needham, Massachusetts.

you would write them on paper. A big difference between a Mathcad worksheet and your paper scratch pad is ***automatic recalculation***: If you make a change to any of the definitions or equations in your worksheet, the rest of the worksheet is automatically updated. This makes it easy to do the "what if" calculations that are so common in engineering. For example, what if the water level rises to 2.8 m? Would the mass in the tank exceed the tank's maximum design value of 40,000 kg?

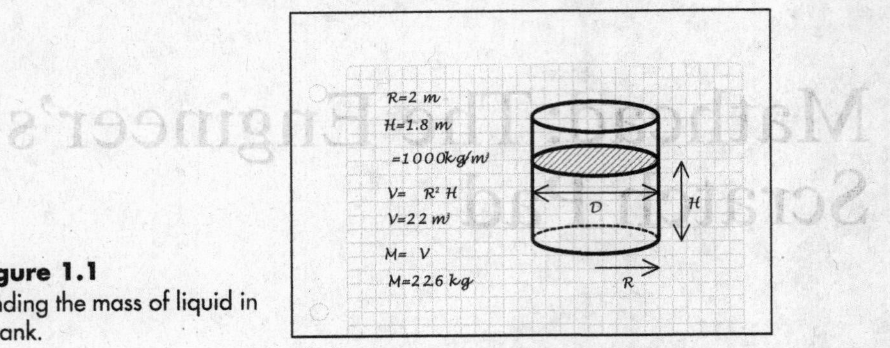

Figure 1.1
Finding the mass of liquid in a tank.

Mathcad - [Tank Volume.xmcd]

File Edit View Insert Format Tools Symbolics Window Help

Normal Arial 10

$R := 2 \cdot m$

$H := 1.8 \cdot m$

$\rho := 1000 \cdot \frac{kg}{m^3}$

$V := \pi \cdot R^2 \cdot H$

$V = 22.619\,m^3$

$M := \rho \cdot V$

$M = 2.262 \times 10^4\,kg$

Press F1 for help. AUTO NUM Page 1

Figure 1.2
The solution in Mathcad.

	A	B	C
1	R:	2	
2	H:	1.8	
3	Rho:	1000	
4			
5	V:	22.6	
6	M:	22619	
7			

Figure 1.3
The solution in a spreadsheet.

```
#include stdio.h
#include math.h

main()
{
      float R, H, Rho, V, M;

      R = 2;
      H = 1.8;
      V = 3.1416 * pow(R,2) * H;
      M = Rho * V;
      printf("V = %f \n M = %f", V,M);
}
```

Figure 1.4
The solution in the C programming language.

Professional Success

Work to develop communication skills as well as technical skills.

An engineer's job is to find solutions to technical problems. Finding a solution requires good technical skills, but the solution must always be communicated to other people. An engineer's communication skills are just as important as her or his technical skills.

Because Mathcad's worksheets are easy to read, they can help you communicate your results to others. You can improve the readability of your worksheets by

- performing your calculations in an orderly way (plan your work),
- adding comments to your worksheet, and
- using units on your variables.

To answer these questions, simply edit the definition of H in the Mathcad worksheet as shown in Figure 1.5. The rest of the equations are automatically updated, and the new result is displayed.

From this calculation, we see that even at a height of 2.8 m, the mass in the tank is still within the design specifications.

The ability to develop a worksheet for a particular case and then vary one or more parameters to observe their impact on the calculated results makes a Mathcad worksheet a valuable tool for evaluating multiple designs.

$R := 2 \cdot m$

$H := 2.8 \cdot m$ << New height

$\rho := 1000 \cdot \frac{kg}{m^3}$

$V := \pi \cdot R^2 \cdot H$

$V = 35.186\ m^3$

$M := \rho \cdot V$

$M = 3.519 \times 10^4\ kg$ << Still below design limit of 40,000 kg

Figure 1.5
The updated solution, with comments.

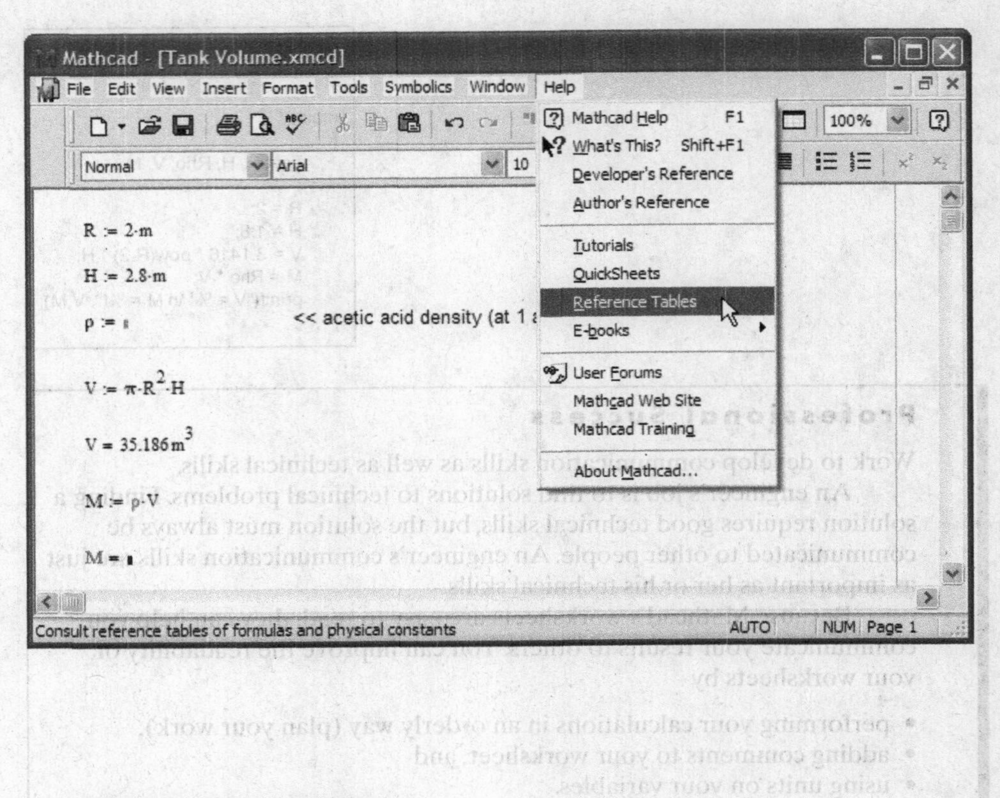

Figure 1.6
Accessing the Mathcad Reference Tables.

1.2.1 Mathcad Reference Tables

Mathcad also provides access to a list of ***reference tables*** that can provide basic information for engineering problems. For example, what if the liquid in the tank was acetic acid rather than water? The density of acetic acid is slightly higher than the density of water, and the value is available in Mathcad's reference tables. To access the reference tables, choose **Reference Tables** from the **Help** menu, as shown in Figure 1.6.

To find the density of acetic acid, select the **Density** option under the **Liquid Properties** section. (See Figure 1.7.)

The densities of a variety of liquids are provided, although only at 1 atm and 300 K (slightly above room temperature). The density value for acetic acid, shown in Figure 1.8, can be copied directly from the reference table and pasted into the Mathcad worksheet.

Once the new density value is inserted into the Mathcad worksheet, the new liquid mass in the tank is automatically calculated, as shown in Figure 1.9.

1.2.2 QuickPlots

Mathcad has another feature, called a ***QuickPlot***, that is very useful for visualizing ***functions***, and this can also speed up the design process. With a QuickPlot, you simply create a graph, put a function on the y-axis, and place a dummy variable on the x-axis. Mathcad evaluates the function for a range of values and displays the graph.

For example, if you wanted to see how the volume of liquid varied with tank radius, you could plot the function for computing the volume ($\pi \cdot RR^2 \cdot H$) against the radius, RR. Mathcad will evaluate the function on the y-axis for a range of RR values

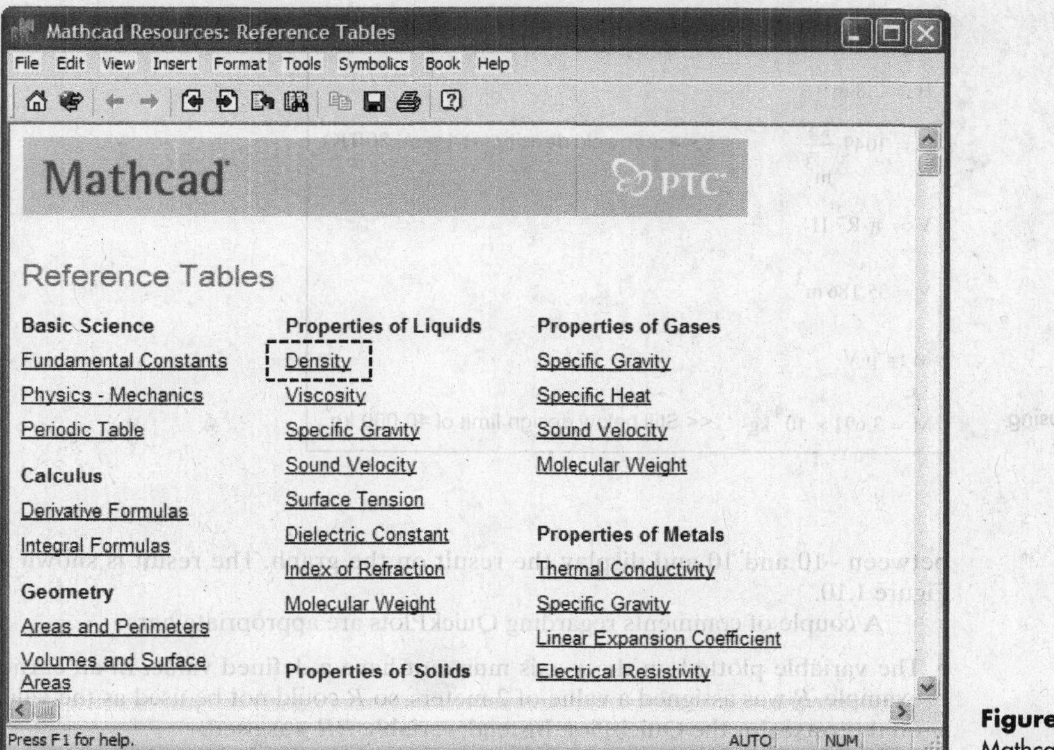

Figure 1.7
Mathcad Reference Tables.

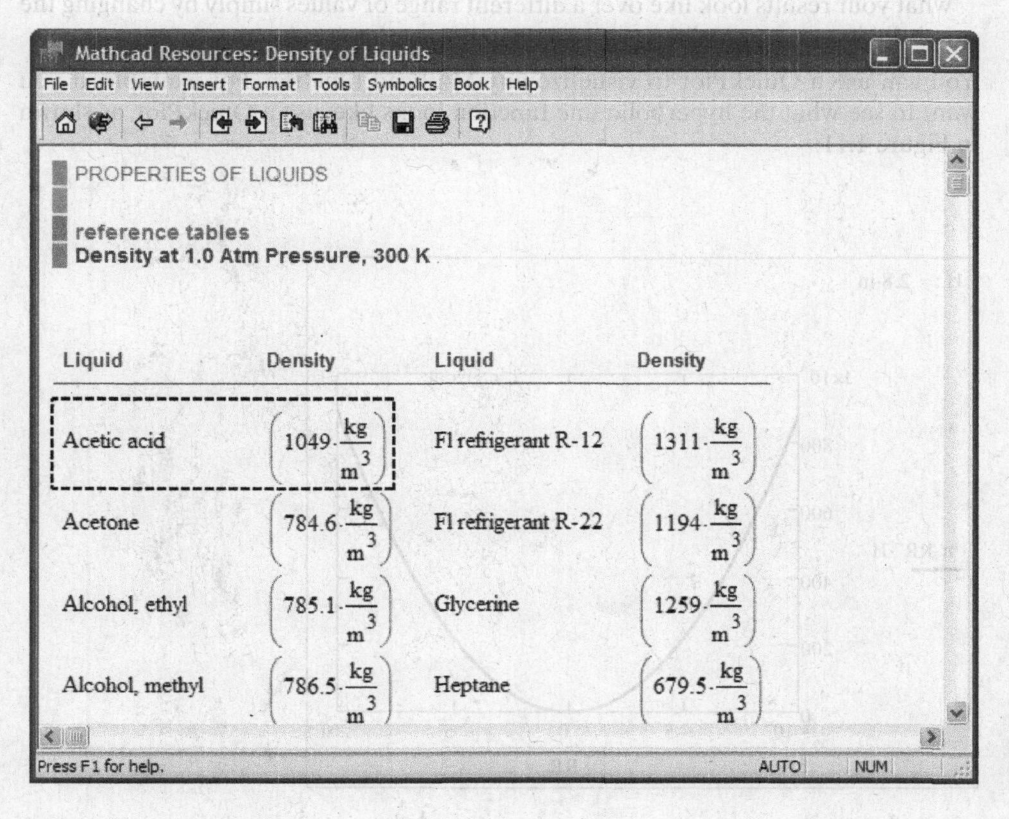

Figure 1.8
Density of acetic acid.

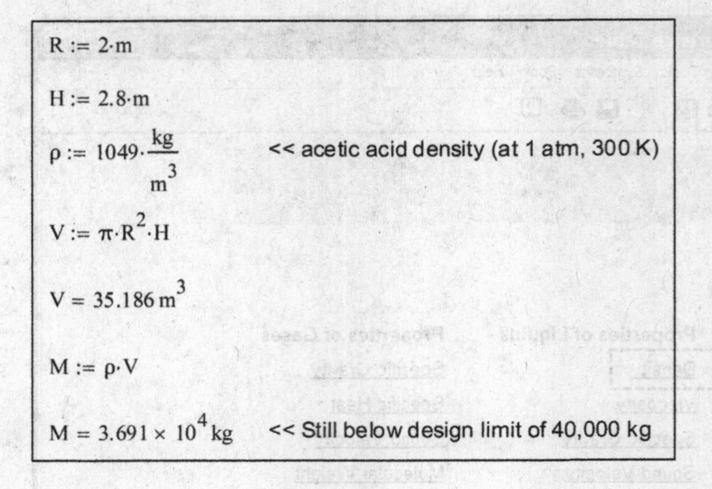

R := 2·m
H := 2.8·m
ρ := 1049·kg/m^3
<< acetic acid density (at 1 atm, 300 K)
V := π·R^2·H
V = 35.186 m^3
M := ρ·V
M = 3.691 × 10^4 kg
<< Still below design limit of 40,000 kg

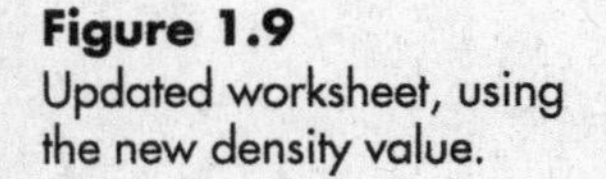

Figure 1.9
Updated worksheet, using the new density value.

between –10 and 10 and display the result on the graph. The result is shown in Figure 1.10.

A couple of comments regarding QuickPlots are appropriate here:

- The variable plotted on the x-axis must not have a defined value. In an earlier example, R was assigned a value of 2 meters, so R could not be used as the value on the x-axis for the QuickPlot. Instead, variable RR was used.
- Mathcad's QuickPlots range from –10 to 10 on the x-axis by default. You can see what your results look like over a different range of values simply by changing the axis limits on the graph.

You can use a QuickPlot to visualize any Mathcad function. For example, if you want to see what the hyperbolic sine function looks like, use a QuickPlot, as shown in Figure 1.11.

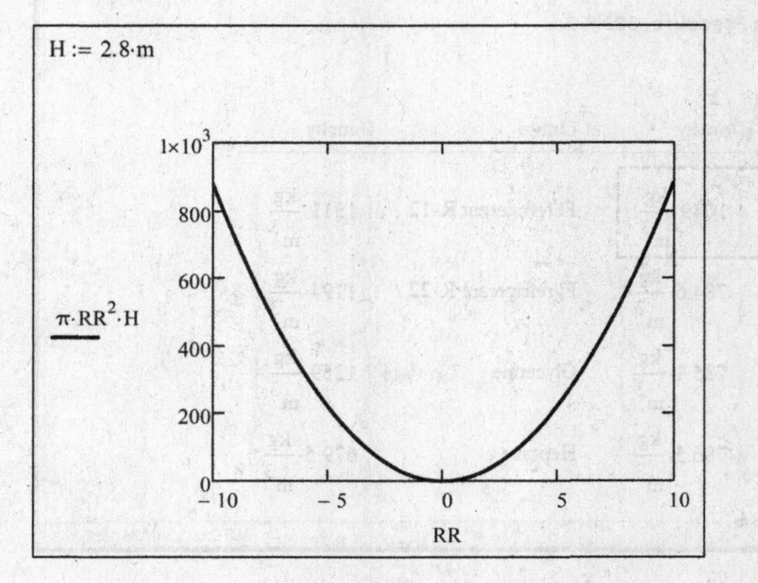

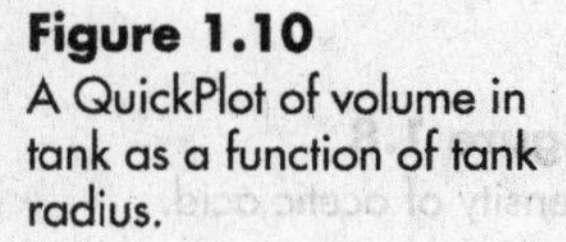

Figure 1.10
A QuickPlot of volume in tank as a function of tank radius.

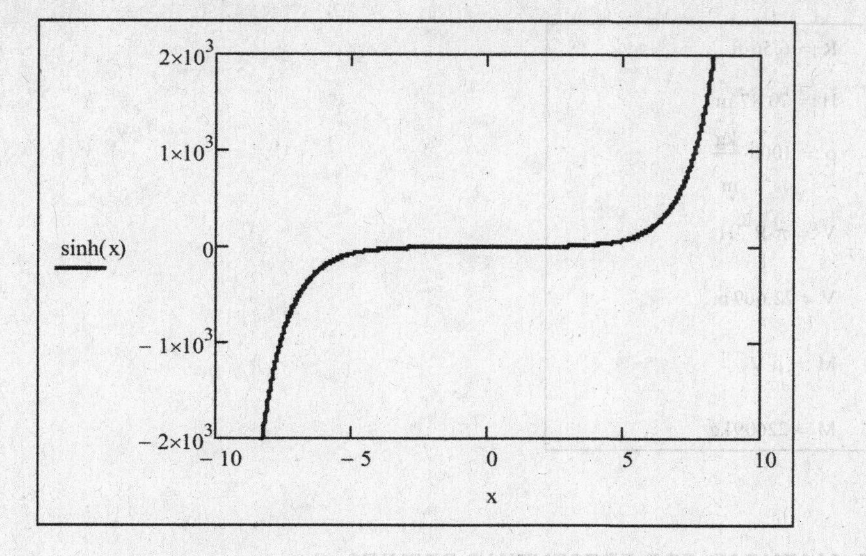

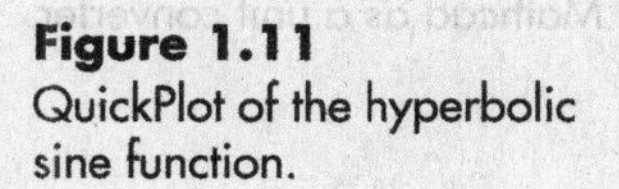
Figure 1.11
QuickPlot of the hyperbolic sine function.

1.3 MATHCAD AS A MATHEMATICAL PROBLEM SOLVER

Mathcad has the ability to compute ***numerical solutions*** (computing a value) or ***symbolic solutions*** (working with the variables directly). It has a large collection of built-in functions for trigonometric calculations, statistical calculations, data analysis (e.g., regression), and matrix operations. Mathcad can calculate derivatives and evaluate integrals, and it can handle many differential equations. It can work with imaginary numbers and handle Laplace transforms. Iterative solutions are tedious by hand, but very straightforward in Mathcad. While it is possible to come up with problems that are beyond Mathcad's capabilities, Mathcad can handle the bulk of an engineer's day-to-day calculations – and do them very well.

Some of these features, such as Laplace transforms and functions for solving differential equations, are beyond the scope of this text, but many of Mathcad's commonly used features and functions will be presented and used.

1.4 MATHCAD AS A UNIT CONVERTER

Mathcad allows you to build ***units*** into most equations. Allowing Mathcad to handle the chore of getting all the values converted to a consistent set of units can be a major time-saver. For example, if the storage tank's radius had been measured in feet and its depth in inches, then

$$R = 6.56 \text{ ft (equivalent to 2 m)}$$

$$H = 70.87 \text{ in (equivalent to 1.8 m)}$$

We could convert the feet and inches to meters ourselves, or we can let Mathcad do the work, as shown in Figure 1.12.

Note: The mass is slightly lower here than in the first example because of rounding of the R and H values.

For complicated problems, and when your input values have many different units, Mathcad's ability to handle the unit conversions is a very nice feature.

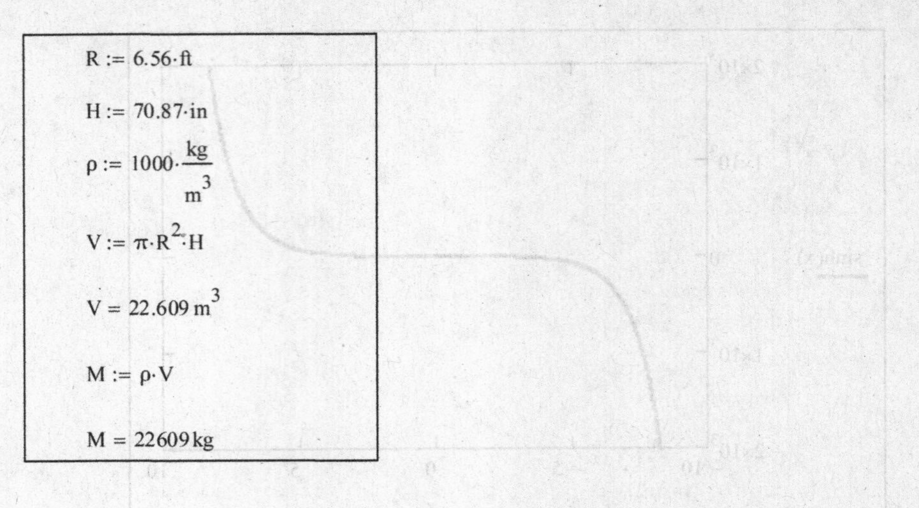

Figure 1.12
Mathcad as a unit converter.

1.5 MATHCAD FOR PRESENTING RESULTS

Engineering and science are both fields in which a person's computed results have little meaning unless they are communicated to someone else. A circuit design has to be passed along to a manufacturer, for example. Getting the results in a useful form can often take as long as computing the results. This is an area where Mathcad can help speed up the process.

Practicing engineers use spreadsheets for many routine calculations, but spreadsheets can be frustrating because they display only the calculated results while hiding the equations. If someone gives you a spreadsheet printout, it probably shows only numbers, and you have to take the person's word for the equations or ask for a copy of the spreadsheet file. If you dig into the spreadsheet to see the equations, they are still somewhat cryptic because they typically use cell references (e.g., B12) rather than variable names.

Computer program listings make it clear how the results were calculated, and recognizable variable names can be used, but long equations on a single line can still be difficult to decipher. Mathcad's ability to show the equations and the results the way people are used to reading them makes a Mathcad worksheet a good way to give your results to someone else. If there is a question about a result, the solution method is obvious. Mathcad's ability to print equations in the order we would write them, but with typeset quality, is a bonus.

Still, there are times when you need to get your results into a more formal report. Mathcad helps out in that area, too. Equations and results (e.g., values, matrices, graphs) on a Mathcad worksheet can be inserted (via copy and paste operations) into other software programs, such as word processors. You don't have to retype the equations in the word processor, which can be a major time-saver. This compatibility with other software also means you can insert a matrix from Mathcad into a spreadsheet or create a Mathcad matrix from a column of values in a spreadsheet.

1.6 MATHCAD'S PLACE IN AN ENGINEER'S TOOL KIT

Spreadsheets, programming languages, and mathematical problem solvers all have their place, and the tools you will use routinely will depend on where your career takes you, or perhaps vice versa. There is no single "right" tool for most problems, but Mathcad seems a logical choice when the requirements of the problem align with Mathcad's strengths, which include

- equations that are displayed in a highly readable form,
- the ability to work with units,
- a symbolic math capability,
- an iterative solution capability, and
- an extensive function library.

Deciding on which software product to use requires an understanding of the various products. For example,

- Spreadsheets can solve equations requiring iteration, but the process is much easier to follow in Mathcad.
- Mathcad can handle lists of numbers (e.g., analyses of experimental data sets), but columns of numbers fit well into the strengths of a spreadsheet.
- Spreadsheets cannot handle symbolic mathematics, so Mathcad (or Maple, or Mathematica)[2] must be used for that type of work.

This text demonstrates some of Mathcad's capabilities; it is my hope that it also will help you learn where the package fits in your engineering tool kit.

1.7 WHAT'S NEW IN MATHCAD 14 AND 15?

Mathcad 14 was the first release of Mathcad by its new owner, the PTC Corporation. The primary change for international users was making Mathcad available in nine languages. Beyond that, new features in Mathcad 14 included:

- A new symbolic math processor with extended support for ordinary differential equation (ODE) solvers.
- More options for formatting XY graphs.
- Inline numeric evaluation.

Inline numeric evaluation of equations will be useful for many people. In the past, you could solve an equation, or you could see the result of the calculation – but not both in the same equation region. Since Mathcad 14 was released, you can include an equal sign at the end of an equation to see the result of the calculation. This makes it easier to check intermediate results as they are calculated. Inline evaluation is illustrated in Figure 1.13.

Prior to Mathcad 14 it took two steps to calculate and display the result:

$Y := \cos(45 \cdot deg)$

$Y = 0.7071$

Now you can calculate a result and display the result on a single line:

$Y := \cos(45 \cdot deg) = 0.7071$

Figure 1.13
Inline numeric evaluation in Mathcad 14.

[2] Maple is a product of Maplesoft, Inc., Ontario, Canada. Mathematica is produced by Wolfram Research, Inc., Champaign, Illinois.

Mathcad 15 has some new features as well:

- Support for Excel 2007.
- New functions to assist in the design of experiments (DOE) – not covered here.

Since Excel 2007 was released after Mathcad 14, Excel's new file format was not supported by Mathcad – until Mathcad 15. With Mathcad 15, it is now possible to save Excel files with the .xlsx file format and work with those files in Mathcad.

1.8 CONVENTIONS USED

Special typefaces are used to make software and math features easier to identify in this text:

Stdev	Function names and variable names in text are shown in a bold font.
[Ctrl-6]	Keystrokes are shown in brackets. This example indicates that the control key and the 6 key should be pressed simultaneously.
File/Save As	Menu selections are listed with the main menu item and submenu items separated by "/".
A: = π·r2	Mathcad examples in text (not figures) are shown in a different font and are indented.
Key term	Key terms are shown in italics the first time they are used.

KEY TERMS

automatic recalculation	inline numeric evaluation	text regions
design	numerical solutions	unit conversion
equations	QuickPlot	variable definitions
functions	reference tables	worksheet
graphs	symbolic solutions	

PROBLEMS

1.1 Use the Mathcad Reference Tables to find the following physical properties at 1 atm, 300 K:

- Density of water
- Density of seawater
- Viscosity of water
- Viscosity of kerosene
- Surface tension of water
- Surface tension of acetone

1.2 Use the Mathcad Reference Tables to answer the following questions:

- Which metal has the higher thermal conductivity, copper or aluminum?
- Which metal has the higher linear expansion coefficient, copper or iron?
- Which metal has the lower modulus of elasticity, gold or silver?
- Which metal has the lower melting point, lead or tin?

1.3 The Mathcad Reference Tables list values at 300 K (27°C, 80°F). When you grab a physical property value from the table, you might introduce a significant error if the property is a strong function of temperature and the temperature of your material is not 300 K. Liquid viscosity, for example, changes dramatically with temperature. The viscosity of water at 350 K is approximately 0.00037 newton · sec/m^2. What percent error would you build into a calculation if you used the 300 K value listed in the Mathcad Reference Tables?

correct value: 0.00037 newton · sec/m^2

incorrect value:_______ newton · sec/m^2 (from Reference Table)

$$\%error = \frac{correct\ value - incorrect\ value}{correct\ value} \cdot 100\%$$

1.4 The Mathcad Reference Tables include useful geometry formulas. Use the reference tables to find the formulas for calculating the

- Area and perimeter of a trapezoid
- Area of a regular polygon with n sides
- Volume of a torus (doughnut shape)

1.5 Use the Mathcad Reference Tables to find the formula for the capacitance between two concentric spheres of radius r_1 and r_2.

ANSWERS TO SELECTED PROBLEMS

Problem 1.1 Density of seawater: 1025 kg/m^3
Viscosity of kerosene: 0.00164 N · s/m^2

Problem 1.4 Area of a trapezoid: $A_{trap} = 12\ h\ (a + b)$, where a and b are the lengths of the parallel sides of the trapezoid.

CHAPTER 2

Mathcad Fundamentals

Objectives

After reading this chapter, you will

- understand how the Mathcad workspace is laid out
- know how to enter an equation
- know how to edit existing equations
- know how to display results and control how they are displayed
- understand how Mathcad determines the order in which to evaluate equations
- know how Mathcad handles unit conversions
- be able to enter and format text regions on a Mathcad worksheet
- know how to save your Mathcad worksheets for future use

2.1 THE MATHCAD WORKSPACE

When you first start Mathcad your computer should display the *Mathcad workspace*, shown in Figure 2.1.

Note: The exact appearance of the screen will depend on the version of Mathcad you are using, the display options that have been selected, and your operating system. Some of the display options will be described later in this chapter. All screen examples in this text are from Mathcad 14 or Mathcad 15 running under Windows XP.

Most of the workspace is blank, white space called the ***worksheet***. This is the area available to you to enter your equations, text, graphs, and so forth. The worksheet scrolls if you need additional space. You can have multiple worksheets open at the same time, similar to working with multiple documents in a word processor.

There is a small crosshair cursor displayed on the worksheet. This is the *edit cursor*, and it indicates where the next equation or ***text region*** will be displayed. Clicking the mouse anywhere on the worksheet moves the edit cursor to the mouse pointer's location.

Below the worksheet, you may see the *Trace Window*. (See Figure 2.1.) The Trace Window is used when debugging Mathcad programs. We won't need the Trace Window at this point, so it can be closed by clicking the "x" at the right side of the Trace Window title bar, as indicated in Figure 2.2.

Several of the features near the top of Figure 2.1 should be pretty familiar to Windows users. The Title Bar and Menu Bar are common to most Windows applications. The *Title Bar* displays the name of the application program, "Mathcad" in Figure 2.1. The *Menu Bar* displays the main menu options. Clicking on any of the main menu options will cause a drop-down menu of additional options to be displayed. Many of the most common menu options are also available as buttons on ***toolbars***, such as the *Standard Toolbar*.

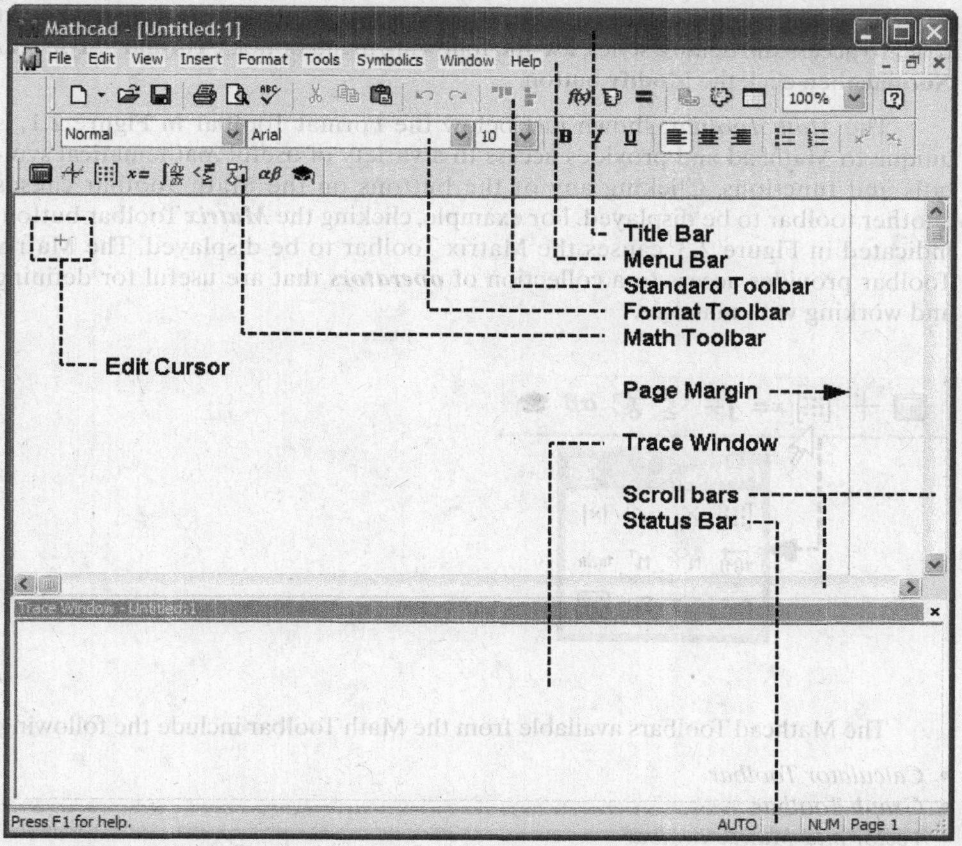

Figure 2.1
The Mathcad workspace.

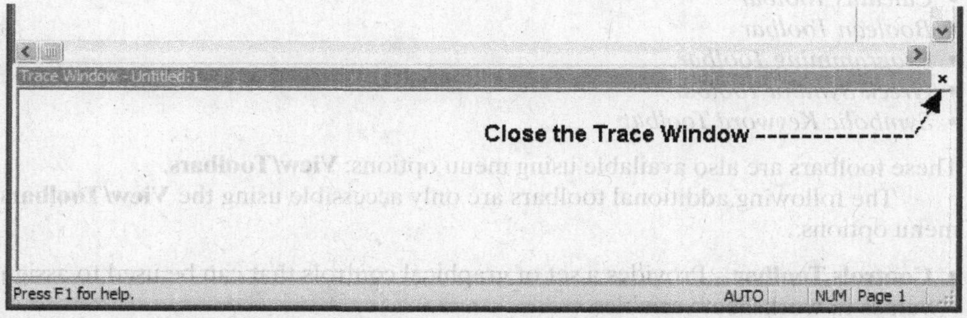

Figure 2.2
Closing the Trace Window.

The *Format Toolbar* is typically located just below the Standard Toolbar, but any of the toolbars in Mathcad can be relocated. The Format Toolbar is very similar to the ones used with word processors, but in Mathcad it displays the font type and size of the ***style*** applied to the item you are editing. There are separate styles defined for constants, variables, and text. When you change the appearance of any constant or variable, you change the display style used to format *all* constants or variables in the worksheet. However, in text regions, only the selected portion of text is changed when you use the Format Toolbar.

Note: To change the way all text regions are displayed, modify the "Normal" text style. To access the default styles, use the following menu options: **Format / Style . . . / Normal**, then click the **Modify** button.

The *Math Toolbar*, shown just below the Format Toolbar in Figure 2.1, is unique to Mathcad and provides access to a variety of useful mathematical symbols and functions. Clicking any of the buttons on the Math Toolbar causes another toolbar to be displayed. For example, clicking the ***Matrix*** Toolbar button, indicated in Figure 2.3, causes the Matrix Toolbar to be displayed. The Matrix Toolbar provides access to a collection of ***operators*** that are useful for defining and working with matrices.

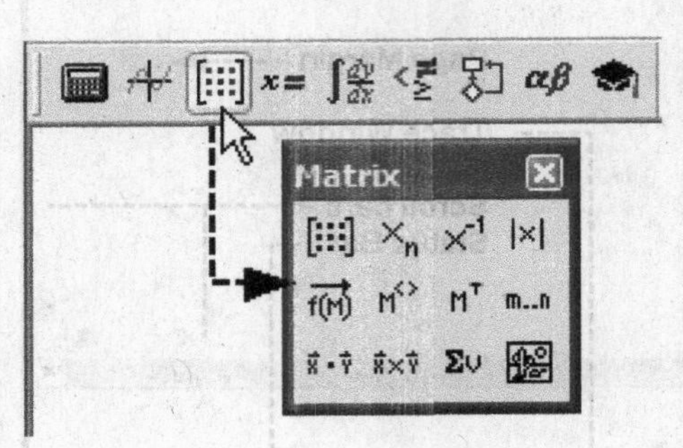

Figure 2.3
Clicking the Matrix Toolbar button causes the Matrix Toolbar to be displayed.

The Mathcad Toolbars available from the Math Toolbar include the following:

- *Calculator Toolbar*
- *Graph Toolbar*
- *Vector and Matrix Toolbar*
- *Evaluation Toolbar*
- *Calculus Toolbar*
- *Boolean Toolbar*
- *Programming Toolbar*
- *Greek Symbol Toolbar*
- *Symbolic Keyword Toolbar*

These toolbars are also available using menu options: **View/Toolbars.**

The following additional toolbars are only accessible using the **View/Toolbars** menu options:

- **Controls Toolbar** – Provides a set of graphical controls that can be used to assign values to worksheet variables.
- **Debug Toolbar** – Used to help debug Mathcad programs.
- **Modifier Toolbar** – A slate of symbolic math modifier keywords.
- **Custom Character Toolbar** – Provides quick access to some hard-to-enter custom characters such as °C and °F.

2.2 CONTROLLING THE ORDER OF SOLVING EQUATIONS IN MATHCAD

It is usually very important to solve a set of equations in a particular order. In Mathcad, you use the placement of equations on the worksheet to control the order of their solution. Mathcad evaluates equations from left to right and top to bottom (see Figure 2.4). When two equations are side by side on the worksheet, the equation on the left will be evaluated first, followed by the equation on the right. When there are no more equations to evaluate on a line, Mathcad moves down the worksheet and continues evaluating equations from left to right.

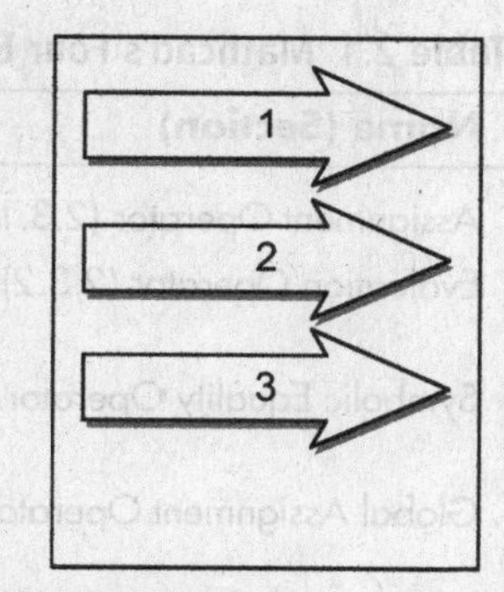

Figure 2.4
Mathcad evaluates equations from left to right, then top to bottom.

Some equations take up a lot more space on the screen than others, so how do you determine which equation will be evaluated first? Mathcad assigns each equation an ***anchor point*** on the screen as shown in Figure 2.5. The anchor point is located to the left of the first character in the equation, at the character baseline. You can ask Mathcad to display the anchor points by selecting **Regions** from the View Menu (**View/Regions**). When **View/Regions** is on, the background of the worksheet is dimmed, and the ***equation regions*** appear as bright boxes. The anchor point is indicated by a black dot.

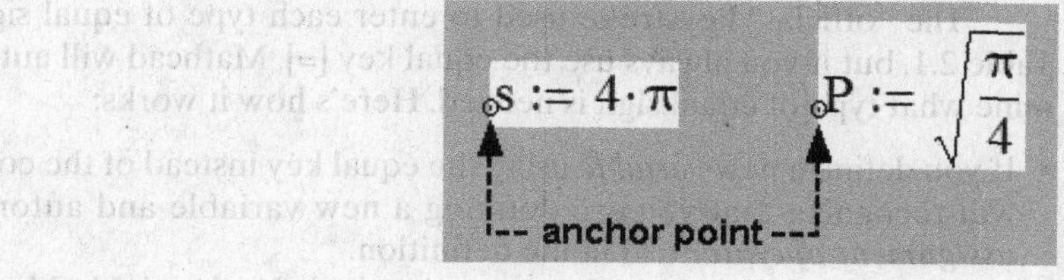

Figure 2.5
Equation anchor points.

In Figure 2.5, the equation on the left would be evaluated first, since the anchor points are at the same distance from the top of the worksheet, but the anchor point on the **s** is to the left of the anchor point on the **P**. For simple variable definitions, like those of **s** and **P** shown here, the anchor point is always located at the bottom-left corner of the variable name.

2.3 FOUR DIFFERENT KINDS OF EQUAL SIGNS!

In high school algebra, you learned that the equal sign indicates that the left and right sides of an equation are equal. Then, in a programming course, you may have seen a statement such as

```
COUNT = COUNT + 1
```

This statement is never an algebraic equality, but it is a valid programming statement because the equal sign, in a programming context, means *assignment*, not algebraic equality. That is, the result of the calculation on the right side is assigned to the variable on the left. Mathcad allows both types of usage (algebraic equality and assignment) within a single worksheet, but different types of equal signs are used to keep things straight.

In Mathcad, four different symbols are available to represent equality or assignment (six, if you count the *symbolic evaluation* and the programming *temporary assignment* symbols). Fortunately, some are much more commonly used than others. The four kinds of equal signs are summarized in Table 2.1, and each will be discussed in the following sections.

Table 2.1 Mathcad's Four Equal Signs

Name (Section)	Symbol	Keystroke	Usage
Assignment Operator (2.3.1)	:=	[:] (colon)	Used to define new variables.
Evaluation Operator (2.3.2)	=	[=] (equal)	Used to display the value assigned to a variable, or the result of a calculation.
Symbolic Equality Operator (2.3.3)	= (bold =)	[Ctrl =]	Used to show the relationship between variables in an equation (algebraic equality).
Global Assignment Operator (2.3.4)	≡	[~] (tilde)	Operates like the regular assignment operator, except global assignments (variable definitions) are performed before evaluating the rest of the worksheet.

The "official" keystroke used to enter each type of equal sign is indicated in Table 2.1, but if you always use the equal key [=], Mathcad will automatically determine what type of equal sign is needed. Here's how it works:

- If you define a new *variable* using the equal key instead of the colon key, Mathcad will recognize that you are defining a new variable and automatically use the ***assignment operator*** (:=) in the definition.
- If you use the equal key after a *previously defined variable*, Mathcad will use the evaluation operator (=) and display the value assigned to the variable.

Four different kinds of equal signs may seem complicated, but you can solve a lot of problems in Mathcad using only the assignment operator (:=) and the evaluation operator (=). A typical use of these two types of equal signs is shown in Figure 2.6.

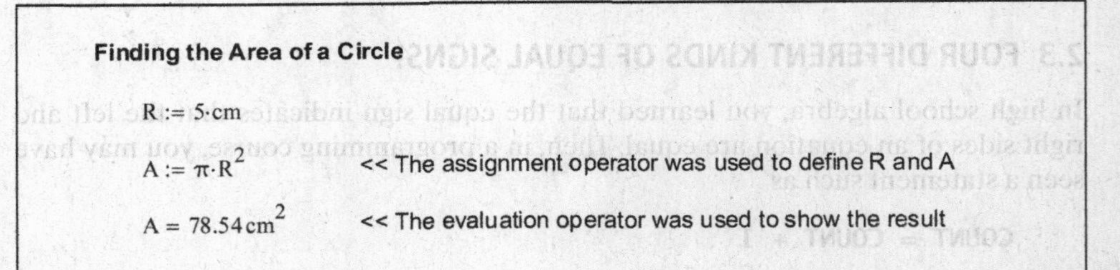

Finding the Area of a Circle
R := 5·cm
A := π·R2
<< The assignment operator was used to define R and A
A = 78.54 cm2
<< The evaluation operator was used to show the result

Figure 2.6
Using the assignment and evaluation types of equal signs.

2.3.1 Assignment (:=)

The most commonly used assignment operator is **:=**, which is entered by using the colon key [**:**]. It is often called the "define as equal to" operator. This type of equal sign was used to assign values to variables **R** and **A** in Figure 2.6.

2.3.2 Display the Value of a Variable or Result of a Calculation (=)

Once a variable has been assigned a value (either directly or by means of a calculation), you can display the value by using the plain equal sign [=], called the

Professional Success

Practice, Practice, Practice!

This does not mean that you should perform the same tasks over and over again (you're not trying to improve your manual dexterity), but you can take conceptual information in a textbook and make it your own by putting it into practice.

The fastest way to learn Mathcad is to use it. Try working through the examples in this text on a computer. There are "Practice!" boxes throughout the text that have been designed to help you learn what Mathcad can do – to help you put this new knowledge into practice.

Solutions to the Practice! problems are available at the end of each chapter.

evaluation operator. Displaying a value is the only usage of the plain equal sign in Mathcad; it is never used for assignment or algebraic equality. To display the value of the calculated variable **s**, you first define the equation used to compute **s** and then display the result using the equal sign. For example, we might have

$$s := 4\cdot\pi$$
$$s = 12.566$$

You can also display a calculated result without assigning it to a variable. This displays the result of the calculation, but you cannot use that result in further calculations without assigning it to a variable. An example is

$$4\cdot\pi = 12.566$$

2.3.3 Symbolic Equality (=)

A ***symbolic equality*** is used to indicate that the combination of variables on the left side of an equation is equal to the combination of variables on the right side of an equation – the high school algebra meaning of the equal sign. Symbolic equality is shown as a heavy boldface equal sign in Mathcad and is entered by pressing [Ctrl =]. (Hold down the control key while pressing the equal key.) An example of a formula that uses this operation is

$$P\cdot V = n\cdot R\cdot T$$

Symbolic equality is used to show a relationship between variables. There is no assignment of a value to any variable when symbolic equality is used. This type of equal sign is used for symbolic math and for solving equations by means of iterative methods.

2.3.4 Global Assignment (≡)

There is one way to override the left-to-right, top-to-bottom evaluation order in Mathcad – use a *global assignment operator*. Mathcad actually evaluates a worksheet in two passes. In the first pass, all ***global assignment*** statements are evaluated (from left to right and top to bottom). Then, in the second pass, all other equations are evaluated. Defining a variable with a global assignment equal sign has the same effect as putting the equation at the top of the worksheet, since both cause the statement to be evaluated first.

Global assignments are not used a lot, but it is fairly common to use them for unit definitions. For example, Mathcad already knows what a year (**yr**) is, but you could define a new unit, **decade**, in terms of years, as

decade ≡ 10·yr

Note: You could also define the decade by using the regular assignment operator, : = . It is common, but not required, to use global assignment for units.

Practice!

What will Mathcad display as the value of x in each of the examples that follow? (Try each in a separate worksheet.)

a. Use of "define as equal to"

```
y := 3
x := 3
x =
```

b. Use of a symbolic equality

```
y = 3
x := y
x =
```

c. With units

```
y := 3 · cm
x := y
x =
```

d. Use of a "global define as equal to"

```
x := y
x =
y ≡ 3
```

2.4 ENTERING AN EQUATION

To enter an equation, you simply position the edit cursor (crosshair) where you want the equation to go and start typing. Mathcad creates an ***equation region*** and displays the equation as you enter it. (Whenever Mathcad is waiting for you to type in the workspace, it waits in equation ***edit mode*** so that you can easily enter a new equation.) To enter the defining equation for the variable **s**, you would type [**s**] [:] [4] [*] [Ctrl-Shift-p][Enter]. To see the result of this calculation, move the cursor to the right or down (or both), and type [s] [=]. Mathcad will display the result after the equal sign:

```
s := 4·π
s = 12.566
```

2.4.1 Predefined Variables

Pi is such a commonly used value that it comes as a ***predefined variable*** in Mathcad. You can get the π symbol either from the Greek Symbols Toolbar or by pressing

[Ctrl-Shift-p], the shortcut used in the previous paragraph. *Pi* comes predefined in Mathcad with a value of 3.1415926. . . . You can redefine *pi* (or any other predefined variable) simply by building it into a new definition:

$\pi := 7$
$s := 4 \cdot \pi$
$s = 28$

However, redefining a commonly used constant is not a good idea in most situations.

Four common predefined values in Mathcad are **π, e, g,** and **%**. These are entered by using [Ctrl-Shift-p], [e], [g], and [%], respectively.

2.4.2 Math Operators and Precedence Rules

The symbols we use to indicate math operations such as addition, division, or exponentiation are called *operators*. In earlier examples we have already used the addition operator (+) the subtraction operator (–) and the multiplication operator (*). A summary of Mathcad's math operators is shown in Table 2.2.

Table 2.2 Standard Math Operators

Symbol	Name	Shortcut Key
+	Addition	+
–	Subtraction	–
*	Multiplication	[Shift 8]
/	Division	/
e^x	Exponentiation	
1/x	Inverse	
x^y	Raise to a Power	[^], or [Shift 6]
n!	Factorial	!
|x|	Absolute Value	
$\sqrt{}$	Square Root	\
$\sqrt[n]{}$	Nth Root	[Ctrl \]

Mathcad also uses standard ***operator precedence rules*** to determine how to evaluate equations. Mathcad generally works from left to right, but operations with higher precedence are evaluated first. The operator precedence rules are summarized in Table 2.3.

Table 2.3 Operator Precedence Rules

Precedence	Operator	Name
First	^	Exponentiation
Second	*, /	Multiplication, Division
Third	+, –	Addition, Subtraction

As an example of the use of operator precedence rules, consider the following equation:

$$x = 3 + 4 * 5 - 2^3$$

The highest precedence is the operation of raising 2 to the third power, so that it is done first:

$$x = 3 + 4 * 5 - \mathbf{8}$$

Next, the multiplication is performed:

$$x = 3 + \mathbf{20} - \mathbf{8}$$

Finally, the addition and subtraction steps are performed. Since addition and subtraction have the same level of precedence, Mathcad simply evaluates from left to right.

$$x = \mathbf{23} - 8$$

$$x = \mathbf{15}$$

Parentheses can be used to override the operator precedence rules, if needed. The math operations within parentheses are carried out first, and if there are multiple operations within a set of parentheses, the operator precedence rules still apply to those operations.

Practice!

Use the operator precedence rules listed in Table 2.3 to predict the result of the following calculations. Then check your predictions using Mathcad.

a. $x = 3 + 4 * 5 - 2^3$
b. $y = (3 + 4) * 5 - 2^3$
c. $z = 3 + 4 * (5 - 2)^3$
d. $w = (3 + 4 * 5) - 2^3$

2.4.3 Entering Complex Numbers

Complex numbers are values that contain both a real and an imaginary part. They are sometimes used in engineering calculations to track two related phenomena, such as voltage and phase changes in AC circuits.

A complex number made up of a real value of 12 and an imaginary value of 4 would be written as

$$12 + 4\mathbf{i}.$$

Mathcad can handle math involving complex numbers. To define a complex number, simply enter the real part, then add on the imaginary part and type an **i** immediately after the numeric value. When complex values are added (see Figure 2.7), the real and imaginary parts are handled separately.

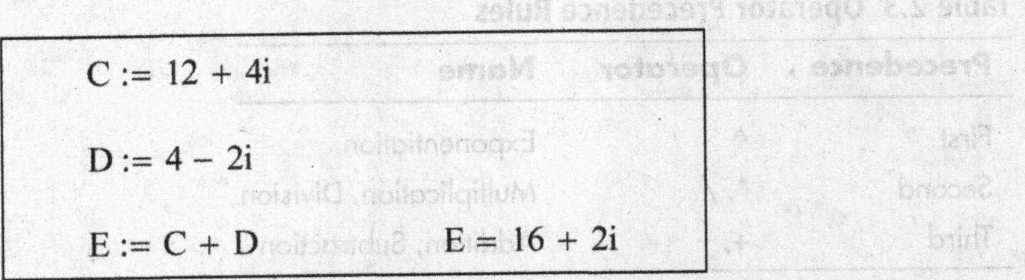

Figure 2.7
Math using complex numbers.

Notes:

- Mathcad recognizes both **i** and **j** as indicators of the imaginary part of a complex number. They are treated identically when performing math operations.
- You cannot append an **i** to a variable name to identify it as imaginary, but you can multiply the variable by 1**i**.

2.4.4 Entering Nondecimal Values

To enter a hexadecimal value when defining a variable, simply add the letter **h** after the value. Hexadecimal values can contain the digits **0** through **9** and the letters **a** through **f**, followed by an **h** to indicate that the value should be interpreted as a hexadecimal value. No decimal points are allowed. Similarly, octal values can be used in variable definitions simply by adding the letter **o** after the value. Only digits **0** through **7** may be used. Mathcad allows binary values (consisting of zeroes and ones only) to be entered by adding a letter **b** at the end of the value. The following examples are representatives:

```
A := 12       A = 12     decimal
B := 12o      B = 10     octal
C := 12h      C = 18     hexadecimal
D := 1011b    D = 11     binary
```

To see a result expressed as an octal, a hexadecimal, or a binary value, double-click the value and change the ***radix*** of the displayed result on the Result Format dialog box on the Display Options panel:

```
E := 201      E = 201
              E = 311o
              E = 0c9h
              E = 11001001b
```

2.4.5 Text Subscripts and Matrix Index Subscripts

Mathcad allows two types of subscripts on variables: ***text subscripts*** and ***matrix index subscripts***. They are very different in usage, and using a text subscript when a matrix index subscript is needed is a common error.

Text Subscripts

Text subscripts, also known as *literal subscripts,* are used to help identify variables. For example, to calculate the total surface area of a cylinder (Figure 2.8) we need to account for the area of each end, plus the area of the side of the cylinder. Just to help identify the areas, we might use subscripts as shown in Figure 2.8.

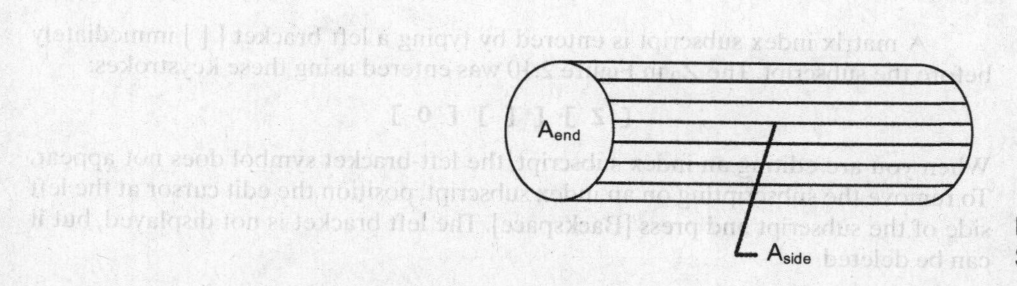

Figure 2.8
Surface area of a cylinder.

This type of subscript is a text subscript and is entered in Mathcad by typing a period [.] just before the subscript text. So, A_{end} is entered using the following keystrokes:

[A] [.] [e] [n] [d]

While you are editing an equation that contains a text subscript, the period is displayed as shown in Figure 2.9. To remove the subscripting, simply delete the period. After you have finished editing the equation, the subscripted text still appears, but the period is hidden.

Mathcad - [Untitled:2]

File Edit View Insert Format Tools Symbolics Window Help

$A_{.end}$ << the period is displayed while editing an equation

A_{side} << the period is not shown unless the equation is being edited

Figure 2.9
The period used to create a text subscript is displayed while editing an equation.

Matrix Index Subscripts

Matrix index subscripts, also known as *vector subscripts* or *array subscripts*, are used to identify particular elements of an *array* (an array can contain a single column or row (i.e., a *vector*), or multiple columns and/or rows (a ***matrix***)). A matrix index subscript may sometimes look like a text subscript, but it is not text. It is either a numeric value or a variable that holds a numeric value.

Mathcad's default array numbering system starts at zero, not one, so the top element of vector **Z** (shown in Figure 2.10) is element zero. This is often referred to as "**Z**-zero" which is written as $\mathbf{Z}_0$. The subscript (0) on the vector name (Z) identifies the particular element of the vector. In the example shown in Figure 2.10, vector **Z** has three elements, and element zero has a value of **2**.

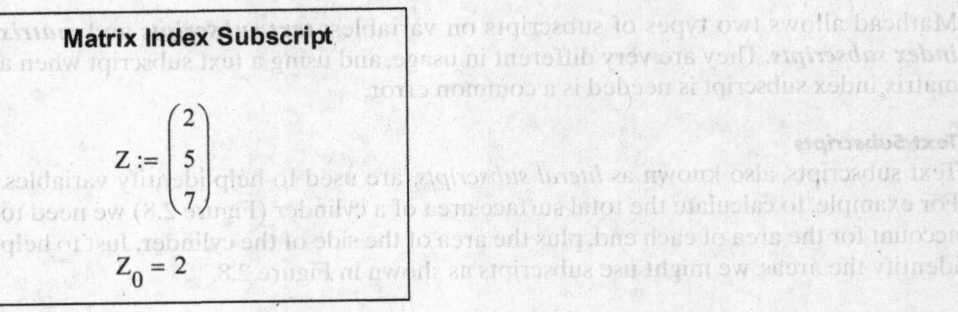
Matrix Index Subscript
Z := (2 5 7)
Z0 = 2

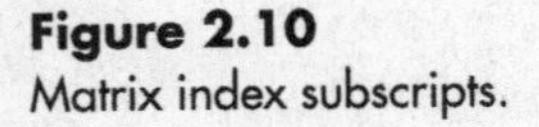

Figure 2.10
Matrix index subscripts.

A matrix index subscript is entered by typing a left bracket [[ ] immediately before the subscript. The Z_0 in Figure 2.10 was entered using these keystrokes:

[Z] [[ ] [0]

When you are editing an index subscript, the left-bracket symbol does not appear. To remove the subscripting on an index subscript, position the edit cursor at the left side of the subscript and press [Backspace]. The left bracket is not displayed, but it can be deleted.

You can use index subscripts to work with specific values in an array. For example, in Figure 2.11, the value of the bottom element (element 2) of the **Z** vector is changed to 12.

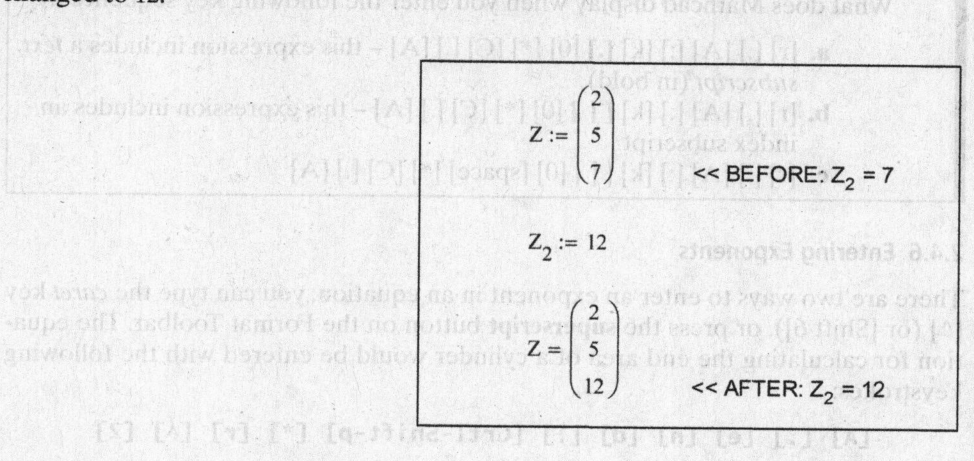

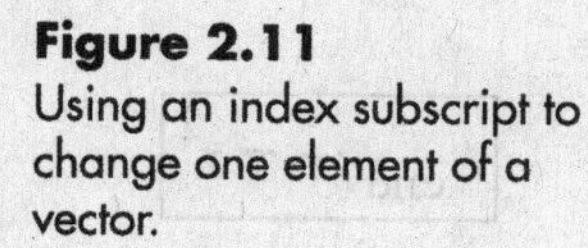

Figure 2.11
Using an index subscript to change one element of a vector.

Mathcad uses row and column numbers to identify elements of a matrix and, by default, calls the top-left element of a matrix element 0, 0. You can use index subscripts to access a specific element of a matrix; just separate the row number from the column number with a comma [,]. In Figure 2.12 index subscripts have been used to identify the top-left element of matrix **M**.

$$M := \begin{pmatrix} 1 & 3 & 5 & 2 \\ 2 & 6 & 9 & 4 \\ 8 & 5 & 3 & 7 \end{pmatrix}$$

$$M_{0,0} = 1$$

Figure 2.12
Accessing one element of a matrix using index subscripts.

Variables can be used in index subscripts, but the value of the variable must be a valid row or column number. An example of this is shown in Figure 2.13.

$$M := \begin{pmatrix} 1 & 3 & 5 & 2 \\ 2 & 6 & 9 & 4 \\ 8 & 5 & 3 & 7 \end{pmatrix}$$

<< row 0
<< row 1
<< row 2
^^^ column 2

$row := 1$

$col := 2$

$M_{row,col} = 9$

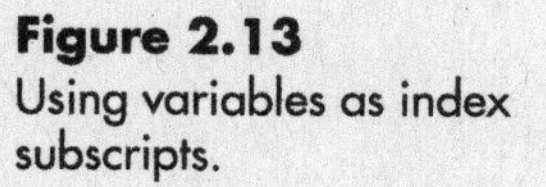

Figure 2.13
Using variables as index subscripts.

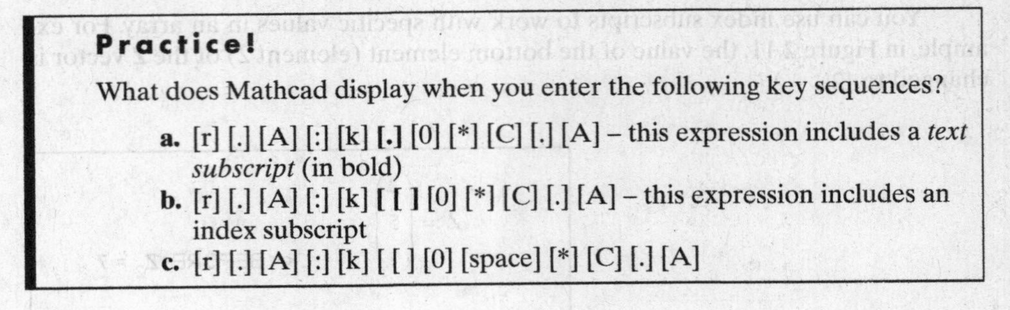

Practice!

What does Mathcad display when you enter the following key sequences?

a. [r] [.] [A] [:] [k] **[.]** [0] [*] [C] [.] [A] – this expression includes a *text subscript* (in bold)

b. [r] [.] [A] [:] [k] **[[ ]** [0] [*] [C] [.] [A] – this expression includes an index subscript

c. [r] [.] [A] [:] [k] [[ ] [0] [space] [*] [C] [.] [A]

2.4.6 Entering Exponents

There are two ways to enter an exponent in an equation: you can type the *caret* key [^] (or [Shift 6]), or press the **superscript** button on the Format Toolbar. The equation for calculating the end area of a cylinder would be entered with the following keystrokes:

```
[A] [.] [e] [n] [d] [:] [Crtl-Shift-p] [*] [r] [^] [2]
```

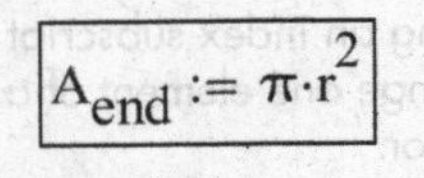

Figure 2.14
Entering an exponent.

The result is shown in Figure 2.14.

When working with exponents you will quickly discover that once the edit cursor has moved up so that you can enter the exponent, it stays in the exponent. That is, if you were to enter a plus sign after the **2** in the preceding equation, you would be adding to the **2** in the exponent, not to the $\mathbf{r^2}$. If you need to add something to the $\mathbf{r^2}$ you must first press the space bar to select all of the $\mathbf{r^2}$ before pressing the plus key. Selecting and modifying equations is the subject of the next section.

2.5 MODIFYING AN EQUATION

When solving problems with any math package, you will invariably generate errors in equations that will have to be corrected. In this section, methods for selecting equations for editing, methods for selecting parts of equations for editing, and methods for changing variables and operators will be presented.

2.5.1 Selecting an Equation

Selecting an equation on a Mathcad worksheet may seem pretty intuitive, but there are actually two ways to do it, depending on whether you want to edit the equation or you are selecting the equation in order to move it or delete it.

Selecting an Equation for Editing

If you click on an equation region, Mathcad will jump into the equation edit mode and allow you to edit the equation. An equation region in the edit mode is enclosed by a solid line, and the horizontal and vertical edit bars are visible.

Selecting an Equation for Moving or Deletion

If you drag-select an equation region by clicking outside the region and dragging the mouse pointer over the equation region, Mathcad will assume you want to move the equation region. You can also select an equation to move or delete by holding the [Shift] key down while clicking on the equation (or equations).

Note: If you need to make room in your worksheet to move an equation, you can add space between two stacked equation regions by clicking between the regions and pressing [Enter].

An equation region selected for moving or deletion will be shown surrounded by a dashed line (see Figure 2.15), and you cannot edit the equation until you click outside the region (to deactivate the "move" mode). You can drag-select one or multiple equation and text regions. If multiple regions are selected, they can be moved together by positioning the mouse pointer over any selected region and dragging.

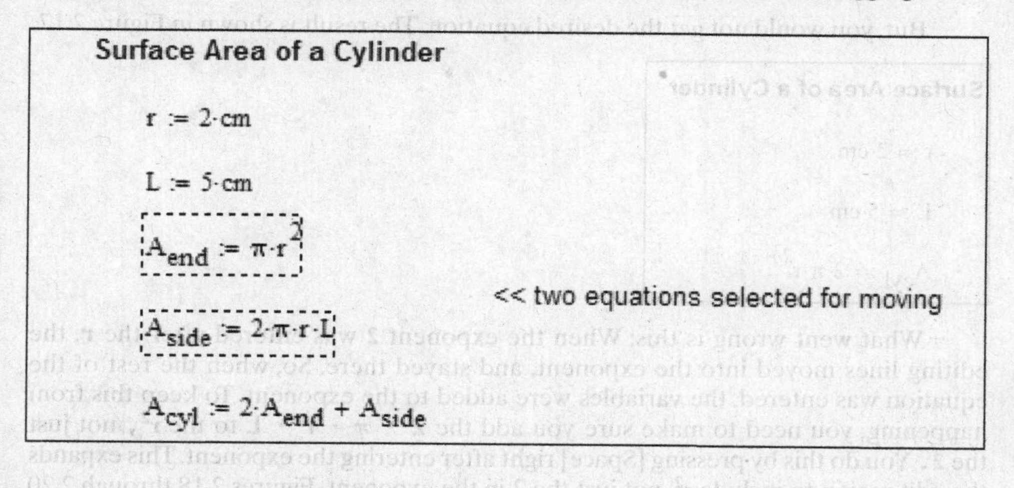

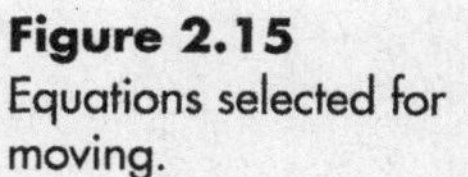

Figure 2.15
Equations selected for moving.

2.5.2 Selecting Part of an Equation

While you are editing an equation, Mathcad displays two ***editing lines***, one horizontal and one vertical. In Figure 2.16, the definition of the Boltzmann Constant has been selected for editing, and the edit lines are visible.

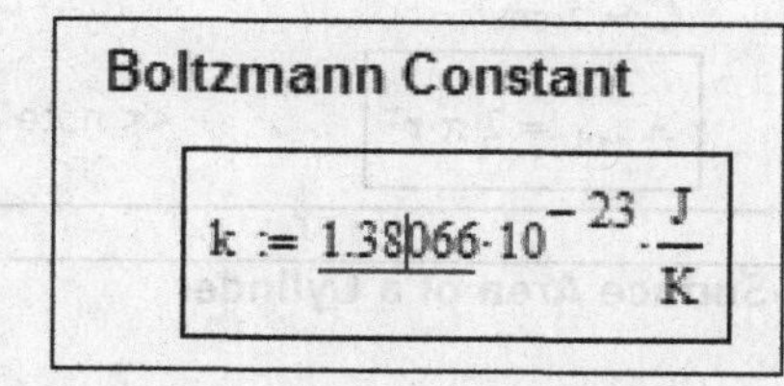

Figure 2.16
Editing an equation.

The *vertical editing line* indicates where any editing will occur. For example, with the vertical editing line between the 8 and the 0 as shown in Figure 2.16, pressing [Backspace] will delete the 8, and pressing [Delete] will delete the 0. You can move the vertical editing line using the arrow keys, or by clicking a different location with the mouse.

The *horizontal editing line* indicates the portion of the equation that is being edited. Pressing [Space] increases the length of the horizontal editing line to include a greater portion of the equation. This is very useful for selecting a part of the equation for editing and can be essential in entering equations correctly. If we consider (again) calculating the total surface area of a cylinder (shown in Figure 2.8), the equation is

$$A_{\text{cyl}} = 2A_{\text{end}} + A_{\text{side}}$$

Substituting geometric formulas for A_{end} and A_{side}, we get

$$A_{cyl} = 2 \cdot \pi \cdot r^2 + 2 \cdot \pi \cdot r \cdot L$$

where r is the radius of the cylinder, and L is the cylinder length.

The following keystrokes might be used to try to get this equation into Mathcad:

```
[A][.][c][y][l][:][2][*][Ctrl-Shift-p][*][r][^][2][+][2][*]
[Ctrl-Shift-p][*][r][*][L]
```

But, you would not get the desired equation. The result is shown in Figure 2.17.

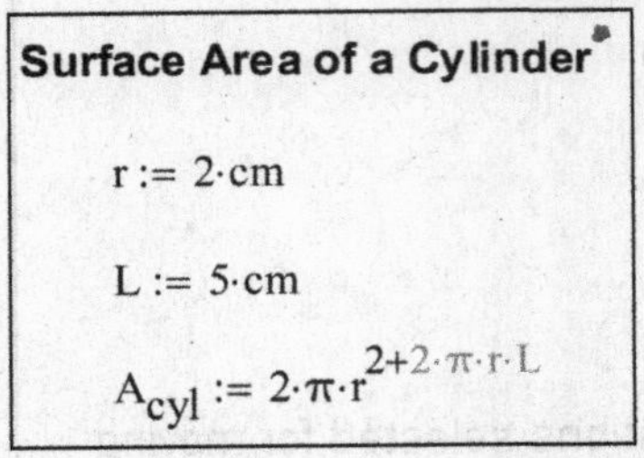

Surface Area of a Cylinder

$r := 2 \cdot cm$

$L := 5 \cdot cm$

$A_{cyl} := 2 \cdot \pi \cdot r^{2+2 \cdot \pi \cdot r \cdot L}$

Figure 2.17
The incorrect equation for surface area.

What went wrong is this: When the exponent **2** was entered after the **r**, the editing lines moved into the exponent, and stayed there. So, when the rest of the equation was entered, the variables were added to the exponent. To keep this from happening, you need to make sure you add the **2 · π · r · L** to the $\mathbf{r^2}$, not just the **2**. You do this by pressing [Space] right after entering the exponent. This expands the edit region to include $\mathbf{r^2}$, not just the 2 in the exponent. Figures 2.18 through 2.20 illustrate this.

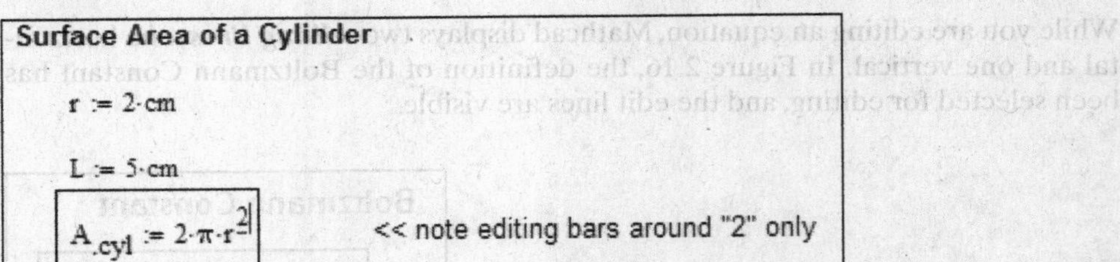
Surface Area of a Cylinder
r := 2·cm
L := 5·cm
A.cyl := 2·π·r2
<< note editing bars around "2" only

Figure 2.18
The editing bars immediately after entering the exponent on *r*.

Surface Area of a Cylinder

$r := 2 \cdot cm$

$L := 5 \cdot cm$

$A_{cyl} := 2 \cdot \pi \cdot r^2$ << note editing bars around "r2"

Figure 2.19
The editing bars after pressing [Space] to expand the edit region to include r^2.

Surface Area of a Cylinder

$r := 2 \cdot cm$

$L := 5 \cdot cm$

$A_{cyl} := 2 \cdot \pi \cdot r^2 + 2 \cdot \pi \cdot r \cdot L$

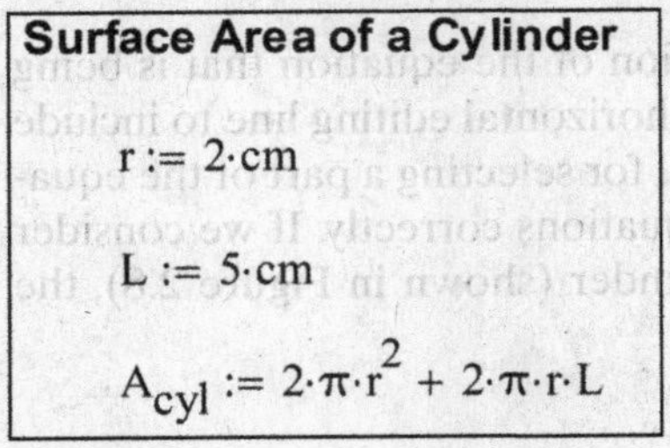

Figure 2.20
The completed equation for surface area of a circle.

Note: You might think that you should select all of $2 \cdot \pi \cdot r^2$ rather than just the r^2 before adding on the second half of the area calculation. That would work, too. You would just press [Space] two more times to select the π and the **2**. To reduce the number of keystrokes required to enter an equation, Mathcad keeps multiplied variables together when you add to the collection. This is just a convenience; it is handy, but it does take some time getting used to.

Practice!

What does Mathcad display when you enter the following key sequences?

a. [P][*][V][Ctrl =][n][*][R][*][T]
b. [P][:][n][*][R][*][T][/][V]
c. [P][:][n] [*][R][*][T][Space][Space][/][V]
d. [r][:][k][1][*][C][A] [^][2][–][k][2][*][C][B]
e. [r][:][k][1][*][C][A] [^][2][Space][–][k][2][*][C][B]

2.5.3 Changing a Value or a Variable Name

When you click in the middle of a value or a variable name, the vertical editing line will be displayed. The editing line indicates where any edits will take place. You can use the left- and right-arrow keys to move the line. The delete key will remove the character to the right of the editing line, while the backspace key will remove the character to the left.

Not all characters that are entered are displayed by Mathcad, but they can still be deleted. For example, the left bracket [[ ] that is used to enter a matrix index subscript is not shown on the screen, but if you position the vertical editing line at the beginning of the subscript and press [Backspace], the left bracket will be removed.

For example, if the ideal gas constant (0.08206 liter atm/mole K) had been entered incorrectly as 0.08506, you would click to the right of the numeral 5, as in the following figure:

$R_{gas} := \underline{0.085|06}$

Then press [Backspace] to remove 5, and press [2] to enter the correct value. The final result would look like this:

$R_{gas} := 0.08206$

2.5.4 Changing an Operator

When you need to change an operator, such as a symbolic equality (=) to an assignment operator (: =), you click just to the right of the operator itself. This puts the editing lines around the character just to the right of the operator. Press [Insert] to move the vertical editing line to the left side of the character (just to the right of the operator):

$s = \underline{|4} \cdot \pi$

Then press [Backspace] to delete the operator. An open placeholder appears, indicating where the new operator will be placed:

$s \;\square\; \underline{|4} \cdot \pi$

Enter the new assignment operator by pressing the colon key [:]:

$s := \lfloor 4 \cdot \pi$

2.5.5 Inserting a Minus Sign

Special care must be taken when you need to insert a minus sign, because the same key is used to indicate both negation and subtraction. Here is how Mathcad decides which symbol to insert when you press the [–] key:

- If the vertical editing line is to the left of a character (to the right of the open placeholder), as in

  $A := 12 \cdot x^2 \ \square \ \lfloor 5 \cdot y^2$

  then the sign of the character to the right of the placeholder is changed (indicating negation):

  $A := 12 \cdot x^2 \ \square \ -5 \cdot y^2$

- If the editing line is to the right of a character (to the left of the open placeholder), as in

  $A := 12 \cdot x^2 \rfloor \square 5 \cdot y^2$

  then the open placeholder is filled by a subtraction operator:

  $A := 12 \cdot x^2 - 5 \cdot y^2$

2.5.6 Highlighting a Region

To make your results stand out, Mathcad allows you to show them in a highlighted region. The highlighting can consist of a border around the result and/or a colored background.

To highlight a result, right-click on it, and select **Properties...** from the pop-up menu as shown in Figure 2.21.

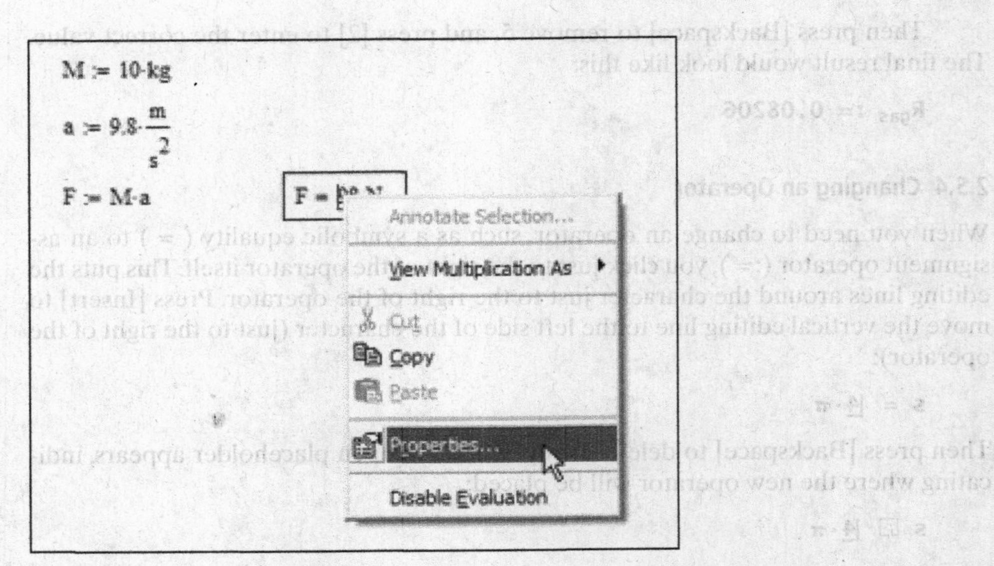

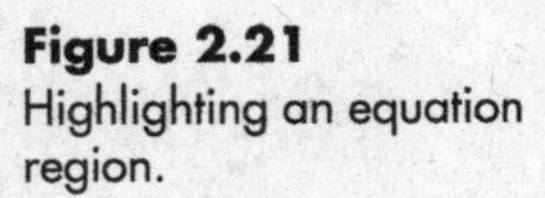

Figure 2.21
Highlighting an equation region.

On the Properties dialog box, select **Highlight Region** and **Choose Color**, or check the **Show Border** box as shown in Figure 2.22.

Figure 2.22
The equation region Properties dialog box.

2.5.7 Changing the Way Operators Are Displayed

Mathcad uses specific symbols for operators, such as the := used to indicate assignment. While these symbols help you understand a Mathcad worksheet, they can confuse people who are not used to working with the software. You can change the way operators are displayed either for a single equation or throughout an entire worksheet.

Changing the Appearance of a Single Operator
If you right-click on an equation region directly over the operator you want to modify, a drop-down menu will appear. For example, right-clicking on the assignment operator (:=) in the definition of `M` displays the pop-up menu shown in Figure 2.23.

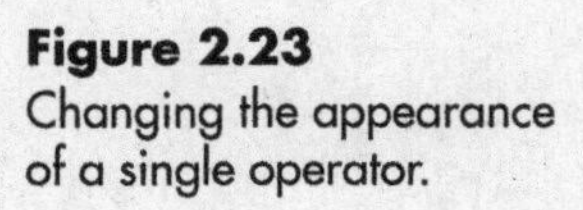

Figure 2.23
Changing the appearance of a single operator.

By selecting **View Definition As** from the menu, a menu of options for the definition operator is displayed. By selecting **Equal** from the options list, you can have the assignment operator displayed as a simple equal sign as shown in Figure 2.24.

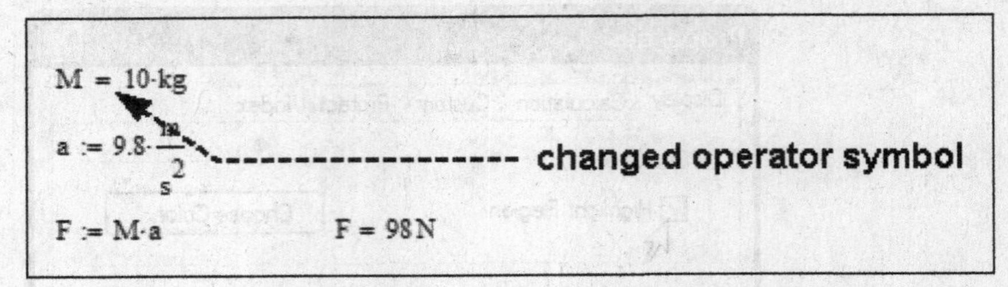

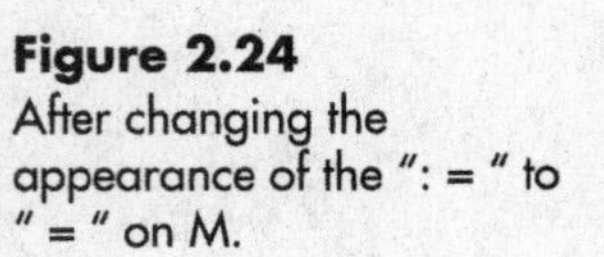
Figure 2.24
After changing the appearance of the ": = " to " = " on M.

Changing the Appearance of All Operators in the Worksheet

To change the appearance of all operators on a worksheet, use the Worksheet Options dialog box (Figure 2.25) that is available from the Tools menu, as **Tools/Worksheet Options . . .**

Worksheet Options

Unit System | Dimensions | Compatibility

Built-In Variables | Calculation | Display

Array Origin (ORIGIN) 0 (0)

Convergence Tolerance (TOL) 0.001 (0.001)

Constraint Tolerance (CTOL) 0.001 (0.001)

Seed value for random numbers 1 (1)

PRN File Settings

Precision (PRNPRECISION) 4 (4)

Column Width (PRNCOLWIDTH) 8 (8)

Restore Defaults

OK Cancel Help

Figure 2.25
Worksheet Options dialog box.

Click on the **Display** tab (see Figure 2.26) to change the way operators are displayed.

There is a drop-down list of display options for each type of operator. The drop-down list for the multiplication operator can be used to change the display of all multiplication operators from "**Narrow Dot**" to "**x**", for example (see Figure 2.27.)

Figure 2.28 shows the force calculation worksheet after changing the appearance of the operators.

Note: Changing the way the operators are *displayed* does not change the way equations are *entered*. The multiplication operator, for example, is still entered using [*], even if it is displayed as "**x**".

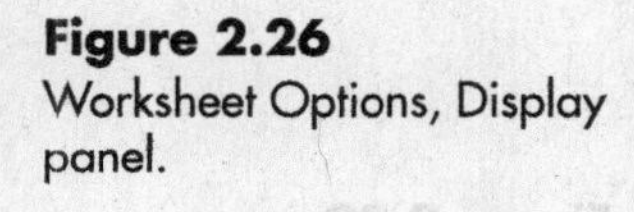

Figure 2.26
Worksheet Options, Display panel.

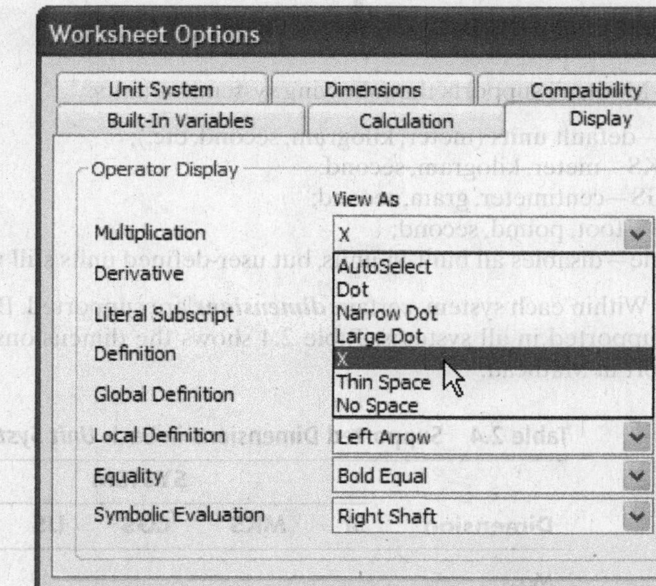

Figure 2.27
Changing the way multiplication operators are displayed.

$$M = 10 \times kg$$

$$a := 9.8 \times \frac{m}{s^2}$$

$$F := M \times a \qquad F = 98\,N$$

Figure 2.28
The worksheet after changing the way operators are displayed.

2.6 WORKING WITH UNITS

Mathcad supports units. Its ability to automatically handle unit conversions is a very nice feature for engineering calculations, since the number of required unit conversions can be considerable. Mathcad handles units by storing all values in a base set of units (International System of Units, or SI, by default, but you can change it).

Values are converted from the units you enter to the base set before the value is stored. (This is illustrated in Figure 2.29.) Then, when a value is displayed, the conversions required to give the requested units are automatically performed. Mathcad's unit handling works very well for most situations, but it does have some limitations, which will be discussed shortly.

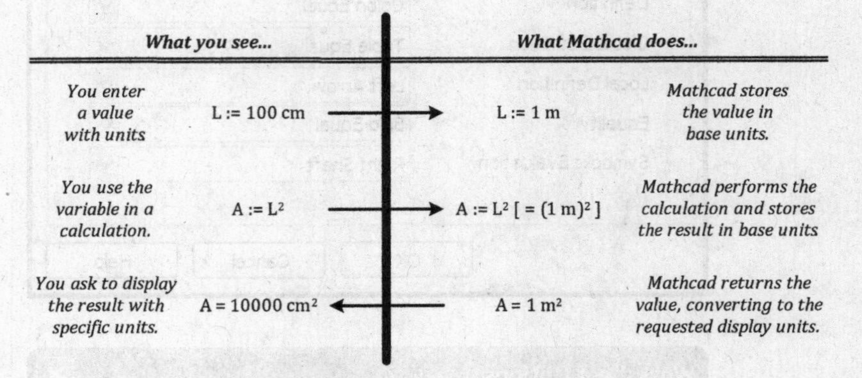

Figure 2.29
How Mathcad handles units.

Mathcad supports the following systems of units:

- SI—default units (meter, kilogram, second, etc.);
- MKS—meter, kilogram, second;
- CGS—centimeter, gram, second;
- US—foot, pound, second;
- none—disables all built-in units, but user-defined units still work.

Within each system, certain ***dimensions*** are supported. But not all dimensions are supported in all systems. Table 2.4 shows the dimensions the various systems support in Mathcad.

Table 2.4 Supported Dimensions in Each *Unit System*.

	SYSTEM				
Dimension	**SI**	**MKS**	**CGS**	**US**	**NONE**
Mass	√	√	√	√	√
Length	√	√	√	√	√
Time	√	√	√	√	√
Current	√	no	no	no	no
Charge	no	√	√	√	√
Temperature	√	√	√	√	√
Luminosity	√	no	no	no	no
Substance	√	no	no	no	no

If you plan to work with current (typical units, amps), luminosity (typical units, candelas) or substance (typical units, moles), then you want to use the default SI units. If these units are not important to you, you have more system options available. You set the system of units to be used as base units from the menu bar: **Tools/Worksheet Options... /Unit System**.

The following are some common unit abbreviations, by category:

mass	kg, gm, lb
length	m, cm, ft
time	sec, hr
current	amp
charge	coul
temperature	K, R
substance	mole
volume	liter, gal, galUK (Imperial gallon)
force	N, dyne, lbf
pressure	Pa, atm, torr, in_Hg, psi
energy and work	joule, erg, cal, BTU
power	watt, kW, hp

Cautions on Unit Names

g – predefined as gravitational acceleration not grams;
gm – unit name for grams;
R – unit name for °R (degrees Rankine), not the gas constant;
mole – gram mole, defined in the SI system only.

You can overwrite the symbols used as predefined unit names by declaring them as variables in a worksheet. For example, if you included the statement

```
m := 1·kg
```

in a worksheet, the symbol *m* would become a variable and no longer represent meters. Redefining unit names is usually not a good idea.

The units used in this chapter represent only a small subset of the predefined units in Mathcad. A list of available units may be obtained by pressing [Ctrl-u]. You can select units from the list or simply type the unit name (using Mathcad's abbreviation).

For example, you could find the area of a circle with a radius of 7 cm by using the following equations:

```
r := 7·cm
Area := π·r²
```

In the first line, the 7 is multiplied by the unit name, *cm*; then the *Area* is computed on the second line. The result is displayed as

```
Area = 0.015 · m² ■
```

The box around this equation and the editing lines around the **Area** variable indicate that we haven't completed editing the equation; that is, we haven't pressed [Enter] yet. As soon as you press [=], Mathcad shows the value currently assigned to the variable **Area** in the base units (SI, by default). A placeholder is also shown to the right of the units (only while the equation is being edited) so that you can request that the value be displayed in different units. If you want the result displayed in cm^2, click on the placeholder and then enter [c] [m] [^] [2] [Enter]. The result is then displayed in the requested units:

$$\text{Area} = 153.938 \cdot \text{cm}^2$$

You can put any defined unit (predefined, or defined part of the worksheet) in the unit placeholder. If the units you enter have the wrong dimensions, Mathcad will make that apparent by showing you the "leftover" dimensions in base units. For example, if you had placed *atm* (atmospheres) in the placeholder, the dimensions would be quite wrong, and the displayed result would make this apparent:

$$\text{Area} = 1.519 \cdot 10^{-7} \cdot \text{kg}^{-1} \cdot \text{m}^3 \cdot \text{s}^2 \cdot \text{atm}$$

2.6.1 Defining a New Unit

New units are defined in terms of predefined units. For example, a commonly used unit of viscosity is the *centipoise*, or *cP*. The *poise* is predefined in Mathcad, but the *centipoise* is not. The new unit can be defined like this:

$$\text{cP} := \frac{\text{poise}}{100}$$

Once the new unit has been defined, it can be used throughout the rest of the worksheet. Here, the viscosity of honey is defined as 10,000 cP, and then the value is displayed by using another common unit for viscosity,

$$\text{visc}_{\text{honey}} := 10000 \cdot \text{cP}$$
$$\text{visc}_{\text{honey}} = 10 \cdot \text{Pa} \cdot \text{s}$$

Practice!

Try these obvious unit conversions:

a. 100 cm → m
b. 2.54 cm → in
c. 454 gm → lb

Now try these less obvious conversions:

a. 1 hp → kw
b. 1 liter · atm → joule
c. 1 joule → watt (This is an invalid conversion. How does Mathcad respond?)

Professional Success

First, make sure your method (or design/computer code/Mathcad worksheet) works correctly on a test case with a known answer. Then try your method (design/computer code/Mathcad worksheet) on a new problem.

In the last "Practice!" box, you were asked to try Mathcad on some obvious cases before trying some less obvious ones. Testing against a known result is a standard procedure in engineering. If you get the right answer in the test case, you have increased confidence (though not total confidence) that your method is working. If you get the wrong answer, it is a lot easier to find out what went wrong using the test problem than it is to try to fix the method and solve the real problem simultaneously.

You'll see this technique used throughout the text.

2.6.2 Editing the Units on a Value or Result

Once you have entered units, simply edit the unit name to change them. Click on the unit name, delete characters as needed, and then type in the desired units.

2.6.3 Converting Temperature Units

Prior to Mathcad 13, describing Mathcad's temperature conversions was easy: Mathcad would convert K (kelvins) to °R (degrees Rankine), or vice versa. That was all. Only absolute temperature scales were supported. But recent versions of Mathcad have added features, and temperature conversions have gotten a little more complicated.

In early versions of Mathcad, unit conversions in Mathcad were based solely on conversion *factors*. This meant that the only unit conversions that Mathcad supported were those involving multiplicative constants. Since a factor of 1.8 is all that is required to convert Kelvins to °R, Mathcad could convert these temperature scales. But conversions such as °C to K or °F to °C require an additive constant (273.15, or 32), so these conversions were once impossible.

Now, Mathcad supports unit conversion *functions*, so both multiplicative and additive conversion constants can be used. But using a unit conversion function is a little different than a simple multiplicative unit conversion. Because of this, we will need to consider the various types of temperature unit conversions separately.

Absolute Temperature Conversions

While the Kelvin temperature scale is by far the most widely used absolute temperature scale, the Rankine scale is also used. Both are absolute temperature scales, meaning that the zero of each scale is absolute zero. But one degree Rankine is the same size as one degree Fahrenheit, which is smaller than a kelvin. The conversion factor is 1.8 °R/K.

Mathcad uses R rather than °R as the symbol for degrees Rankine, but it handles unit conversions between kelvins and degrees Rankine like any other unit conversion.

Celsius and Fahrenheit Temperature Conversions

Temperature conversions between °C, °F, K, and °R are now built into Mathcad – but using them can be a little tricky.

- If you want to *display a calculated value* (i.e., a result) with units of °C or °F, simply place the °C or °F symbol in the unit placeholder that Mathcad provides after displaying any calculated result. Mathcad will perform the unit conversion and display the result in the requested units. So, displaying a calculated result in °C or °F is straightforward and easy. The °C and °F symbols are available on the Custom Characters Toolbar to make them easier to access.[1]

[1]The degree symbol is also available as alternate character 0176. This means you can get the character by holding down the [Alt] key while pressing 0, 1, 7, and 6 from the numeric keypad (not the numbers across the top of the keyboard).

- If you want to *define a variable* with units of °C or °F, you cannot use the usual procedure of entering the value and multiplying that value by the appropriate units. Because Mathcad, by default, stores temperature values with units of kelvins, there will automatically be a unit conversion from °C or °F to kelvins, and that is not a simple multiplication because an additive constant is also required (273.15 for °C to K, for example). Because a unit conversion function (rather than factor) is needed, you must use Mathcad's *postfix operator*[2] to define a temperature variable using °C or °F.

The postfix operator is on the Evaluation Toolbar[3] and has the symbol **xf**. To define a variable with units of °C or °F, enter the definition as usual, but use the postfix operator (instead of the multiplication operator) before the units.

An example of a calculation using a temperature in degrees Fahrenheit is shown in Figure 2.30. In the example, the volume of a gas at standard temperature and pressure is calculated. The temperature is specified in °F, but an absolute temperature is required for ideal gas calculations. This calculation works, because Mathcad converts all defined variables to base units, which are kelvins for temperature.

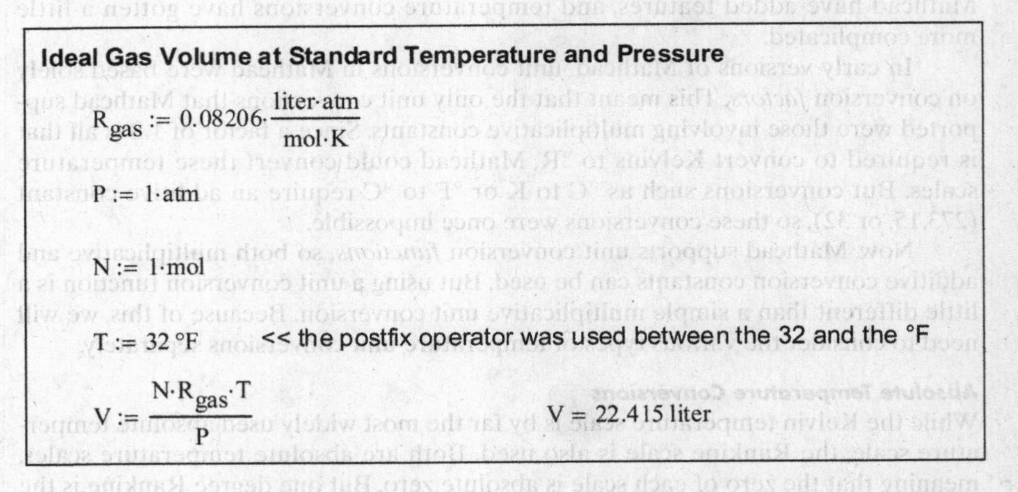

Ideal Gas Volume at Standard Temperature and Pressure
$R_{gas} := 0.08206 \cdot \frac{liter \cdot atm}{mol \cdot K}$
$P := 1 \cdot atm$
$N := 1 \cdot mol$
T := 32 °F
<< the postfix operator was used between the 32 and the °F
$V := \frac{N \cdot R_{gas} \cdot T}{P}$
V = 22.415 liter

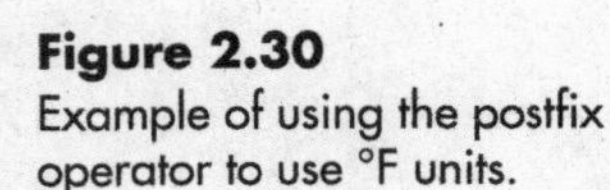

Figure 2.30
Example of using the postfix operator to use °F units.

Units on Temperature Changes

Another source of confusion is that we typically use the same temperature symbols for both temperatures (T) and changes in temperature (ΔT), but the conversion process is different for T's and ΔT's when working with the Fahrenheit or Celsius scales. ΔT's are more common in engineering than you might realize, since they are built into some common property values such as heat capacity (J/kg K) and thermal conductivity (W/m K).

For the absolute temperature scales, the conversion process is the same for temperatures and changes in temperature:

- to convert $T_{°R}$ to T_K, use $T_{°R} = T_K \cdot 1.8$
- to convert $\Delta T_{°R}$ to ΔT_K, use $\Delta T_{°R} = \Delta T_K \cdot 1.8$

---

[2]The postfix operator causes a function that is entered just after an operand to be evaluated using the operand. In this context, we are asking the unit conversion function implied by °C or °F to use the temperature just to the left of the function.

[3]The keyboard shortcut for the postfix operator is [Ctrl-Shift-x].

So, Mathcad's absolute temperature conversions work for either T or ΔT. However, the conversions for the Fahrenheit or Celsius scales are different:

- to convert $T_{°F}$ to $T_{°C}$, use $T_{°F} = (T_{°C} \cdot 1.8) + 32$
- to convert $\Delta T_{°F}$ to $\Delta T_{°C}$, use $\Delta T_{°F} = \Delta T_{°C} \cdot 1.8$

Mathcad provides unit conversions functions for temperature changes in °C and °F:

- Δ°C—specifies a temperature change in °C
- Δ°F—specifies a temperature change in °F

These are available from the Units dialog box (press [Ctrl-u] to open the Units dialog box) in the **Temperature** section. The Mathcad example in Figure 2.31 illustrates how to use these unit conversion functions.

Using Mathcad's Units for Temperature Change (Δ°C and Δ°F)

$T_{boil} := 212\,°F$

$T_{freeze} := 32\,°F$

$\Delta T := T_{boil} - T_{freeze}$

$\Delta T = 180\,\Delta°F$ << correct result using Δ°F

$\Delta T = -279.67\,°F$ << incorrect result using °F

$\Delta T = 100\,K$ << correct result using K

$\Delta T = 100\,\Delta°C$ << correct result using Δ°C

Figure 2.31
Using units with temperature changes.

2.6.4 Limitations to Mathcad's Units Capabilities

- Some of Mathcad's built-in functions do not support, or do not fully support, units. For example, the linear regression function **`linfit`** does not accept values with units. The iterative solver (**`given-find`** solve block) does allow units, but if you are solving for two or more variables simultaneously, all of the variables must have the same units.
- Mathcad's built-in graphics always display the values in the base (stored) units.
- Only the SI system of units fully supports moles. In the other systems, Mathcad allows you to use the term "mole" as a unit but does not consider it as a dimension and does not display it when presenting units. This means that you should use the SI system if you plan to work with moles.
- The mole defined in Mathcad's SI system is the gram-mole. You can make this obvious by defining a new unit, the "gmol":

```
gmol := mole
```

The other commonly used molar units can also be defined:

```
kmol := 1000 · mole
lbmol := 453.593 · mole
```

With these definitions, Mathcad can convert between the various types of moles (but only if you are using the SI system of units).

APPLICATION

Determining the Current in a Circuit

Consider a simple circuit containing a 9-volt battery and a 90-ohm resistor, as shown in Figure 2.32. What current would flow in the circuit? To solve this problem, you need to know Ohm's law,

$$V = iR,$$

and Kirchhoff's law of voltage,

For a closed (loop) circuit, the algebraic sum of all changes in voltage must be zero.

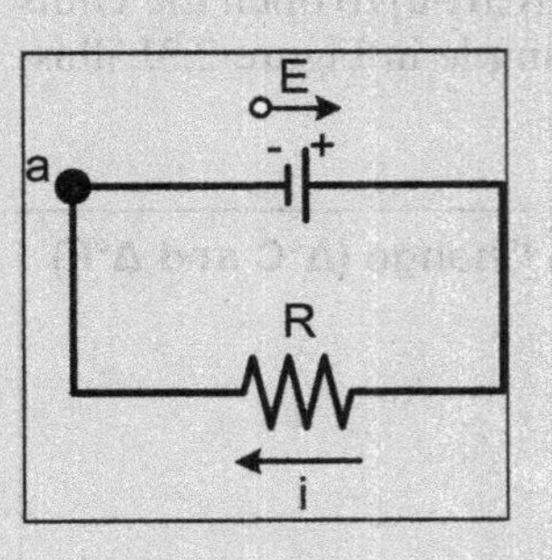

Figure 2.32
Closed loop circuit.

Kirchhoff's law is perhaps easier to understand if you consider the following analogy:

If you are hiking in the hills and you end up back at the same spot you started from (a loop trail), then the sum of the changes in elevation must be zero: You must end up at the same elevation at which you started.

The electrons moving through the circuit from point a may have their electric potential (volts) increased (by the battery) or decreased (by the resistor), but if they end up back where they started from (in a loop circuit), they must end up with the same potential they started with. We can use Kirchhoff's law to write an equation describing the voltage changes through the circuit. Starting at an arbitrary point a and moving through the circuit in the direction of the current flow the voltage is first raised by the battery ($V_B = E$) and then lowered by the resistor (V_R). These are the only two elements in the circuit, so

$$V_B - V_R = 0$$

by Kirchhoff's law. Now, we know that the battery raises the electric potential by 9 volts, and Ohm's law relates the current flowing through the resistor (and the rest of the circuit) to the resistance, *90 ohms*. Thus,

$$V_B - iR = 0.$$

The Mathcad equations needed to calculate the current through the circuit look like as shown in Figure 2.33:

Determining the Current in a Circuit

$V_B := 9 \cdot \text{volt}$

$R := 90 \cdot \text{ohm}$

$i := \frac{V_B}{R}$ $\quad i = 0.1\,\text{amp}$

Figure 2.33
Mathcad solution.

2.7 CONTROLLING HOW RESULTS ARE DISPLAYED

The Result Format dialog box allows you to control the way individual numbers, matrices, and units are displayed. If you have not selected a result when you change a setting in the Result Format dialog box, you will change the way all results on the worksheet will be displayed. However, if you select a value before you open the Result Format dialog box; the changes you make in the Result Format dialog box will affect only that one result.

The Result Format dialog box can be opened in two ways:

1. from the menu, using **Format/Result...** or
2. by double-clicking on a displayed result.

2.7.1 Controlling the Way Numbers Are Displayed

Mathcad tries to present results in a readable form, with only a few decimal places displayed. You may want to change the way results are displayed, in any of the following ways:

1. Format: You can use a **General, Decimal, Scientific, Engineering**, or **Fraction** format. (The default is **General** format, which displays small values without exponents, but switches to scientific notation for larger values.)
2. Number of decimal places displayed (three by default).
3. Whether or not to display trailing zeros. (By default, they are not displayed.)
4. Exponential threshold (for **General** format only): You can indicate how large values can get before Mathcad switches to scientific notation.

You can open the Result Format dialog box by double-clicking on a displayed result. As an example, we'll use **B** = 1/700, a very small value with lots of decimal places. The Mathcad defaults are shown with the Result Format dialog box in Figure 2.34:

- **General** format
- three decimal places
- trailing zeros not shown
- exponential threshold of 3 (i.e., 999 displayed as decimal, 1000 displayed in scientific notation)

$B := \frac{1}{700}$ $B = 1.429 \times 10^{-3}$

Figure 2.34
Result Format dialog box.

If you were to select the **Decimal** format, the result would be displayed in standard notation, but still with only three decimal places:

$$B := \frac{1}{700}$$

$$B = 0.001$$

To see more decimal places, change the **Number of decimal places** from 3 to some larger value, say, 7:

$$B := \frac{1}{700}$$

$$B = 0.0014286$$

The following table compares **General, Scientific,** and **Engineering** formats:

DEFINITION	GENERAL FORMAT	SCIENTIFIC FORMAT	ENGINEERING FORMAT
$x := \begin{pmatrix} 10 \\ 100 \\ 1000 \\ 10000 \\ 100000 \\ 1000000 \end{pmatrix}$	$x := \begin{pmatrix} 10.000 \\ 100.000 \\ 1.000 \times 10^3 \\ 1.000 \times 10^4 \\ 1.000 \times 10^5 \\ 1.000 \times 10^6 \end{pmatrix}$	$x := \begin{pmatrix} 1.000 \times 10^1 \\ 1.000 \times 10^2 \\ 1.000 \times 10^3 \\ 1.000 \times 10^4 \\ 1.000 \times 10^5 \\ 1.000 \times 10^6 \end{pmatrix}$	$x := \begin{pmatrix} 10.000 \times 10^0 \\ 100.000 \times 10^0 \\ 1.000 \times 10^3 \\ 10.000 \times 10^3 \\ 100.000 \times 10^3 \\ 1.000 \times 10^6 \end{pmatrix}$

General format displays small values without a power but switches large values (greater than the exponential threshold) to scientific notation. **Scientific** format uses scientific notation for values of any size. **Engineering** format is similar to scientific notation, except that the powers are always multiples of 3 (−6, −3, 0, 3, 6, 9, etc.)

2.7.2 Controlling the Way Matrices Are Displayed

By default, Mathcad shows small matrices (Mathcad calls them *arrays*) in their entirety, but switches to scrolling tables when the matrices get large. You can change the way a matrix result is displayed by double-clicking on the matrix (Figure 2.35).

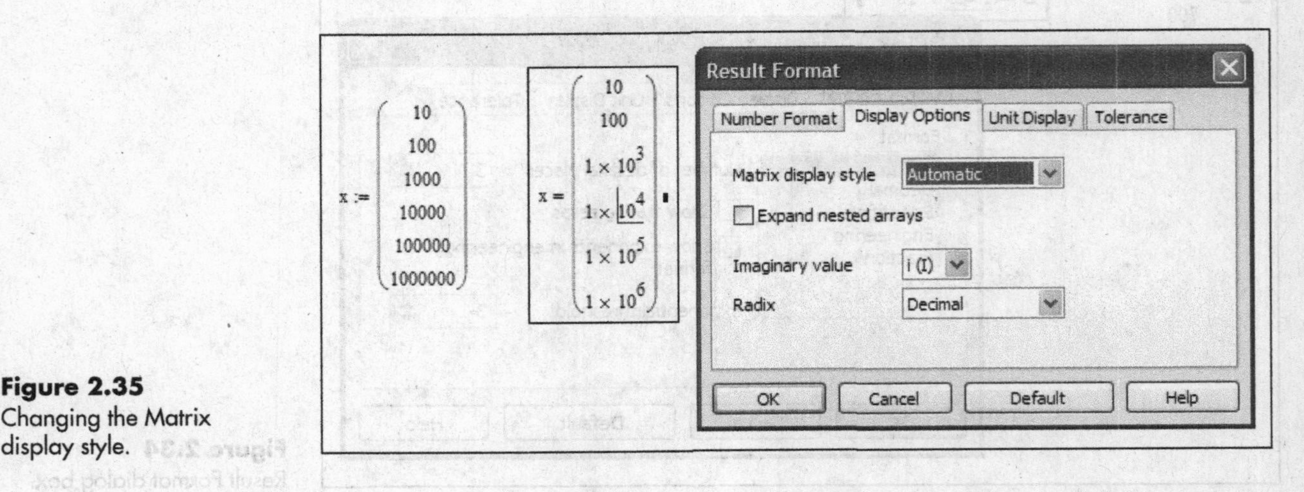

Figure 2.35
Changing the Matrix display style.

The **x** matrix used in this example is a pretty small matrix, so the **Automatic** selection (default) displays the result as a matrix. You can change the display style using the drop-down list and selecting **Table** as shown in Figure 2.36.

When you click on the **OK** button, the **x** matrix is displayed as a scrolling table (Figure 2.37).

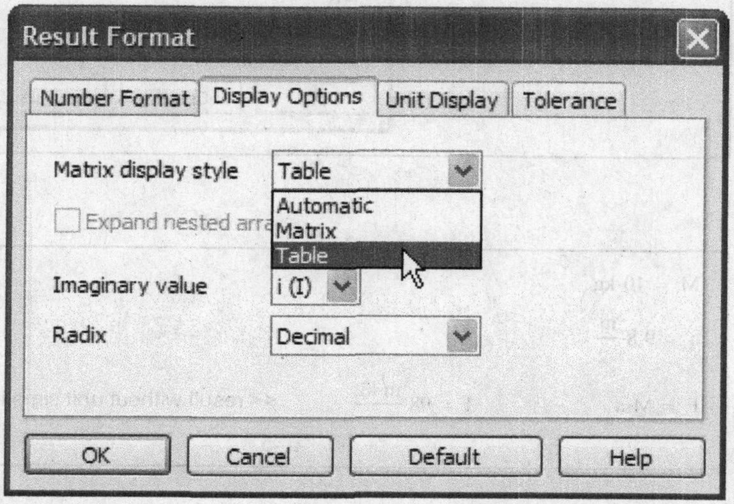

Figure 2.36
Changing the Matrix display style to Table.

Figure 2.37
Matrix x displayed as a scrolling table.

2.7.3 Controlling the Way Units Are Displayed

By default, Mathcad tries to simplify units whenever possible. Because of this, the units on the result in Figure 2.38 are *newtons* rather than $kg \cdot m/s^2$.

If you do not want Mathcad to simplify the units for a particular result, double-click on that result to bring up the Result Format dialog box. Then, clear the **Simplify units when possible** check box. When you leave the dialog box, the result will be displayed without simplifying the units (Figure 2.39).

The **Format units** check box (see Figure 2.38) indicates that units should be displayed in common form. If the **Format units** check box is cleared, the result would be displayed as

$$F = 98.000\ kg\ m\ s^{-2}$$

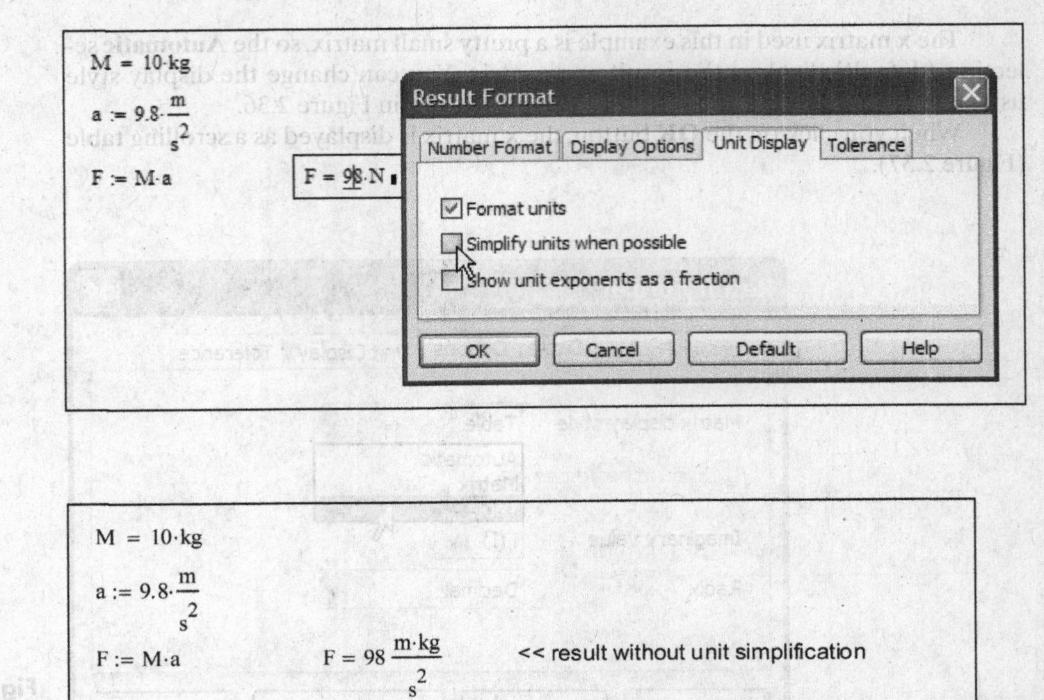

Figure 2.38
Controlling how units are displayed.

$M = 10 \cdot kg$

$a := 9.8 \cdot \frac{m}{s^2}$

$F := M \cdot a$ $\qquad F = 98 \frac{m \cdot kg}{s^2}$ << result without unit simplification

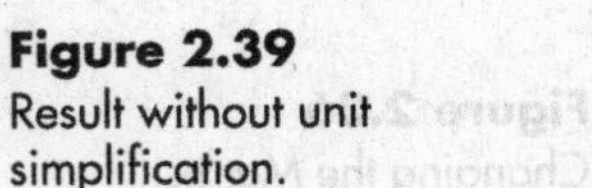

Figure 2.39
Result without unit simplification.

2.8 ENTERING AND EDITING TEXT

Mathcad defaults to equation edit mode, so if you just start typing, Mathcad will try to interpret your entry as an equation. If you type a series of letters and then a space, Mathcad will recognize that you are entering text and will switch to *text edit mode* and create a *text region*. Or, you can tell Mathcad you want to enter text by pressing the quote key ["].

To create a text region, position the edit cursor (crosshair) in the blank portion of the worksheet where you want the text to be placed, and then press ["] (the quote key). A small rectangle with a vertical line (called the *insert bar*) appears on the worksheet, indicating that you are creating a text region. The text region will automatically expand as you type, until you reach the page margin (the vertical line at the right side of the worksheet). Then the text will automatically wrap to the next line. This is illustrated in Figure 2.40.

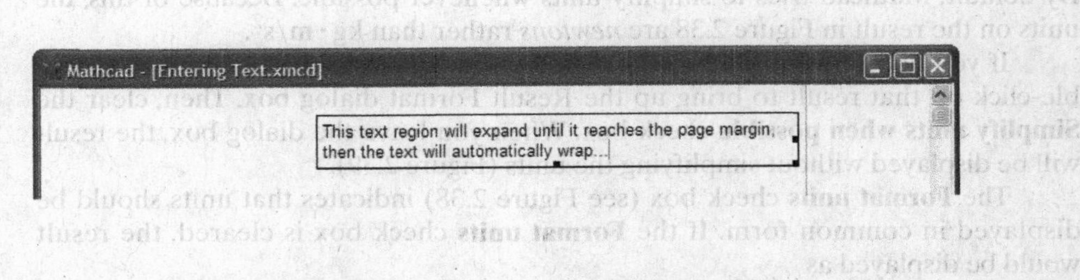

Figure 2.40
Entering text.

You can change the size or location of an existing text region by clicking on the text to select the region and display the border. Then, drag the border to move the region, or drag one of the *handles* (the small boxes on the border) to change the size of the region. If you change the width of a text region, the text will wrap as necessary to fit in the new width.

Once you have a block of text in a text region, you can select all or part of the text with the mouse and then use the formatting buttons on the Format Toolbar. These buttons allow you to make the font boldface, add italics, or underline the text. For example, to create a heading for some climate change calculations, you might type in some text and then select "Climate Change" as illustrated in Figure 2.41.

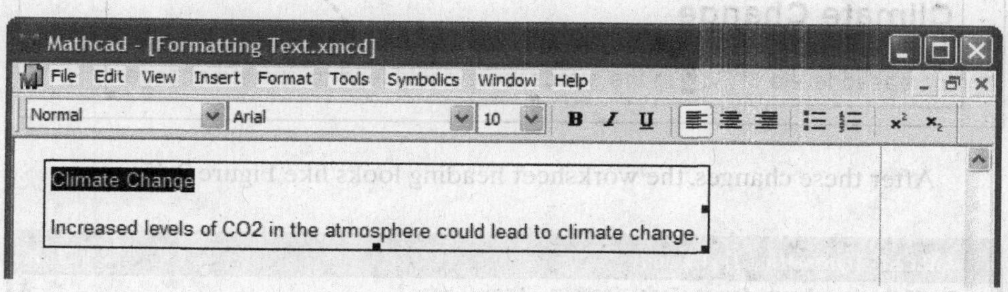

Figure 2.41
Selecting text for formatting.

Then increase the font size of the selected text by using the Format Toolbar (Figure 2.42).

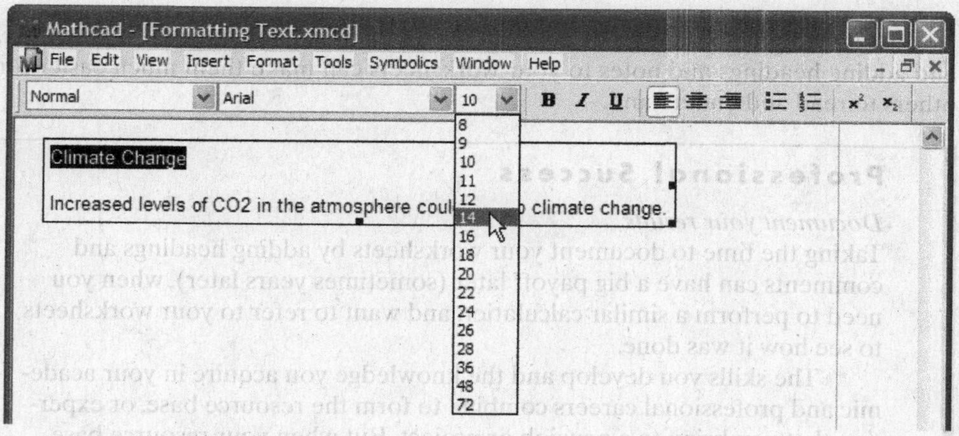

Figure 2.42
Changing font size.

Finally, make the font in the heading boldface by using the Format Toolbar (Figure 2.43).

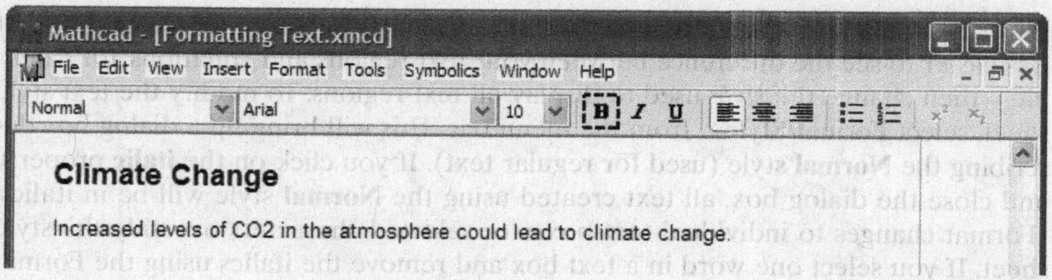

Figure 2.43
Changing text attributes.

You can also create sub- or superscripts within the text region by selecting the characters to be raised or lowered; then press the **Subscript** or **Superscript** buttons on the Format Toolbar (Figure 2.44).

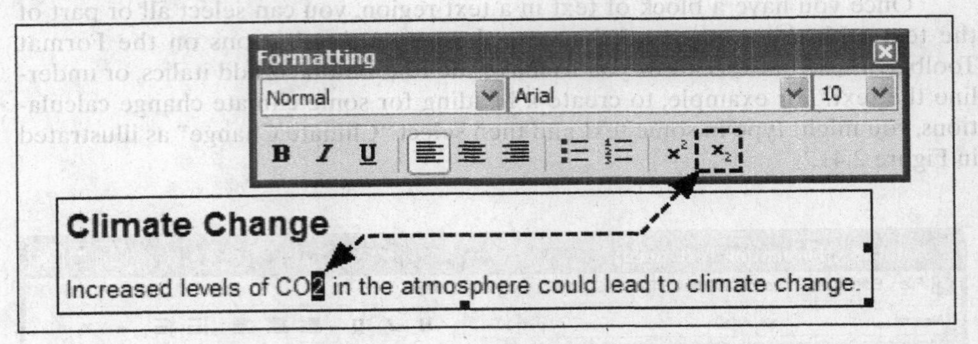

Figure 2.44
Subscripting selected text.

After these changes, the worksheet heading looks like Figure 2.45.

Mathcad - [Formatting Text.xmcd]
File Edit View Insert Format Tools Symbolics Window Help
Climate Change
Increased levels of CO_2 in the atmosphere could lead to climate change.

Figure 2.45
The formatted result.

Mathcad gives you a great deal of control over the appearance of your text, and adding headings and notes to your worksheets can make them much easier for others to read and understand.

> **Professional Success**
>
> *Document your results.*
> Taking the time to document your worksheets by adding headings and comments can have a big payoff later (sometimes years later), when you need to perform a similar calculation and want to refer to your worksheets to see how it was done.
>
> The skills you develop and the knowledge you acquire in your academic and professional careers combine to form the resource base, or expertise, that you bring to a new job or project. But when your resource base gets stale, you can spend a lot of time relearning things you once knew. Plan ahead and make the relearning as easy as possible by making your worksheets easy to understand.

Note: If you want to make a change to all text regions – italicizing all the text so that it is easier to see the difference between your text regions and equations, for example – then change the *style* used to display all text regions. To modify the text style sheet, select **Format/Style...** from the menu bar. This will bring up a dialog box describing the **Normal** style (used for regular text). If you click on the **italic** property and close the dialog box, all text created using the **Normal** style will be in italics. (Format changes to individual text boxes or selected characters override the style sheet. If you select one word in a text box and remove the italics using the Format

Bar, that individual change would override the style sheet, and the italics would be removed from that word.)

2.8.1 Sizing, Moving, and Deleting a Text Region

Just like selecting equation regions, there are two ways to select text regions: You can select for editing or select for moving.

Selecting a Text Region for Editing

To select a text region for editing, simply click on the text. When you click on a text region, Mathcad automatically switches to text editing mode. Click outside of the text region when you have finished editing.

To resize a text region, first select the region for editing. When a text region has been selected for editing, three black markers appear on the border; these are called *handles* and are used to resize the text region. Use the mouse pointer to grab a handle and drag the handle and stretch the border to a new location. The mouse pointer will change to a double-headed arrow when the mouse is positioned over a handle.

To move a text region that has been selected for editing, position the mouse pointer over the border, but not over a handle. The mouse pointer will change to a small hand when positioned over the border. Grab the border and drag the entire region to the desired location.

Selecting Text Regions for Moving or Deleting

To select one or more regions for moving, either hold the shift key down when you click on the region(s), or drag-select the region(s) by clicking outside of any region and dragging the mouse pointer over the region(s) you want to select. Regions selected for moving are shown with a dashed line border. When regions have been selected for moving, simply position the mouse over any selected region. The mouse pointer will change to a small hand. Drag the mouse pointer to move all selected regions to a new location.

A region that has been selected for moving cannot be edited. When you have finished moving the region(s), click on the worksheet away from all regions to release the selection.

Regions selected for moving can also be quickly deleted just by pressing the [Delete] key after selecting the regions.

2.8.2 Inserting Equations inside Text Regions

You can move an existing equation into a text region so that the equation is read as part of the text. When equations are moved inside text regions, they are still functioning equations in the Mathcad worksheet and are evaluated just like any other equation.

As an example, consider a force calculation with a paragraph describing Newton's second law, as shown in Figure 2.46.

Newton's Second Law

$M = 10 \cdot kg$

$a := 9.8 \cdot \frac{m}{s^2}$

The relationship between force, mass, and acceleration was defined by Isaac Newton. The commonly used version of this equation is called Newton's Second Law.

$F := M \cdot a$

$F = 98\,N$

Figure 2.46
Example before moving equation.

The equation for calculating force can be moved inside the paragraph, and it will still calculate the force. The equation must still be placed on the worksheet after **M** and **a** have been defined and before the result is displayed. (See Figure 2.47.)

Newton's Second Law

$M = 10 \cdot kg$

$a := 9.8 \cdot \frac{m}{s^2}$ *The relationship between force, mass, and acceleration was defined by Isaac Newton. The commonly used version of this equation,* $F := M \cdot a$ *, is called Newton's Second Law.*

$F = 98\,N$

Figure 2.47
Example of an equation inside a text region.

2.9 A SIMPLE EDITING SESSION

We will use one of the examples mentioned earlier, namely, determining the surface area of a cylinder, to demonstrate how to solve a simple problem using Mathcad. Even in this simple example, the following Mathcad features will be used:

- text editing and formatting;
- variable definitions (**r** and **L**);
- equation definition;
- exponentiation ($\mathbf{r}^2$);
- displaying the result (=); and
- working with units.

Statement of the Problem
Determine the surface area of a cylinder with a radius of 7 cm and a length of 21 cm.

The solution will be presented with a step-by-step commentary. For this simple example, the complete Mathcad worksheet will be shown as it develops at each step.

Step 1: Use a Text Region to Describe the Problem

1. Position the edit cursor (crosshair) near the top of the blank worksheet and click the left mouse button.
2. Press ["] to create a text region.
3. Enter the statement of the problem.
4. Click outside of the text region when you are done entering text.

The result might look something like Figure 2.48. (Italic text is not the Mathcad default, but adding italics can help differentiate text regions from equations.)

Statement of the Problem

Determine the surface area of a cylinder with a radius of 7 cm and a length of 21 cm.

Figure 2.48
Worksheet development, Step 1.

Step 2: Enter the Known Values of Radius and Length, with Appropriate Units

1. Position the edit cursor below the text region and near the left side of the worksheet. Click the left mouse button.

2. Define the value of **r** by typing [r] [:] [7] [*] [c] [m] [Enter].
3. Similarly, position the edit cursor, and enter the value of **L** (21 cm).

The positions of the **r** and **L** equations are arbitrary, but they might look like the example in Figure 2.49.

Statement of the Problem

Determine the surface area of a cylinder with a radius of 7 cm and a length of 21 cm.

$r := 7 \cdot cm$

$L := 21 \cdot cm$

Figure 2.49
Worksheet development, Step 2.

Step 3: Enter the Equation for Computing the Surface Area of a Cylinder

1. Position the edit cursor below or to the right of the definitions of **r** and **L**.
2. Enter the area variable name, $\mathbf{A_{cyl}}$, by using a text subscript, as [A] [.] [c] [y] [l].
3. Press [:] to indicate that you are defining the $\mathbf{A_{cyl}}$ variable.
4. Enter the first term on the right side ($\mathbf{2\pi r^2}$) as [2] [*] [Ctrl-Shift-p] [*] [r][^][2].
5. Press [Space] to select the $\mathbf{r^2}$ in the $\mathbf{2\pi r^2}$ term.
6. Press [+] to add to the selected term.
7. Enter the second term on the right side ($\mathbf{2\pi rL}$) as [2][*][Ctrl-Shift-p][*][r][*][L].
8. Press [Enter] to conclude the equation entry.

The result is shown in Figure 2.50.

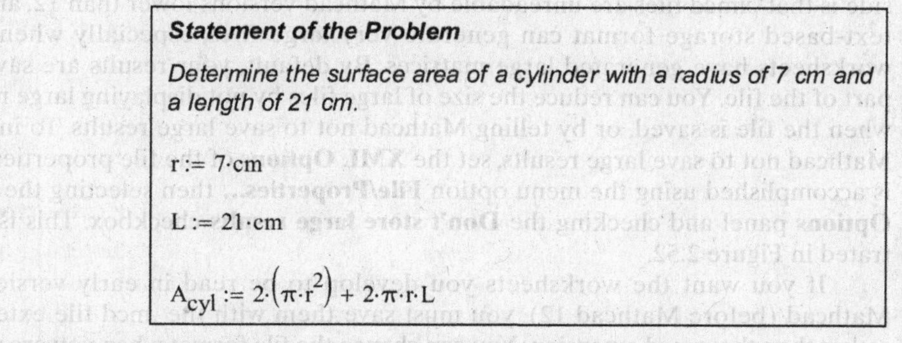
Statement of the Problem
Determine the surface area of a cylinder with a radius of 7 cm and a length of 21 cm.
r := 7·cm
L := 21·cm
A_cyl := 2·(π·r^2) + 2·π·r·L

Figure 2.50
Worksheet development, Step 3.

Step 4: Display the Result in the Desired Units

1. Position the edit cursor below or to the right of the $\mathbf{A_{cyl}}$ equation.
2. Ask Mathcad to display the result of the calculation by typing [A] [.] [c] [y] [l] [=].
3. Click on the units placeholder (to the right of the displayed base units).
4. Enter the desired units (e.g., cm^2).
5. Press [Enter] to conclude the entry.

The worksheet will look like the one shown in Figure 2.51.

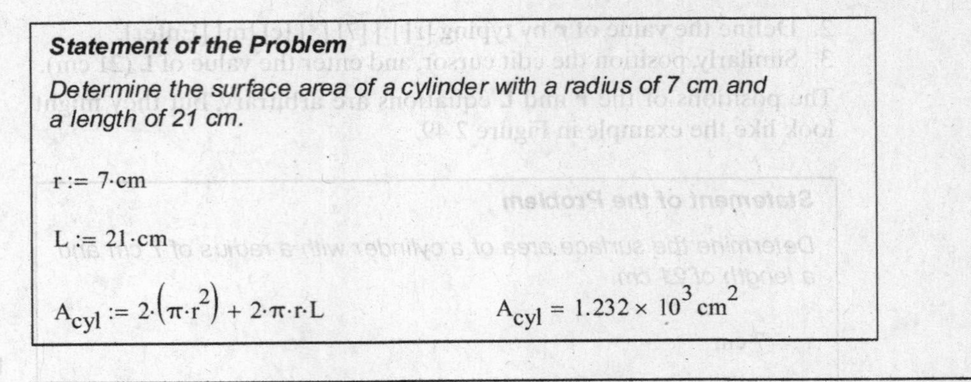
Statement of the Problem
Determine the surface area of a cylinder with a radius of 7 cm and a length of 21 cm.
r := 7·cm
L := 21·cm
A_cyl := 2·(π·r^2) + 2·π·r·L
A_cyl = 1.232 × 10^3 cm^2

Figure 2.51
Worksheet development, Step 4.

> **Practice!**
>
> Calculate the surface area and volume of
>
> **a.** a cube 1 cm on a side.
> **b.** a sphere with a radius of 7 cm.
>
> Equations for the sphere are $A = 4\pi r^2$ and $V = (4/3)\pi r^3$.

2.10 SAVING YOUR WORKSHEETS

To save your worksheet, click the **Save** button on the Standard toolbar, or use the equivalent menu options: **File/Save**. Saving a file under a new file name can be useful when solving several related problems. To save a file with a new name, use **File/Save As...**.

By default, Mathcad saves worksheets in XML format using the file extension. xmcd. The text-based XML format is supposed to improve accessibility to your files across operating systems and allows improved search capabilities. The downside is that .xmcd files are unreadable by Mathcad versions lower than 12, and the text-based storage format can generate very large files, especially when your worksheets have generated large matrices. By default, your results are saved as part of the file. You can reduce the size of large files by not displaying large results when the file is saved, or by telling Mathcad not to save large results. To instruct Mathcad not to save large results, set the **XML Options** of the file properties. This is accomplished using the menu option **File/Properties...** then selecting the **XML Options** panel and checking the **Don't store large results** checkbox. This is illustrated in Figure 2.52.

If you want the worksheets you develop to be read in early versions of Mathcad (before Mathcad 12), you must save them with the .mcd file extension rather than the .xmcd extension. You can change the file format when you are saving the worksheet using the Save as type: drop-down list on either the **Save** or **Save As...** dialog box.

2.10.1 Autosave Feature

Mathcad provides an ***Autosave*** feature to protect your work from program or system crashes. When Autosave is activated, your worksheet is saved periodically so that if something goes wrong (e.g., a power outage) all of your work prior to the last Autosave is preserved.

Figure 2.52
Setting XML options.

To activate the Autosave feature, use menu options **Tools/Preferences...** then select the **Save** panel and check the **Autosave** checkbox shown in Figure 2.53. By default, files are saved after every five minutes, but this can also be changed on the **Save** panel.

Figure 2.53
Activating Autosave.

If you exit Mathcad without errors (i.e., no system crash, power loss, etc.), then the Autosave files are deleted. The only time the Autosave files will be available is after a serious problem has abnormally stopped the program. When you restart Mathcad after a crash, you will have an opportunity to open the recovered Autosave files, as shown in Figure 2.54.

Figure 2.54
Recovering Autosave files.

APPLICATION

Nanotechnology for Computer Chips

The field of *nanotechnology* is relatively, but is growing rapidly. In general terms, it is the science of the very small. (One nanometer is 1×10^{-9} meter.) The goal of this new field of study is to develop the ability to manufacture on a molecular level, arranging individual atoms to create structures and devices with particular characteristics, properties, or abilities. Computer chip manufacturers are hoping that nanotechnology will provide a solution for problems that are heading their way...

The number of transistors on computer chips has been growing exponentially for the past thirty years, doubling approximately every eighteen months. This has fueled the rapid increase in computer power and the simultaneous decrease in costs that we have witnessed over the past decade. But those transistors are formed on chips using photolithographic methods; essentially they are printed on silicon using light. The wavelength of the light source imposes a fundamental limitation on the minimum size of the transistors, and chip manufacturers have been switching to higher frequency, shorter wavelength light sources to get the printed line width down to 0.14 μm (or 140 nm). But chip manufacturers know that they are rapidly running out of opportunities to obtain ever-smaller line widths using photolithography. The search is on for radically different approaches to chip building.

One of the radical new approaches that shows promise is the use of semiconducting carbon nanotubes as transistors. These nanotubes are cylinders made of carbon atoms, approximately 5 to 10 atoms in diameter. Nanotubes are currently being produced and are being studied for a wide range of applications, including potential use as transistors.

As a quick (and very superficial) estimate of the potential for nanotubes to increase the power of computer chips, let's calculate how many nanotubes could fit on a 220 mm^2 computer chip.

First, we need to define millimeters, micrometers, and nanometers in Mathcad:

$$mm := 10^{-3} \cdot m$$
$$\mu m := 10^{-6} \cdot m$$
$$nm := 10^{-9} \cdot m$$

Nanotubes are approximately 1 nm in diameter and perhaps 200 to 500 nm in length. If the nanotube is attached to the surface of a chip, it would occupy an area of up to 500 nm^2. Thus, we have

$$D := 1 \cdot nm$$
$$L := 500 \cdot nm$$
$$A_{nt} := D \cdot L \qquad A_{nt} = 500\ nm^2$$

A recent, commonly produced computer chip held about 50 million transistors on a silicon chip with an area of 220 mm^2. If you pack the surface with nanotubes, you could fit 440 billion nanotubes in the same area, a result determined as follows:

$$A_{chip} := 220 \cdot mm^2 \qquad A_{chip} = 2.2 \times 10^{14} nm^2$$
$$N_{tubes} := \frac{A_{chip}}{A_{nt}} \qquad N_{tubes} = 440 \times 10^{9}$$

Although, you probably would not be able to pack the nanotubes so tightly, if you used just a tenth of the tubes in the same area, you would still increase the number of transistors per chip by a factor of nearly 1,000 compared with today's chips.

APPLICATION

AC Circuits and Complex Numbers

When an AC (alternating current) circuit has multiple voltage sources, you must account for the phase of each voltage source as well as the voltage to calculate the combined voltage and phase angle. (We are assuming that all sources are at the same frequency.)

The circuit shown in Figure 2.55 has three AC voltage sources in series, but the three sources have different magnitudes and phases, as shown in the figure. What is the combined voltage and phase for the three sources?

The voltage and angle of the three sources can be described as vectors where the voltage is related to the length of the vector and the phase is the vector's angle. As vectors, the three sources can be depicted as shown in Figure 2.56.

25 V at -65° E_1
10 V at 75° E_2
15 V at -10° E_3

Figure 2.55
Three AC voltage sources in series.

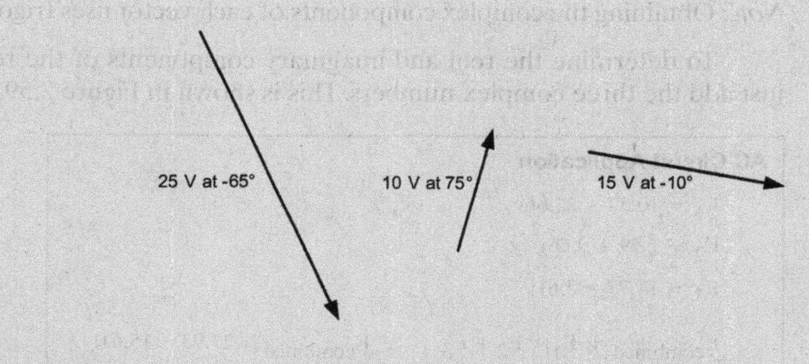

Figure 2.56
The three AC voltage sources as vectors.

Because the voltage sources are connected in series, the vectors can be connected end-to-end to find the resultant vector (the arrow from the tail of the first vector to the tip of the third vector). The length and direction of the resultant vector tells us the voltage and phase angle of the combined AC voltage sources. The resultant vector is shown with a dashed line in Figure 2.57.

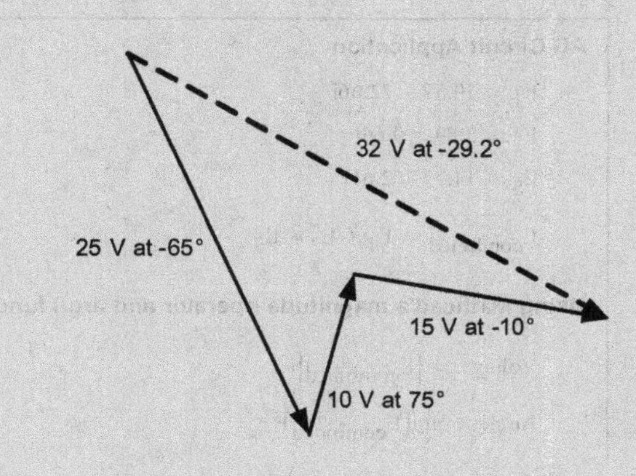

Figure 2.57
The combined vectors and the resultant vector.

From the resultant vector shown in Figure 2.57, we expect the combined voltage to be 32V at an angle of $-29.2°$, but we need to find a way to calculate the result without drawing a picture. One way to do so is to rewrite each voltage-angle vector in terms of horizontal and vertical components. These components can be written as complex numbers. The real part of the complex number represents the horizontal component of the vector, and the imaginary part represents the vertical component, as shown in Figure 2.58.

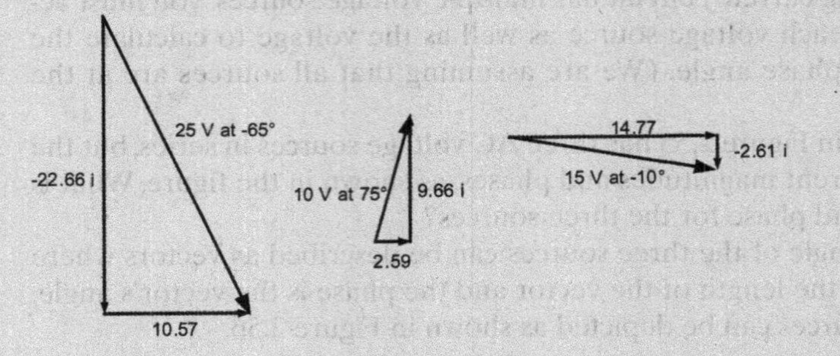

Figure 2.58
The three AC voltage sources as vectors, with complex components indicated.

Note: Obtaining the complex components of each vector uses trigonometry functions.

To determine the real and imaginary components of the resultant vector, we just add the three complex numbers. This is shown in Figure 2.59.

AC Circuit Application

$E_1 := 10.57 - 22.66i$

$E_2 := 2.59 + 9.66i$

$E_3 := 14.77 - 2.61i$

$E_{combined} := E_1 + E_2 + E_3$ $\quad E_{combined} = 27.93 - 15.61i$

Figure 2.59
Using Mathcad to find the components of the resultant vector, or the combined AC voltage.

The combined voltage can be written as 27.93 − 15.61i volts, but this is equivalent to $E_{combined} = 32$ V at −29.2°. Obtaining the magnitude and phase angle form uses functions that are presented in later chapters, but the Mathcad solution is shown in Figure 2.60 as a preview of things to come.

AC Circuit Application

$E_1 := 10.57 - 22.66i$

$E_2 := 2.59 + 9.66i$

$E_3 := 14.77 - 2.61i$

$E_{combined} := E_1 + E_2 + E_3$ $\quad E_{combined} = 27.93 - 15.61i$

Using Mathcad's magnitude operator and arg() function...

$Voltage := |E_{combined}|$ $\quad Voltage = 31.996$

$Angle := arg(E_{combined})$ $\quad Angle = -29.201\ deg$

Or, using Mathcad's square root operator and atan() function...

$Voltage := \sqrt{Re(E_{combined})^2 + Im(E_{combined})^2}$ $\quad Voltage = 31.996$

$Angle := atan\left(\frac{Im(E_{combined})}{Re(E_{combined})}\right)$ $\quad Angle = -29.201\ deg$

Figure 2.60
The completed Mathcad solution.

SUMMARY

In this chapter, we learned the basics of working with Mathcad: how the screen is laid out, how to enter equations and text, and how Mathcad handles units.

MATHCAD SUMMARY

Four Kinds of Equal Signs

:=	Assigns a value or the result of a calculation to a variable.
=	Displays a value or the result of a calculation.
=	Symbolic equality; shows the relationship between variables.
≡	Global assignment; these are evaluated before the rest of the worksheet.

Predefined Variables

π	3.141592...	Press [Ctrl-Shift-p] or choose π from the Greek Symbols Toolbar.
e	2.718281...	
g	9.8 m/s^2	
%	multiplies the displayed value by 100 and displays the percent symbol.	

Entering Equations

+,–	addition and subtraction.	
*, /	multiplication and division.	Press [Shift-8] for the multiplication symbol.
^	exponentiation.	Press [Shift-6] for the caret symbol.
[Space]	enlarges the currently selected region – used after typing in exponents and denominators.	
[Insert]	moves the vertical editing line between the beginning and end of a selected region – used when you need to delete an operator to the left of a selected region.	
[.]	text subscript.	
[ ]	index subscript – used to indicate a particular element of an array.	

Operator Precedence

^	Exponentiation is performed before multiplication and division.
*, /	Multiplication and division are performed before addition and subtraction.
+, –	Addition and subtraction are performed last.

Text Regions

["]	Creates a text region – if you type in characters that include a space (i.e., two words), Mathcad will automatically create a text region around these characters.

KEY TERMS

- anchor point
- assignment operator
- Autosave
- dimension
- editing lines
- equation region
- global assignment
- index (array) subscript
- matrix
- operator
- operator precedence
- placeholder
- predefined variable
- radix
- style
- symbolic equality
- text region
- text subscript
- toolbar
- unit system
- worksheet

PROBLEMS

2.1 Unit Conversions

(a) speed of light in a vacuum — 2.998×10^8 m/sec to miles per hour.

(b) density of water at room temperature — 62.3 lb/ft^3 to kg/m^3.

(c) density of water at 4°C — 1,000 kg/m^3 to lb/gal.

(d) viscosity of water at room temperature (approx.) — 0.01 poise to lb/ft sec. 0.01 poise to kg/m sec.

(e) ideal gas constant — 0.08206 L · atm/mole · K to joules/mole · K.

Note: The "lb" in parts b, c, and d is a pound mass, and "mole" in part e is a gram·mole.

2.2 Volume and Surface Area of a Sphere

Calculate the volume and surface area of a sphere with a radius of 3 cm as shown in Figure 2.61.

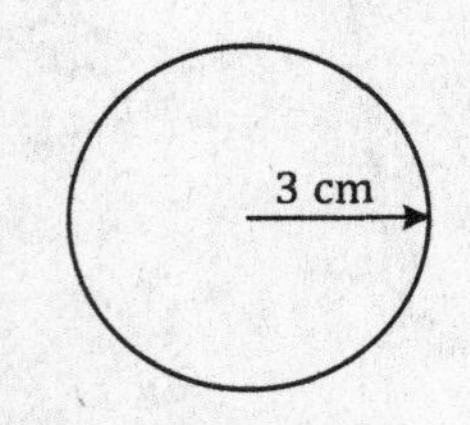

Figure 2.61
Volume and surface area of a sphere.

Use

$$V = \frac{4}{3}\pi r^3$$

and

$$A = 4\pi r^2.$$

2.3 Volume and Surface Area of a Torus

A doughnut is shaped roughly like a torus, as illustrated in Figure 2.62.

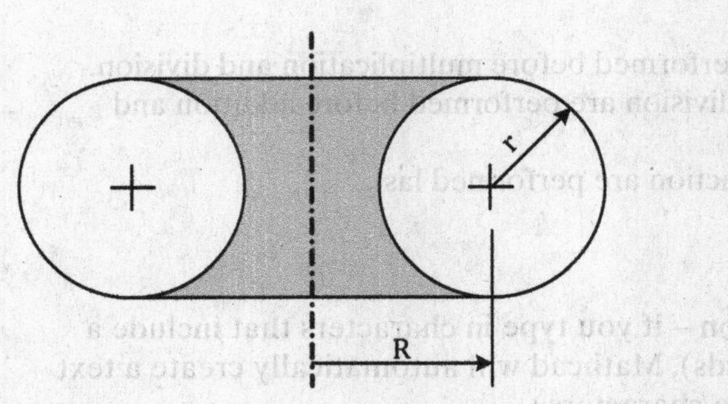

Figure 2.62
Volume and surface area of a torus.

The surface area and volume functions for a torus are

$$A = 4\pi^2 Rr$$

and

$$V = 2\pi^2 Rr^2,$$

respectively. Calculate the surface area and volume of a doughnut with **R** = 3 cm and **r** = 1.5 cm.

2.4 Ideal Gas Behavior, I

(a) A glass cylinder fitted with a movable piston (see Figure 2.63) contains 5 *grams* of chlorine gas. When the gas is at room temperature (25°C), the piston is 2 cm from the bottom of the container. The pressure on the gas is *1 atm*. What is the volume of gas in the glass cylinder (in liters)?

(b) The gas is heated at constant pressure until the piston is 5 cm from the bottom of the container. What is the final temperature of the chlorine gas (in kelvins)? Assume chlorine behaves as an ideal gas under these conditions. The molecular weight of chlorine is 35.45 grams per mole.

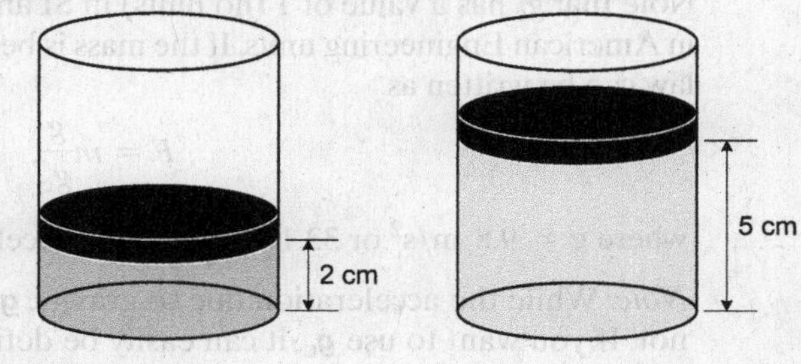

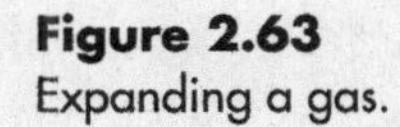

Figure 2.63
Expanding a gas.

The ideal gas equation is written as

$$PV = NRT,$$

where **P**, **V**, and **T** represent the pressure, volume, and temperature of **N** moles of gas. **R** is the ideal gas constant. The value of **R** depends on the units used on the various variables. A commonly used value for **R** is *0.08206 liter·atm/mole·***K**.

2.5 Ideal Gas Behavior, II

(a) Chlorine gas is added to a glass cylinder of *2.5 cm* in radius, fitted with a movable piston (illustrated in Figure 2.64). Gas is added until the piston is lifted *5 cm* off the bottom of the glass. The pressure on the gas is *1 atm*, and the gas and cylinder are at room temperature (25°C). How many moles of chlorine were added to the cylinder?

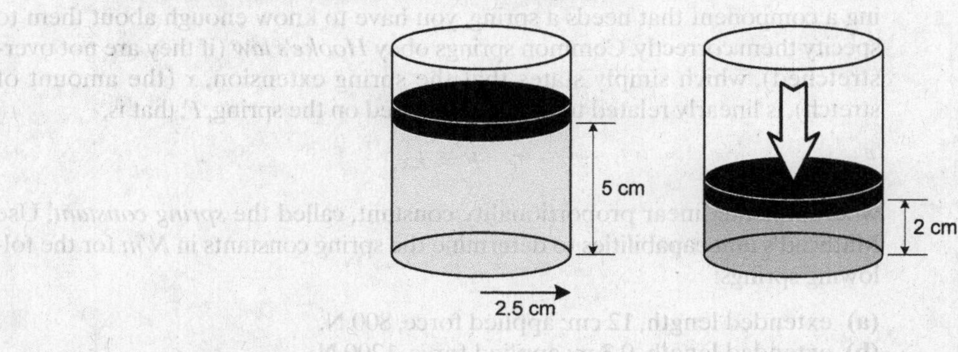

Figure 2.64
Compressing a gas.

(b) Pressure is applied to the piston, compressing the gas until the piston is *2 cm* from the bottom of the cylinder. When the temperature returns to *25°C*, what is the pressure in the cylinder (in atm)?

Assume that chlorine behaves as an ideal gas under these conditions. The molecular weight of chlorine is 35.45 grams per mole.

2.6 Relating Force and Mass

Fundamental to the design of the bridge is the relation between the mass of the deck and the force on the wires supporting the deck.

The relationship between force and mass is called *Newton's law*, and in physics courses it is usually written as

$$F = ma.$$

In engineering courses, we often build in the gravitational constant, g_c to help keep the units straight:

$$F = m\frac{a}{g_c}.$$

Note that g_c has a value of 1 (no units) in SI and a value of 32.174 $\text{ft} \cdot \text{lb}/\text{lb}_f \cdot \text{s}^2$ in American Engineering units. If the mass is being acted on by gravity, Newton's law can be written as

$$F = m\frac{g}{g_c},$$

where $g = 9.8\ \text{m/s}^2$ or $32.174\ \text{ft/s}^2$ is the acceleration due to gravity.

Note: While the acceleration due to gravity, **g**, is predefined in Mathcad, $\mathbf{g_c}$ is not. If you want to use $\mathbf{g_c}$, it can easily be defined in your worksheet with the use of the SI value, then let Mathcad take care of the unit conversions. The definition is simply

$$g_c := 1$$

(a) If a 150-kg mass is hung from a hook by a fine wire (of negligible mass) as illustrated in Figure 2.65, what force (*N*) is exerted on the hook?

(b) If the mass in part (a) were suspended by two wires, the force on the hook would be unchanged, but the tension in each wire would be halved. If the duty rating on the wire states that the tension in any wire should not exceed 300 N, how many wires should be used to support the mass?

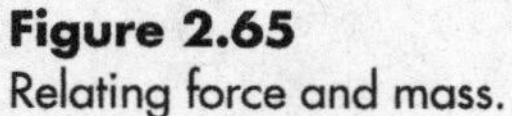
Figure 2.65
Relating force and mass.

2.7 Spring Constants

Springs are so common that we hardly even notice them, but if you are designing a component that needs a spring, you have to know enough about them to specify them correctly. Common springs obey *Hooke's law* (if they are not overstretched), which simply states that the spring extension, x (the amount of stretch), is linearly related to the force exerted on the spring, F; that is,

$$F = kx,$$

where **k** is the linear proportionality constant, called the *spring constant.* Use Mathcad's unit capabilities to determine the spring constants in *N/m* for the following springs:

(a) extended length, 12 cm; applied force, 800 N.

(b) extended length, 0.3 m; applied force, 1200 N.

(c) extended length, 1.2 cm; applied force, 100 dynes.
(d) extended length, 4 inches; applied force, 2000 lb_f.

Note: The preceding equation is actually only a simplified version of Hooke's law that is applicable when the force is in the direction of motion of the spring. Sometimes you will see a minus sign in Hooke's law. It all depends on whether the force being referred to is being applied to the spring or is within the spring, restraining the applied force.

2.8 Specifying a Spring Constant

The backrest of a chair is to be spring loaded to allow the chair to recline slightly. The design specifications call for a deflection of no more than 2 inches when a 150-lb person leans 40% of his or her body weight on the backrest. (Assume that there are no lever arms between the backrest and the spring to account for in this problem.)

(a) Use Newton's law (described in Problem 2.6) to determine the force applied to the backrest when a 150-lb person leans 40% of his or her body weight against it.
(b) Determine the constant required for the spring.
(c) If a 200-lb person puts 70% of his or her body weight on the backrest, what spring extension would be expected? (Assume that the applied force is still within the allowable limits for the spring.)

2.9 Simple Harmonic Oscillator

If a 50-gram mass is suspended on a spring (as shown in Figure 2.66) and the spring is stretched slightly and released, the system will oscillate.

The period, **T**, and natural frequency, $\mathbf{f_n}$, of this simple harmonic oscillator can be determined using the formulas

$$T = \frac{2\pi}{\sqrt{k/m}}$$

and

$$f_n = \frac{1}{T},$$

where **k** is the spring constant and **m** is the suspended mass. (This equation assumes that the spring has negligible mass.)

(a) If the spring constant is 100 N/m, determine the period of oscillation of the spring.
(b) What spring constant should be specified to obtain a period of 1 second?

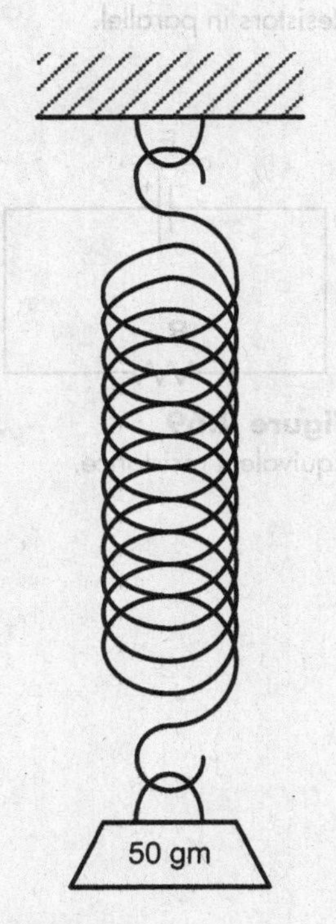

Figure 2.66
Simple harmonic oscillator.

2.10 Determining the Current in a Circuit

A real battery has an internal resistance, often shown by combining a cell symbol and a resistor symbol, as illustrated in Figure 2.67.

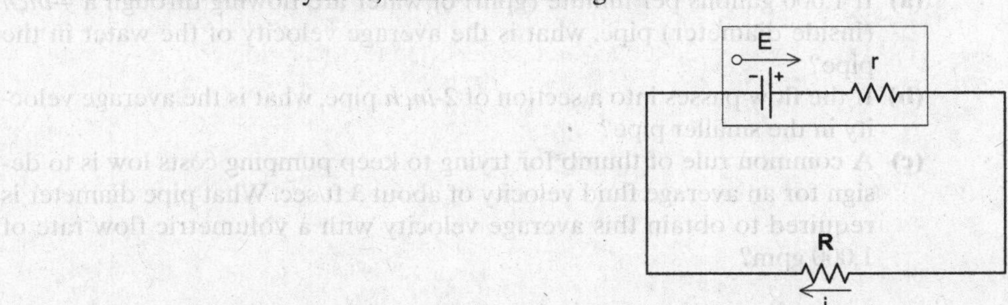

Figure 2.67
Determining current in a circuit.

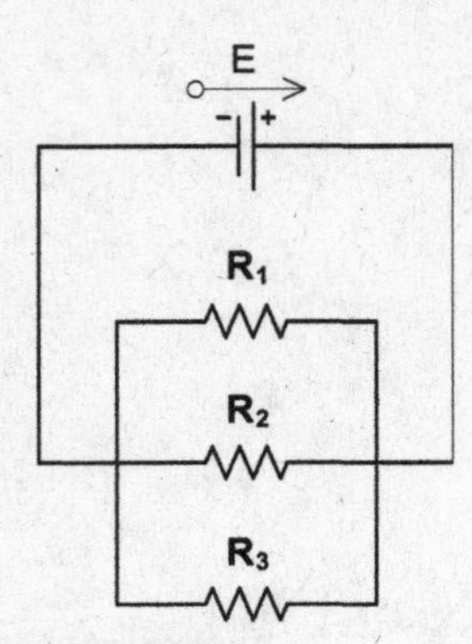

Figure 2.68
Resistors in parallel.

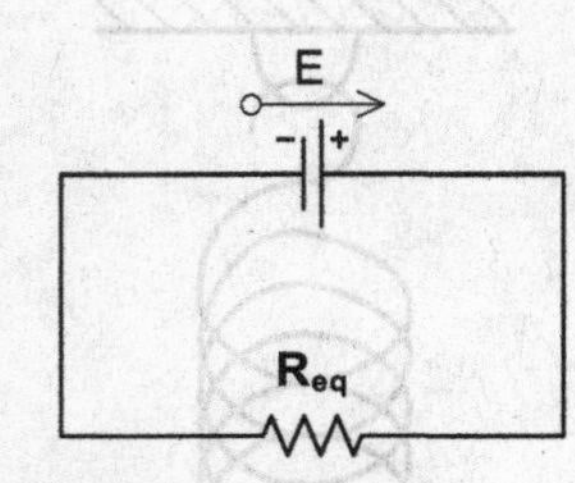

Figure 2.69
Equivalent resistance.

If a 9-volt battery has an internal resistance, **r**, of 15 ohms and is in a circuit with a 90-ohm resistor, **R**, what current would flow in the circuit?

Note: When resistors are connected in series, the combined resistance is simply the sum of the individual resistance values.

2.11 Resistors in Parallel

When multiple resistors are connected in parallel (see Figure 2.68), the equivalent resistance of the collection of resistors can be computed as

$$\frac{1}{R_{eq}} = \sum_{i=1}^{N} \frac{1}{R_i},$$

where **N** is the number of resistors connected in parallel.

Compute the equivalent resistance, $\mathbf{R_{eq}}$, in the circuit (illustrated in Figure 2.69), and determine the current that would flow through the equivalent resistor.

The data are as follows:

$$E = 9 \text{ volts};$$
$$R_1 = 20 \text{ ohms};$$
$$R_2 = 30 \text{ ohms};$$
$$R_3 = 40 \text{ ohms}.$$

2.12 Building Nanotubes

Carbon has an atomic radius of 77.2 pm (*pm* means picometers, 1 pm = 1×10^{-12} m). Approximately, how many carbon atoms are required to build a nanotube of 1 nm in diameter by 300 nm long?

2.13 Average Fluid Velocity

When a fluid is flowing in a pipe, the rate and direction of flow of the fluid – the fluid velocity – is not the same everywhere in the pipe. The fluid near the wall moves more slowly than the fluid near the center of the pipe, and things that cause the flow field to bend, such as obstructions and bends in the piping, cause some parts of the flow to move faster and in different directions than other parts of the flow. For many calculations, the details of the flow pattern are not important, and an *average fluid velocity* can be used in design calculations. The average velocity, $\mathbf{V_{avg}}$, can be determined by measuring the volumetric flow rate, Q, and dividing that by the cross-sectional area of the pipe, $\mathbf{A_{flow}}$.

$$V_{avg} = \frac{Q}{A_{flow}}$$

(a) If 1,000 gallons per minute (gpm) of water are flowing through a *4-inch* (inside diameter) pipe, what is the average velocity of the water in the pipe?

(b) If the flow passes into a section of *2-inch* pipe, what is the average velocity in the smaller pipe?

(c) A common rule of thumb for trying to keep pumping costs low is to design for an average fluid velocity of about 3 ft/sec. What pipe diameter is required to obtain this average velocity with a volumetric flow rate of 1,000 gpm?

Note: You may need to take a square root in part c. You can use either an exponent of 0.5 or Mathcad's square root operator, which is available by pressing [\] (backslash).

2.14 Orifice Meter

An *orifice meter* (see Figure 2.70) is a commonly used flow measuring device. The flow must squeeze through a small opening (the orifice), generating a pressure drop across the orifice plate. The higher the flow rate, the higher the pressure drop, so if the pressure drop is measured, it can be used to calculate the flow rate through the orifice.

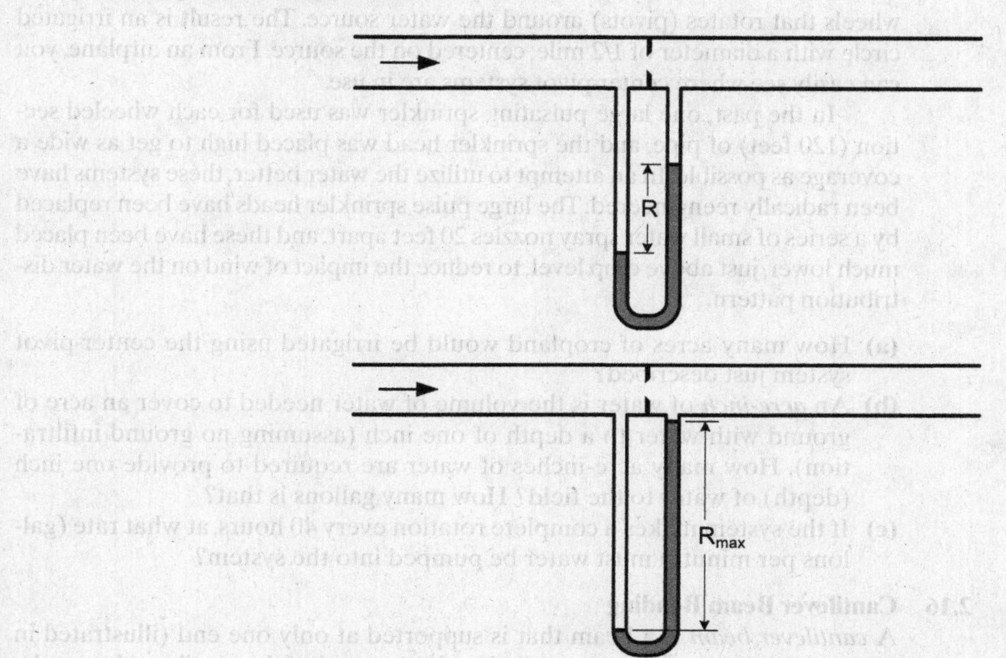

Figure 2.70
Orifice meter.

In the past, mercury manometers were often used to measure the pressure drop across the orifice plate. The manometer reading, **R**, can be related to the pressure drop, **ΔP**, by the differential manometer equation

$$\Delta P = (\rho_m - r)\frac{g}{g_c}R,$$

where ρ_m is the density of the manometer fluid; ρ is the density of the process fluid in the pipe; **g** is the acceleration due to gravity; g_c, is the gravitational constant (used to help manage units in the American Engineering system); and **R** is the difference in height of the manometer fluid on the two sides of the manometer, which is called the manometer reading.

(a) Given the following data, what is the pressure drop (atm) across the manometer when the manometer reading is `R = 32 cm`?

$$\rho_m = 13{,}600 \text{ kg/m}^3;$$
$$\rho = 1{,}000 \text{ kg/m}^3;$$
$$g = 9.80 \text{ m/sec}^2;$$
$$g_c = 1 \text{ (SI)}.$$

(b) If the flow rate gets too high, the mercury in the manometer can be washed out of the device into the pipe downstream of the orifice plate. To avoid this, the orifice must be designed to produce a pressure drop smaller than the maximum permissible **ΔP** for the manometer. If $\mathbf{R_{max}}$ = *70 cm*, what is the maximum permissible **ΔP** (atm) for the manometer?

2.15 Designing an Irrigation System

In arid areas, farmers cannot rely on periodic rains to water crops, so many of them depend on irrigation systems. Center-pivot irrigation systems are a type of system that is commonly used. These systems have a 1/4-mile-long water pipe on wheels that rotates (pivots) around the water source. The result is an irrigated circle with a diameter of 1/2 mile, centered on the source. From an airplane, you can easily see where center-pivot systems are in use.

In the past, one large pulsating sprinkler was used for each wheeled section (120 feet) of pipe, and the sprinkler head was placed high to get as wide a coverage as possible. In an attempt to utilize the water better, these systems have been radically reengineered. The large pulse sprinkler heads have been replaced by a series of small water spray nozzles 20 feet apart, and these have been placed much lower, just above crop level, to reduce the impact of wind on the water distribution pattern.

(a) How many acres of cropland would be irrigated using the center-pivot system just described?

(b) An *acre-inch* of water is the volume of water needed to cover an acre of ground with water to a depth of one inch (assuming no ground infiltration). How many acre-inches of water are required to provide one inch (depth) of water to the field? How many gallons is that?

(c) If the system makes a complete rotation every 40 hours, at what rate (gallons per minute) must water be pumped into the system?

2.16 Cantilever Beam Bending

A *cantilever beam* is a beam that is supported at only one end (illustrated in Figure 2.71). When a force is applied to the free end of the cantilever beam, the beam bends.

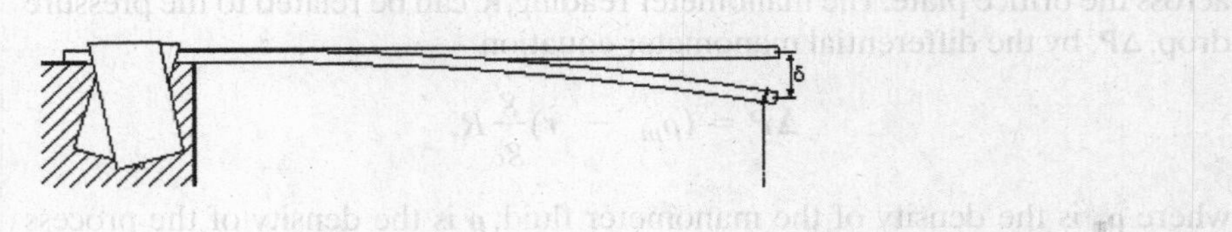

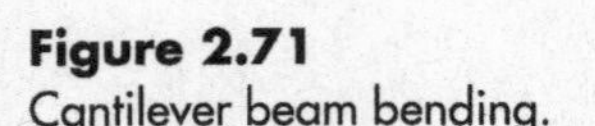

Figure 2.71
Cantilever beam bending.

The amount of bending depends on the size and shape of the beam, as well as the material from which the beam was made. The shape can be accounted for through a quantity known as the *moment of inertia*. For a cylindrical beam, the moment of inertia, **I**, can be calculated as

$$I = \frac{\pi}{4} R^4$$

where **R** is the beam radius. The type of material comes into play through a material property called the *modulus of elasticity*, **E**.

A simple experiment can be used to determine the modulus of elasticity for a material. In this experiment, two aluminum rods are strapped tightly to a table, as shown in Figure 2.60. A known mass is hung from one of the beams (not shown), which bends by an amount, δ. (The second rod is used to make it easy to measure δ.)

The deflection of a cantilever beam is related to the beam length, L; the modulus of elasticity, **E**; and the applied force, **F**, by the equation

$$\delta = \frac{L^3}{3EI}F$$

where **L** is the beam length from the edge of the table to the point where the force, **F**, is applied. If you have experimental data for **L**, **F**, and δ, it is possible to calculate the value of **E** for the rod material from this equation.

Experimental Data		
R	1	cm
L	38	cm
F	3550	N
δ	12	cm

(a) Calculate the moment of inertia and the modulus of elasticity from the experimental data.

(b) Compare your calculated modulus of elasticity with the value reported in the Mathcad Reference Table for aluminum. Does it appear that the rods are pure aluminum?

To access the Reference Tables, use menu options **Help/Reference Tables/ Modulus of Elasticity.**

2.17 Specific Heat of Marble

A cube of marble *8 cm* on each side is placed in a 200°C oven and left there for a long time to ensure that the entire cube reaches a temperature of 200°C. Meanwhile, a beaker is filled with a mixture of ice and water with a combined volume of two liters. The ice is allowed to melt in the beaker while the cube is in the oven.

At the moment that the last piece of ice melts (i.e., the ice becomes pure water at 0°C), the cube is moved from the oven and dropped in the water. After letting the system equilibrate, the final temperature of the water and the marble cube is 25.1°C. The equation that describes the energy required to warm or cool a mass is

$$Q = M \cdot C_P \cdot (T_{\text{final}} - T_{\text{initial}})$$

where **Q** is the amount of energy, **M** is the mass, and $\mathbf{C_p}$ is the specific heat (or heat capacity) of the material.

(a) Given the properties of liquid water listed in Table 2.5, determine the following:

1. the mass of water in the beaker.
2. the energy required to warm the water from 0°C to 25.1°C.

Table 2.5 Properties of Liquid Water

Density, ρ	1000 kg/m^3
Specific Heat, C_P	4.19 Joule/gram · K

(b) Given the properties of solid marble listed in Table 2.6, and assuming that all of the energy required to warm to water came from cooling the marble, determine:

1. the mass of the marble cube.
2. the specific heat of marble.

Table 2.6 Properties of Solid Marble

Density, ρ	2720 kg/m^3

(c) Compare your calculated heat capacity for marble with the value reported in the Mathcad Reference Table. Do the values agree?

To access the Reference Tables, use menu options **Help/Reference Tables/Specific Heat.**

2.18 Volume in a Partially Filled Spherical Tank

Use the Mathcad Reference Tables to find the equations for the volume of a spherical cap. Then, use the formula to determine the volume in a 12-foot diameter spherical tank that is

(a) filled to a depth of 3 feet.
(b) filled to a depth of 10 feet.

Express your results in both gallons and m^3.

To access the Reference Tables, use menu options **Help/Reference Tables/Volumes and Surface Areas.**

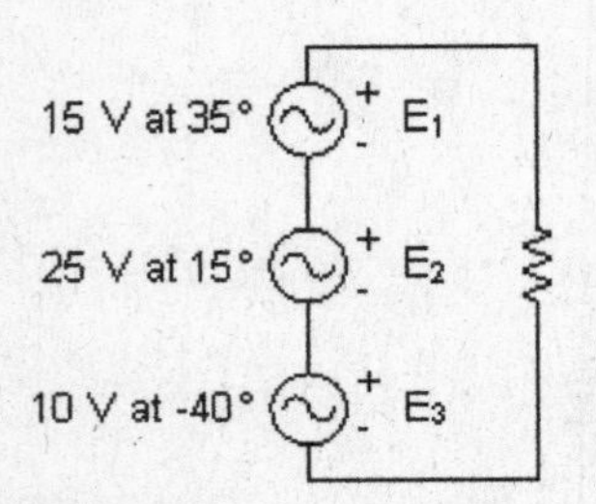

Figure 2.72
Three AC voltage sources in series.

2.19 **AC Circuit**

The circuit shown in Figure 2.72 has three AC voltage sources in series, but the three sources have different magnitudes and phases, as shown in the figure. All three sources are operating at the same frequency.

Vectors can be used to describe the three voltage sources as shown in Figure 2.73.

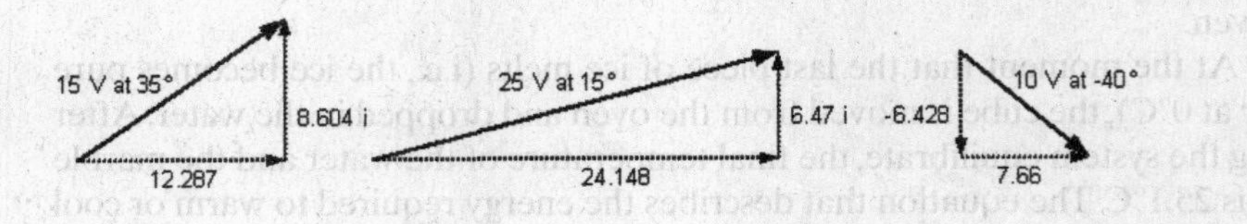

Figure 2.73
Vector representation of the voltage sources, including components.

What is the combined voltage and phase for the three sources, expressed as a complex voltage?

2.20 **Area Under a Curve**

Sometimes it is necessary to know the area under a curve. One way to approximate the area is to break up the actual area into a series of trapezoids. The area of each trapezoid can be calculated, and then the total area under the curve determined from the sum of the trapezoid areas.

$$A_{\text{trap}} = \frac{y_{\text{left}} + y_{\text{right}}}{2} \cdot (x_{\text{right}} - x_{\text{left}})$$

The area under the curve in Figure 2.74 has been subdivided into five trapezoids. The (x, y) coordinates of the upper vertices of each trapezoid are indicated in the figure.

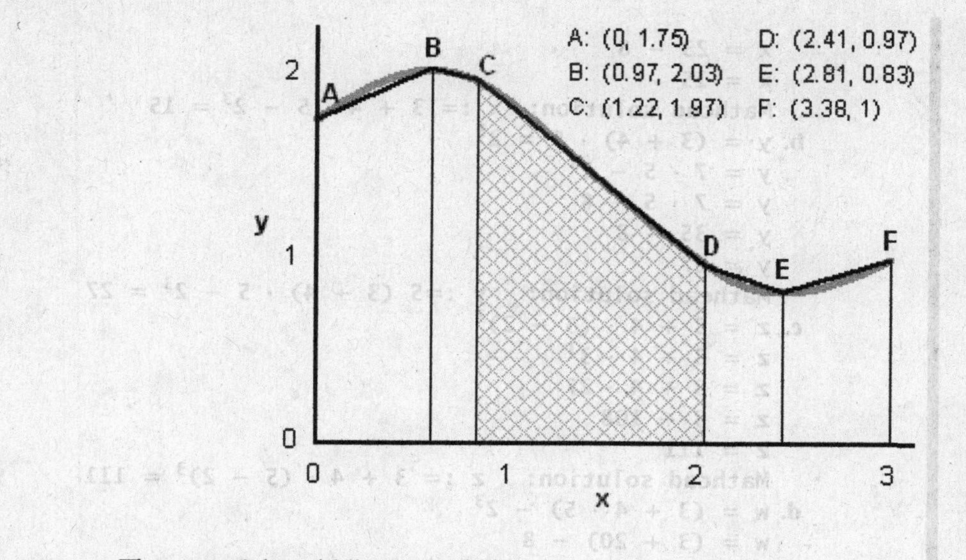

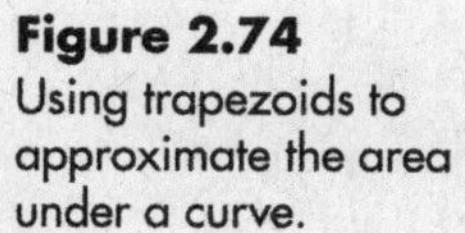

Figure 2.74
Using trapezoids to approximate the area under a curve.

The area of the middle trapezoid is

$$A_{\text{mid}} = \frac{1.97 + 0.97}{2} \cdot (2.41 - 1.22) = 1.75$$

Calculate the area of all five trapezoids and sum them to find the area under the curve.

SOLUTIONS TO PRACTICE! PROBLEMS

Section 2.3 Practice!

```
a. y := 3
   x := y
    x = 3
b. y  = 3
   x := y   << Mathcad indicates an error since y was not
   x := ■      assigned a value by the symbolic equality
c. y := 3 cm
   x := y
   x = 0.03 m   << the value of x is reported in base
                   units (SI by default)
d. x := y
   x = 3
   y ≡ 3
```

Section 2.4.2 Practice!

```
a. x = 3 + 4 : 5 − 2³
   x = 3 + 4 · 5 − 8
   x = 3 + 20 − 8
```

$x = 23 - 8$
$x = 15$
Mathcad solution: $x := 3 + 4 \cdot 5 - 2^3 = 15$

b. $y = (3 + 4) \cdot 5 = 2^3$
$y = 7 \cdot 5 - 2^3$
$y = 7 \cdot 5 - 8$
$y = 35 - 8$
$y = 27$
Mathcad solution: $y :=5\ (3 + 4) \cdot 5 - 2^3 = 27$

c. $z = 3 + 4 \cdot (5 - 2)^3$
$z = 3 + 4 \cdot 3^3$
$z = 3 + 4 \cdot 27$
$z = 3 + 108$
$z = 111$
Mathcad solution: $z := 3 + 4 \cdot (5 - 2)^3 = 111$

d. $w = (3 + 4 \cdot 5) - 2^3$
$w = (3 + 20) - 8$
$w = 23 - 8$
$w = 15$
Mathcad solution: $w := (3 + 4 \cdot 5) - 2^3 = 15$

Section 2.4.5 Practice!

a. $r_A := k_0 \cdot C_A$

b. $r_A := k_{0 \cdot C_A}$ ≪ the C_A is part of the subscript on k

c. $r_A := k_0 \cdot C_A$

Section 2.5.2 Practice!

a. $P \cdot V = n \cdot R \cdot T$ ≪ symbolic equality

b. $P := n \cdot R \cdot \frac{T}{V}$

c. $P := \frac{n \cdot R \cdot T}{V}$

d. $r := k1 \cdot CA^{2 - k2 \cdot CB}$

e. $r := k1 \cdot CA^2 - k2 \cdot CB$

Section 2.6.1 Practice!

<u>**Obvious Conversions**</u>

a. $100 \cdot cm = 1\ m$

b. $2.54 \cdot cm = 1\ in$

c. $454 \cdot gm = 1.001\ lb$

Less Obvious Conversions

a. $1 \cdot hp = 0.746\ kw$

b. $1 \cdot liter \cdot atm = 101.325\ J$

c. $1 \cdot joule = 1\ s\ watt$ << Mathcad shows the "leftover" seconds

Section 2.9 Practice!

Cube

$L := 1 \cdot cm$ ≪ each side

$A_{1side} := L^2$

$A_{cube} := 6 \cdot A_{1side}$ $A_{cube} = 6 \times 10^{-4} m^2$

$A_{cube} = 6\ cm^2$

$V_{cube} := L^3$ $V_{cube} = 1 \times 10^{-6} m^3$

$V_{cube} = 1\ cm^3$

Sphere

$R := 7 \cdot cm$

$A_{sphere} := 4 \cdot \pi \cdot R^2$ $A_{sphere} = 0.062\ m^2$

$A_{sphere} = 615.752\ cm^2$

$V_{sphere} := \frac{4}{3} \cdot \pi \cdot R^3$ $V_{sphere} = 1.437 \times 10^{-3}\ m^3$

$V_{sphere} = 1436.8\ cm^3$

ANSWERS TO SELECTED PROBLEMS

Problem 1	speed of light in a vacuum	2.998×10^8 m/sec → 6.706×10^8 mi/hr
	viscosity of water at room temperature	0.01 poise → 6.72×10^{-4} lb/ft · s → 0.001 kg/m³s
Problem 3	V = 133 cm³	
Problem 4	a. V = 3.489 L b. T = 745 K, or 472 °C	
Problem 7	b. 4000 N/m d. 8.76×10^4 N/m	
Problem 9	b. 1.97 N/m	
Problem 13	a. 7.8 m/s, or 25.5 ft/s	
Problem 15	a. 126 acres b. 3.4×10^6 gal	
Problem 17	a. ii. 2.1×10^5 joules	
Problem 19	44.1 + 8.6i volts	

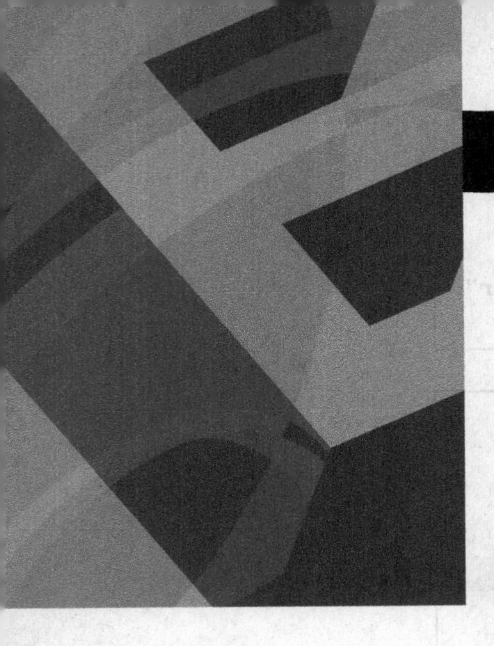

3

Mathcad Functions

Objectives

After reading this chapter, you will

- understand what a function is, and how functions are used in Mathcad
- be able to use Mathcad's built-in functions
- know what functions Mathcad provides for handling logarithms
- know what trigonometric functions are available as built-in functions in Mathcad
- know which Mathcad functions will be helpful when working with complex numbers
- be aware of a few advanced math functions that Mathcad provides for specialized applications
- know that Mathcad provides functions for handling ***character strings***
- know how to write your own functions, when necessary

3.1 MATHCAD FUNCTIONS

In a programming language, the term ***function*** is used to describe a piece of the program dedicated to a particular calculation. Many advanced calculators provide built-in functions to handle specific calculations. A function accepts input from a list of ***parameters*** (the values needed to perform the calculation), performs calculations, and then returns a value or a set of values. Functions are used whenever you want to

- perform the same calculations multiple times using different input values,
- reuse the function in another program without retyping it, or
- make a complex program easier to comprehend.

Mathcad's functions work the same way and serve the same purposes. They receive input from a parameter list, perform a calculation, and return a value or a set of values. Mathcad's functions are useful when you need to perform the same calculation multiple times. You can also cut and paste a function from one worksheet to another.

Professional Success

When do you use a calculator to solve a problem, and when should you use a computer?

These three questions will help you make this decision:

1. Is the calculation long and involved?
2. Will you need to perform the same calculation numerous times?
3. Do you need to document the results for the future, either to give them to someone else or for your own reference?

A "yes" answer to any of these questions suggests that you should consider using a computer. Moreover, a "yes" to the second question suggests that you may want to write a reusable function to solve the problem.

Mathcad provides a wide assortment of built-in functions, as well as allowing you to write your own functions. While most Mathcad functions *can* accept arrays as input, some Mathcad functions *require* data arrays as input. For example, the **mean** function needs a column of data values in order to compute the average value of the data set. In this chapter, we discuss functions that can take single-valued (scalar) inputs.

The following are the commonly used scalar functions:

- elementary mathematical functions and ***operators***
- ***trigonometric*** functions
- advanced mathematical functions
- ***string functions***

3.2 ELEMENTARY MATHEMATICAL FUNCTIONS AND OPERATORS

Many of the functions available in programming languages are implemented as operators in Mathcad. For example, to take the square root of four in FORTRAN, you would use **SQRT(4)**. **SQRT** is FORTRAN's square-root function. In Mathcad, the square-root symbol is an operator available on the Calculator ***Toolbar*** (see Figure 3.1), and shows up in a Mathcad worksheet just as you would write it on paper.

Figure 3.2 shows a couple of examples of Mathcad's square-root operator. Notice that the square-root symbol changes size as necessary as you enter your equation.

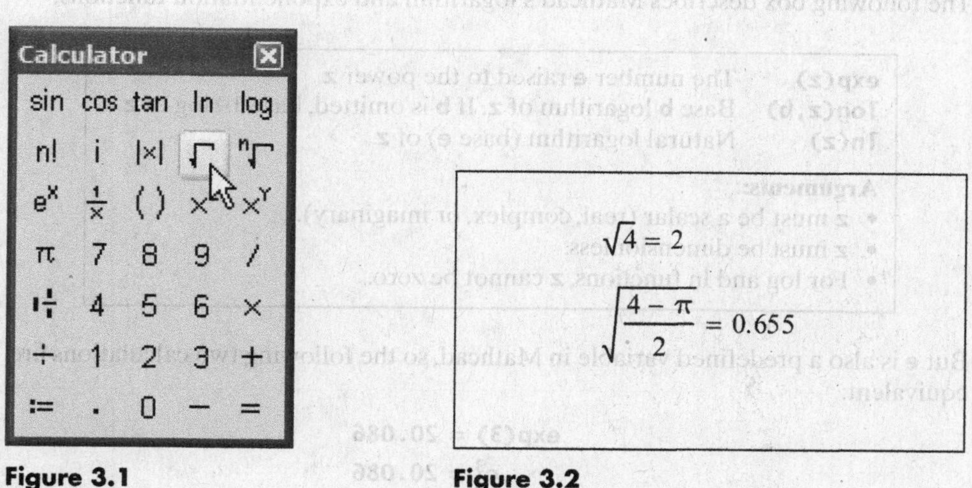

Figure 3.1
The Calculator Toolbar.

Figure 3.2
Using the square-root operator.

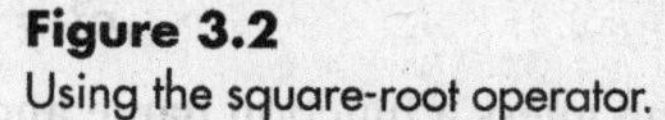

3.2.1 Common Mathematical Operators

The table on page 68 presents some common mathematical operators in Mathcad.

Many additional mathematical operations are available as built-in functions such as:

- ***logarithm*** and ***exponentiation*** functions
- ***round-off*** and ***truncation*** functions

Table 3.1 Mathcad Operators

Operator	Mathematical Operation	Source Toolbar	Alternate Keystroke
Square Root	$\sqrt{}$	Calculator Toolbar	[] (backslash)
n^{th} Root	$\sqrt[n]{}$	Calculator Toolbar	[Ctrl-\]
Absolute Value	$\|x\|$	Calculator Toolbar	[|] (vertical bar)
Factorial	$x!$	Calculator Toolbar	[!]
Summation	Σx	Calculator Toolbar	[Shift-4]
Product	$\prod x$	Calculus Toolbar	[Shift-3]
NOT	$\neg A$	Boolean Toolbar	[Ctrl-Shift-1]
AND	$A \wedge B$	Boolean Toolbar	[Ctrl-Shift-7]
OR	$A \vee B$	Boolean Toolbar	[Ctrl-Shift-6]
XOR	$A \oplus B$	Boolean Toolbar	[Ctrl-Shift-5]

Note: Mathcad's help files include descriptions of every built-in function, including information about requirements on arguments or parameters. Excerpts from the Mathcad help files are shown in text boxes.

3.2.2 Logarithm and Exponentiation Functions

The following box describes Mathcad's logarithm and exponentiation functions:

exp(z)	The number **e** raised to the power **z**
log(z,b)	Base **b** logarithm of **z**. If **b** is omitted, base 10 log of **z**.
ln(z)	Natural logarithm (base **e**) of **z**

Arguments:

- **z** must be a scalar (real, complex, or imaginary).
- **z** must be dimensionless.
- For log and ln functions, **z** cannot be zero.

But **e** is also a predefined variable in Mathcad, so the following two calculations are equivalent:

$$\text{exp}(3) = 20.086$$

$$e^3 = 20.086$$

If you know or can guess a function name and want to see the **Help** text file for information on the function in order to find out what arguments are required, type the function name on the worksheet, and while the edit lines are still around the function name, press the [F1] key. If Mathcad recognizes the function name you entered, it will display information on that function. For example, if you type **log** and press [F1], the information shown in the preceding text box will be displayed.

If you don't know the function name, use the Help menu and search the Mathcad index for your subject. For example, you can access the index list from the Help menu by clicking **Help/Mathcad Help/Index**. A search box will appear. When you type the word "functions," you will receive a lot of information about Mathcad's functions.

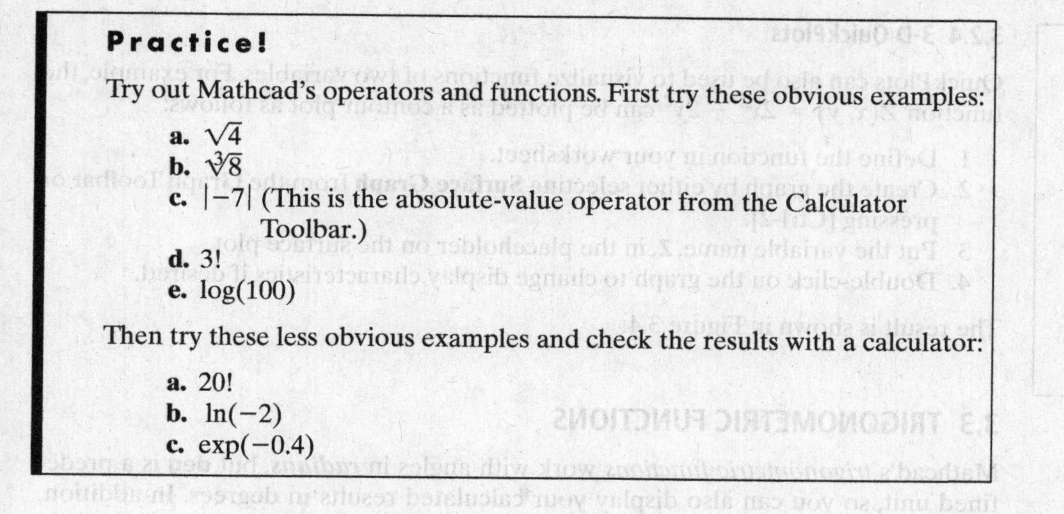

Practice!

Try out Mathcad's operators and functions. First try these obvious examples:

a. $\sqrt{4}$
b. $\sqrt[3]{8}$
c. $|-7|$ (This is the absolute-value operator from the Calculator Toolbar.)
d. 3!
e. log(100)

Then try these less obvious examples and check the results with a calculator:

a. 20!
b. ln(−2)
c. exp(−0.4)

3.2.3 Using QuickPlots to Visualize Functions

Mathcad has a feature called a ***QuickPlot*** that produces the graph of a function. For example, to obtain a visual display of the natural logarithm function, **ln**, you would follow these steps:

1. First, create a graph by either selecting **X-Y Graph** from the Graph Toolbar or pressing [Shift-2].
2. Enter **ln(x)** in the *y*-axis placeholder. (The **x** can be any unused variable, but you need to use the same variable in step 3.)
3. Enter **x** in the *x*-axis placeholder.

The result will look like the graph shown in Figure 3.3.

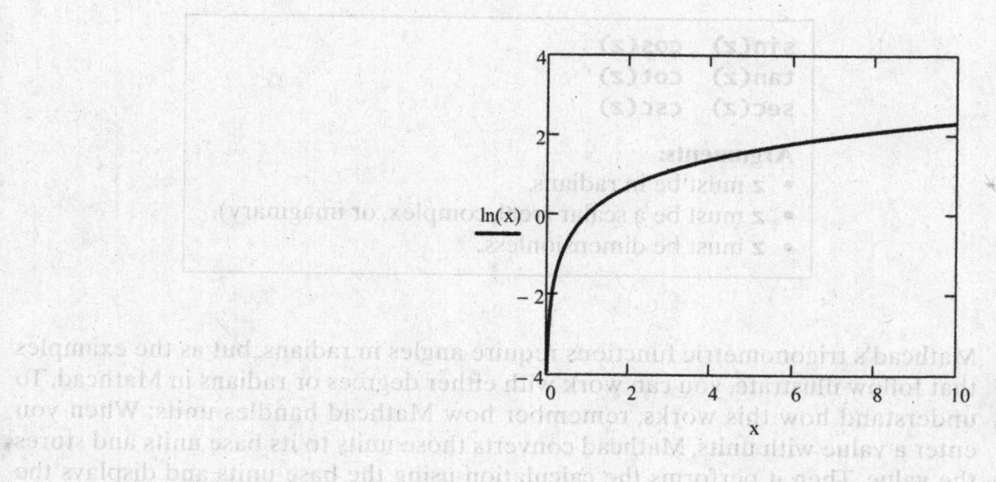

Figure 3.3
QuickPlot of ln(x).

Mathcad shows what the function looks like over an arbitrary range of **x** values. The default range is −10 to 10 but the natural logarithm is not defined for negative numbers, so Mathcad used only positive numbers for this function. If you want to see the plot over a different range, just click on the *x*-axis and change the plot limits.

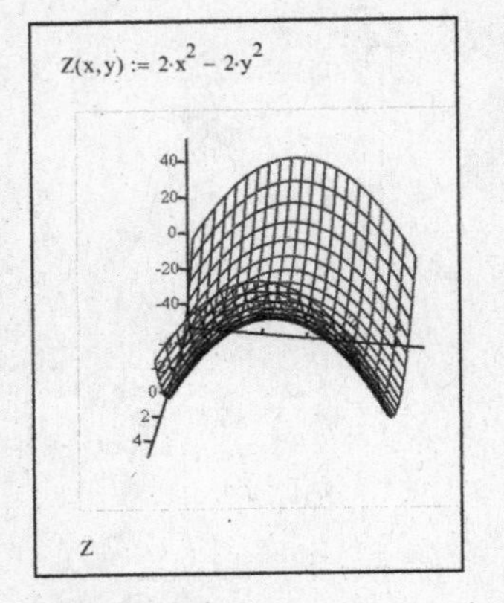

Figure 3.4
3-D QuickPlot.

3.2.4 3-D QuickPlots

QuickPlots can also be used to visualize functions of two variables. For example, the function $Z(x, y) = 2x^2 - 2y^2$ can be plotted as a contour plot as follows:

1. Define the function in your worksheet.
2. Create the graph by either selecting **Surface Graph** from the Graph Toolbar or pressing [Ctrl-2].
3. Put the variable name, **Z**, in the placeholder on the surface plot.
4. Double-click on the graph to change display characteristics if desired.

The result is shown in Figure 3.4.

3.3 TRIGONOMETRIC FUNCTIONS

Mathcad's *trigonometric functions* work with angles in ***radians***, but **deg** is a predefined unit, so you can also display your calculated results in degrees. In addition, there is a predefined unit **rad.** You probably won't need to use it, since Mathcad assumes radians if the unit is not specified, but you can include the **rad** unit label to make your worksheet easier to read. The following Mathcad example shows how the **rad** and **deg** units can be used.

```
angle := 2·π
angle = 6.283
angle = 6.283 rad
angle = 360 deg
```

3.3.1 Standard Trigonometric Functions

Mathcad's basic trigonometric functions are:

```
sin(z)  cos(z)
tan(z)  cot(z)
sec(z)  csc(z)
```

Arguments:
- **z** must be in radians.
- **z** must be a scalar (real, complex, or imaginary).
- **z** must be dimensionless.

Mathcad's trigonometric functions require angles in radians, but as the examples that follow illustrate, you can work with either degrees or radians in Mathcad. To understand how this works, remember how Mathcad handles units: When you enter a value with units, Mathcad converts those units to its base units and stores the value. Then it performs the calculation using the base units and displays the result.

In the example that follows, Mathcad first converts the units that were entered (30 degrees) to base units (radians) and stores the value. Then the sine was computed using the stored value in base units (radians), and the result was displayed.

```
sin(30·deg) = 0.5
```

In the next example, no units were entered with the *π argument*, so Mathcad used the default unit, radians, and computed the result.

```
cos(π) = -1
```

Remember: You can work with degrees with Mathcad's trigonometric functions, but you must include the **deg** unit abbreviation on each angle. When you do so, Mathcad automatically converts the units to radians before performing the calculation.

Professional Success

Validate your functions and worksheets as you develop them.

If you are "pretty sure" that the tangent of an angle is equal to the sine of the angle divided by the cosine of the angle, test the **tan** function before building it into your worksheet. Learning to devise useful tests is a valuable skill. In this case, you could test the relationship between the functions like this:

```
tan(30·deg) = 0.577
```

$$\frac{\sin(30\cdot deg)}{\cos(30\cdot deg)} = 0.577$$

It takes only a second to test, and building a number of "pretty sure" items into a worksheet will quickly lead to a result with low confidence.

3.3.2 Inverse Trigonometric Functions

The following box shows Mathcad's inverse trigonometric functions:

```
asin(z)
acos(z)
atan(z)
```

Arguments:

- **z** must be a scalar.
- **z** must be dimensionless.

Values returned are angles in radians between 0 and 2π. Values returned are from the principal branch of these functions.

The angles returned by these functions will be in radians. You can display the result in degrees by using the **deg** unit abbreviation. You can also append the **rad** unit abbreviation; doing so will not change the displayed angle, but it might help someone else understand your results. The following Mathcad example illustrates the use of the **asin** function with and without the **deg** and **rad** units.

```
asin(0.5) = 0.524
asin(0.5) = 0.524 rad
asin(0.5) = 30 deg
```

APPLICATION

Resolving Forces

If one person pulls on a rope connected to a hook imbedded in a floor with a force of 400 N and another person pulls on the same hook with a force of 200 N, what is the total force on the hook?

Answer: Can't tell—there's not enough information.

What's missing in the statement is some indication of the direction of the applied forces, since force is a vector. If two people are pulling in the same direction, then the combined force is 600 N. But if they are pulling in different directions, it's a little tougher to determine the net force on the hook. To help find the answer, we often resolve the forces into horizontal and vertical components.

For example, if one person pulls to the right on a rope connected to a hook imbedded in a floor with a force of 400 N at an angle of 20° from the horizontal (as illustrated in Figure 3.5), and another person pulls to the left on the same hook with a force of 200 N at an angle of 45° from the horizontal, what is the net force on the hook?

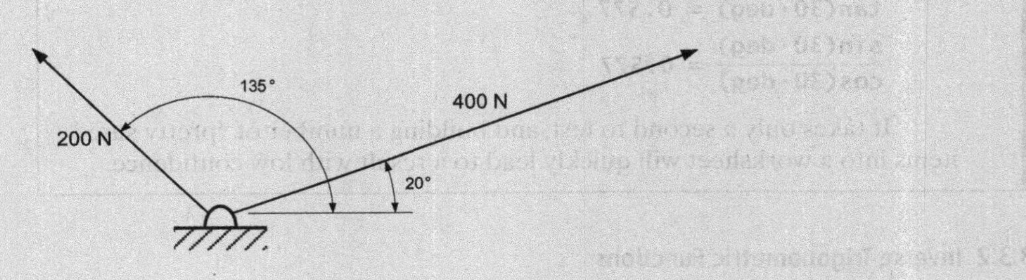

Figure 3.5
Two forces acting on a point.

Since both people are pulling upward, their vertical contributions add. But since one is pulling left and the other right, they are (in part) counteracting each other's efforts. To quantify this distribution of forces, we can calculate the horizontal and vertical components of the force being applied by each person, as shown in Figure 3.6. Mathcad's trigonometric functions are helpful for these calculations.

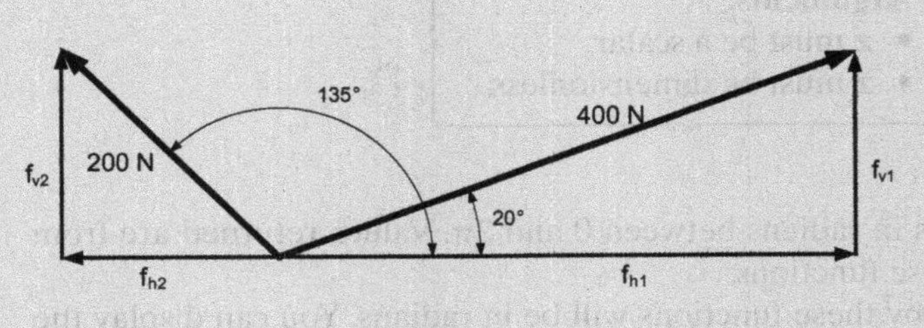

Figure 3.6
Resolving the forces into horizontal and vertical components.

The 400-N force from person 1 resolves into a vertical component $\mathbf{f_{v1}}$ and a horizontal component $\mathbf{f_{h1}}$. The magnitudes of these force components can be calculated as shown in the following Mathcad example.

```
F1 := 400·N
fv1 := F1·sin(20·deg)      fv1 = 136.8 N
fh1 := F1·cos(20·deg)      fh1 = 375.9 N
```

Similarly, the 200-N force from person 2 can be resolved into the following component forces:

$$F_2 := 200 \cdot N$$

$$f_{v2} := F_2 \cdot \sin(45 \cdot deg) \qquad f_{v2} = 141.4\ N$$

$$f_{h2} := F_2 \cdot \cos(45 \cdot deg) \qquad f_{h2} = 141.4\ N$$

Actually, force component f_{h2} would usually be written as $f_{h2} = -141.421\ N$, since it is pointed in the $-x$ direction. If all angles had been measured from the same position (usually the three o'clock angle is called 0°), the angle on the 200-N force would have been 135° and the signs on the force components would have taken care of themselves:

$$F_2 := 200 \cdot N$$

$$f_{v2} := F_2 \cdot \sin(135 \cdot deg) \qquad f_{v2} = 141.4\ N$$

$$f_{h2} := F_2 \cdot \cos(135 \cdot deg) \qquad f_{h2} = -141.4\ N$$

Once the force components have been computed, the net force in the horizontal and vertical directions can be determined. (Force f_{h2} has a negative value in this calculation.)

$$f_{v\_net} := f_{v1} + f_{v2} \qquad f_{v\_net} = 278.2\ N$$

$$f_{h\_net} := f_{h1} + f_{h2} \qquad f_{h\_net} = 234.5\ N$$

The net horizontal and vertical components can be recombined as illustrated in Figure 3.7 to find a combined net force and angle on the hook.

$$F_{net} := \sqrt{f_{h\_net}^2 + f_{v\_net}^2} \qquad F_{net} = 363.8\ N$$

$$\theta := \operatorname{atan}\left(\frac{f_{v\_net}}{f_{h\_net}}\right) \qquad \theta = 49.9\ deg$$

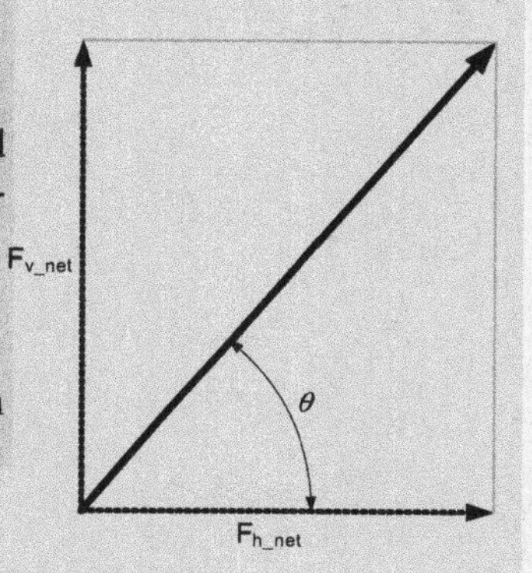

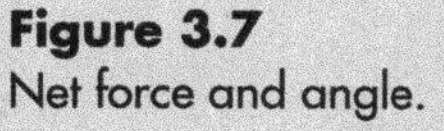

Figure 3.7
Net force and angle.

3.3.3 Hyperbolic Trigonometric Functions

Mathcad's hyperbolic trigonometric functions are:

```
sinh(z)  cosh(z)
tanh(z)  csch(z)
sech(z)  coth(z)
```

Arguments:

- z must be in radians.
- z must be a scalar.
- z must be dimensionless.

3.3.4 Inverse Hyperbolic Trigonometric Functions

The following box describes the inverse hyperbolic trigonometric functions in Mathcad:

```
asinh(z)
acosh(z)
atanh(z)
```

Arguments:

- z must be a scalar.
- z must be dimensionless.

Values returned are from the principal branch of these functions.

The angles returned by the inverse hyperbolic trigonometric functions will be in radians, but you can put **deg** in the unit's placeholder to convert the displayed units.

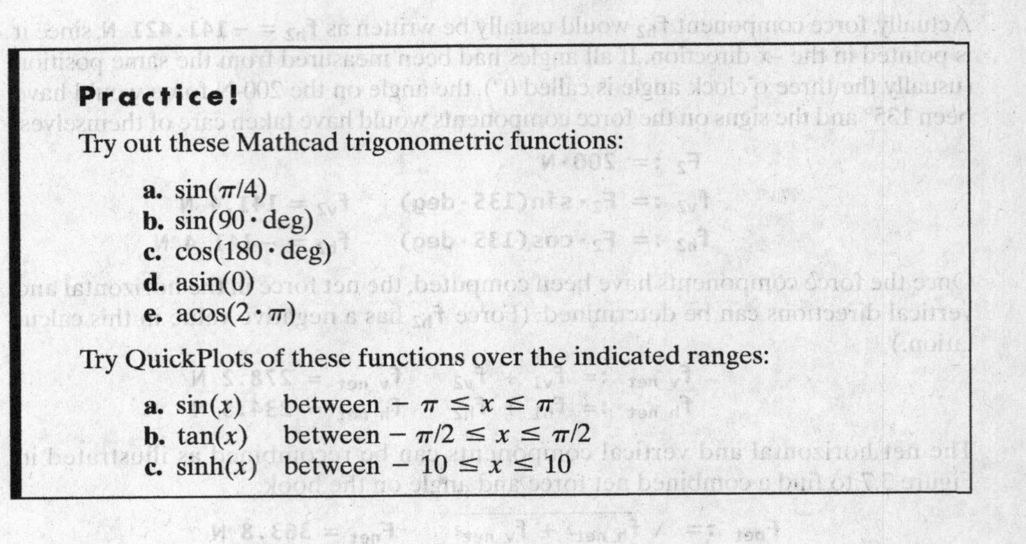

Practice!

Try out these Mathcad trigonometric functions:

a. $\sin(\pi/4)$
b. $\sin(90 \cdot \text{deg})$
c. $\cos(180 \cdot \text{deg})$
d. $\text{asin}(0)$
e. $\text{acos}(2 \cdot \pi)$

Try QuickPlots of these functions over the indicated ranges:

a. $\sin(x)$ between $-\pi \leq x \leq \pi$
b. $\tan(x)$ between $-\pi/2 \leq x \leq \pi/2$
c. $\sinh(x)$ between $-10 \leq x \leq 10$

3.4 FUNCTIONS FOR HANDLING COMPLEX NUMBERS

Complex numbers are used routinely by electrical engineers to keep track of amplitudes and phase angles of signals, and occasionally by engineers from other disciplines.

In Mathcad, complex numbers can be defined simply by appending an `i` or `j` after the imaginary part. For example, the complex number N, with value $12 + 3i$ is entered by typing the variable name, the colon [:] to enter the define as equal to symbol (:=), then the 12 and a plus sign, and then the `3i`. Once the complex numbers are entered, all of Mathcad's complex number mathematics capabilities are available.

Mathcad provides a few functions to help make working with complex numbers easier.

- **`Re(z)`** returns the real part of a complex number
- **`Im(z)`** returns the imaginary part of a complex number
- **`arg(z)`** returns the principal argument of **`z`**, between $-\pi$ and π.

Mathcad will also calculate the magnitude of **`z`**, but it does so using the *magnitude operator*, |z|, rather than a function. The magnitude operator is available on the Calculator Toolbar, where it is called the absolute value operator. (The behavior of the operator changes depending on the nature of the operand in the placeholder. When the placeholder contains a complex number, the operator returns the magnitude.) The magnitude operator can also be entered by pressing the [|] (vertical bar) key.

Figure 3.8 illustrates the use of these functions with complex numbers.

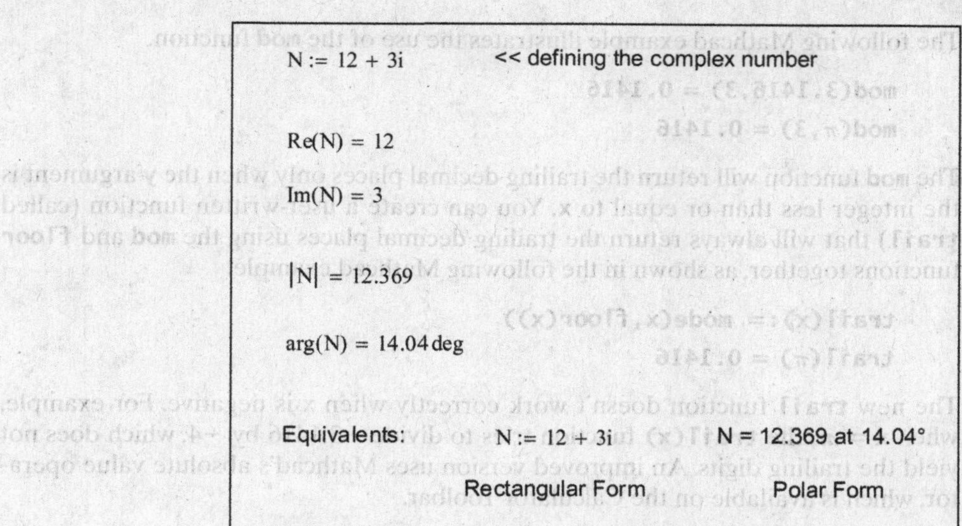

Figure 3.8
Using Mathcad's functions for handling complex numbers.

3.5 ADVANCED MATHEMATICAL FUNCTIONS

Mathcad provides a wide range of advanced math functions; only a few will be presented here.

- Round-off and truncation functions
- Logical functions
- ***Discontinuous functions***
- Functions for ***random numbers***, distributions, and histograms

3.5.1 Round-Off and Truncation Functions

Mathcad provides two functions that determine the nearest integer value. The **floor(x)** function returns the largest integer that is less than or equal to **x**, while the **ceil(x)** function returns the smallest integer that is greater than or equal to **x**. To remove the decimal places following a value, use the **floor** function. For example, to find the greatest integer less than or equal to π, either of the following will work:

```
floor(3.1416) = 3
floor(π) = 3
```

The second example used Mathcad's predefined variable for π.

If you want the values after the decimal point (i.e., the *mantissa*), use the **mod** function, which performs ***modulo division***, returning the remainder after the division.

mod(x,y)	**This function returns the remainder on dividing *x* by *y*. The result has the same sign as *x*.**

Arguments:
- **x** and **y** should both be real scalars.
- **y** must be nonzero.

The following Mathcad example illustrates the use of the **mod** function.

```
mod(3.1416,3) = 0.1416
mod(π,3) = 0.1416
```

The **mod** function will return the trailing decimal places only when the **y** argument is the integer less than or equal to **x**. You can create a user-written function (called **trail**) that will always return the trailing decimal places using the **mod** and **floor** functions together, as shown in the following Mathcad example:

```
trail(x):= mode(x,floor(x))
trail(π) = 0.1416
```

The new **trail** function doesn't work correctly when x is negative. For example, when **x** = π, the **trail(x)** function tries to divide -3.1416 by -4, which does not yield the trailing digits. An improved version uses Mathcad's absolute value operator, which is available on the Calculator Toolbar.

```
trail(x):= mod(|x|,floor(|x|))
trail(π) = 0.1416
trail(-π) = 0.1416
```

3.5.2 The if( ) Function

Mathcad provides an **if** function for making logical decisions. The **if** function is described as:

if(cond, Tval, Fval)	This function returns one of two values, depending on the value of a logical condition, *cond*.

Arguments:

- **cond** is usually an expression involving a logical operator. For example, you can use
  - **(i < 2)**
  - **(x < 1) * (x > 0) for an AND gate**
  - **(x > 1) + (x < 0) for an OR gate**
- **Tval** is the value returned when **cond** is true.
- **Fval** is the value returned when **cond** is false.

The value that is returned can be a simple scalar value, a computed value, or even a text string. For example, the **if** function can be used to determine whether water would be a liquid or a solid by checking the temperature of the water:

```
state(Temp) := if(Temp -> 0,"liquid","solid")
state(25)    = "liquid"
state(-3)    = "solid"
```

You probably noticed that the user-written **state** function requires that **Temp** be in °C, without actually accepting units on the variable. This function can be improved by using units, as shown in Figure 3.9.

Defining the Function

$\text{State}(\text{Temp}) := \text{if}(\text{Temp} > 0\,°\text{C}, \text{"liquid"}, \text{"solid"})$

Using the Function

$\text{State}(25\,°\text{C}) = \text{"liquid"}$

$\text{State}(-3\,°\text{C}) = \text{"solid"}$

$\text{State}(30\,°\text{F}) = \text{"solid"}$

Figure 3.9
The new State function, with units.

Notes:

- The name of the function in Figure 3.9 was changed by capitalizing the `S` in `State`.
- Units were brought into the function by including °C on the value in the condition inside the `if` function.
- Notice in the fourth line in Figure 3.6 that you do not have to use °C with the `State` function, but you do have to use valid temperature units.
- Every time temperature units were used in Figure 3.6, they were connected to a value using the postfix operator, not multiplication.

In the `if` function, the `Tval` and `Fval` "variables" can actually be formulas. Thus, we can create an absolute value function by using `if` and changing the sign on `x`, if `x` is not greater than `0`.

```
abs(x) := if(x ≥ 0, x, -x)
abs(12) = 12
abs(-6) = 6
```

Practice!

Write an expression that returns a value of 1 if the volume in a tank is greater than or equal to 5,000 gallons and a value of 0 otherwise. An expression like this might be used to control a valve through which a liquid flows into the tank, shutting the valve when the tank is full. Test your expression in Mathcad.

3.5.3 Creating AND and OR Gates

The condition (logical test) in the `if` function can be extended to include two (or more) logical tests.

- An AND gate requires two logical tests, and both individual tests must evaluate to **true** in order for the AND gate to evaluate **true**.
- An OR gate also requires two logical tests, but if either of the individual tests evaluates to **true**, then the OR gate evaluates **true**.

These gates sound more complicated than they really are. To help clarify how these gates can be used, consider these common automotive situations:

- Is your car due for an oil change?

  Oil change businesses tell us that our cars' oil should be changed every 3,000 miles or three months, whichever comes first. There are two logical tests to perform, the mileage test and the time test, but only one has to be **true** to decide that an oil change is due. This is an example of an OR gate since a **true** on either test is enough to make the OR gate return **true**.

- Is the excessive tread wear on your tires covered by the warranty?

  Two conditions must be satisfied if your tire warranty is going to come into play: the tread wear must be excessive (must be greater than a tread wear criteria), and the mileage on the tires must be less than the warranty mileage. Since both of the tests (tread wear and mileage) must evaluate **true** in order for the warranty to kick in, this is an example of an AND gate.

The oil change (OR gate) example is shown in Figure 3.7. In this example it is assumed that the car has been driven 2,300 miles since the most recent oil change, but that oil change was 3.2 months ago. So, the mileage test will fail, but the time test will evaluate **true**. Since a **true** on either test is enough to cause the OR gate to evaluate **true**, the oil change test will tell us that it is time to change the oil.

Figure 3.7 illustrates that you can construct an OR gate by adding two logical tests inside the `if` function's condition argument. You can also use the **OR** operator from the Boolean Toolbar, as shown in Figure 3.10.

Example of an OR gate...

Change the oil if "mileage > 3000" or "time since last oil change > 3 months".

Specify the Criteria Unit Definition: $mo := \frac{yr}{12}$

$distance_{crit} := 3000 \cdot mi$

$time_{crit} := 3 \cdot mo$

Specify Actual Values

$distance_{act} := 2300 \cdot mi$

$time_{act} := 3.2 \cdot mo$

Specify Messages

DoMsg := "Change Oil"

DontMsg := "Oil Change Not Yet Required"

Define Test Function (Using the OR operator from the Boolean Toolbar)

$test_{oil} := if\left[\left(distance_{act} \geq distance_{crit}\right) \vee \left(time_{act} \geq time_{crit}\right), DoMsg, DontMsg\right]$

$test_{oil}$ = "Change Oil"

Figure 3.10
Example of an OR gate used to determine if a car is due for an oil change.

The Tire Warranty (AND gate) example is shown in Figure 3.8. In this example it is assumed that the tire wear is excessive (the 11 mm or actual wear exceeds the 9 mm criteria), but the mileage on the tires (53,000 miles) exceeds the warranty mileage (50,000 miles) so the mileage test will evaluate **false**. Because both the tread wear and mileage tests must evaluate **true** for the AND gate to evaluate **true** and the warranty to apply, the tire warranty test will tell us that the warranty does not apply.

Figure 3.8 illustrates how you can construct an AND gate by multiplying two logical tests inside the `if` function's condition argument. You can also use the **AND** operator from the Boolean Toolbar, as shown in Figure 3.11.

Example of an AND gate...

The tire warranty applies if tread wear exceeds the threshold and mileage since the tire purchase is less than the warranty mileage. Both conditions must be met for the warranty to apply.

Specify the Criteria

$distance_{crit} := 50000 \cdot mi$

$wear_{crit} := 9 \cdot mm$

Specify Actual Values

$distance_{act} := 53000 \cdot mi$

$wear_{act} := 11.mm$

Specify Messages

YesMsg := "Warranty Applies"

NoMsg := "Warranty Does Not Apply"

Define Test Function (Using the AND operator from the Boolean Toolbar)

$test_{tire} := \text{if}\left[\left(wear_{act} > wear_{crit}\right) \wedge \left(distance_{act} \le distance_{crit}\right), YesMsg, NoMsg\right]$

$test_{tire}$ = "Warranty Does Not Apply"

Figure 3.11
Example of an AND gate used to determine if a tire warranty applies.

3.5.4 Discontinuous Functions

Mathcad supports the following discontinuous functions:

KRONECKER DELTA

δ(m,n) This function returns **1** if **m = n** and returns **0** otherwise.

Argument:

- Both **m** and **n** must be integers without units.

HEAVISIDE STEP FUNCTION

ϕ(x) This function returns **0** if **x** is negative and returns **1** otherwise.

Argument:

- **x** must be a real scalar.

Both of these functions can be used to activate and deactivate terms in an equation, and have practical applications in mathematical modeling. As a very simple example, the Heaviside function was used in Figure 3.9 in a mathematical model of a tank filling and draining. The Heaviside function caused the fill pump to be activated at **t = 5** (time units) at a rate of **100** (volume units/time), and the drain pump to activate at **t = 10** (time units) at a rate of **125** (volume units/time) as illustrated in Figure 3.12.

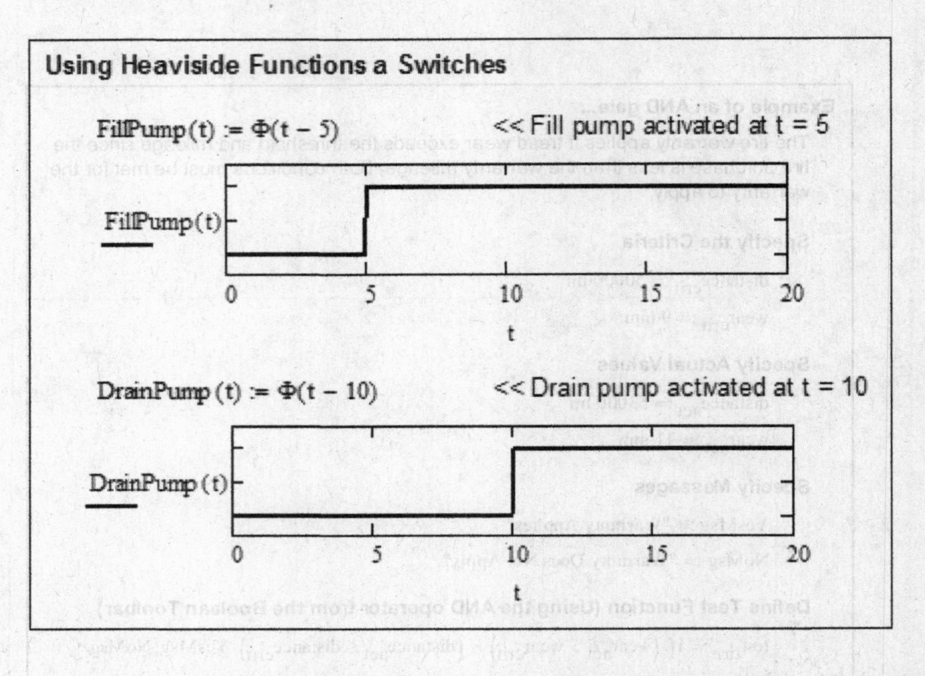

Figure 3.12
Using a Heaviside function as a switch in mathematical modeling.

The tank volume stays at zero until the fill pump is activated then increases until the drain pump is activated, as shown in Figure 3.13.

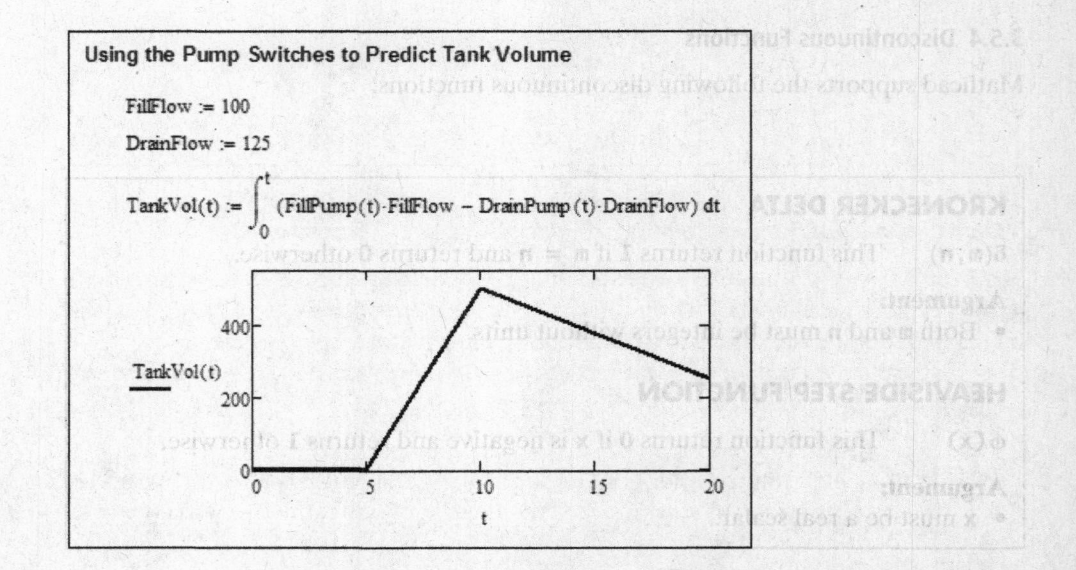

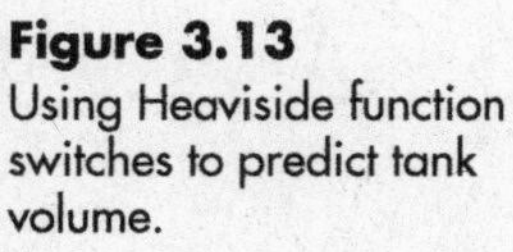

Figure 3.13
Using Heaviside function switches to predict tank volume.

3.5.5 Boolean (Logical) Operators

Mathcad provides the following ***Boolean (logical) operators*** on the Boolean Toolbar.

Name	Symbol	Shortcut	Description
NOT Operator	$\neg A$	[Ctrl-Shift-1]	Returns 1 if A is zero; returns 0 otherwise.
AND Operator	$A \wedge B$	[Ctrl-Shift-7]	Returns 1 if A and B both have nonzero values; returns 0 otherwise.
OR Operator	$A \vee B$	[Ctrl-Shift-6]	Returns 1 if A or B is nonzero; returns 0 otherwise.
XOR Operator (Exclusive OR)	$A \oplus B$	[Ctrl-Shift-5]	Returns 1 if A or B, but not both, is nonzero; returns 0 otherwise.

Examples using the AND and OR operators were shown in Figures 3.10 and 3.11. Figure 3.14 repeats the oil change example to highlight the usage of the OR operator.

Example of an OR gate -- Revisited

Change the oil if "mileage > 3000" or "time since last oil change > 3 months".

Specify the Criteria Unit Definition: $mo := \frac{yr}{12}$

$distance_{crit} := 3000 \cdot mi$

$time_{crit} := 3 \cdot mo$

Specify Actual Values

$distance_{act} := 2300 \cdot mi$

$time_{act} := 3.2 \cdot mo$

Test the Distance Travelled Since Last Oil Change

$Test_{dist} := distance_{act} \geq distance_{crit}$

$Test_{dist} = 0$ Returned 0, False

Test the Time Since Last Oil Change

$Test_{time} := time_{act} \geq time_{crit}$

$Test_{time} = 1$ Returned 1, True

Perform the OR Comparison of the Two Test Results

$Test_{dist} \vee Test_{time} = 1$ Returned 1, True -- time to change the oil

Figure 3.14
Using the OR operator.

3.5.6 Functions for Random Numbers, Distributions, and Histograms

Mathcad provides a number of functions for generating sets of random numbers with specific characteristics. Only a few of the more commonly used functions will be described here; the others are described in the Mathcad help system.

The random number generating functions are summarized first, and then described in the following paragraphs.

- **`rnd(x)`** returns one random number between **0** and **x**.
- **`runif(m, a, b)`** returns **m** uniformly distributed random numbers between **a** and **b**.
- **`rnorm(m, μ, σ)`** returns **m** normally distributed random numbers with a mean, **μ**, and standard deviation, **σ**.

To generate a single random number, the **rnd(x)** function is commonly used. For example, to build uncertainty into a board game, it is common to roll dice. To simulate rolling a die (one of a pair of dice), you might use **rnd(6)**—except that it won't work correctly because **rnd** returns noninteger values and you can't roll a 4.32145 with dice. A quick modification gives only integer values between 1 and 6:

```
roll := ceil(rnd(6))
```

The **ceil** function (see Section 3.5.1) always rounds up to the nearest integer, which is just what is needed to simulate rolling dice.

Theoretically, if you roll a die many times, you should roll as many 2s as 5s or any other value between 1 and 6. That is, each face of the die has an equal probability of landing face up. The result of many rolls of a single die should be a *uniform distribution*. Rather than type in the **rnd** function many times, we can use the **runif(*m*,*a*,*b*)** function to create a uniform distribution containing m values. If we still want integer values between 1 and 6, we can use **runif** like this:

```
data := ceil(runif(1000,0,6))
```

Notice that the range is 0 to 6, not 1 to 6. Since the **ceil** function rounds up, we need to allow values less than 1 so that the **ceil** function can generate rolls of 1. If you used the range 1 to 6 you would be excluding nearly all rolls of 1 from the data set.

To see what the distribution of values in data looks like, the **histogram (intervals, data)** function can be used. The **histogram** function returns the number of **data** values that fall into each interval. The **intervals** argument can either be a vector of interval endpoints or an integer indicating how many intervals are desired. The **data** argument is the vector of values being analyzed. For the die example, we want six intervals, so we will use **histogram(6, data)**. This is illustrated in Figure 3.15.

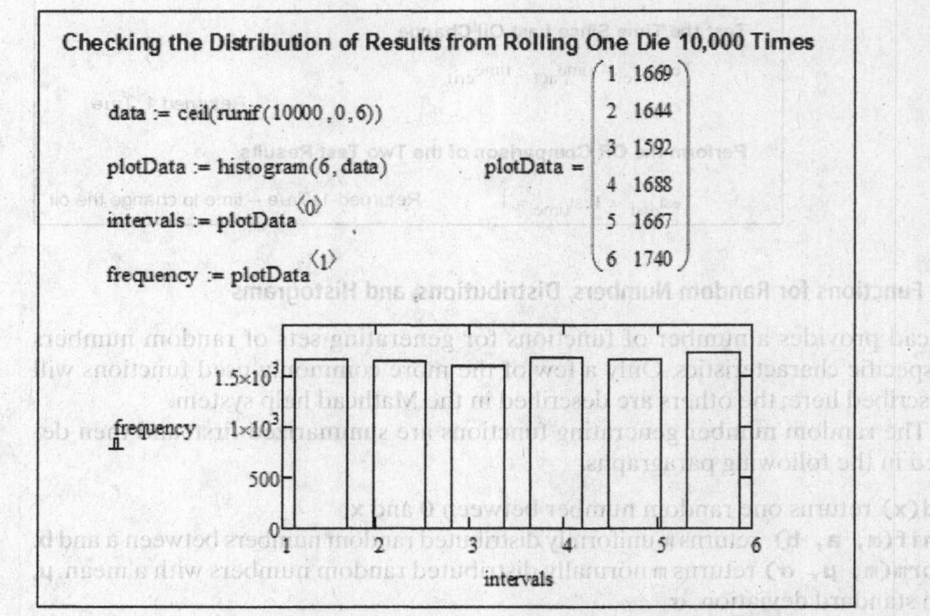

Figure 3.15
Checking the distribution of a die simulator.

The **`histogram`** function returns all of the information needed to create a histogram plot, but it is in the form of a matrix with two columns (as shown in Figure 3.11). The left column contains the intervals used, and the right column contains the number of values in each interval (often called the *frequency*). We need to get the information out of the matrix and assigned to variables that can be used for plotting. The equations used to accomplish this are highlighted in the following Mathcad example.

`intervals := plotData`$^{\langle 0 \rangle}$
`frequency := plotData`$^{\langle 1 \rangle}$

The < > shown at the right end of the *plotData* array is called the *column operator* [Ctrl-6]. It is used to grab one column from a matrix. Once the plot information has been assigned to the variables, **`intervals`** and **`frequency`**, the distribution can be plotted, as shown in Figure 3.11. The distribution is not perfectly uniform, but it is pretty close. (If the distribution was perfectly uniform, you might question whether the random numbers were truly "random.")

The situation changes when you roll two or more dice, because the probabilities of each possible result change. For example, when you roll three dice there is only one way to roll a 3 (three ones), but there are numerous ways to roll a 10 (e.g., 6,3,1; 6,2,2; 5,4,1; . . .). The probability of rolling a 10 is much higher than the probability of rolling a 3. This result is shown in Figure 3.16.

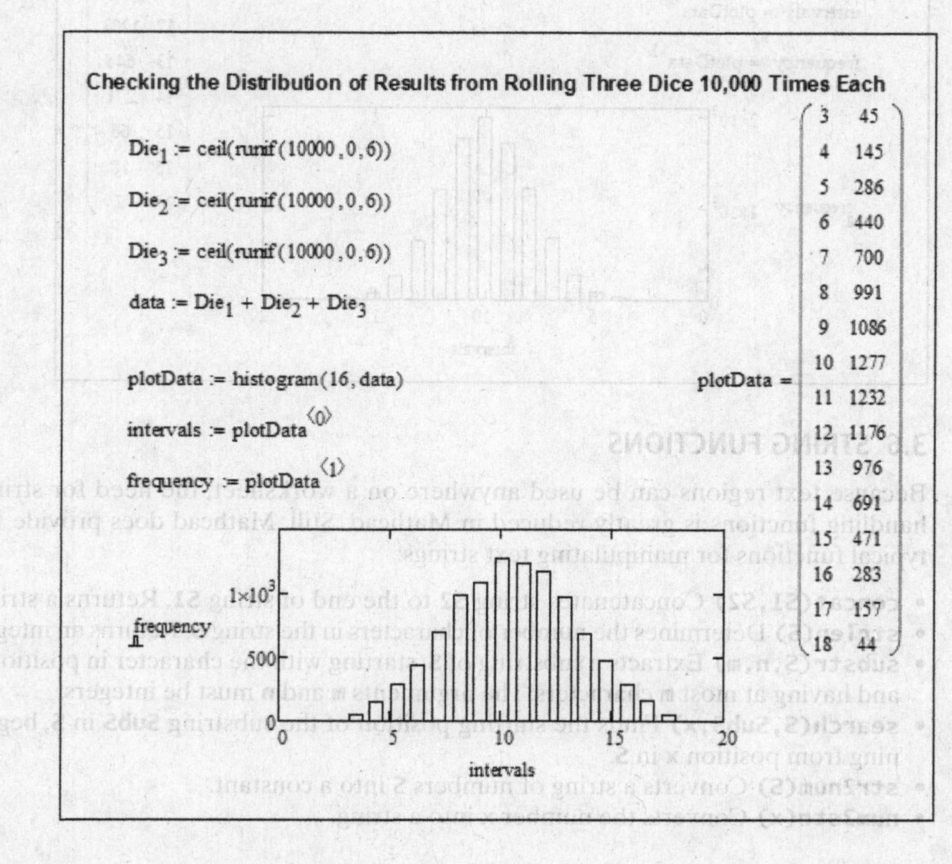

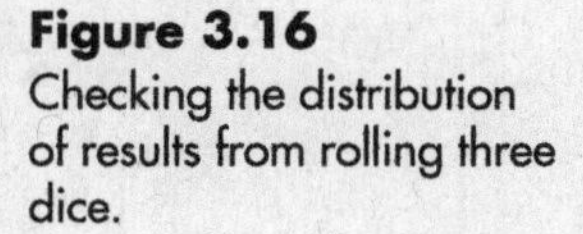
Figure 3.16
Checking the distribution of results from rolling three dice.

Mathcad provides another random number function, called **rnorm(m, μ, σ)**, that is designed to generate sets of *normally distributed* (i.e., *bell curve*) values. The arguments are as follows:

- **m** the function returns **m** values,
- **μ** the distribution is centered at the mean value, **μ**,
- **σ** the spread of the distribution around the mean value is controlled by **σ**, the standard deviation.

Figure 3.17 shows the distribution returned by **rnorm(10000,10,2)**—a normally distributed data set containing 10,000 values, centered at 10, with a standard deviation of 2.

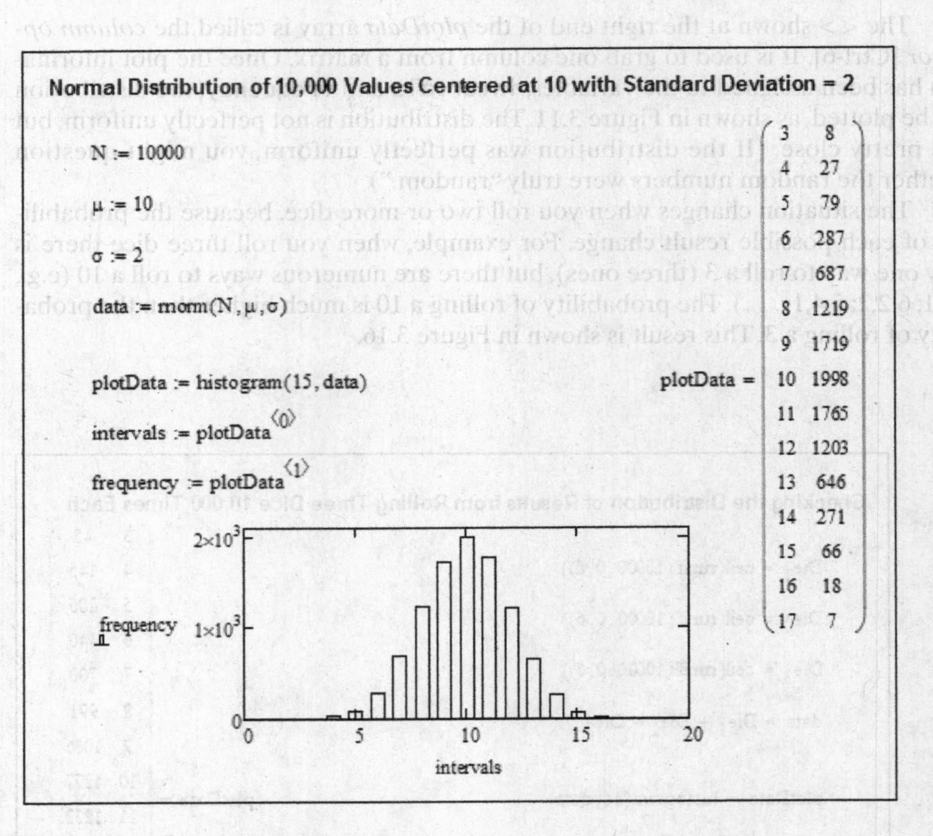

Figure 3.17
Using rnorm to generate normally distributed random numbers.

3.6 STRING FUNCTIONS

Because text regions can be used anywhere on a worksheet, the need for string-handling functions is greatly reduced in Mathcad. Still, Mathcad does provide the typical functions for manipulating text strings:

- **concat(S1,S2)** Concatenates string **S2** to the end of string **S1**. Returns a string.
- **strlen(S)** Determines the number of characters in the string **S**. Returns an integer.
- **substr(S,n,m)** Extracts a substring of **S**, starting with the character in position **n** and having at most **m** characters. The arguments **m** and **n** must be integers.
- **search(S,SubS,x)** Finds the starting position of the substring **SubS** in **S**, beginning from position **x** in **S**.
- **str2num(S)** Converts a string of numbers **S** into a constant.
- **num2str(x)** Converts the number **x** into a string.

The example in Figure 3.18 illustrates the use of the **search** function to find the word **Mathcad** in the string called **MyString**. The position of the **M** in **Mathcad** is returned by the function. Since Mathcad says the first character of the string (the **T** in **This**) is in position zero, the **M** in **Mathcad** is in position 10 (the eleventh character of the string).

```
MyString := "This is a Mathcad example."

pos := search(MyString, "Mathcad", 0)        pos = 10

Next, check to see if Mathcad string search is case sensitive...

pos := search(MyString, "mathcad", 0)        pos = -1        << search failed
```

Figure 3.18
Searching strings in Mathcad.

The last two lines in Figure 3.18 illustrate that Mathcad differentiates between uppercase and lowercase letters. That is, Mathcad is ***case sensitive***. The **search** function returns a **-1** when the search string is not found.

3.7 USER-WRITTEN FUNCTIONS

When Mathcad does not provide a built-in function, you can always write your own. The ability to use Mathcad's functions inside your own function definitions greatly increases the power of Mathcad to solve engineering problems and makes writing functions much simpler. Several user-written functions have been utilized in this chapter, including the following:

- **trail(x) := mod(|x|,floor(|x|))**
- **state(Temp) := if (Temp> 0, "liquid", "solid")**
- **abs(x) := if(x≥0,x,-x)**

Each of these user-written functions has a name (e.g., **trail**). a parameter list on the left, and a mathematical expression of some sort on the right side. User-written functions must always have one or more arguments. In these examples, only a single argument was used in each parameter list, but multiple arguments are allowed and are very common. For example, a function to calculate the surface area of a cylinder could be written as

$$A_{cyl}(r, L) := 2 \cdot \pi \cdot r^2 + 2 \cdot \pi \cdot r \cdot L$$

Once the function has been defined, it can be used multiple times in the worksheet. In Figure 3.19, the function is employed four times, illustrating how functions can be used with and without units and with both values and variables.

Recall that the A_{cyl} function was defined using **r** and **L** for arguments. In a function definition, the arguments are *dummy variables*. That is, they show what argument is used in which location in the calculation, but neither **r** nor **L** needs to have defined values when the function is declared. They are just *placeholders*. When the function is used, the values in the **r** and **L** placeholders are put into the appropriate spots in the equation, and the computed area is returned.

In a function, you can use arguments with or without units, and the arguments can be values or defined variables. In the last calculation, **r** and **L** were used for the

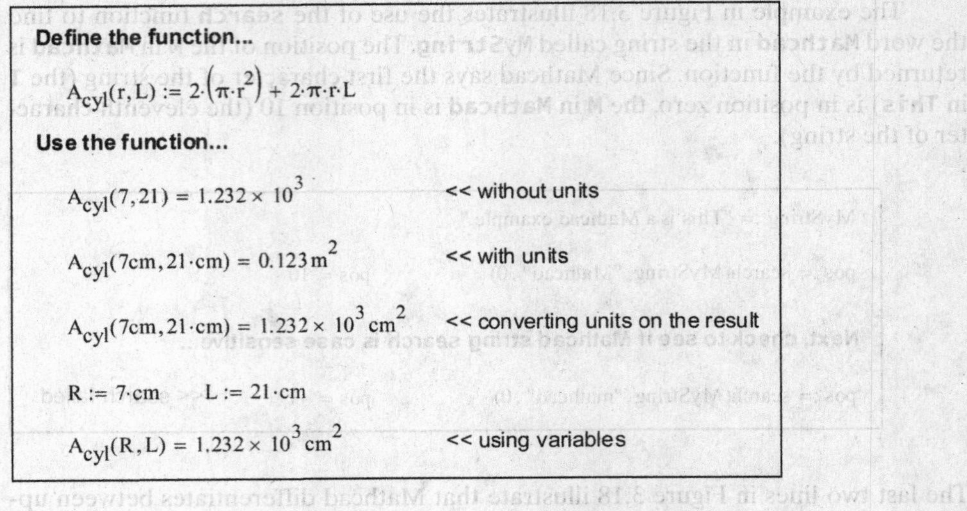

Figure 3.19
Using a user-written function.

radius and length of the cylinder, respectively. This choice of variable names is reasonable but entirely irrelevant to Mathcad. Variables such as **Q** and **Z** would work equally well, but they would make the worksheet harder to understand. For example, we might have

$$A_{cyl}(Q, Z) := 2 \cdot \pi \cdot Q^2 + 2 \cdot \pi \cdot Q \cdot Z$$
$$Q := 7 \cdot cm$$
$$Z := 21 \cdot cm$$
$$A_{cyl}(Q, Z) := 1 \cdot 232 \times 10^3 cm^2$$

Practice!

Define a function for computing the volume of a rectangular box, given the height, width, and length of the box. Then use the function to compute the volumes of several boxes:

a. H = 1, W = 1, L = 1
b. H = 1, W = 1, L = 10
c. H = 1 cm, W = 1 cm, L = 10 cm
d. H = 1 cm, W = 1 m, L = 10 ft

APPLICATION

Fiber Optics

The use of optics can be dated back at least to the time when someone noticed that putting a curved piece of metal behind a lamp could concentrate the light. Or perhaps it goes back to the time when someone noticed that the campfire felt warmer if it was built next to a large rock. In any case, while the use of optics might be nearly as old as humanity itself, optics is a field that has seen major changes in the past twenty years. And these changes are affecting society in a variety of ways:

- Fiber-optic communication systems are now widely used and will become even more commonplace in the future.
- Lasers have moved from laboratory bench tops to supermarket checkout scanners and CD players.
- Traffic signs are much more visible at night now that they incorporate microreflective optical beads.

This application considers a feature of fiber-optic systems: *total internal reflection.*

According to the *law of refraction*, the *angle of incidence*, θ_i, is related to the *angle of refraction*, θ_r, by the index of refraction n, for the two materials through which light passes (this is illustrated in Figure 3.20). That is,

$$n_1 \sin(\theta_r) = n_2 \sin(\theta_2).$$

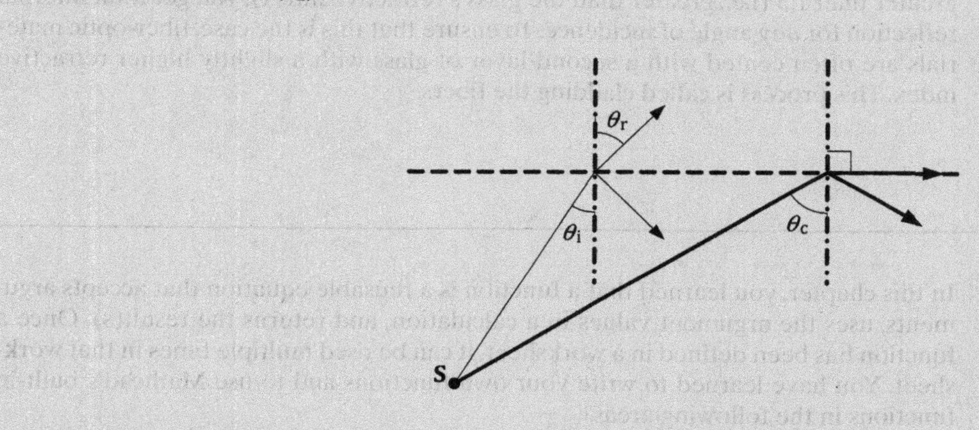

Figure 3.20
Reflection and refraction.

For light in a glass fiber ($n_1 \approx 1.5$) reflecting and refracting at the air ($n_2 \approx 1$) interface with an angle of incidence of 20°, what is the angle of refraction?

To solve this problem, we will use Mathcad's **sin** and **asin** functions and the built-in unit **deg**:

$$\theta_r := \mathrm{asin}\left(\frac{1.5 \cdot \sin(20 \cdot deg)}{1}\right)$$

$$\theta_r = 30.866 \cdot deg$$

Note: The "1" in the first equation is there just to show how the index of refraction of air enters into the equation. It could be left out with no impact on the result.

When the angle of reflection is 90°, the angle of incidence is called the *critical angle.* For the fiber-optic material in air, what is the critical angle?

The critical angle is

$$\theta_r := \mathrm{asin}\left(\frac{1.5 \cdot \sin(90 \cdot deg)}{1.5}\right)$$

$$\theta_c = 41.81 \cdot deg$$

The critical angle is important because if you keep the angle of incidence greater than the critical angle, you get total internal reflection – there is no refracted ray, and thus no light is lost because of refraction. This is extremely important if the fiber-optic material is to be used for transmitting signals, since refracted light shortens the potential transmission distance.

FIBER CLADDING

If the medium outside the fiber is something other than air, the critical angle changes. For example, if the fiber is in water ($n_2 \approx 1.33$), the critical angle is higher:

$$\theta_r := \text{asin}\left(\frac{1.33 \cdot \sin(90 \cdot \text{deg})}{1.5}\right)$$

$$\theta_c = 62.457 \cdot \text{deg}$$

By increasing the refractive index of the material on the outside of the fiber, the angle of incidence can be closer to perpendicular to the wall of the fiber and still have total internal reflection. If the material outside the fiber has a refractive index greater than 1.5 (i.e., greater than the glass's refractive index), you get total internal reflection for *any* angle of incidence. To ensure that this is the case, fiber-optic materials are often coated with a second layer of glass with a slightly higher refractive index. This process is called cladding the fibers.

SUMMARY

In this chapter, you learned that a function is a reusable equation that accepts arguments, uses the argument values in a calculation, and returns the result(s). Once a function has been defined in a worksheet, it can be used multiple times in that worksheet. You have learned to write your own functions and to use Mathcad's built-in functions in the following areas.

Logarithm and Exponentiation	
`exp(x)`	Raises **e** to the power **x**.
`ln(x)`	Returns the natural logarithm of **x**.
`log(x)`	Returns the base-10 logarithm of **x**.

Trigonometric Functions

The units on the angle **a** in these functions must be radians. If you want to work in degrees, replace **a** by **a · deg**, where **deg** is a predefined unit for degrees:

`sin(a)`	Returns the sine of **a**.
`cos(a)`	Returns the cosine of **a**.
`tan(a)`	Returns the tangent of **a**.
`sec(a)`	Returns the secant of **a**.
`csc(a)`	Returns the cosecant of **a**.
`cot(a)`	Returns the cotangent of **a**.

Inverse Trigonometric Functions	
These functions take a value between −1 and 1 and return the angle in radians. You can display the result in degrees by using **deg**.	
`asin(x)`	Arcsine—returns the angle that has a sine value equal to **x**.
`acos(x)`	Arccosine—returns the angle that has a cosine value equal to **x**.
`atan(x)`	Arctangent—returns the angle that has a tangent value equal to **x**.

Hyperbolic Trigonometric Functions	
`sinh(a)`	`cosh(a)`
`tanh(a)`	`csch(a)`
`sech(a)`	`coth(a)`

Inverse Hyperbolic Trigonometric Functions
`asinh(x)`
`acosh(x)`
`atanh(x)`

Round-off and Truncation Functions	
`floor(x)`	Returns the largest integer that is less than or equal to **x**.
`ceil(x)`	Returns the smallest integer that is greater than or equal to **x**.
`mod(x,y)`	Returns the remainder after dividing **x** by **y**.

Working with Complex Numbers	
`Re(z)`	Returns the real part of complex number **z**.
`Im(z)`	Returns the imaginary part of complex number **z**.
`arg(z)`	Returns the principal argument of **z**, between $-\pi$ and π.
`|z|`	Returns the magnitude of complex number **z**. This operator is available on the Calculator Toolbar, or it can be entered using the vertical bar [|] key.

Advanced Mathematical Functions	
`iif(cond,Tval,Fval)`	Returns **Tval** or **Fval**, depending on the value of the logical condition *cond*.
`δ(m,n)`	Kronecker delta—returns 1 if **m = n** and returns 0 otherwise.
`ϕ(x)`	Heaviside step function—returns 0 if **x** is negative and returns 1 otherwise.
`rnd(x)`	Returns a random number (not an integer) between 0 and **x**.
`runif(m,a,b)`	Returns a vector of **m** uniformly distributed values between **a** and **b**.
`rnorm(m,μ,σ)`	Returns a vector of **m** normally distributed (i.e., bell curve) values with a mean **μ** and standard deviation **σ**.

Boolean (Logical) Operators	
¬A	NOT Operator—returns 1 if A is zero; returns 0 otherwise.
A∧B	AND Operator—returns 1 if A and B both have nonzero values; otherwise returns 0.
A∨B	OR Operator—returns 1 if A or B is nonzero; otherwise returns 0.
A ⊕ B	XOR (Exclusive OR) Operator—returns 1 if A or B, but not both, is nonzero; otherwise returns 0.

String Functions	
`concat(S1,S2)`	Concatenates string **S2** to the end of string **S1**. Returns a string.
`strlen(S)`	Determines the number of characters in the string **S**. Returns an integer.
`substr(S,n,m)`	Extracts a substring of **S**, starting with the character in position **n** and having at most **m** characters. The arguments **m** and **n** must be integers.
`search(S,SubS,x)`	Finds the starting position of the substring **SubS** in **S**, beginning from position **x** in **S**.
`str2num(S)`	Converts a string of numbers **S** into a constant.
`num2str(x)`	Converts the number **x** into a string.

KEY TERMS

Boolean (logical) operator
case sensitive
character string
discontinuous function
exponentiation function
logarithm
modulo division
operator
parameter
QuickPlot
radian
random number
round-off
string function
toolbar
trigonometric
truncation

PROBLEMS

3.1 Force and Pressure

A force of 800 N is applied to a piston with a surface area of 24 cm^2 as illustrated in Figure 3.21.

(a) What pressure (Pa) is applied to the fluid? Use

$$P = \frac{F}{A}.$$

(b) What is the diameter (cm) of a piston with a (circular) surface area of 24 cm^2?

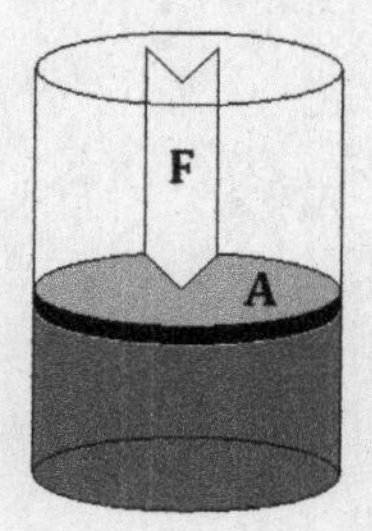

Figure 3.21
Force exerted on a piston.

3.2 Calculating Bubble Size

Very small gas bubbles (e.g., air bubbles less than 1 mm in diameter in water) rising through liquids (see Figure 3.22) are roughly spherical in shape.

(a) What volume of air (mm^3) is contained in a spherical bubble 1 mm in diameter?

(b) What is the surface area (mm^2) of a spherical bubble 1 mm in diameter?

(c) Calculate the diameter of a spherical bubble containing 1.2×10^{-7} mole of nitrogen at 25°C at 1 atm. Use the formulas

$$V = \frac{4}{3}\pi R^3$$

and

$$A = 4\pi R^3.$$

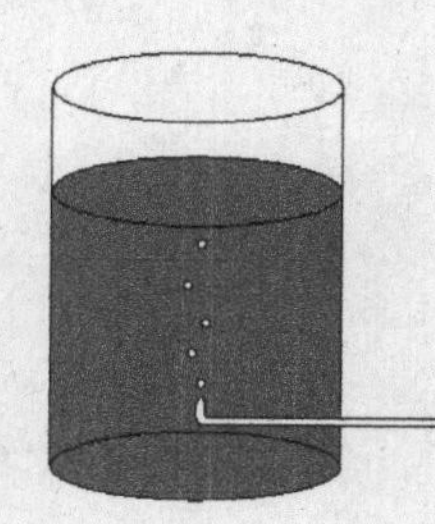

Figure 3.22
Calculating bubble size.

3.3 Relationships Between Trigonometric Functions

Devise a test to demonstrate the validity of the following common trigonometric formulas:

(a) $\sin(A + B) = \sin(A)^3 \cos(B) + \cos(A)^3 \sin(B)$
(b) $\sin(2 \cdot A) = 2^3 \sin(A)^3 \cos(A)$
(c) $\sin^2(A) = 1/2 - 1/2^3 \cos(2^3 A)$

What values of *A* and *B* should be used to thoroughly test these functions?

Note: In Mathcad, `sin²(A)` should be entered as `sin(A)²`. This causes `sin(A)` to be evaluated first and that result to be squared.

3.4 Calculating Acreage

An odd-shaped corner lot (see Figure 3.23) is up for sale. The "going rate" for property in the area is $3.60 per square foot.

(a) Determine the corner angle, α (in degrees).
(b) What is the area of the lot in square feet? In acres?
(c) How much should the seller ask for the property?

3.5 Using the Angle of Refraction to Measure the Index of Refraction

A laser in air ($n_2 \approx 1$) is aimed at the surface of a liquid with an angle of incidence of 45°. A photodetector is moved through the liquid until the beam is located at an angle of refraction of 32°. What is the index of refraction of the liquid?

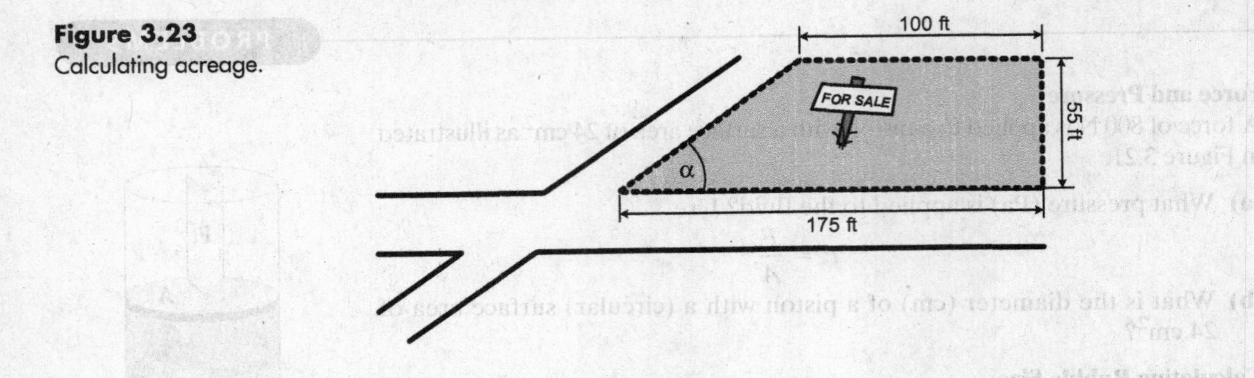

Figure 3.23
Calculating acreage.

3.6 Measuring Heights

A surveyor's *transit* is a telescope-like device mounted on a protractor that allows the surveyor to sight distant points and then read the angle of the scope off the scale. With a transit it is possible to get accurate measurements of large objects.

In this problem, a transit is being used to measure the height of a tree (see Figure 3.24). The transit is set up 80 feet from the base of the tree ($L = 80$ ft), and the sights are set on the top of the tree. The angle reads 43.53° from the horizontal. Then the sights are set at the base of the tree, and the second angle is found to be 3.58°. Given this information, how tall is the tree?

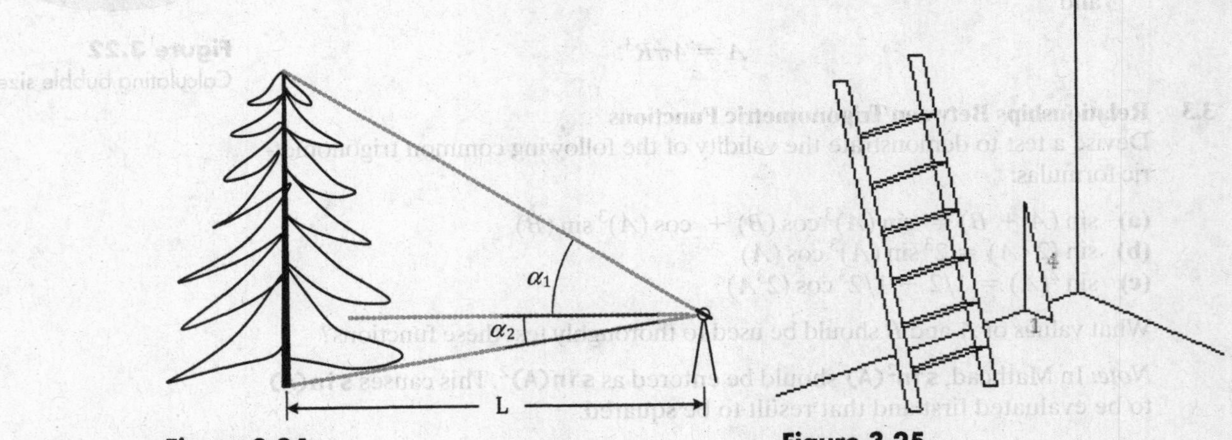

Figure 3.24
Measuring heights.

Figure 3.25
Ladder safety.

3.7 Ladder Safety

Ladders are frequently used tools, but they are also the source of many injuries. Improperly positioning a ladder is a common cause of falls.

Ladders are supposed to be set against walls at an angle such that the base of the ladder stands away from the wall 1 foot for every 4 feet of working length of the ladder. So the base of a 12-foot ladder with the top resting against a wall (as shown in Figure 3.25) should be 3 feet away from the wall. At this angle, most people can stand up straight on a rung and reach out and easily grab another rung with their hands. The 1-in-4 rule is for comfort and for safety—and it is written into Occupational Safety and Health Administration (OSHA) regulations (OSHA Standards, 29.b.5.i).

(a) What is the angle θ between the floor and the ladder if the 1-in-4 rule is used?

(b) If the 1-in-4 rule is used, at what height does the top of a 12-foot ladder touch the wall?

3.8 Solving Quadratic Equations

A quadratic equation can be written in the form

$$ax^2 + bx + c = 0 \ (a \neq 0).$$

These equations have two solutions, which can be found by using the formula

$$x = \frac{-b \pm \sqrt{b^2 - 4ac}}{2a}$$

Find the solutions to the following equations:

(a) $-2x^2 + 3x + 4 = 0.$ (Expect real roots, since $b^2 - 4ac > 0$.)
(b) $-4x^2 + 4x - 1 = 0.$ (Expect real, equal roots, since $b^2 - 4ac = 0$.)
(c) $3x^2 + 2x + 1 = 0$ (Expect imaginary roots, since $b^2 - 4ac < 0$.)

3.9 Heating a Hot Tub

A 500-gallon hot tub (see Figure 3.26) has been unused for some time, and the water in the tub is at 65°F.

Figure 3.26
Heating a hot tub.

Hot water (130°F) will be added to the tub at a rate of 5 gallons per minute. The temperature in the hot tub can be calculated with the equation

$$T = T_{IN} - (T_{IN} - T_{START})e^{\frac{-Q}{V}t},$$

where

T	is the temperature of the water already in the tub,
T_{IN}	is the temperature of the water flowing into the tub (130°F),
T_{START}	is the initial temperature of the water in the hot tub (65°F),
Q	is the hot water flow rate (5 gpm).
V	is the volume of the tub (500 gallons), and
T	is the elapsed time since the hot water started flowing.

(a) What will be the temperature of the water in the hot tub after one hour?

(b) How long will it take (the water in the tub) to reach a temperature of 110°F?

(c) If the temperature of the water flowing into the tub could be raised to 150°F, how long would it take to warm the water to 110°F?

3.10 Vapor Pressure

Knowing the vapor pressure is important in designing equipment that will or might contain boiling or highly volatile liquids. When the pressure in a vessel is at or below the liquid's vapor pressure, the liquid will boil. If you are designing a boiler, then designing it to operate at the liquid's vapor pressure is a pretty good idea. But if you are designing a waste solvent storage facility, you want the pressure to be significantly higher than the liquid's vapor pressure. In either case, you need to be able to calculate the liquid's vapor pressure.

Antoine's equation is a common way to calculate vapor pressures for common liquids. It requires three coefficients that are unique to each liquid. If the coefficients are available, Antoine's equation[1] is easy to use. The equation is

$$\log(p_{vapor}) = A - \frac{B}{T + C},$$

where

p_{vapor}	is the vapor pressure of the liquid in mm Hg,
T	is the temperature of the liquid in °C, and
A, B, C	are coefficients for the particular liquid (found in tables).

In using Antoine's equation, keep the following important considerations in mind:

1. Antoine's equation is a *dimensional equation;* that is, specific units must be used to get correct results.
2. The `log` function is not supposed to be used on a value with units, but that's the way Antoine's equation was written. Mathcad, however, will not allow you to take the logarithm of a value with units, so you must work without units while using Antoine's equation.
3. There are numerous forms of Antoine's equation, each requiring specific units. The coefficients are always designed to go along with a particular version of the equation. You cannot use Antoine coefficients from one text in a version of Antoine's equation taken from a different text.

Test Antoine's Equation

Water boils at 100°C at 1 atm pressure. So the vapor pressure of water at 100°C should be 1 atm, or 760 mm Hg. Check this using Antoine's equation. For water,

$$A = 7.96671 \quad B = 1668.21 \quad C = 228.0.$$

Use Antoine's Equation

(a) A chemistry laboratory stores acetone at a typical room temperature of 25°C (77°F) with no problems. What would happen if on a hot summer day the air-conditioning failed and a storage cabinet warmed by the sun reached 50°C (122°F)? Would the stored acetone boil?

(b) At what temperature would the acetone boil at a pressure of 1 atm? For acetone:

$$A = 7.02447 \quad B = 1161.0 \quad C = 224.$$

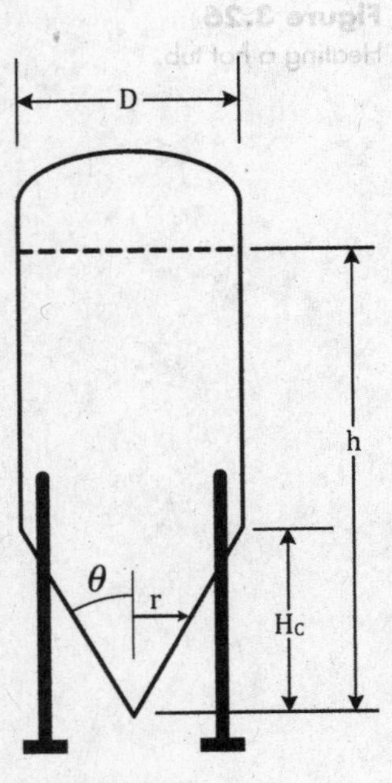

Figure 3.27
Volume of a storage tank.

3.11 Calculating the Volume and Mass of a Substance in a Storage Tank

Vertical tanks with conical base sections are commonly used for storing solid materials such as grain, gravel, salt, catalyst particles, and so forth. The sloping sides help prevent plugging as the material is withdrawn. (See Figure 3.27.)

If the height of solid in the tank is less than the height of the conical section, H_c, then the volume is computed by using the formula for the volume of a cone:

$$V = \frac{1}{3}\pi r^2 h.$$

[1]This version of Antoine's equation is from *Elementary Principles of Chemical Processes*, by R. M. Felder and R. W. Rousseau (New York: Wiley, 1978).

But if the tank is filled to a depth greater than H_C, then the volume is the sum of the filled conical section, V_{cone}, and a cylindrical section, V_{cyl}, of height $h - H_C$:

$$V = V_{cone} + V_{cyl}$$

$$V = \frac{1}{3}\pi R^2 H_C + \pi R^2(h - H_C).$$

Here, $R = D/2$.

The radius of the tank depends on h if $h < H_C$, but has a value of R when $h = H_C$. The `if( )` function can be used to return the correct value of r for any h:

```
r(h):= if (h < Hc,h·tan(θ),R)
```

where θ is the angle of the sloping walls. (Variables **H**, **R**, and $\boldsymbol{\theta}$ must be defined before using this function.)

When working with solids, you often want to know the mass as well as the volume. The mass can be determined from the volume using the *apparent density*, ρ_A, which is the mass of a known volume of the granular material (including the air in between the particles) divided by the known volume. Since the air between the particles is much less dense than the solid particles themselves, the apparent density of a granular solid is typically much smaller than the true density of any individual particle.

(a) Use Mathcad's `if` function to create a function that will return the volume of solids for any height.

(b) Use the function from part **(a)** to determine the volume and mass of stored material (ρ_A = 20 lb/ft^3) in a tank (D = 12 ft, θ = 30□) when the tank is filled to a depth h = 21 ft.

3.12 Force Components and Tension in Wires

A 150-kg mass is suspended by wires from two hooks. (See Figure 3.28.) The lengths of the wires have been adjusted so that the wires are each 50° from horizontal. Assume that the mass of the wires is negligible.

(a) Since two hooks support the mass equally, the vertical component of force exerted by either hook will be equal to the force resulting from 75 kg being acted on by gravity. Calculate this vertical component of force, $\mathbf{f_v}$, on the right hook. Express your result in newtons.

(b) Compute the horizontal component of force, $\mathbf{f_h}$, by using the result obtained in problem **(a)** and trigonometry.

(c) Determine the force exerted on the mass in the direction of the wire $\mathbf{F_R}$, (equal to the tension in the wire).

(d) If you moved the hooks farther apart to reduce the angle from 50° to 30°, would the tension in the wires increase or decrease? Why?

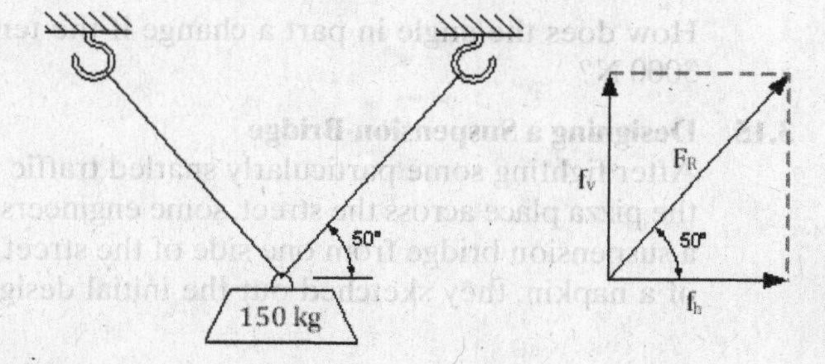

Figure 3.28
Mass supported equally by two wires.

3.13 Multiple Loads

If two 150-kg masses are suspended on a wire as shown in Figure 3.29, such that the section between the loads (wire *B*) is horizontal, then wire *B* is under tension, but it is doing no lifting. The entire weight of the 150-kg mass on the right is being held up by the vertical component of the force in wire *C*. In the same way, the mass on the left is being supported entirely by the vertical component of the force in wire *A*.

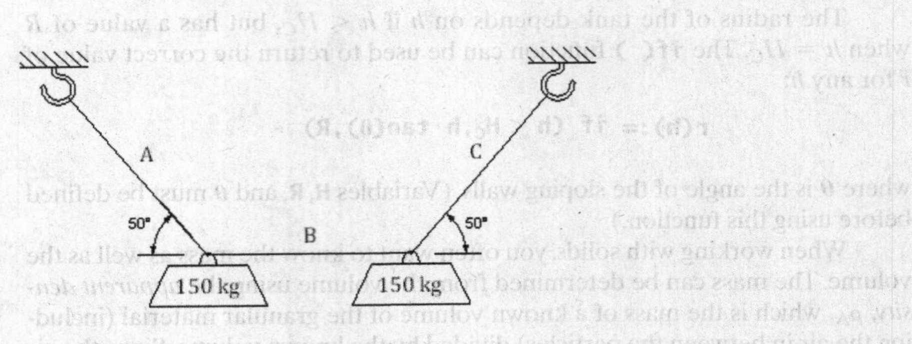

Figure 3.29 Multiple loads supported by two wires.

Calculate

(a) the vertical force component in wires *A* and *C*.

(b) the horizontal force component in each wire.

(c) the tension in wire *A*.

3.14 Tension and Angles

If the hooks are pulled farther apart (see Figure 3.30), the tension in wire *B* will change, and the angle of wires *A* and *C* with respect to the horizontal will also change.

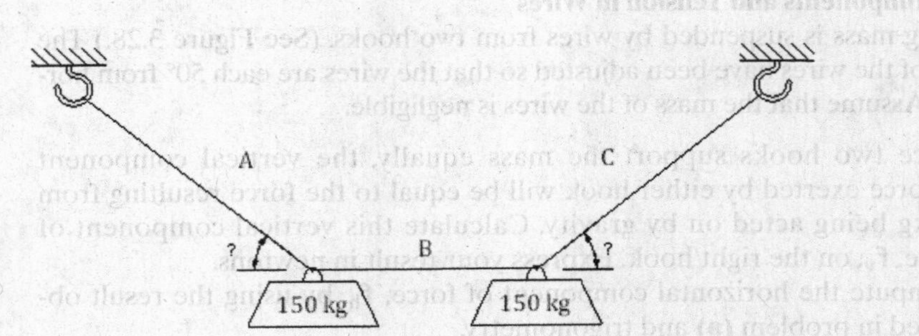

Figure 3.30 Effect of angle on tension in wires.

If the hooks are pulled apart until the tension in wire *B* is 2000 N, determine

(a) the angle between the horizontal and wire *C*, and

(b) the tension in wire *C*.

How does the angle in part a change if the tension in wire *B* is increased to 3000 N?

3.15 Designing a Suspension Bridge

After fighting some particularly snarled traffic to get from their apartment to the pizza place across the street, some engineers decided that they should build a suspension bridge from one side of the street to the other. On the back side of a napkin, they sketched out the initial design shown in the accompanying

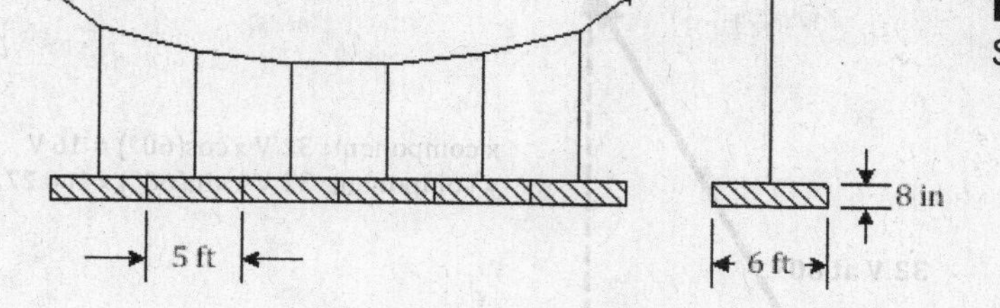

Figure 3.31
Suspension bridge design.

figure. The plan calls for a single cable to be hung between the buildings, with one support wire to each deck section. The deck is to be 6 feet wide to allow pedestrian traffic on either side of the support wires. The bridge will be 30 feet long and made up of six 5-foot sections as illustrated in Figure 3.31. The engineers want their design to be environmentally friendly, so the design calls for the decking to be made of a material constructed of used tires, with a density of 88 lb/ft^3.

Note: To help keep things straight, we use the term "wire" for the vertical wires connected to the deck sections and the term "cable" for the main suspension between the buildings.

(a) What is the mass of each deck section?
(b) What force is a deck section imparting to its suspension wire?

The engineers had quite a discussion about how much tension to impose on the bridge. One wanted to minimize the stress on the buildings by keeping the tension low, but others thought that would produce a "droopy" bridge. They finally decided that the tension in the center (horizontal) section would be five times the force that any deck section put on its suspension wire.

(c) What is the horizontal component of force in any of the cable sections?
(d) Calculate all the forces and angles in the main cable segments.

Hint: Since the center segment of the cable has only a horizontal force component and you can calculate the tension in that segment, start in the center and work your way toward the outside. Also, since the bridge is symmetrical, you need only to solve for the angles and forces in one-half of the bridge.

At this point, the waiter walked up and made a comment that upset all of their plans. The result was a lively discussion that went on into the night, but no follow-through on the original design.

So What Did the Waiter Say?

"As soon as someone stands on that bridge, the deck will tip sideways and dump the person off."

The engineers spent the rest of the evening designing solutions to this problem. Do you have any ideas?

3.16 AC Circuit I

The voltage and phase angle of an AC voltage source can be written in polar form or rectangular form. An example of polar form notation is 32 V at 60°. The same voltage can be presented in rectangular form as a complex number: (16 + 27.7i) volts.

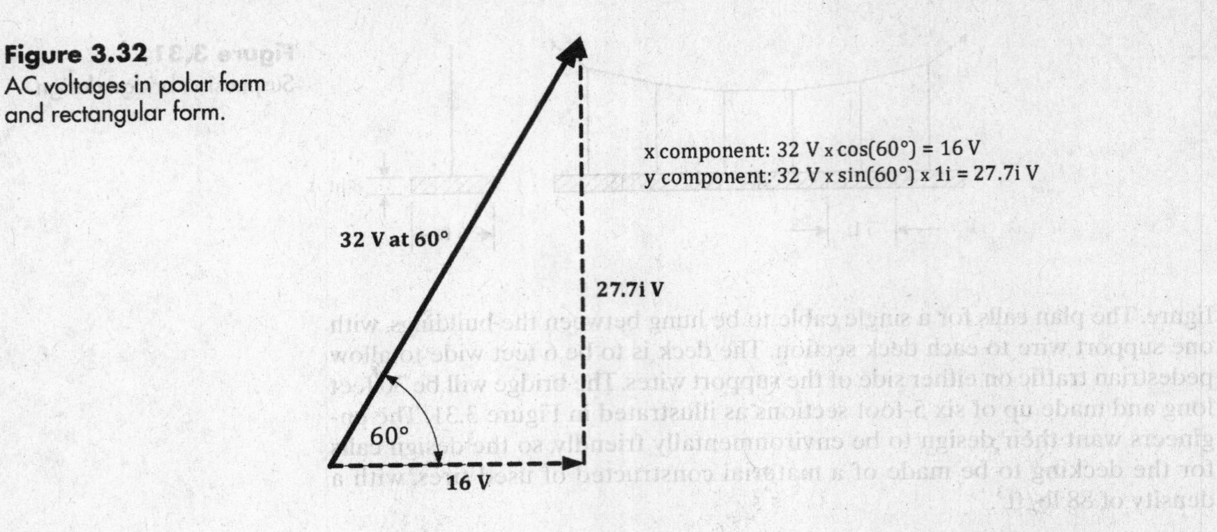

Figure 3.32 AC voltages in polar form and rectangular form.

When viewed in terms of a vector, the polar form presents the magnitude and direction (angle) of the vector. The rectangular form reports the *x*- and *y*-components of the vector reported as the real and imaginary parts of a complex number. Both forms are illustrated in Figure 3.32.

(a) Write a function that receives a polar form voltage and phase angle as arguments, then returns the rectangular form as a complex number.
Note: You cannot multiply something by "i"; Mathcad will think "i" is a variable. But you can multiply by "1i" or $\sqrt{-1}$.

(b) Test your function using the example shown in Figure 3.27.

32 V at 60° ans: (16 + 27.7i)V

(c) Use your function to determine the rectangular forms of the following AC voltages:

25 V at − 65°
10 V at 75°
15 V at −10°

(d) Use your function to determine the rectangular forms of the following AC voltages:

110 V at 30°
220 V at 60°
60 V at 45°

3.17 AC Circuit II

When the voltage and phase angle of an AC voltage source are reported in rectangular form, such as VAC = (6.339 − 13.595i)V, the theorem of Pythagoras can be used to find the magnitude of the AC voltage, and the **atan** function can be used to determine the phase angle. Also, to separate the real and imaginary parts of the complex number, Mathcad's **Re** and **Im** functions can be used. Alternatively, Mathcad's magnitude operator |■|, and **arg** function can be used to determine the magnitude and phase angle of the voltage.

Determine the magnitude and phase angle of the following AC voltages:

(a) (6.339 − 13.595i)V
(b) (7.015 + 12.25i)V
(c) (12.25 + 12.25i)V
(d) (1.30 − 20.76i)V

3.18 Angles of Incidence and Refraction

If you want to try spear fishing, you have to learn that the fish you see under the water is not exactly where it appears to be because of the refraction of light as it passes between water and the air. The angle of refraction, θ_r, depends on the angle of incidence, θ_i, and the indices of refraction for the two materials.

$$n_{water}\theta_i = n_{air}\theta_r$$

The spear fishing example is illustrated in Figure 3.33. The index of refraction for air is about 1, while the index for water is 1.33.

(a) If the angle of refraction is 44°, as shown in Figure 3.28, what is the angle of incidence, θ_i?

(b) If the fish is two vertical feet below the water, what is the horizontal distance between where the fish appears to be, and the true location of the fish (marked as distance Δx in Figure 3.33)?

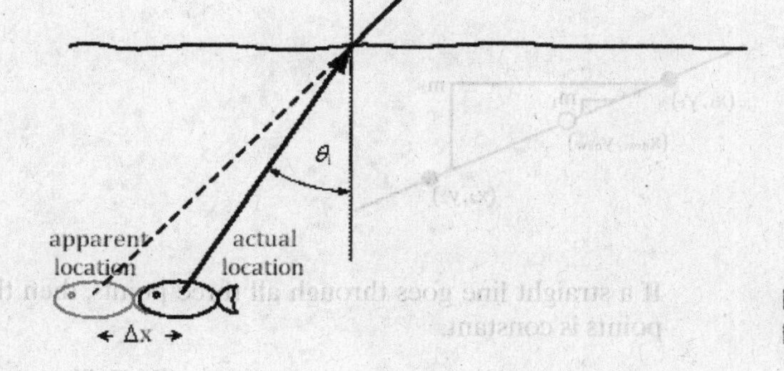

Figure 3.33
Refraction of light.

3.19 Refractometer

A *refractometer* is a device used to measure the refractive index of a material, and it is often used to determine the composition of a sample. A schematic of a possible refractometer design is shown in Figure 3.34.

With this design, the sample is placed in a V-shaped glass sample holder. A laser light passes through the glass and the sample, and then exits into the air. The light enters and leaves the glass at 90°, so the only angle of incidence that

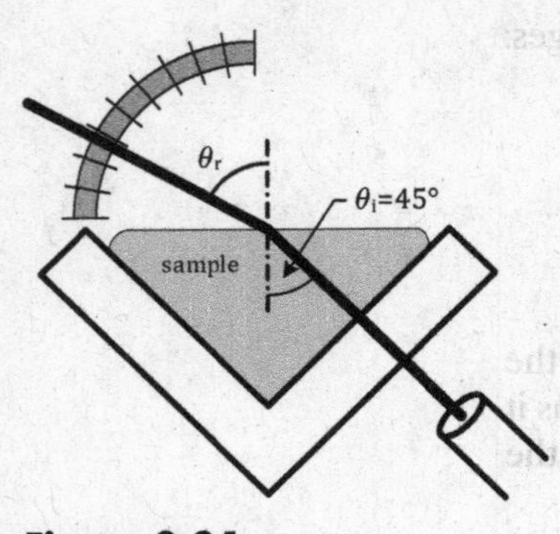

Figure 3.34
Refractometer.

must be considered is the 45° angle that the light makes with the sample surface, marked as θ_i. (The index of refraction of air is very close to $n_{air} = 1.0$.) The angle of refraction, θ_r, is read off the scale shown behind the exiting ray in Figure 3.29.

To prepare to determine the concentration of a water–ethanol mixture, the angles of refraction of some water and ethanol standards were first measured. The compositions are reported as mole percentages.

0% ethanol (100% water)	n = 1.33	$\theta_r = 59.85°$
50% ethanol		$\theta_r = 60.52°$
100% ethanol	n = 1.36	$\theta_r = 61.20°$

(a) Prepare a plot, on paper, of ethanol concentration (mole%) on the x-axis vs. the angle of refraction (°) on the y-axis. Does it look as though there is a linear relationship between the molar composition and angle of refraction?

When the unknown sample was tested, the angle of refraction was found to be 60.05°.

(b) What is the refractive index of the unknown mixture?
(c) What is the composition (mole%) of the unknown mixture?

3.20 A Linear Interpolation Function

Interpolating between tabulated values is a very common chore for engineers. Linear interpolation is the most common interpolation method. Mathcad provides a linear interpolation function, called `linterp`, but it requires that the known data values be entered as vectors. For this problem you will write a linear interpolation function that does not require vectors.

The assumption behind linear interpolation is that a straight line connects the two known data values, and that the desired value between those two points will lie on the same line. This is illustrated in Figure 3.35. The solid dots represent known values from a data table. The open dot represents the intermediate value that we want to determine.

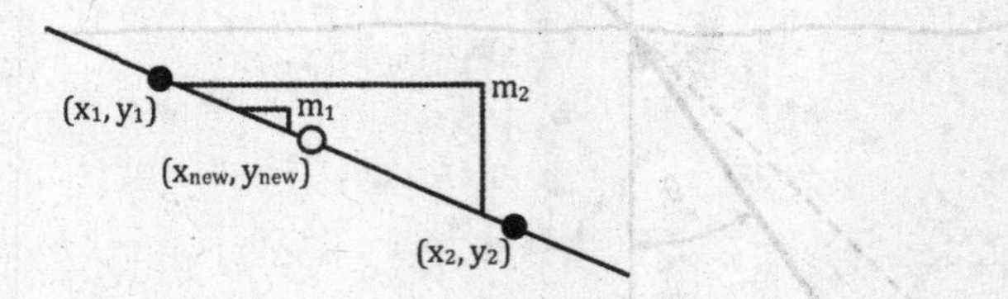

Figure 3.35
Graphical illustration of linear interpolation.

If a straight line goes through all three points, then the slope between any two points is constant.

$$m_1 = m_2$$

$$\frac{y_{new} - y_1}{x_{new} - x_1} = \frac{y_2 - y_1}{x_2 - x_1}$$

In the previous equation, (x_1, y_1) and (x_2, y_2) are known (tabulated) data values, and x_{new} is the value at which we want to find y_{new}. Every value in the equation is known, except for y_{new}. Solving for y_{new} yields

$$y_{new} = y_1 + \left[\frac{y_2 - y_1}{x_2 - x_1}\right] \cdot (x_{new} - x_1)$$

(a) Write a function called `linear` to perform linear interpolation using the equation for y_{new}. Five values (x_1, x_2, y_1, y_2, x_{new}) will need to be passed into the function as arguments. Then your function should solve for y_{new} and return that value.

Test your function using these values:

$x_1 = 10 \quad y_1 = 25$
$x_2 = 5 \quad y_2 = 20$
$x_{new} = 7$ answer: $y_{new} = 22$

(b) Experimental data on warming a hot tub are shown in Table 3.2. The heater was turned on at time zero. Use your interpolation function to back calculate the time at which the temperature was 40°C (104°F), which is as warm as recommended for hot tubs. In this problem, the temperatures will be the x values, and the times will be the y values.

Table 3.2 Hot Tub Temperature Data

Time (min.)	Temp. (°C)
0	4.4
5	17.1
10	26.4
15	33.2
20	38.1
25	41.8
30	44.4
35	46.4
40	47.8
45	48.8

SOLUTIONS TO PRACTICE! PROBLEMS

Section 3.2.2 Practice!

More Obvious Examples

a. $\sqrt{4} = 2$
b. $\sqrt[3]{8} = 2$
c. $|-7| = 7$
d. $3! = 6$
e. $\log(100) = 2$

Less Obvious Examples

a. $20! = 2.433 \times 10^{18}$
b. $\ln(-2) = 0.693 + 3.142i$
c. $\exp(-0.4) = 0.67$

Section 3.4 Practice!

Mathcad Trig Functions

a. $\sin\left(\frac{\pi}{4}\right) = 0.7071$

b. $\sin(90 \cdot \text{deg}) = 1$

c. $\cos(180 \cdot \text{deg}) = -1$

d. $\text{asin}(0) = 0$

e. $\text{acos}(2 \cdot \pi) = 2.525i$

QuickPlots

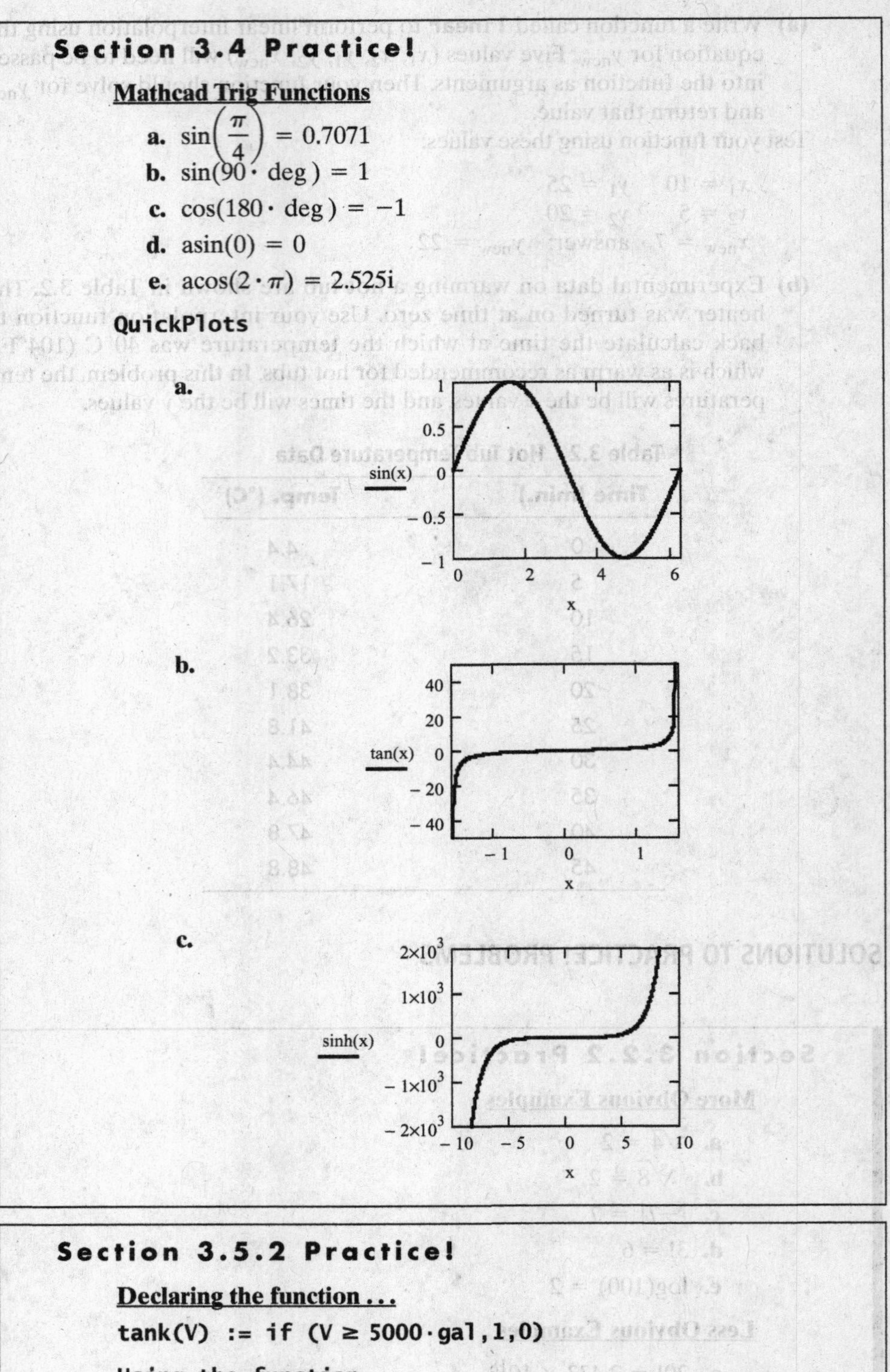

Section 3.5.2 Practice!

Declaring the function . . .

```
tank(V) := if (V ≥ 5000·gal,1,0)
```

Using the function...

```
tank(4900·gal) = 0
tank(5100·gal) = 1
tank(22000·liter) = 1
```

Section 3.7 Practice!

Declaring the function . . .

`boxArea(H,W,L) := H W L`

Using the function...

a. `boxArea(1,1,1) = 1`

b. `boxArea(1,1,10) = 10`

c. boxArea(1·cm, 1·cm 10·cm) = 10 cm^3

d. `boxArea(1·cm, 1·m, 10·ft) = 30.48 liter`

ANSWERS TO SELECTED PROBLEMS

Problem 1	b. 5.53 cm
Problem 3	(value of A is to be chosen by student; results will vary) $A = \pi/8, \sin(2A) = 0.707, 2\sin(A)\cos(A) = 0.707$ $A = 5\pi/8, \sin(2A) = -0.707, 2\sin(A)\cos(A) = 0.707$
Problem 7	b. 11.6 ft
Problem 8	a. −0.851, 2.351
Problem 11	b. $V = 1590\ ft^3$
Problem 12	c. 960 N
Problem 18	a. 33°
Problem 19	c. 16% ethanol

CHAPTER 4

Working with Matrices

Objectives

After reading this chapter, you will

- understand how Mathcad uses the terms "array," "matrix," and "vector"
- know several ways to create a matrix and fill it with values
- be able to add and subtract matrices using Mathcad
- be able to use Mathcad to multiply matrices
- be able to perform element-by-element multiplication of matrices with Mathcad
- be able to use Mathcad to transpose matrices
- be able to invert matrices, if possible, using Mathcad
- be able to find the determinant of a matrix to determine whether inversion is possible
- be aware of Mathcad's built-in functions to manipulate matrices
- be able to use matrix math to solve systems of linear equations

4.1 MATHCAD'S MATRIX DEFINITIONS

A ***matrix*** is a collection of numbers, called ***elements***, that are related in some way. We commonly use matrices to hold data sets. For example, if you recorded the temperature of the concrete in a structure over time as the concrete set, the time and temperature values would form a data set of related numbers. This data set would be stored in a computer as a matrix.

Common usage in mathematics calls a single column or row of values a ***vector***. If the temperature and time values were stored separately, we would have both a time vector and a temperature vector. A matrix is a collection of one or more vectors. That is, a matrix containing a single row or column would also be a vector. On the other hand, a matrix containing three rows and two columns would be called a matrix, not a vector. (But you could say the matrix is made up of three row vectors or two column vectors.) So every vector is a matrix, but only single-row and single-column matrices are called vectors.

Mathcad modifies these definitions slightly by adding the term ***array***. In Mathcad, a vector has only one row or one column, and a matrix always has at least two rows or two columns. That is, there is no overlap in Mathcad's definitions of vectors and matrices. Mathcad uses the term array to mean a collection of related values that could be either a vector or a matrix. Mathcad uses the new term to indicate what type of parameter must be sent to functions that operate on vectors and matrices. For example, you can send either a vector or a matrix to the **`rows(A)`** function, so Mathcad's help files show an array, **`A`**, as the function's parameter to indicate that either a vector or a matrix is acceptable.

Mathcad Definitions:

`A`	Array argument—either a matrix or a vector.
`M`	Matrix argument—an array with two or more rows or columns.
`v`	Vector argument—an array containing a single row or column.

The `length(v)` function requires a vector and will not accept multiple rows or columns. Mathcad indicates this vector-only restriction in its help files by showing a `v` as the `length` function's parameter. The ***determinant*** operator, |`M`|, requires a *square matrix* (same number of rows as columns) and will not work on a vector. To show that a matrix is required, Mathcad's help files display an `M` as the operator's parameter.

4.1.1 Array Origin

By default, Mathcad refers to the first element in a vector as element zero. For a two-dimensional matrix, the ***array origin*** is 0,0 by default. If you would rather have Mathcad start counting elements at another ***index*** value, you can change the default by using the `ORIGIN` variable. The example in Figure 4.1 illustrates how changing the origin from 0,0 to 1,1 changes the row and column numbering of the array elements.

Mathcad defaults to (0,0) for the top-left element

$$\text{MyArray} := \begin{pmatrix} 12 & 15 & 17 \\ 23 & 25 & 19 \end{pmatrix}$$

$\text{MyArray}_{0,0} = 12$ $\quad$ $\text{MyArray}_{0,1} = 15$ $\quad$ ...

$\text{MyArray}_{1,0} = 23$ $\quad$ ...

The ORIGIN variable can be used to change array indexing

$\text{ORIGIN} := 1$ << from this point on, start array indexing at 1, not 0

$\text{MyArray}_{1,1} = 12$ $\quad$ $\text{MyArray}_{1,2} = 15$ $\quad$ ...

$\text{MyArray}_{2,1} = 23$ $\quad$ ...

Figure 4.1
Using the `ORIGIN` variable to change array indexing.

Notice that the array in Figure 4.1 was not changed by reassigning the `ORIGIN` variable; only the way the matrix elements were accessed changed. Before the `ORIGIN` variable was redefined, the top-left element of `MyArray` was called element 0,0. After assigning `ORIGIN` a value of 1, the top-left element is called element 1,1. Notice that reassigning the `ORIGIN` variable in the middle of a worksheet changes the way array elements are addressed from that point down in the worksheet. Changing the way an array is addressed in the middle of a worksheet is confusing to a reader, so the `ORIGIN` variable, when used, is typically used at the top of the worksheet.

You can also change the array origin for the entire worksheet from the Tools menu using **Tools/Worksheet Options. . . .** The `Array Origin` value can be set on the **Built-In Variables** panel.

4.1.2 Maximum Array Size

There are two constraints on array sizes in Mathcad:

- If you enter arrays from the keyboard, the arrays may have no more than 600 elements. You can use ***input tables*** to work around this limitation.

- The total number of elements in all arrays is dependent on the amount of memory in your computer. According to the Mathcad help system, "For most systems, the array limit is at least 1 million elements."

4.2 DEFINING A MATRIX

Before a matrix can be used, it must be *initialized*, that is, it must be defined and filled with values. There are several ways to initialize a matrix in Mathcad. You can

- type in the values from the keyboard,
- read the values from a file,
- use an input table to fill the matrix,
- compute the values by using a function or a ***range variable***, or
- copy and paste the values from another Windows program.

In order to begin working with matrices more quickly, most of these methods will be discussed later in the chapter (Section 4.7). For now, we will simply create matrices directly from the keyboard.

You can define one element of an array using the standard Mathcad assignment operator (:=). For example, typing

```
[G][[][3][,][2][:][6][4]
```

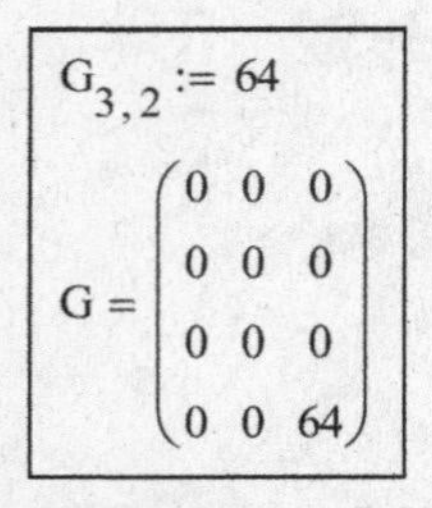

Figure 4.2
Defining a matrix element causes Mathcad to create the matrix.

defines element 3,2 of a matrix called G and assigns it a value of 64. This is illustrated in Figure 4.2. If the G matrix has not been previously defined, then Mathcad will create a matrix just large enough to have an element at position 3,2 and will fill the undefined elements with zeroes.

In Figure 4.2, notice that the default array indexing starting at 0 has been used (no **ORIGIN** statement), so that element 3,2 is actually in the fourth row and third column.

Filling a matrix by defining every element would be extremely slow. Fortunately, Mathcad provides a better way by allowing you to use the Insert Matrix dialog box to define an array of the desired number of rows and columns, then allowing you to fill in the element values as a second step.

Step 1: Create the Empty Array

To create a new array, begin by choosing a variable name and using the assignment operator (:=). In the example in Figure 4.3, we again use the variable **G** and create the left side of a definition by typing **[G] [:]**. Then, open the Insert Matrix dialog box in one of the following (equivalent) ways:

1. Use the keyboard shortcut [Ctrl-M]
2. Use menu options **Insert/Matrix**
3. Use the **Matrix** or **Vector** button on the Matrix Toolbar.

Use the Insert Matrix dialog box to tell Mathcad how many rows and columns the **G** matrix should contain, and then click **OK** to close the dialog box. Mathcad will create an array of placeholders with the requested number of rows and columns. (See Figure 4.4.)

Note: The Insert Matrix dialog box will not create matrices with more than 600 elements. For very large arrays, the alternative methods for creating matrices described in Section 4.7 may be used.

Mathcad - [Ch04 Images.xmcd]

G := ▮

Insert Matrix

Rows: 4

Columns: 3

OK

Insert

Delete

Cancel

Figure 4.3
Using the Insert Matrix dialog box to create a matrix.

Step 2: Fill the Placeholders to Assign a Value to Each Matrix Element
Once the array of placeholders has been created, click on each placeholder and enter the element value. You can use the [Tab] key to move from one placeholder to the next to quickly enter values. The resulting matrix might look like Figure 4.5.

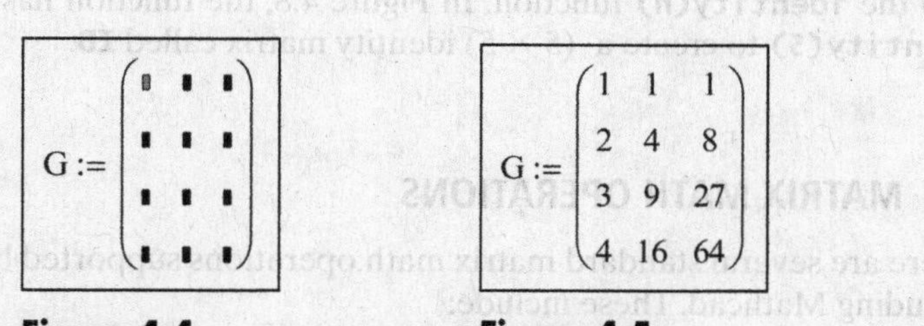

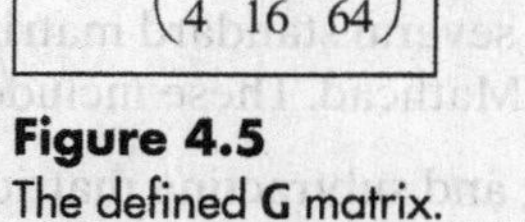

Figure 4.4
The array of placeholders created using the Insert Matrix dialog box.

Figure 4.5
The defined G matrix.

Practice!

Use the Insert Matrix dialog box to create the following matrices.

$$t := (2\ 4\ 6\ 8\ 10)$$

$$W := \begin{pmatrix} 1 & 2 & 3 & 4 \\ 3 & 1 & 5 & 7 \end{pmatrix}$$

$$Y := \begin{pmatrix} 1\cdot sec & 2\cdot min & 3\cdot hr \\ 4\cdot min & 5\cdot min & 6\cdot min \end{pmatrix}$$

4.2.1 Using Units with Matrices

Matrix elements can have units, but every element of a matrix must have the same dimensions. If all of the elements of an array have the same units, you can multiply the entire array by the units, as shown in Figure 4.6.

If the matrix elements have different units (but the same dimensions), you can multiply each element by the appropriate units, as shown in Figure 4.7.

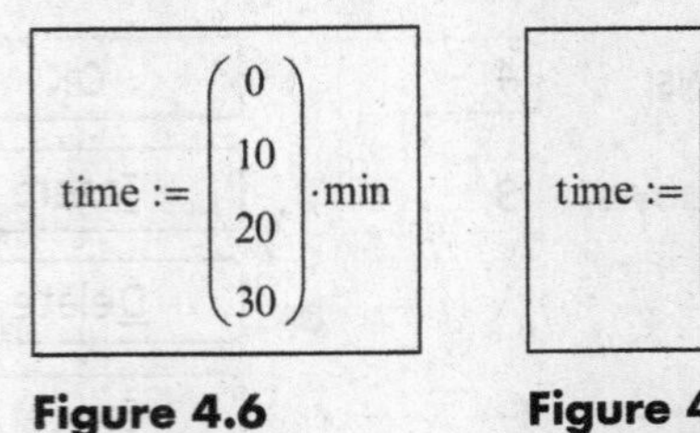

Figure 4.6
Applying units to all matrix elements.

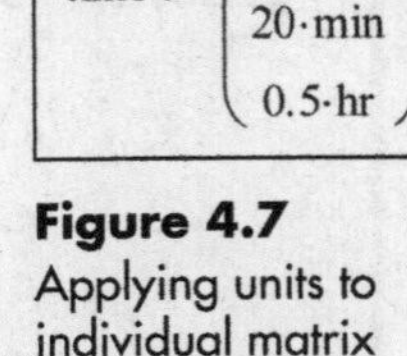

Figure 4.7
Applying units to individual matrix elements.

4.2.2 Creating an Identity Matrix

An ***identity matrix*** is a square matrix filled with 1s along the diagonal and 0s everywhere else. Mathcad's **`identity(n)`** function creates identity matrices. Since the number of rows is always equal to the number of columns in an identity matrix, you need to specify just one or the other so that the **`identity`** function can create a matrix with the correct dimensions. The desired number of rows, *n*, is the argument sent into the **`identity(n)`** function. In Figure 4.8, the function has been used as **`ID := identity(5)`** to create a (5×5) identity matrix called **`ID`**.

$$ID := \text{identity}(5)$$

$$ID = \begin{pmatrix} 1 & 0 & 0 & 0 & 0 \\ 0 & 1 & 0 & 0 & 0 \\ 0 & 0 & 1 & 0 & 0 \\ 0 & 0 & 0 & 1 & 0 \\ 0 & 0 & 0 & 0 & 1 \end{pmatrix}$$

Figure 4.8
Creating an identity matrix.

4.3 MATRIX MATH OPERATIONS

There are several standard matrix math operations supported by all math programs, including Mathcad. These include:

- Adding and subtracting matrices
- Multiplying matrices
- Multiplying elements of matrices
- Transposing a matrix
- Inverting a matrix
- Calculating the determinant of a matrix

These matrix math operations often have certain requirements that must be met, and each follows a prescribed mathematical procedure. As each of these matrix math operations is presented, the requirements and procedure will be described.

4.3.1 Matrix Addition

Requirement: The arrays to be added must be the same size.

Procedure: Each element of the first array is added to (or subtracted from) the same element of the second array.

We can see how Mathcad adds array by using Mathcad's symbolic math capability. While symbolic math is not discussed in this chapter, we will use it here simply to illustrate the procedure. In Figure 4.9, two matrices are added symbolically using the symbolic evaluation operator, →

$$\begin{pmatrix} A & B \\ C & D \\ E & F \end{pmatrix} + \begin{pmatrix} J & K \\ L & M \\ N & O \end{pmatrix} \rightarrow \begin{pmatrix} A+J & B+K \\ C+L & D+M \\ E+N & F+O \end{pmatrix}$$

Figure 4.9
Symbolic matrix addition to illustrate the procedure.

Figure 4.10 illustrates how matrix addition is typically performed.

$$M_A := \begin{pmatrix} 1 & 2 & 3 \\ 2 & 3 & 4 \\ 3 & 4 & 5 \end{pmatrix} \qquad M_B := \begin{pmatrix} 7 & 8 & 9 \\ 8 & 9 & 10 \\ 10 & 11 & 12 \end{pmatrix}$$

$$Sum := M_A + M_B \qquad Sum = \begin{pmatrix} 8 & 10 & 12 \\ 10 & 12 & 14 \\ 13 & 15 & 17 \end{pmatrix}$$

Figure 4.10
Matrix addition.

Matrix Subtraction

Mathcad will also subtract two matrices, as illustrated in Figure 4.11. The requirement is the same as for matrix addition; the matrices must be the same size.

$$Diff := Sum - M_B \qquad Diff = \begin{pmatrix} 1 & 2 & 3 \\ 2 & 3 & 4 \\ 3 & 4 & 5 \end{pmatrix}$$

Figure 4.11
Matrix subtraction.

4.3.2 Matrix Multiplication

Requirement: The inside dimensions of the arrays to be multiplied must be equal. The outside dimensions determine the size of the product matrix. As an example, if we want to multiply a 2×3 matrix and a 3×2 matrix, this is allowed because the inside dimensions (the 3s) are equal. The outside dimensions tell us that the result will be a 2×2 matrix.

Procedure: Working across the columns of the first array and down the rows of the second array and then add the product of each pair of elements. This is easier to do than to describe. Figure 4.12 uses symbolic math to show how Mathcad multiplies matrices.

$$\begin{pmatrix} A & B & C \\ D & E & F \end{pmatrix} \cdot \begin{pmatrix} P & Q \\ R & S \\ T & U \end{pmatrix} \rightarrow \begin{pmatrix} A \cdot P + B \cdot R + C \cdot T & A \cdot Q + B \cdot S + C \cdot U \\ D \cdot P + E \cdot R + F \cdot T & D \cdot Q + E \cdot S + F \cdot U \end{pmatrix}$$

Figure 4.12
Symbolic matrix multiplication to illustrate the procedure.

Multiplying matrices in Mathcad is illustrated in Figure 4.13.

$$M_D := \begin{pmatrix} 1 & 2 & 3 \\ 4 & 5 & 6 \end{pmatrix} \qquad M_E := \begin{pmatrix} 10 & 11 \\ 12 & 13 \\ 14 & 15 \end{pmatrix}$$

$$Prod := M_D \cdot M_E \qquad Prod = \begin{pmatrix} 76 & 82 \\ 184 & 199 \end{pmatrix}$$

Figure 4.13
Matrix multiplication.

The operator used to multiply matrices is the same operator used for multiplying scalars; the asterisk [Shift-8]. However, unlike multiplying scalars, the order in which the matrices are multiplied makes a difference.

$$M_D \cdot M_E \neq M_E \cdot M_D$$

Practice!

The order in which two matrices are multiplied can change both the numerical results and the size of the product matrix.

a. What is the size of the matrix that results from multiplying ME · MD?
b. What numerical result would you expect from multiplying ME · MD?

4.3.3 Element-by-Element Multiplication

Sometimes you don't want true ***matrix multiplication***. Instead, you want each element of the first matrix multiplied by the corresponding element of the second matrix. ***Element-by-element multiplication*** is available in Mathcad by means of the ***vectorize operator*** on the Matrix Toolbar. (You can also use the shortcut key [Ctrl -]; hold the Control key down and press the minus key.) The vectorize operator is indicated by an arrow above the vectorized operation, as shown in Figure 4.14.

Requirement: The arrays must be the same size for element-by-element multiplication.

Procedure: Multiply each individual element of the first matrix by the corresponding element of the second matrix. Figure 4.15 shows how element-by-element multiplication works, and Figure 4.16 shows how to use element-by-element multiplication with numeric arrays.

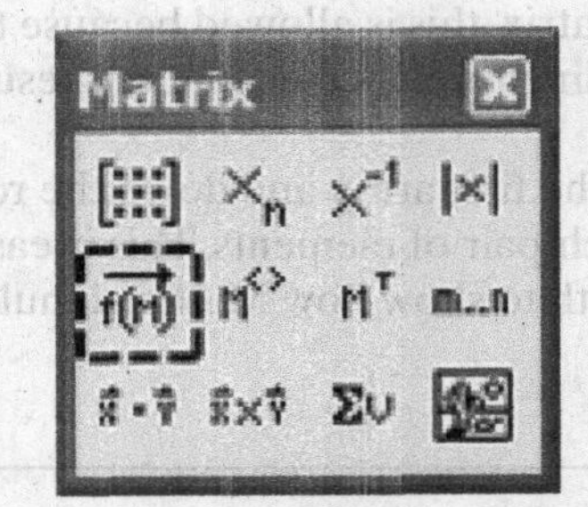

Figure 4.14
The vectorize operator is available on the Matrix Toolbar.

$$\overrightarrow{\left[\begin{pmatrix} A & B \\ C & D \\ E & F \end{pmatrix} \cdot \begin{pmatrix} J & K \\ L & M \\ N & O \end{pmatrix}\right]} \rightarrow \begin{pmatrix} A \cdot J & B \cdot K \\ C \cdot L & D \cdot M \\ E \cdot N & F \cdot O \end{pmatrix}$$

Figure 4.15
Element-by-element multiplication procedure.

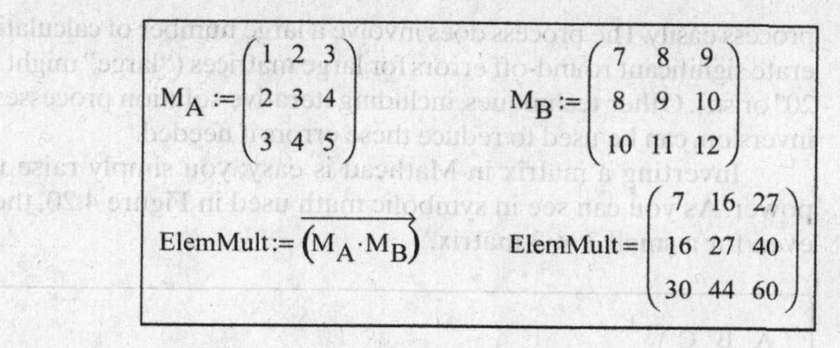

$M_A := \begin{pmatrix} 1 & 2 & 3 \\ 2 & 3 & 4 \\ 3 & 4 & 5 \end{pmatrix}$
$M_B := \begin{pmatrix} 7 & 8 & 9 \\ 8 & 9 & 10 \\ 10 & 11 & 12 \end{pmatrix}$
$ElemMult := \overrightarrow{(M_A \cdot M_B)}$
$ElemMult = \begin{pmatrix} 7 & 16 & 27 \\ 16 & 27 & 40 \\ 30 & 44 & 60 \end{pmatrix}$

Figure 4.16
Element-by-element multiplication.

Practice!

Using the matrices defined in Figure 4.16, compare the results of the following multiplications.

a. $M_A \cdot M_B$
b. $\overrightarrow{(M_A \cdot M_B)}$

4.3.4 Transposing a Matrix

Transposing a matrix interchanges rows and columns. The ***transpose operator*** is available on the Matrix Toolbar (see Figure 4.17) or with keyboard shortcut [Ctrl-1].

Requirement: Any array can be transposed.

Procedure: Interchange row and column elements. $Trans_{j,i} = C_{i,j}$. This is illustrated using symbolic math in Figure 4.18 and using numeric arrays in Figure 4.19.

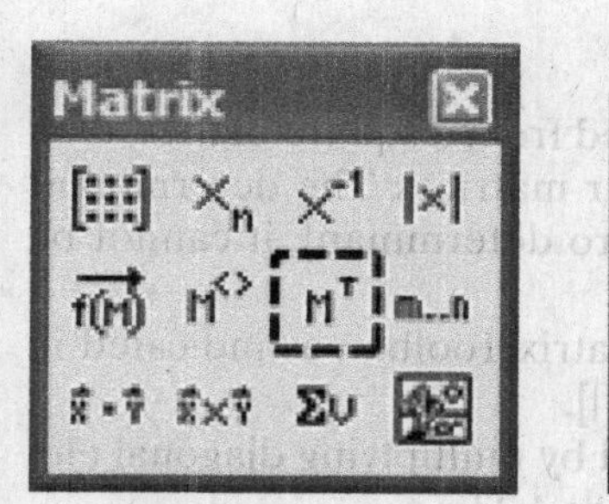

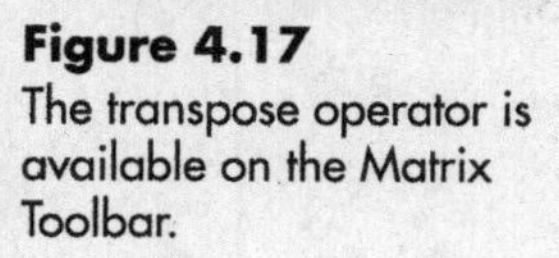

Figure 4.17
The transpose operator is available on the Matrix Toolbar.

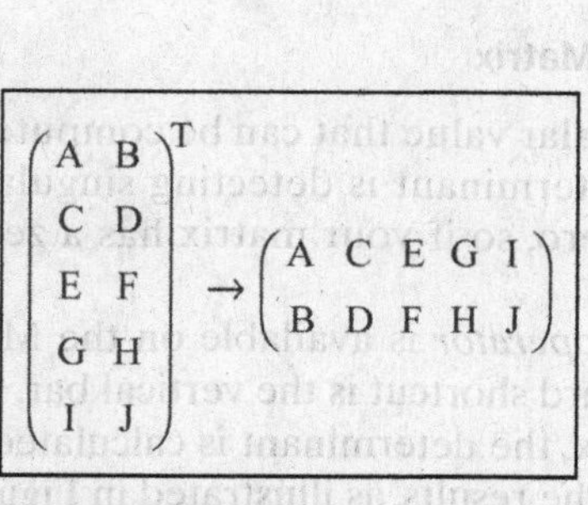

$\begin{pmatrix} A & B \\ C & D \\ E & F \\ G & H \\ I & J \end{pmatrix}^T \rightarrow \begin{pmatrix} A & C & E & G & I \\ B & D & F & H & J \end{pmatrix}$

Figure 4.18
Matrix transposition procedure.

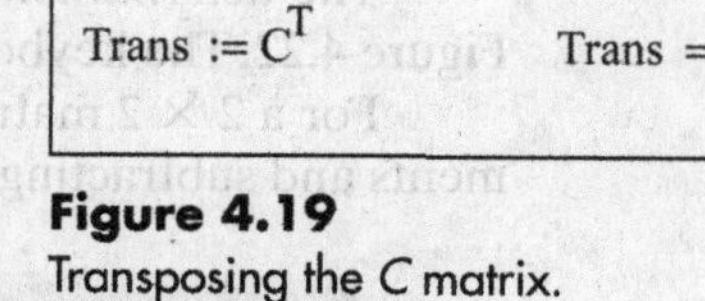

$C := \begin{pmatrix} 1 & 1 \\ 2 & 8 \\ 3 & 27 \\ 4 & 64 \\ 5 & 125 \end{pmatrix}$
$Trans := C^T$
$Trans = \begin{pmatrix} 1 & 2 & 3 & 4 & 5 \\ 1 & 8 & 27 & 64 & 125 \end{pmatrix}$

Figure 4.19
Transposing the C matrix.

4.3.5 Inverting a Matrix

When you ***invert*** a scalar, you divide 1 by the scalar. The result is a value that, when multiplied by the original value, yields 1 as a product. Similarly, when you multiply an inverted matrix by the original matrix, you obtain an identity matrix as the result.

Requirement: Only square matrices can be inverted, and the matrix must be *nonsingular*. In the context of simultaneous linear equations, the term nonsingular implies that there is a solution to the set of equations.

Procedure: The procedure for inverting a matrix is quite involved. The process is described in detail in matrix math texts. The good news here is that Mathcad handles the process easily. The process does involve a large number of calculation steps and can generate significant round-off errors for large matrices ("large" might be "greater than 20 × 20" or so). Other techniques, including iterative solution processes and symbolic matrix inversion, can be used to reduce these errors if needed.

Inverting a matrix in Mathcad is easy; you simply raise the matrix to the −1 power. As you can see in symbolic math used in Figure 4.20, the process is involved, even for a small 3 × 3 matrix.

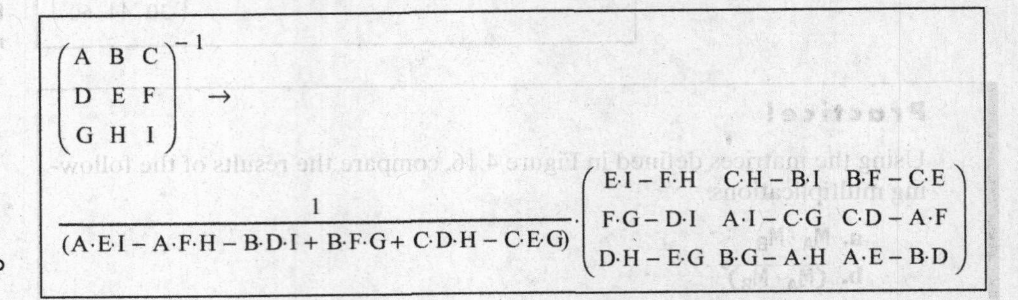

Figure 4.20
Symbolic matrix inversion to illustrate the procedure.

Fortunately, all of the calculation steps are handled by Mathcad, which shows only the final result, as illustrated in Figure 4.21. If Mathcad cannot invert the matrix, it will tell you that the matrix is singular.

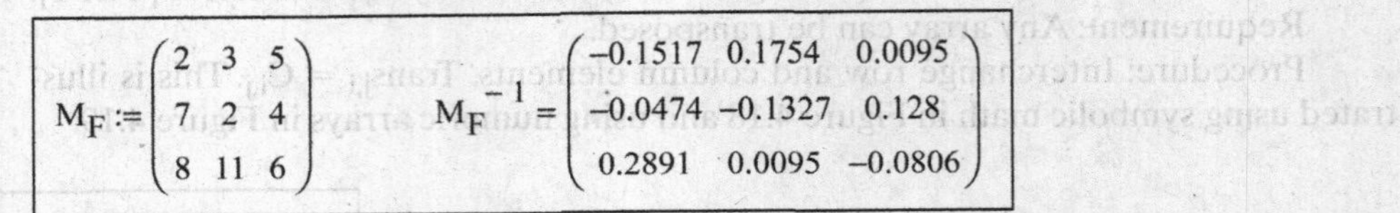

Figure 4.21
Inverting a matrix.

4.3.6 Determinant of a Matrix

The *determinant* is a scalar value that can be computed from a square matrix. One common use of the determinant is detecting singular matrices. The determinant of singular matrix is zero, so if your matrix has a zero determinant, it cannot be inverted.

The *determinant operator* is available on the Matrix Toolbar, as indicated in Figure 4.22. The keyboard shortcut is the vertical bar, [|].

For a 2 × 2 matrix, the determinant is calculated by multiplying diagonal elements and subtracting the results, as illustrated in Figure 4.23.

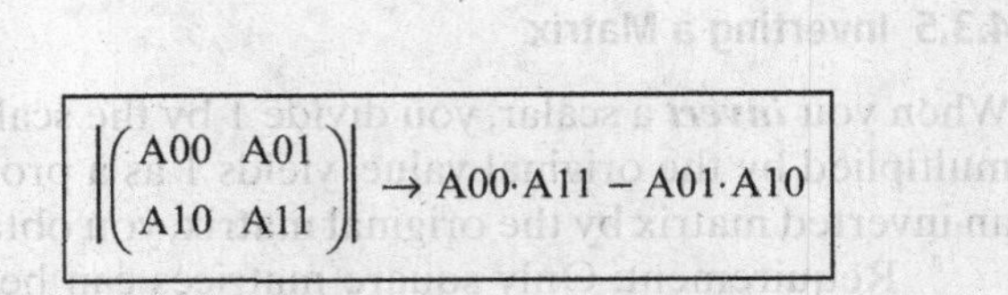

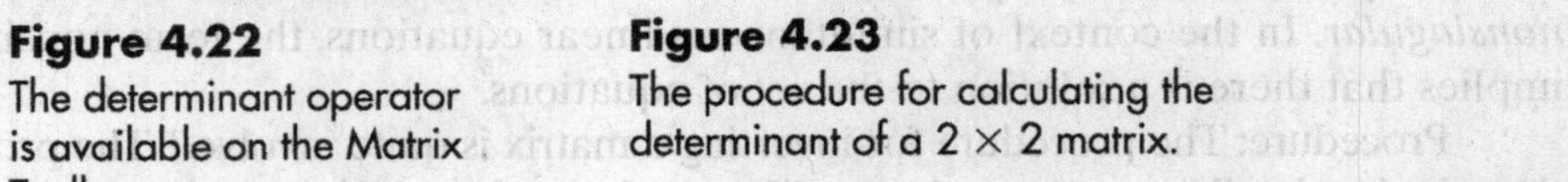

Figure 4.22
The determinant operator is available on the Matrix Toolbar.

Figure 4.23
The procedure for calculating the determinant of a 2 × 2 matrix.

Things get a bit more complicated for a 3×3 matrix, as shown in Figure 4.24.

$$\left|\begin{pmatrix} A00 & A01 & A02 \\ A10 & A11 & A12 \\ A20 & A21 & A22 \end{pmatrix}\right| \rightarrow$$

$$A00 \cdot A11 \cdot A22 - A00 \cdot A12 \cdot A21 - A01 \cdot A10 \cdot A22 + A01 \cdot A20 \cdot A12 + A10 \cdot A02 \cdot A21 - A02 \cdot A11 \cdot A20$$

Figure 4.24
The procedure for calculating the determinant of a 3×3 matrix.

The process gets increasingly involved as the matrices get larger. Fortunately, Mathcad handles the process and simply reports the result. An example is shown in Figure 4.25.

$$M_F := \begin{pmatrix} 2 & 3 & 5 \\ 7 & 2 & 4 \\ 8 & 11 & 6 \end{pmatrix}$$

$$\text{Determ} := |M_F| \qquad \text{Determ} = 211$$

Figure 4.25
Calculating the determinant of a matrix.

4.4 SOLVING SYSTEMS OF LINEAR ALGEBRAIC EQUATIONS

A common use of matrix math is solving systems of linear algebraic equations. Sets of *simultaneous linear equations* occur fairly commonly in many areas of engineering including linear programming, material balance calculations, and current calculations in DC circuits.

A system of linear algebraic equations is shown in Figure 4.26.

$$8 \cdot x_1 + 4 \cdot x_2 - 3 \cdot x_3 = 14$$

$$6 \cdot x_1 + 2 \cdot x_2 - 4 \cdot x_3 = -4$$

$$4 \cdot x_1 - 3 \cdot x_2 + 6 \cdot x_3 = 32$$

Figure 4.26
A system of three simultaneous linear equations.

The unknowns are x_1, x_2, and x_3, which can be written as an *unknown vector*, **x**, as shown in Figure 4.27.

Note: Because the equations have been written in terms of x_1, x_2, and x_3 rather than x_0, x_1, and x_2, this set of equations was solved in Mathcad with **ORIGIN := 1** to start array indexing at 1 rather than 0 (Mathcad's default index origin).

The coefficients multiplying the unknowns in the three equations can be collected in a *coefficient matrix*, **C**. This is illustrated in Figure 4.28.

$$x := \begin{pmatrix} x_1 \\ x_2 \\ x_3 \end{pmatrix}$$

Figure 4.27
The unknowns written as vector **x**.

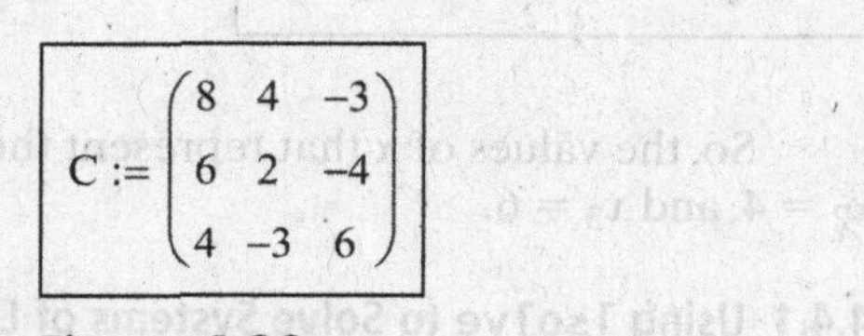

$$C := \begin{pmatrix} 8 & 4 & -3 \\ 6 & 2 & -4 \\ 4 & -3 & 6 \end{pmatrix}$$

Figure 4.28
The coefficient matrix, **C**.

Checking to See If a Solution Exists

The determinant of the coefficient matrix can be calculated to see if a solution exists. If the determinant of the coefficient matrix is zero, there is no solution to the set of equations. Calculating the determinant is illustrated in Figure 4.29. Since the determinant is nonzero, there is a solution to the system of equations.

$$\text{determ} := |C| \qquad \text{determ} = -130$$

Or, simply

$$|C| = -130$$

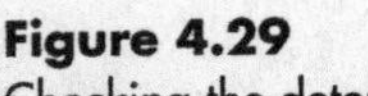

Figure 4.29
Checking the determinant to see if a solution exists.

Multiplying the coefficient matrix and the unknown vector, **C·x**, generates the left-hand side of the system of equations (Figure 4.30). This is not a necessary calculation step, but it demonstrates that the **C** and **x** arrays have been correctly configured.

$$C\cdot x \rightarrow \begin{pmatrix} 8\cdot x_1 + 4\cdot x_2 - 3\cdot x_3 \\ 6\cdot x_1 + 2\cdot x_2 - 4\cdot x_3 \\ 4\cdot x_1 - 3\cdot x_2 + 6\cdot x_3 \end{pmatrix}$$

Figure 4.30
The result of multiplying matrix *C* and vector *x* (symbolic evaluation).

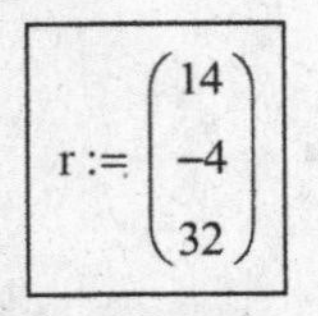

Figure 4.31
The right-hand-side vector, **r**.

The values on the right side of the equations in Figure 4.31 can also be collected in a vector. This vector, shown in Figure 4.31, is typically called the *right-hand-side vector*, **r**.

The set of equations shown in Figure 4.26 can be written in matrix form as **C·x = r**. This matrix equation can be solved for the unknown vector, **x**.

$$C\cdot x = r$$
$$C^{-1}\cdot C\cdot x = C^{-1}\cdot r$$
$$I\cdot x = C^{-1}\cdot r$$
$$x = C^{-1}\cdot r$$

You can determine the element values of the unknown vector by inverting the coefficient matrix and multiplying the result with the right-hand-side vector. This is illustrated in Figure 4.32.

$$x := C^{-1}\cdot r \qquad x = \begin{pmatrix} 2 \\ 4 \\ 6 \end{pmatrix}$$

Figure 4.32
Solving the system of linear equations for the unknown vector, **x**.

So, the values of **x** that represent the solution to the three equations are $x_1 = 2$, $x_2 = 4$, and $x_3 = 6$.

4.4.1 Using `lsolve` to Solve Systems of Linear Algebraic Equations

While the process of inverting the coefficient matrix and multiplying with the right-hand-side vector is pretty straightforward. Mathcad provides a function that makes solving systems of simultaneous equations even easier. The `lsolve(C, r)` function receives the coefficient matrix and right-hand-side vector as arguments and returns the solution vector, **x**. The solution using `lsolve` is shown in Figure 4.33.

$$C := \begin{pmatrix} 8 & 4 & -3 \\ 6 & 2 & -4 \\ 4 & -3 & 6 \end{pmatrix} \qquad r := \begin{pmatrix} 14 \\ -4 \\ 32 \end{pmatrix}$$

$$x := \text{lsolve}(C, r) \qquad x = \begin{pmatrix} 2 \\ 4 \\ 6 \end{pmatrix}$$

Figure 4.33
Solving the system of linear equations using the `lsolve` function.

The `lsolve` function is actually a better way to solve simultaneous linear equations than inverting the coefficient matrix directly. This is because the `lsolve` function does not actually invert the coefficient matrix; it uses a different algorithm (LU decomposition) that allows the solution to be determined with fewer calculations. Fewer calculations mean less round-off error. This can be significant when solving large systems of equations.

Practice!

Each of the two systems of equations that follow has no unique solution. Verify that the determinant of each coefficient matrix is zero.

a. The second and third equations are identical.

$$2x_1 + 3x_2 + 1x_3 = 12$$
$$1x_1 + 4x_2 + 7x_3 = 16$$
$$1x_1 + 4x_2 + 7x_3 = 16$$

b. The third equation is the sum of the first and second equations.

$$2x_1 + 3x_2 + 1x_3 = 12$$
$$1x_1 + 4x_2 + 7x_3 = 16$$
$$3x_1 + 7x_2 + 8x_3 = 28$$

The two systems of equations that follow might have solutions. For each system, use the determinant to find out if there is a solution and, if so, use the `lsolve()` function to find the solution.

c.

$$2x_1 + 3x_2 + 1x_3 = 12$$
$$1x_1 + 4x_2 + 7x_3 = 16$$
$$4x_1 + 1x_2 + 3x_3 = 9$$

d.

$$2x_1 + 3x_2 + 1x_3 = 12$$
$$1x_1 + 4x_2 + 7x_3 = 16$$
$$7x_1 + 18x_2 + 23x_3 = 28$$

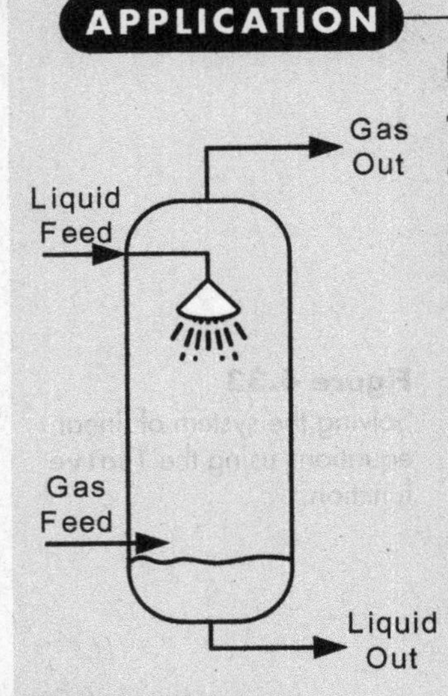

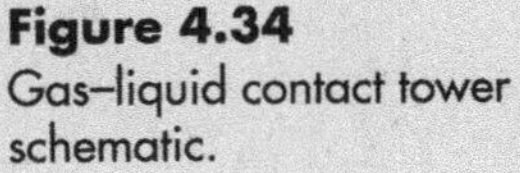

Figure 4.34
Gas–liquid contact tower schematic.

Removal of CO_2 from a Gas Stream

There are various solutions that can be used to remove CO_2 from gas streams, including ethanolamine solutions and solutions of potassium carbonate. Removing CO_2 from fossil fuel burning processes is technically possible, but it is not yet economical for most processes. Still, the process can be used to demonstrate how material balances can be solved using the techniques presented in this section.

Consider a gas–liquid contact tower (see Figure 4.34) in which a gas stream enters the bottom of the tower and flows upward, leaving the top of the tower. A liquid stream enters at the top and flows down, leaving at the bottom. The goal is to move CO_2 from the gas stream to the liquid stream inside the tower.

If we know the composition and flow rate of the gas feed steam and measure the composition of the other three streams, we will have enough information to determine the flow rates of each of the remaining streams. The known values are listed in Table 4.1.

Table 4.1 Data for Material Balance over the Gas–Liquid Contact Tower

	Concentrations		
Stream	**CO_2**	**Other Gases**	**Amine**
Gas Feed, 100 moles	10%	90%	—
Liquid Feed, s_1	2%	—	98%
Liquid Out, s_2	12%	1%	87%
Gas Out, s_3	1%	99%	—

Three material balances can be written, one for each type of concentration.

CO_2 Balance

$$CO_2 \text{ in Gas Feed} + CO_2 \text{ in Liquid Feed} = CO_2 \text{ in Liquid Out} + CO_2 \text{ in Gas Out}$$

$$0.10 \cdot 100 \text{ moles} + 0.02 \cdot s_1 = 0.12 \cdot s_2 + 0.01 \cdot s_3$$

Other Gases (OG) Balance

$$\text{OG in Gas Feed} + \text{OG in Liquid Feed} = \text{OG in Liquid Out} + \text{OG in Gas Out}$$

$$0.90 \cdot 100 \text{ moles} + 0.00 \cdot s_1 = 0.01 \cdot s_2 + 0.99 \cdot s_3$$

Amine (AM) Balance

$$\text{AM in Gas Feed} + \text{AM in Liquid Feed} = \text{AM in Liquid Out} + \text{AM in Gas Out}$$

$$0.00 \cdot 100 \text{ moles} + 0.98 \cdot s_1 = 0.87 \cdot s_2 + 0.00 \cdot s_3$$

The three material balances can be written as a system of simultaneous linear algebraic equations. If we rearrange the equations to put the known values on the right side, we get

$$\begin{aligned}
0.02 \cdot s_1 - 0.12 \cdot s_2 - 0.01 \cdot s_3 &= -10 \text{ moles} \\
0.00 \cdot s_1 - 0.01 \cdot s_2 - 0.99 \cdot s_3 &= -90 \text{ moles} \\
0.98 \cdot s_1 - 0.87 \cdot s_2 - 0.00 \cdot s_3 &= 0 \text{ moles}
\end{aligned}$$

The Mathcad solution for s_1, s_2, and s_3 is shown in Figure 4.35.

$$C := \begin{pmatrix} 0.02 & -0.12 & -0.01 \\ 0.00 & -0.01 & -0.99 \\ 0.98 & -0.87 & 0.00 \end{pmatrix} \qquad r := \begin{pmatrix} -10 \\ -90 \\ 0 \end{pmatrix} \cdot \text{mol}$$

$$s := \text{lsolve}(C, r) \qquad s = \begin{pmatrix} 79 \\ 89 \\ 90 \end{pmatrix} \text{mol}$$

Figure 4.35
Mathcad solution for the stream flow rates.

Solving the three material balance equations simultaneously, we learn that we have 79 moles of liquid feed (s_1), 89 moles of liquid out (s_2), and 90 moles of gas out (s_3) for every 100 moles of gas fed to the column.

4.5 ARRAY FUNCTIONS

Mathcad provides a number of functions to help you to work with vectors and matrices. We will take a look at a number of functions that allow you to get information about matrices or allow you to reorder matrices.

4.5.1 Functions That Return Information about Matrices

Mathcad's **max(A)** and **min(A)** functions can be used to find the maximum and minimum values in an array (either a vector or a matrix), as illustrated in Figure 4.36.

$$C := \begin{pmatrix} 1 & 1 \\ 2 & 8 \\ 3 & 27 \\ 4 & 64 \\ 5 & 125 \end{pmatrix} \qquad \text{MaxVal} := \max(C) \qquad \text{MaxVal} = 125$$

$$\text{MinVal} := \min(C) \qquad \text{MinVal} = 1$$

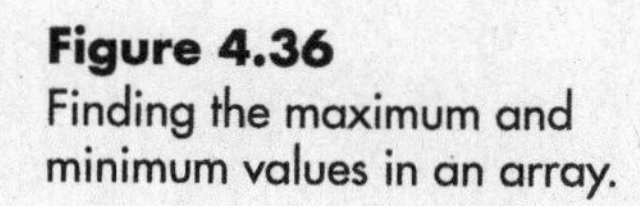

Figure 4.36
Finding the maximum and minimum values in an array.

Four additional functions return information about the size of an array.

- **cols(A)** Returns the number of columns in array **A**.
- **rows(A)** Returns the number of rows in array **A**.
- **length(v)** Returns the number of elements in vector **v** (vectors only).
- **last(v)** Returns the index number of last element in vector **v** (vectors only).

The use of these functions is illustrated in Figure 4.37.

Be careful with the **last(v)** function because the returned value will change depending on the value of **ORIGIN**. This is illustrated in Figure 4.38.

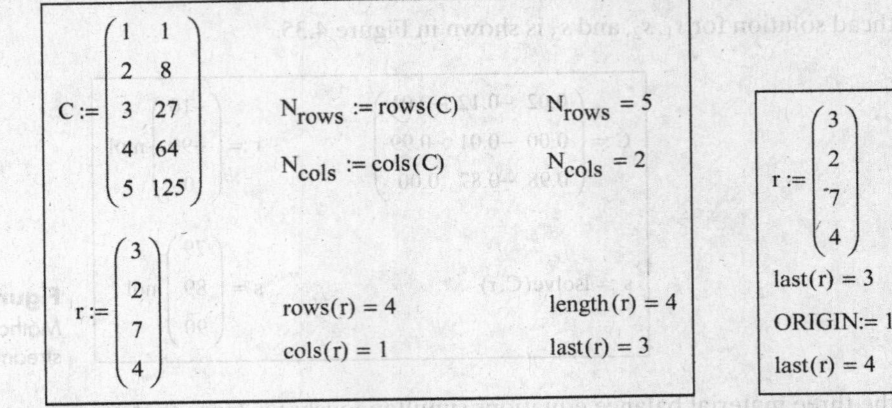

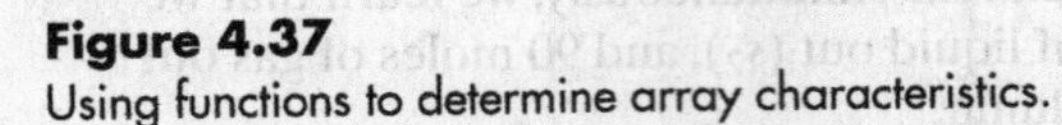
Figure 4.37
Using functions to determine array characteristics.

$r := \begin{pmatrix} 3 \\ 2 \\ 7 \\ 4 \end{pmatrix}$

last(r) = 3

ORIGIN:= 1

last(r) = 4

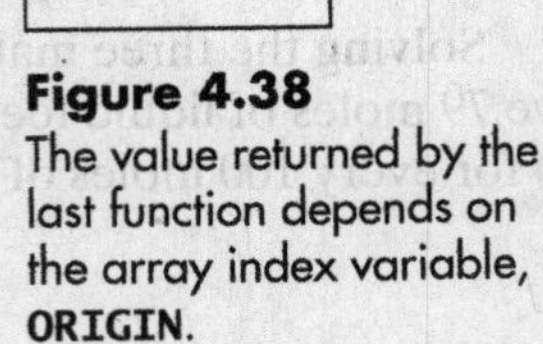
Figure 4.38
The value returned by the last function depends on the array index variable, **ORIGIN**.

4.5.2 Functions for Sorting Vectors and Arrays

There are two functions used for sorting vectors: **sort(v)** and **reverse(v)**. The **sort(v)** function arranges the elements of the vector in ascending order. The **reverse(v)** function is not actually a sorting function at all; it simply reverses the order of elements in a vector. However, the **reverse(v)** function can be used after the **sort(v)** function to generate a vector with elements in descending order. The use of these vectors is illustrated in Figure 4.39.

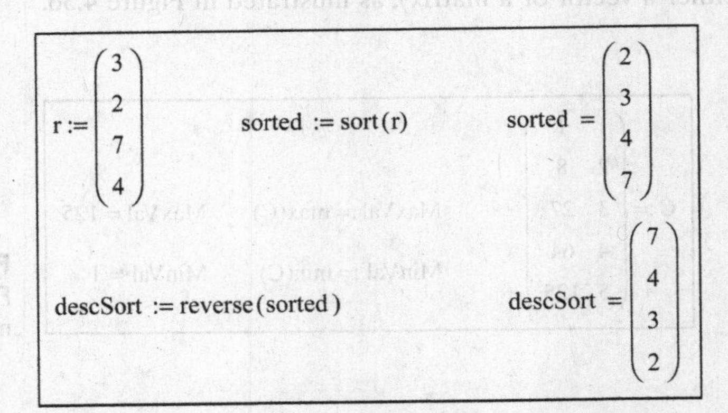

Figure 4.39
Sorting vectors.

You could combine the sorting and reversing steps in a single calculation, as shown in Figure 4.40.

descSort := reverse(sort(r))

$descSort = \begin{pmatrix} 7 \\ 4 \\ 3 \\ 2 \end{pmatrix}$

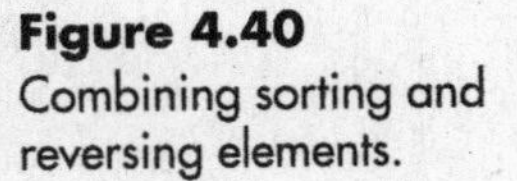
Figure 4.40
Combining sorting and reversing elements.

There are two sorting functions for arrays:

- **`csort(A,n)`** Sorts array **A** so that the elements in column **n** are in ascending order.
- **`rsort(A,n)`** Sorts array **A** so that the elements in row **n** are in ascending order.

An example of the use of the **`csort`** function is shown in Figure 4.41. Note that the **`ORIGIN`** variable was included as a reminder that the numbering of the columns in an array depends on the value assigned to **`ORIGIN`**.

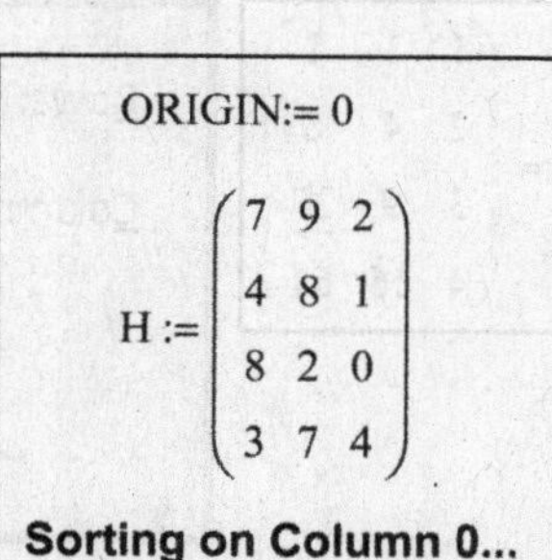

ORIGIN:= 0

$$H := \begin{pmatrix} 7 & 9 & 2 \\ 4 & 8 & 1 \\ 8 & 2 & 0 \\ 3 & 7 & 4 \end{pmatrix}$$

Sorting on Column 0...

$$\text{csort}(H, 0) = \begin{pmatrix} 3 & 7 & 4 \\ 4 & 8 & 1 \\ 7 & 9 & 2 \\ 8 & 2 & 0 \end{pmatrix}$$

Sorting on Column 2...

$$\text{csort}(H, 2) = \begin{pmatrix} 8 & 2 & 0 \\ 4 & 8 & 1 \\ 7 & 9 & 2 \\ 3 & 7 & 4 \end{pmatrix}$$

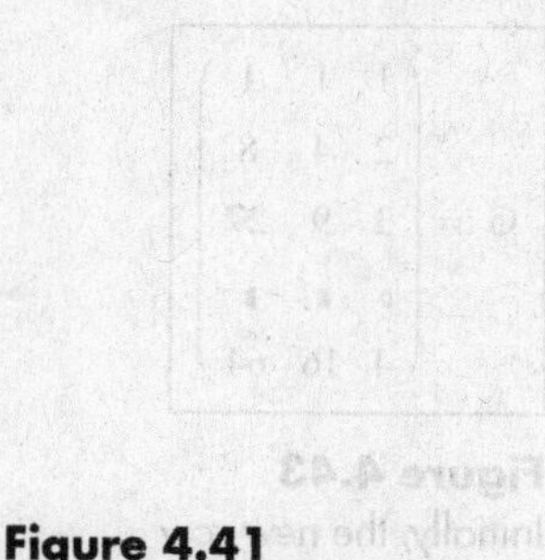

Figure 4.41
Sorting arrays.

4.6 MODIFYING MATRICES

Suppose you have just entered a large array from the keyboard and notice that you accidentally left out one row. You don't have to start over again; you can insert one or more rows or columns into an existing array so that you can correct the entry error. Mathcad also allows you to delete one or more rows or columns from arrays.

You can also join two arrays together to form a larger array. To join two arrays together side to side (***augment*** them), use the **`augment`** function. To put one array on top of another (***stack*** them), use the **`stack`** function. Finally, you can select a portion of an array and assign it to another variable. You can select one column using the ***column operator***, <*n*>, (access using keyboard shortcut [Ctrl-6]), or a different part of an array using the **`submatrix`** function.

Each of these ways to modify a matrix is presented in the following sections.

4.6.1 Inserting a Row or Column into an Existing Array

Use the Insert Matrix dialog box [Ctrl-M] to insert a row or column (or both) into an existing array. The Insert Matrix dialog box will always insert to the right of the selected column, and below the selected row, so that you can control where the new row or column will be positioned by selecting a matrix element before opening the Insert Matrix dialog box.

For example, to insert a new row just below the row containing the 3, 9, and 27 in the **G** matrix shown in Figure 4.42, click anywhere on that row; either on the 3, 9, or the 27.

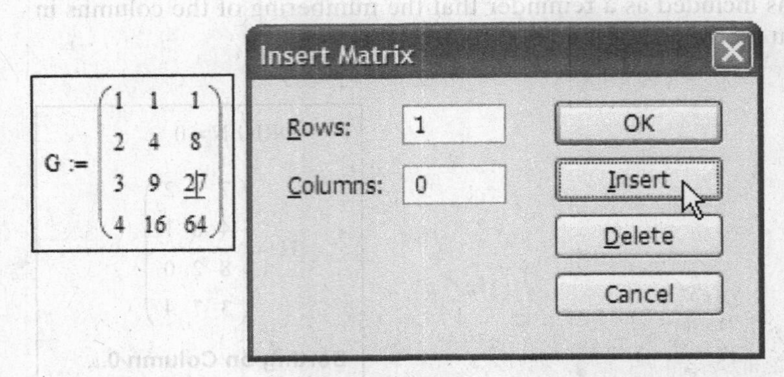

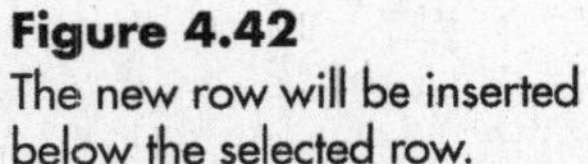

Figure 4.42
The new row will be inserted below the selected row.

Then, open the Insert Matrix dialog box using the keyboard shortcut [Ctrl-M], menu options **Insert/Matrix. . .** or the **Matrix** or **Vector** button on the Matrix Toolbar. To insert only one row, set the **Rows:** field to 1, and the **Columns:** field to 0 in the dialog box (as shown in Figure 4.42).

Click **OK** (or **Insert**) to insert the new row into the **G** matrix. The new row will contain empty placeholders (see Figure 4.43); click on each placeholder to enter values. *Note:* The difference between the **OK** button and the **Insert** button on the Insert Matrix dialog box is this: the Insert button leaves the dialog box open so that you can make additional changes to the array; the **OK** button will insert the row and close the dialog box.

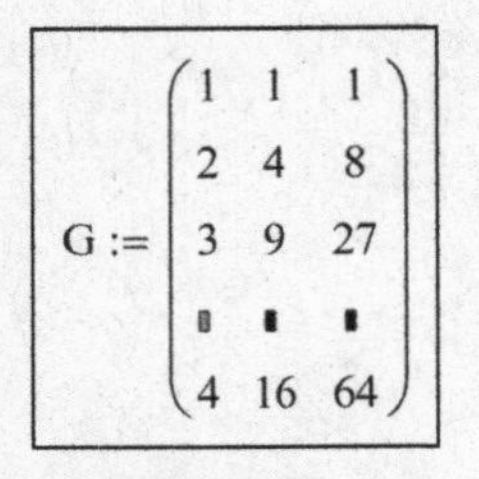

Figure 4.43
Initially, the new row contains only placeholders.

If you wanted to insert two new columns to the right side of the original **G** matrix, you would follow these steps:

1. Click on any element in the third column. This tells Mathcad to insert new columns to the right of the third column. (See Figure 4.44.)

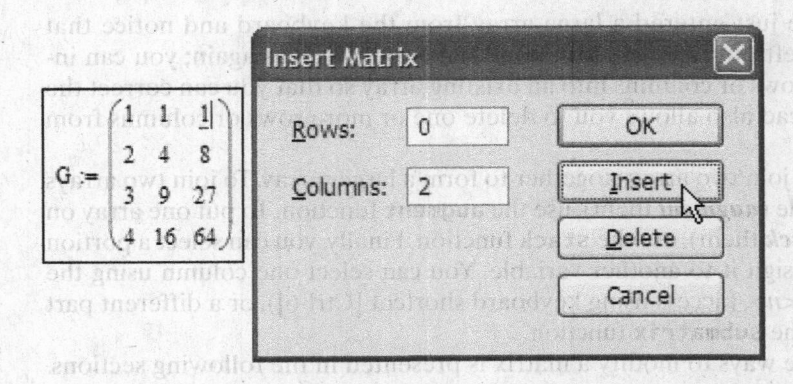

Figure 4.44
Inserting two columns into matrix **G**.

2. Open the Insert Matrix dialog box by pressing [Ctrl-M]. Indicate that you want to add only two columns by setting the **Rows:** field to 0 and the **Columns:** field to 2. (See Figure 4.44.)
3. Click **OK** to insert the two new columns of placeholders and close the dialog box. The result is shown in Figure 4.45.

$$G := \begin{pmatrix} 1 & 1 & 1 & \blacksquare & \blacksquare \\ 2 & 4 & 8 & \blacksquare & \blacksquare \\ 3 & 9 & 27 & \blacksquare & \blacksquare \\ 4 & 16 & 64 & \blacksquare & \blacksquare \end{pmatrix}$$

Figure 4.45
Two new columns of placeholders have been added to the right side of the **G** matrix.

In order to insert columns to the left side of an existing array, or insert rows to the top of an array, you must select the entire array (see Figure 4.46) before opening the Insert Matrix dialog box. In this example, we add both a new row and a new column to the original **G** matrix in a single step. The result is shown in Figure 4.47.

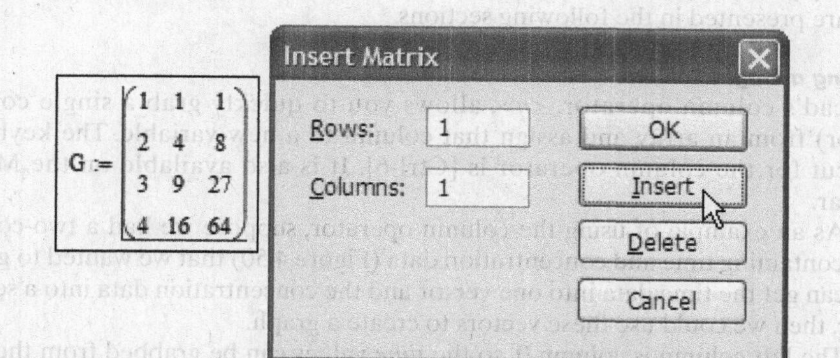

Figure 4.46
Select the entire array to insert columns to the left or rows on top.

$$G := \begin{pmatrix} \blacksquare & \blacksquare & \blacksquare & \blacksquare \\ \blacksquare & 1 & 1 & 1 \\ \blacksquare & 2 & 4 & 8 \\ \blacksquare & 3 & 9 & 27 \\ \blacksquare & 4 & 16 & 64 \end{pmatrix}$$

Figure 4.47
A new row and new column have been added to the **G** matrix.

4.6.2 Deleting Rows or Columns

The procedure used to delete rows or columns is similar to that used to insert them:

1. Select an element in the row or column to be deleted, then
2. Open the Insert Matrix dialog box [Ctrl-M].
3. Indicate how many rows or columns you want to delete.
   - The selected row or column will always be deleted.
   - If you ask to delete more than one row, the additional deleted rows will be below the selected row.
   - If you ask to delete more than one column, the additional columns will be to the right of the selected column.
4. Click the **Delete** button on the Insert Matrix dialog box to delete rows or columns. (This leaves the dialog box open.)
5. Click the **Close** button to close the Insert Matrix dialog box.

Figures 4.48 and 4.49 illustrate the process of deleting the middle two rows of the original **G** matrix.

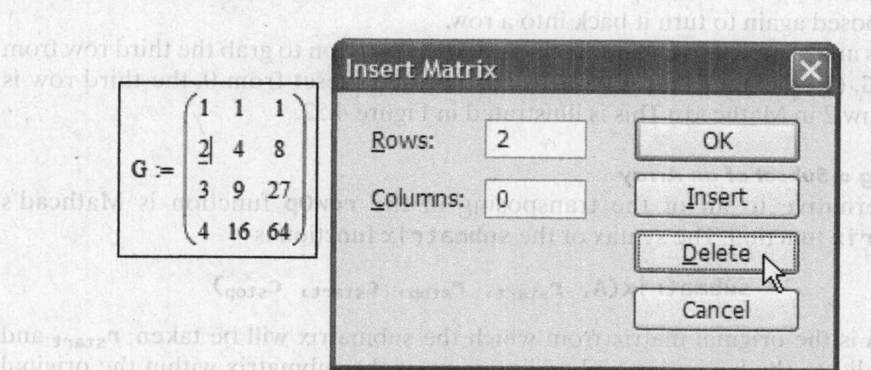

Figure 4.48
An element in the second row of the **G** matrix is selected before opening the Insert Matrix dialog box.

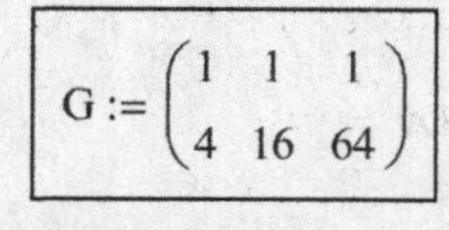

Figure 4.49
The middle two rows (the selected row and the row just below the selected row) have been deleted.

4.6.3 Selecting a Portion of an Array

There are times when you want to use just a part of an array. Mathcad provides a column operator, *<n>*, to grab a single column, *n*, from an array. When the portion of the array you need is not a column, you can use the **submatrix** function. These options are presented in the following sections.

$$\text{data} = \begin{pmatrix} 0 & 50 \\ 5 & 46 \\ 10 & 42.3 \\ 15 & 38.9 \\ 20 & 35.8 \\ 25 & 33 \\ 30 & 30.3 \\ 35 & 27.9 \\ 40 & 25.7 \\ 45 & 23.6 \\ 50 & 21.7 \end{pmatrix}$$

Figure 4.50
Time (left column) and concentration (right column) combined as a data matrix.

Selecting a Single Column

Mathcad's column operator, *<n>*, allows you to quickly grab a single column (vector) from an array and assign that column to a new variable. The keyboard shortcut for the column operator is [Ctrl-6]. It is also available on the Matrix Toolbar.

As an example of using the column operator, suppose we had a two-column array containing time and concentration data (Figure 4.50) that we wanted to graph. If we can get the time data into one vector and the concentration data into a second vector, then we could use these vectors to create a graph.

The left column is column 0, so the *time* values can be grabbed from the data array using

$$\texttt{time} := \texttt{data}^{\langle 0 \rangle}$$

where the column operator was entered using [Ctrl-6].

Similarly, the concentration data can be assigned to variable **conc** as

$$\texttt{conc} := \texttt{data}^{\langle 1 \rangle}$$

Figure 4.51 shows how column operators were used to define **time** and **conc**, what the new vectors look like, and what the graph of concentration versus time looks like.

Note: The graph was used here because extracting vectors for plotting is a common use of the column operator.

Selecting a Single Row

Mathcad does not provide a row operator to grab a single row out of an array. You could create one as a user-written function, such as

$$\texttt{rowOp(A,r)} := \left(A^{T}\right)^{\langle r \rangle T}$$

This function takes array **A** and transposes it. That turns all of the original rows into columns. Then the column operator selects column **<r>**. Finally, the selected column is transposed again to turn it back into a row.

As an example, we will use the new **rowOp** function to grab the third row from matrix **G**. (Remember that if **ORIGIN** has not been reset from 0, the third row is called row 2 in Mathcad.) This is illustrated in Figure 4.52.

Choosing a Subset of an Array

An alternative to all of the transposing in the **rowOp** function is Mathcad's **submatrix** function. The syntax of the **submatrix** function is

$$\texttt{submatrix}(A,\ r_{start},\ r_{stop},\ c_{start},\ c_{stop})$$

where **A** is the original matrix from which the submatrix will be taken, r_{start} and r_{stop} indicate the beginning and ending rows of the submatrix within the original matrix, and c_{start} and c_{stop} indicate the beginning and ending columns of the submatrix within the original matrix.

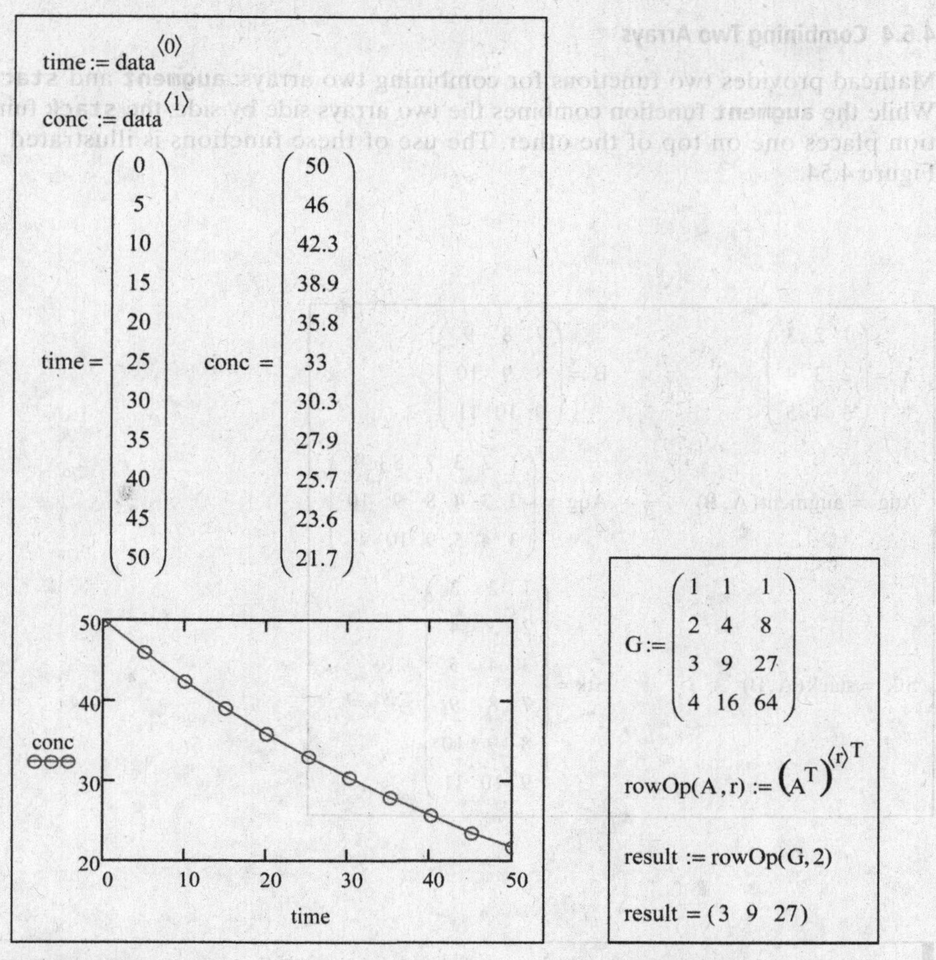

Figure 4.51
The time vs. concentration graph created after extracting the time and concentration data from the original array.

Figure 4.52
Using the user-written function **rowOp** to get one row from the **G** matrix.

Figure 4.53 illustrates how the **submatrix** function could be used to extract the third row from the **G** matrix. Since we want only row 2 (the third row), both $\mathbf{r_{start}}$ and $\mathbf{r_{stop}}$ are equal to **2**. To get all three columns, set $\mathbf{c_{start} = 0}$ and $\mathbf{c_{stop} = 2}$.

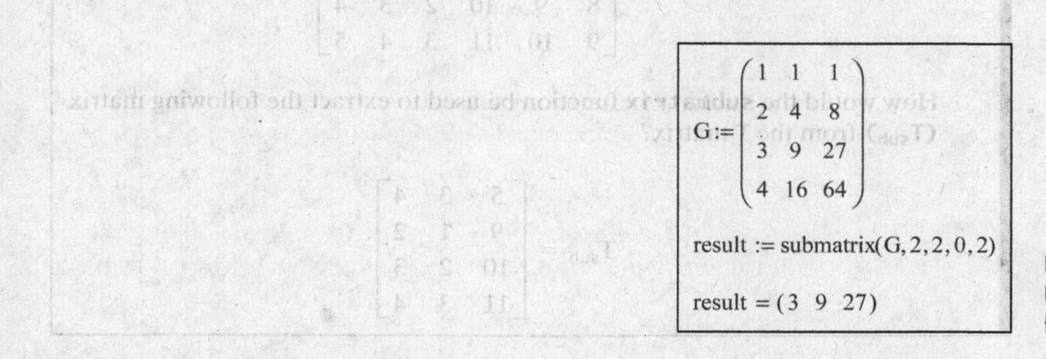

Figure 4.53
Using the submatrix function.

4.6.4 Combining Two Arrays

Mathcad provides two functions for combining two arrays: **augment** and **stack**. While the **augment** function combines the two arrays side by side, the **stack** function places one on top of the other. The use of these functions is illustrated in Figure 4.54.

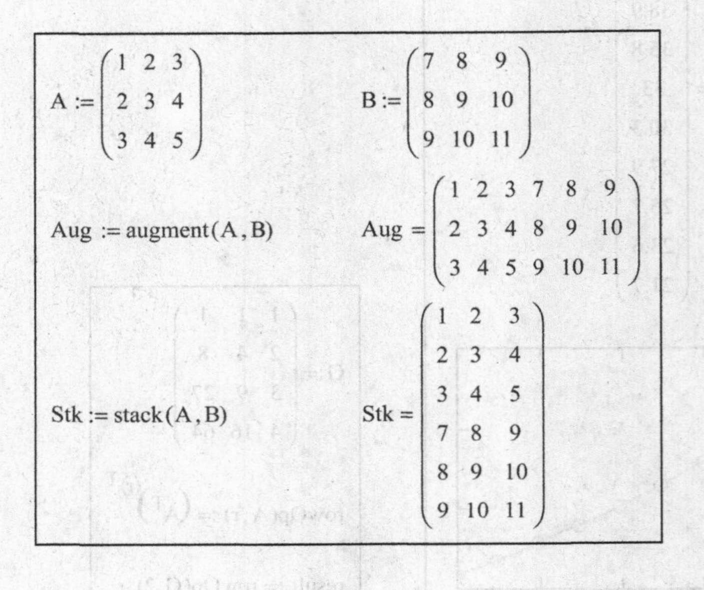

A := (1 2 3; 2 3 4; 3 4 5)
B := (7 8 9; 8 9 10; 9 10 11)
Aug := augment(A, B)
Aug = (1 2 3 7 8 9; 2 3 4 8 9 10; 3 4 5 9 10 11)
Stk := stack(A, B)
Stk = (1 2 3; 2 3 4; 3 4 5; 7 8 9; 8 9 10; 9 10 11)

Figure 4.54
Using the augment and stack functions to combine matrices.

Practice!

In what order would you use the **stack** and **augment** functions to create the following **T** matrix from matrices **A** and **B** shown in Figure 4.54?

$$T = \begin{bmatrix} 1 & 2 & 3 & 1 & 2 & 3 \\ 2 & 3 & 4 & 2 & 3 & 4 \\ 3 & 4 & 5 & 3 & 4 & 5 \\ 7 & 8 & 9 & 1 & 2 & 3 \\ 8 & 9 & 10 & 2 & 3 & 4 \\ 9 & 10 & 11 & 3 & 4 & 5 \end{bmatrix}$$

How would the **submatrix** function be used to extract the following matrix (T_{sub}) from the **T** matrix?

$$T_{sub} = \begin{bmatrix} 5 & 3 & 4 \\ 9 & 1 & 2 \\ 10 & 2 & 3 \\ 11 & 3 & 4 \end{bmatrix}$$

4.7 ADDITIONAL WAYS TO INITIALIZE ARRAYS

So far, every array we have used has been created using the Insert Matrix dialog box. That is a convenient way to create small matrices, but Mathcad provides several other ways to create and fill (or, *initialize*) matrices. Knowing what other methods are available can save a lot of typing.

The ways to initialize matrices can be separated by the source of the array values.

1. Element values can be entered from the keyboard:
   - Use the Insert Matrix dialog box to create the array, and then type values into the placeholders.
   - Use an *Input Table* to create the array (Section 4.7.2).
2. Element values can be imported from other programs:
   - You can import data from a file (such as an Excel file) into an Input Table. (See Section 4.7.2.)
   - You can "cut and paste" values from a spreadsheet into an array definition. (See Section 4.7.1.)
3. Element values can be read from files:
   - You can read data from a file (such as a text file) into an array definition. (See section 4.7.3.)
4. Element values can be calculated:
   - You can use *range variables* to calculate individual elements of arrays. (See Section 4.7.4.)
   - You can use the `matrix` function to calculate element values and assign them to matrices. (See Section 4.7.5.)

There are a lot of ways to create arrays in Mathcad, and you may not want to spend a lot of time learning each method. But knowing that the various methods exist can save you a lot of typing if you know that you can import the data rather than retyping it.

4.7.1 Copying and Pasting Values from a Spreadsheet

Spreadsheets are very commonly used to store data, and one of the easiest ways to get the data from a spreadsheet into Mathcad is to "copy and paste" the data. More specifically, the data values are copied from the spreadsheet to the Windows clipboard, then pasted from the Windows clipboard into Mathcad.

Note: A spreadsheet is also a very handy way to type in a large set of values. Then, once the data is available in the spreadsheet, it can easily be moved to Mathcad using "copy and paste" operations.

Assuming the data already exists in an Excel spreadsheet (or worksheet), the process of getting the values into an array in Mathcad is as follows:

1. In Mathcad, begin the definition of a new array by entering the array name (variable) and the assignment operator (: =).
2. In the spreadsheet, select the values to be copied.
3. In the spreadsheet, copy the values to the Windows clipboard using menu options **Edit/Copy** or the keyboard shortcut [Ctrl-C].
4. In Mathcad, click on the placeholder on the right side of the assignment operator in the new matrix definition.
5. In Mathcad, paste the values from the Windows clipboard into the placeholder. To paste, either use menu options **Edit/Paste** or the keyboard shortcut [Ctrl-V].

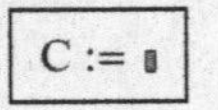

Figure 4.55
Defining the new matrix variable in Mathcad.

These steps will be used in the following example.

Step 1: We start by defining variable **C** in Mathcad, as shown in Figure 4.55. This is the variable that will hold the new matrix. The placeholder will be filled later.

Step 2: In the spreadsheet, select the range of cells that contain the values we want to move to Mathcad. This is illustrated in Figure 4.56.

	A	B	C	D
1	1	1		
2	2	8		
3	3	27		
4	4	64		
5	5	125		
6				

Figure 4.56
Select the Excel cell range that contains the data to be moved to Mathcad.

Step 3: In the spreadsheet, copy the selected cells to the Windows clipboard using [Ctrl-C].

Step 4: In Mathcad, click the placeholder in the new matrix definition. (See Figure 4.55.)

Step 5: In Mathcad, paste the values from the Windows clipboard into the placeholder. To paste, use either menu options **Edit/Paste** or the keyboard shortcut [Ctrl-V]. The result is shown in Figure 4.57.

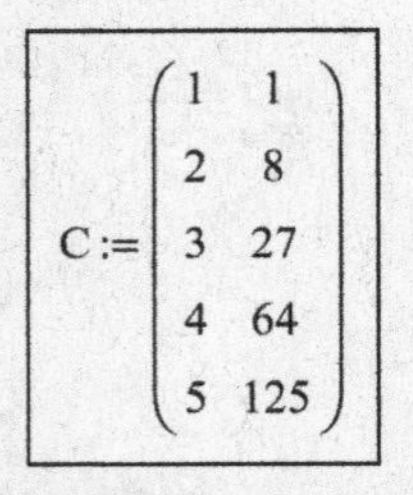

Figure 4.57
The spreadsheet values have been pasted into matrix **C** in Mathcad.

Note: In the spreadsheet, formulas may have been used to calculate some of the selected values. When the values were pasted into Mathcad, only the values (the formula results) were pasted, not the formulas themselves.

> **Professional Success**
>
> *Let your software do the work.*
>
> Retyping data, equations, and results into a report is tedious and tends to introduce errors. The past few generations of software products have had the ability to exchange data, and the capabilities are improving with each new version. Mathcad can share information with other mathematics packages, such as Excel and Matlab, and word processors such as Microsoft Word.[1] If you find yourself frequently reentering information into multiple software products, you might want to see if there is a better way.

4.7.2 Using an Input Table to Create an Array

An *input table* is a spreadsheet-like interface for entering array values in Mathcad. To use an input table, click in the spot on the Mathcad worksheet where you want the table to be inserted, and then use the following menu options: **Insert/Data/Table**. A 2×2 input table will be inserted on the worksheet on the right side of an assignment operator (:=). The left side of the assignment operator is an empty placeholder, as shown in Figure 4.58.

[1]Microsoft Word and Excel are products of the Microsoft Corporation, Redmond, WA. Matlab is produced by Mathworks, Inc., Natick, MA.

■ :=

	0	1
0	0	
1		

Figure 4.58
The empty input table, with a placeholder for the matrix name.

Once the input table has been inserted on the worksheet, the next thing to do is to fill the placeholder with the variable name for the matrix. In this example, we will call the matrix **A**. The matrix name has been entered into the placeholder in Figure 4.59.

A :=

	0	1
0	0	
1		
2		
3		
4		
5		
6		
7		
8		

Figure 4.59
The input table is now attached to variable **A** and has been enlarged to allow more values to be displayed.

An input table can be enlarged to display more values. To enlarge the input table, click on the table to select it. When selected, eight *handles* (small black squares on the boundary of the table) will appear, as shown in Figure 4.59. Grab any handle with the mouse and drag it to change the displayed size of the input table.

All that is left to do is to fill the table with values. This can be done in three ways:

1. Type values into the input table.
2. Paste values from another program (such as a spreadsheet) into the input table.
3. Import values from a file (such as a text file, database file, or spreadsheet file).

The process used to copy and paste values was presented in Section 4.7.1, and that process can be used with input tables as well.

Importing Values from Files

Before you can import values from a file, the values must exist in a file. The spreadsheet shown in Figure 4.56 has been saved as C:\ArrayData.xls. We will import values from this Excel file to fill the input table.

Note: Mathcad 14 input tables can import Excel data using the Excel 2003 format (.xls), but not the Excel 2007 format (.xlsx). Mathcad 15 does support Excel 2007.

To import values from a file into the input table, right-click on any cell of the input table; a pop-up menu will appear as illustrated in Figure 4.60.

When you select the **Import...** option on the pop-up menu, the File Options dialog box will open as shown in Figure 4.61. In Figure 4.61, Microsoft Excel has been

A :=
0
1
0
1
2
3
4
5
6
7
8
0
Undo
Cut
Copy
Paste
Paste Table
Import...
Insert Cells...
Delete Cells...

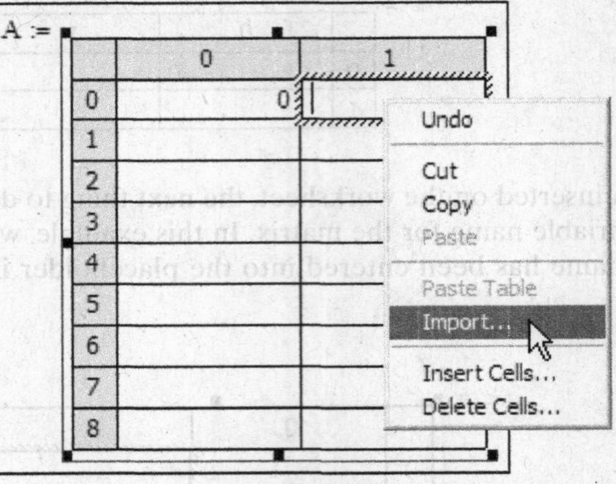

Figure 4.60
Using the Import... option on the pop-up menu to fill the input table.

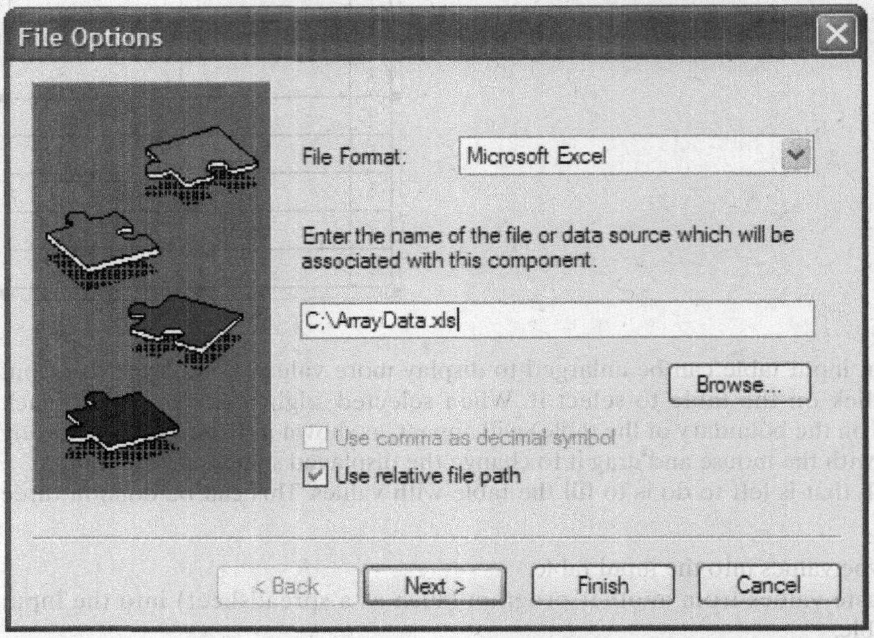

File Options
File Format:
Microsoft Excel
Enter the name of the file or data source which will be associated with this component.
C:\ArrayData.xls
Browse...
Use comma as decimal symbol
Use relative file path
< Back
Next >
Finish
Cancel

Figure 4.61
The File Options dialog box.

selected as the File Format, and the location of the file (C:\ArrayData.xls) has been provided.

If you click **Finish**, all of the data on the first worksheet of the workbook will be imported into the input table. That is adequate for this example, and the result is shown in Figure 4.62.

A :=

	0	1
0	1	1
1	2	8
2	3	27
3	4	64
4	5	125

Figure 4.62
The input table filled with values from the Excel file.

The **Next>** button on the File Options dialog box (shown in Figure 4.63) can be used to open the Data Range dialog box (shown in Figure 4.63) to select a particular worksheet from the Excel file or a particular range of cells. While the Data Range dialog box was not needed for this simple example, it can be very useful when selecting a small data set from a large spreadsheet.

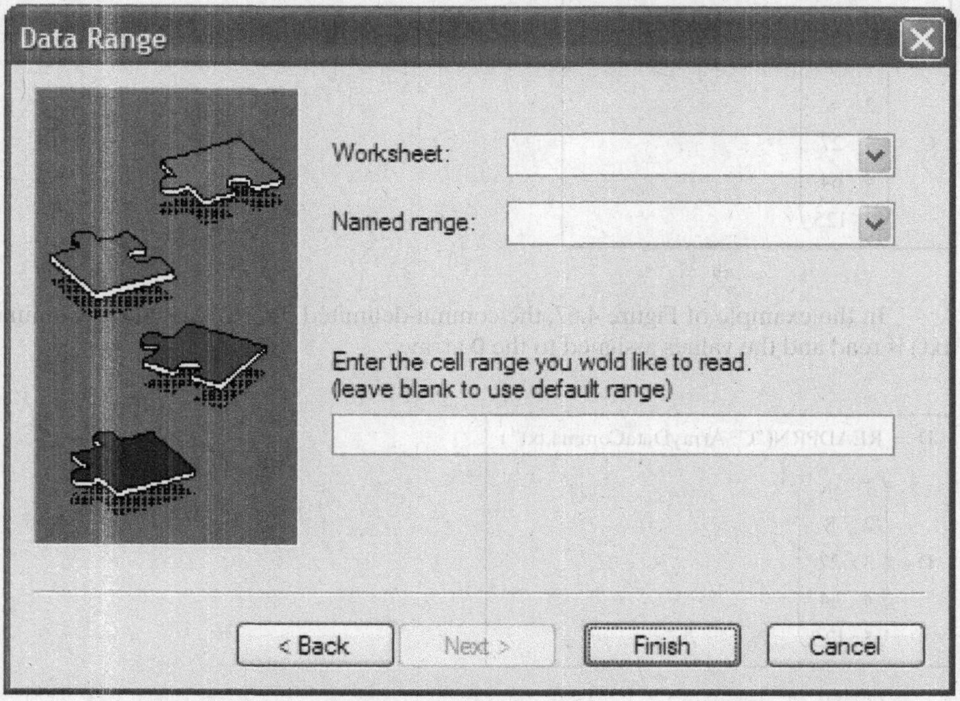

Figure 4.63
The Data Range dialog box.

4.7.3 Reading Data Values from Text Files

There are two ways to get values from text files into an array:

1. You can use an Input Table and import the values. This was shown in Section 4.7.2, using an Excel file. The same process can be used with text files, except that you need to indicate that you are using a text file in the **File Format** field of the File Options dialog box. (See Figure 4.61.)
2. You can read values directly into an array definition using the **READPRN** function. The use of the **READPRN** function will be shown here.

The **READPRN(path)** function receives the **path** name of the file, where the path is a text string or string variable that represents the drive letter, any directories or folders, and the file name with extension. In this example, we will read values from file ArrayData.txt stored in the root directory of the C: drive. The *path* is then "C:\ArrayData.txt".

The file contains the same data we have been importing in previous examples, but it has been stored as a *tab-delimited text file*. (See Figure 4.64.) Tab-delimited means there is a [Tab] character between the values on each line. The [Tab] separates the values so that it is clear where one value ends and the next begins.

An alternative text file format is the *comma-delimited* format, which separates values with commas rather than tabs. Figure 4.65 shows the same data in comma-delimited form.

```
1    1
2    8
3    27
4    64
5    125
```

Figure 4.64
The array values from file ArrayData.txt, in tab-delimited form.

```
1,1
2,8
3,27
4,64
5,125
```

Figure 4.65
The array values from file ArrayDataComma.txt, in comma-delimited form.

The **READPRN** function can handle either tab-delimited or comma-delimited data. We will illustrate both in the following Mathcad examples. In Figure 4.66, the tab-delimited file (C:\ArrayData.txt) is read and the values assigned to variable **C**.

C := READPRN("C:\ArrayData.txt")

$$C = \begin{pmatrix} 1 & 1 \\ 2 & 8 \\ 3 & 27 \\ 4 & 64 \\ 5 & 125 \end{pmatrix}$$

Figure 4.66
Reading a tab-delimited file with the **READPRN** function.

In the example of Figure 4.67, the comma-delimited file (C:\ArrayDataComma.txt) is read and the values assigned to the **D** array.

D := READPRN("C:\ArrayDataComma.txt")

$$D = \begin{pmatrix} 1 & 1 \\ 2 & 8 \\ 3 & 27 \\ 4 & 64 \\ 5 & 125 \end{pmatrix}$$

Figure 4.67
Reading a comma-delimited file with the **READPRN** function.

Other than the file names, the examples in Figures 4.66 and 4.67 look the same because the **READPRN** function automatically determines which delimiter has been used.

You can also assign the path name to a variable, and then use the variable name as the argument for the **READPRN** function, as shown in Figure 4.68.

pathName := "C:\ArrayData.txt"

C := READPRN(pathName)

$$C = \begin{pmatrix} 1 & 1 \\ 2 & 8 \\ 3 & 27 \\ 4 & 64 \\ 5 & 125 \end{pmatrix}$$

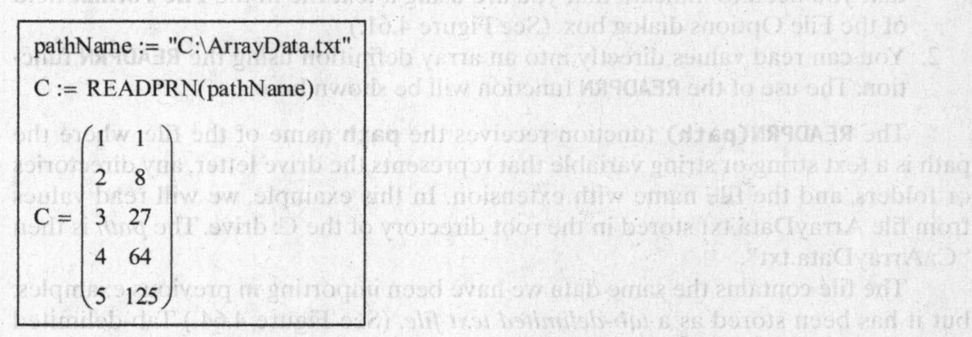

Figure 4.68
Using a variable to hold the file path.

Writing Array Values to a File

You can also write array values to a text file using Mathcad's **WRITEPRN(path)** function. This is illustrated in Figure 4.69. In this example, the **C** array is first transposed, then the values are written to file **NewArrayData.txt**.

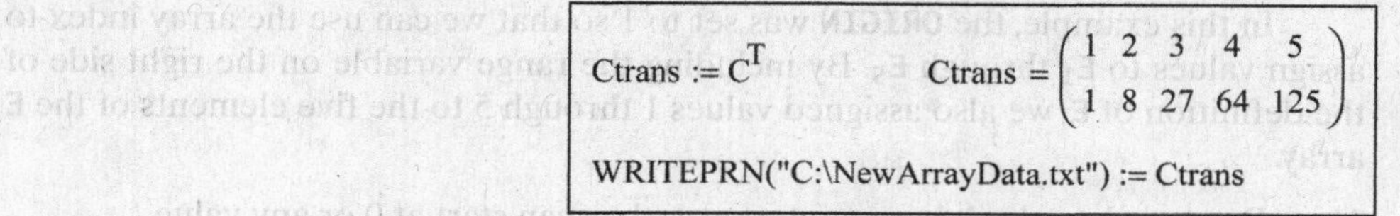

Ctrans := C^T
Ctrans = (1 2 3 4 5 / 1 8 27 64 125)
WRITEPRN("C:\NewArrayData.txt") := Ctrans

Figure 4.69
Writing array values to a file.

The contents of file **NewArrayData.txt** are shown in Figure 4.70. Note that Mathcad tab-delimits data saved using the **WRITEPRN** function.

1	2	3	4	5
1	8	27	64	125

Figure 4.70
The array values saved as a text file.

4.7.4 Using Range Variables to Assign Calculated Values to Array Elements

Mathcad provides an unusual type of variable called a *range variable* that takes on an entire series, or range, of values (sequentially) whenever it is encountered in a Mathcad worksheet. This type of variable tends to confuse people the first time they encounter one, but it can be very powerful. One common use of range variables is assigning calculated values to arrays.

The range of values for a range variable is defined using an ellipsis, which is entered by pressing the semicolon key [;]. The definition of a range variable includes the first value, the second value, an *ellipsis*, and the final value. If you omit the second value, Mathcad will assume you want to step through the range with a step size of 1. In the example in Figure 4.71, the range variable **i** will take on values from 1 to 5.

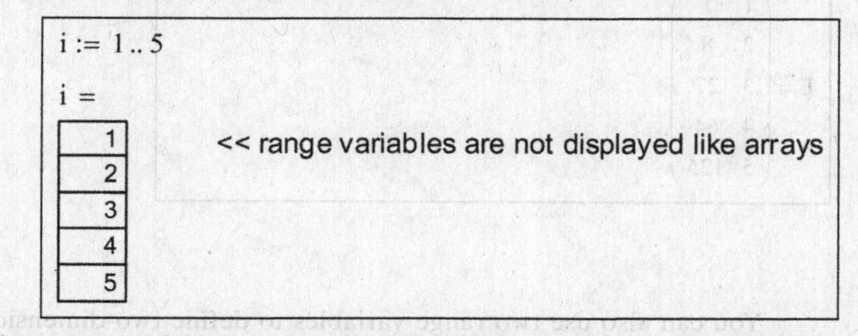

Figure 4.71
Defining a range variable.

Notice that range variables are not displayed like arrays; this is because they aren't arrays. The range variable **i** is not a collection of values but rather a single variable that takes on a series of values, one after another, whenever it is encountered in the worksheet. Any time the range variable defined in Figure 4.71 is used, the equation in which it is used will be evaluated five times: first, with `i = 1`, then with `i = 2`, and so on until `i = 5`. If the **i** appears as an index subscript, it can be used to assign values to an array. An example of this is shown in Figure 4.72.

ORIGIN:= 1

i := 1..5

$E_i := i$ << The range variable, *i*, appears in the index subscript and on the right side of the equation

$$E = \begin{pmatrix} 1 \\ 2 \\ 3 \\ 4 \\ 5 \end{pmatrix}$$

Figure 4.72
Using a range variable to assign values to an array.

In this example, the **ORIGIN** was set to 1 so that we can use the array index to assign values to $\mathbf{E_1}$ through $\mathbf{E_5}$. By including the range variable on the right side of the definition of **E**, we also assigned values 1 through 5 to the five elements of the **E** array.

Note: Range values don't have to start at 1; they can start at 0 or any value.

We can do more sophisticated math with range variables to create more complex arrays. In Figure 4.73 the second column of array **E** was calculated by cubing the range variable.

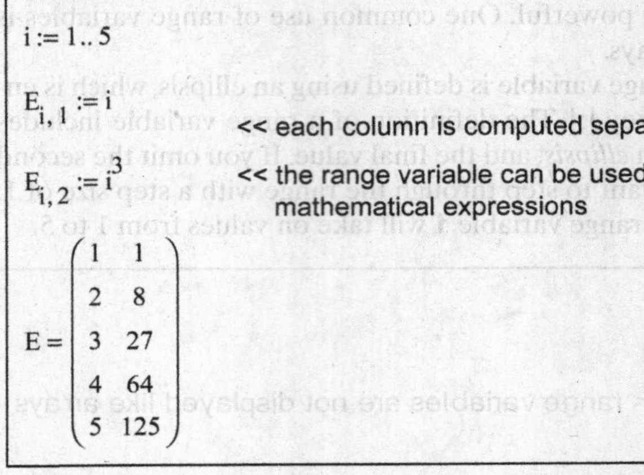

ORIGIN:= 1
i:= 1.. 5
E i, 1 := i
<< each column is computed separately
E i, 2 := i 3
<< the range variable can be used in mathematical expressions
E =
1 1
2 8
3 27
4 64
5 125

Figure 4.73
Using a range variable in a math expression to assign values to an array.

You can also use two range variables to define two-dimensional arrays. In Figure 4.74, range variables **r** and **c** have been used to keep track of the row and column index when defining array **S**. Notice that the **ORIGIN** was returned to the Mathcad default of **0** for this example.

ORIGIN:= 0

r:= 0.. 4 c:= 0.. 2

$S_{r,c} := r^2 + c^2$

$$S = \begin{pmatrix} 0 & 1 & 4 \\ 1 & 2 & 5 \\ 4 & 5 & 8 \\ 9 & 10 & 13 \\ 16 & 17 & 20 \end{pmatrix}$$

Figure 4.74
Using two range variables to create a two-dimensional array.

Practice!

What will array **M** look like when defined by the following expressions?

1. $i := 0 .. 4$
   $M_i := 2 \cdot i$
2. $i := 0 .. 4$
   $j := 0 .. 4$
   $M_{i,j} := (3 + 2 \cdot i - j)$

4.7.5 Using the `matrix` Function to Assign Calculated Values to Array Elements

Mathcad's **matrix** function allows you to fill an array with computed values without explicitly declaring range variables. In effect, Mathcad declares the necessary range variables based on the information you send to the **matrix** function.

$\mathrm{matrix}(N_{rows}, N_{cols}, f)$

N_{rows}	is the number of rows in the new array
N_{cols}	is the number of columns in the new array
f	is the name of a function you have written that defines how to calculate the array elements

For example, the **S** array in Figure 4.74 could have been created using the **matrix** function as shown in Figure 4.75.

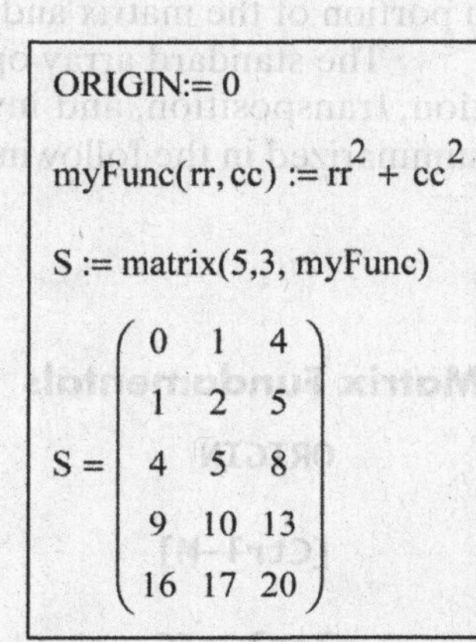

ORIGIN:= 0
myFunc(rr, cc) := rr^2 + cc^2
S := matrix(5,3, myFunc)
S =
0 1 4
1 2 5
4 5 8
9 10 13
16 17 20

Figure 4.75
Using the matrix function to assign calculated values to an array.

Notes:

1. The **ORIGIN** statement was not needed in the example shown in Figure 4.75, but it was included to make it clear that array indexing is starting at **0** in this example.
2. Variable names **rr** and **cc** were used in Figure 4.75 to make it clear that they are unrelated to the range variables **r** and **c** used in Figure 4.74. (Because **rr** and **cc** are dummy variables in a function definition, the variable names **r** and **c** could have been used, and they still would have been unrelated to the range variables used in Figure 4.74.)

Practice!

What would the matrix returned by **`matrix(4, 4, f)`** look like if **`f(r, c)`** was defined as follows?

1. $f(r,c) := 2 + r + c$
2. $f(r,c) := 0.5 \cdot r + c^2$

The matrix **`M`** shown in Figure 4.76 was created using the **`matrix`** function.

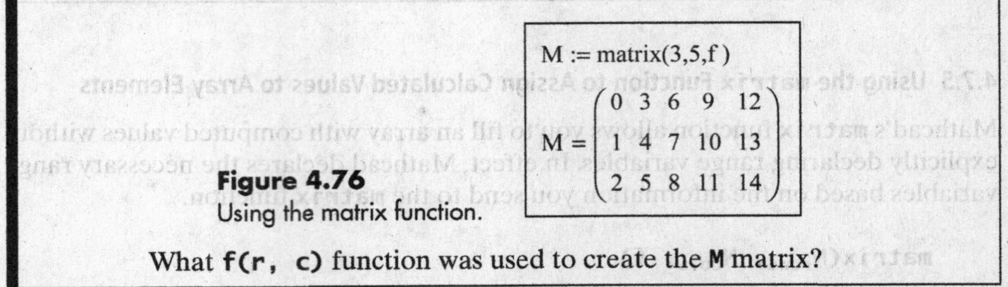

Figure 4.76
Using the matrix function.

What **`f(r, c)`** function was used to create the **`M`** matrix?

SUMMARY

In this chapter, we learned about how to work with matrices in Mathcad. Matrix values can be entered from the keyboard, computed from equations, read from data files, or copied from other programs. Once a matrix exists, there are a variety of ways it can be modified and manipulated to insert or delete rows or columns or to choose a portion of the matrix and assign it to another variable.

The standard array operations were described, including addition, multiplication, transposition, and inversion. The way Mathcad handles these operations is summarized in the following two lists:

MATHCAD SUMMARY

Matrix Fundamentals

`ORIGIN`	Changes the starting value of the first array element (0 by default).
`[Ctrl-M]`	Opens the Matrix dialog box to create a matrix and to insert or delete rows or columns.
`lsolve(C, r)`	Solves a set of simultaneous linear algebraic equations. **C** is the coefficient matrix and **r** is the right-hand-side vector for the system of equations.
`matrix(r,c,f)`	Creates a matrix with **r** rows and **c** columns, using function **f**. Function **f** is user defined and must be a function of two variables.
`identity(n)`	Creates an identity matrix with **n** rows and columns.
`[Ctrl-6]`	Selects a single column of a matrix.
submatrix $(A, r_{start}, r_{stop}, c_{start}, c_{stop})$	Extracts a portion of matrix **A**.

`augment (A1, A2)`	Combines arrays **A1** and **A2** side by side.
`stack (A1, A2)`	Stacks array **A1** on top of array **A2**.
`READPRN (path)`	Reads array values from a text file.
`WRITEPRN (path)`	Writes array values to a text file.

Matrix Operations

`+`	Addition of matrices.
`[Shift-8]`	Matrix multiplication.
→	Vectorize (from the Matrix Toolbar)—used when you want element-by-element multiplication instead of matrix multiplication.
`T`	Transpose operator (from the Matrix Toolbar).
`[Shift-6]`	Invert a matrix.
`|M|`	Determinant operator (from the Matrix Toolbar).
`sort(v)`	Sorts vector **v** into ascending order.
`length(v)`	Returns the number of elements in vector **v**.
`last(v)`	Returns the index number of the last element of vector **v**. The current value of the array **ORIGIN** impacts the value returned by the **last** function.
`reverse(v)`	Reverses the order of vector **v**—this operation is often used after **sort(v)** to get a vector sorted into descending order.
`csort(A,n)`	Sorts array **A** so that the values in column **n** are in ascending order.
`rsort(A,n)`	Sorts array **A** so that the values in row **n** are in ascending order.

KEY TERMS

array	element-by-element multiplication	matrix multiplication
array origin	identity matrix	range variable
augment	index	stack
column operator	input table	transpose
determinant	invert	vector
element	matrix	vectorize

PROBLEMS

4.1 Matrix Operations

Given the matrices and vectors

$$I = \begin{bmatrix} 1 & 0 & 0 & 0 \\ 0 & 1 & 0 & 0 \\ 0 & 0 & 1 & 0 \\ 0 & 0 & 0 & 1 \end{bmatrix},\ a = \begin{bmatrix} 1 \\ 2 \\ 3 \\ 4 \end{bmatrix},\ b = [2 \quad 4 \quad 6 \quad 8], \text{ and } C = \begin{bmatrix} 2 & 3 & 7 & 11 \\ 1 & 4 & 3 & 9 \\ 0 & 6 & 5 & 1 \\ 1 & 8 & 4 & 2 \end{bmatrix}.$$

which of the following matrix operations are allowed?

(a) I^T transpose the identity matrix
(b) $|a|$ find the determinant of vector a
(c) a^{-1} invert the a vector
(d) $|C|$ find the determinant of matrix C
(e) C^{-1} invert the C matrix
(f) $I \cdot a$ multiply the identity matrix by vector a
(g) $a \cdot b$ multiply the a vector by vector b
(h) $b \cdot a$ multiply the b vector by vector a
(i) $b \cdot a$ multiply the inverse of the C matrix by vector a

For each operation that can be performed, what is the result?

4.2 Simultaneous Equations, I

The arrays that follow represent coefficient matrices and right-hand-side vectors for sets of simultaneous linear equations written

$$[C][x] = [r]$$

in matrix form. Calculate the determinant of these coefficient matrices to see whether each set of simultaneous equations can be solved, and if so, solve for the solution vector [x]:

$$C = \begin{bmatrix} 3 & 1 & 5 \\ 2 & 3 & -1 \\ -1 & 4 & 0 \end{bmatrix}, \quad r = \begin{bmatrix} 20 \\ 5 \\ 7 \end{bmatrix};$$

$$C = \begin{bmatrix} 4 & 2 & 1 \\ 2 & 3 & 0 \\ 0 & 4 & -1 \end{bmatrix}, \quad r = \begin{bmatrix} 18 \\ 6 \\ -2 \end{bmatrix};$$

$$C = \begin{bmatrix} 7 & 3 & 1 \\ 2 & -5 & 6 \\ 1 & 5 & 1 \end{bmatrix}, \quad r = \begin{bmatrix} 108 \\ -62 \\ 56 \end{bmatrix};$$

$$C = \begin{bmatrix} 4 & 2 & 1 & 0 \\ 2 & 3 & 0 & 1 \\ 0 & 4 & -1 & 3 \\ 2 & 1 & 4 & 2 \end{bmatrix}, \quad r = \begin{bmatrix} 13 \\ 8 \\ 4 \\ 19 \end{bmatrix}.$$

4.3 Simultaneous Equations, II

Write the following sets of simultaneous equations in matrix form, and solve (if possible):

(a) $3x_1 + 1x_2 + 5x_3 = 20$
$2x_1 + 3x_2 - 1x_3 = 5$
$-1x_1 + 4x_2 = 7$

(b) $6x_1 + 2x_2 + 8x_3 = 14$
$x_1 + 3x_2 + 4x_3 = 5$
$5x_1 + 6x_2 + 2x_3 = 7$

(c) $4y_1 + 2y_2 + 1y_3 + 5y_4 = 52.9$
$3y_1 + y_2 + 4y_3 + 7y_4 = 74.2$
$2y_1 + 3y_2 + y_3 + 6y_4 = 58.3$
$3y_1 + y_2 + y_3 + 3y_4 = 34.2$

4.4 Element-by-Element Matrix Mathematics

Figure 4.77 is a diagram of a heat exchanger; cold fluid flows through the inside tube and is warmed from T_{C_in} to T_{C_out} by energy from the hot fluid surrounding the inside tube.

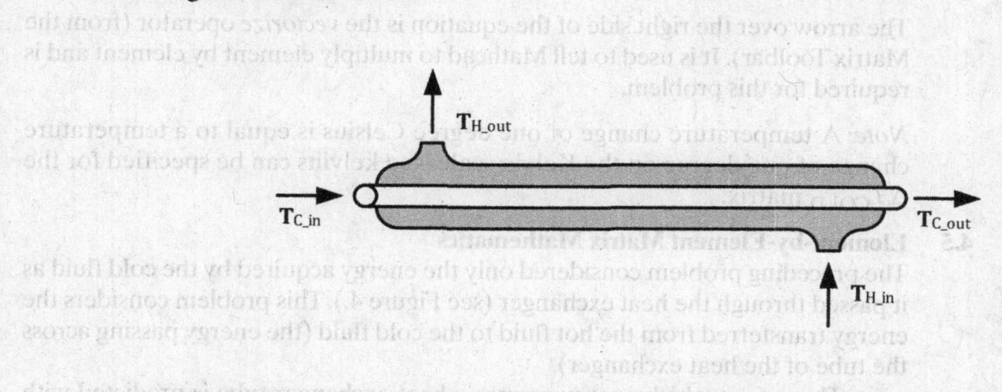

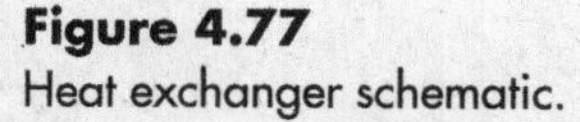

Figure 4.77
Heat exchanger schematic.

The temperature change of the cold fluid depends on the amount of energy the fluid picks up during the time it is flowing through the exchanger—that is, the rate at which energy is acquired by the cold fluid. The energy and the change in temperature are, respectively,

$$q_{COLD} = \dot{m} C_p \, \Delta T_{COLD}$$

and

$$\Delta T_{COLD} = T_{C\_out} - T_{C\_in}.$$

If you performed a series of experiments on this heat exchanger, you might vary the flow rate of the cold fluid to determine the effect on the energy acquired by the cold fluid. Suppose you did this and got the following results:

Experiment	Cold Fluid Rate	Heat Capacity	T_{C_in}	T_{C_out}	q_{COLD}
	kg/minute	joule/kg K	°C	°C	watts
1	2	4187	6	62	
2	5	4187	6	43	
3	10	4187	6	26	
4	15	4187	6	20	
5	20	4187	6	14	

You could solve for the five q_{COLD} values one at a time, but if you create vectors containing the cold fluid rate values and the ΔT_{COLD} values, Mathcad can solve for all of the q_{COLD} values at once by using element-by-element matrix mathematics.

Create two matrices like the following:

$$\text{flow} := \begin{bmatrix} 2 \\ 5 \\ 10 \\ 15 \\ 20 \end{bmatrix} \cdot \frac{\text{kg}}{\text{min}} \qquad \Delta T_{COLD} := \begin{bmatrix} 62 - 6 \\ 43 - 6 \\ 26 - 6 \\ 20 - 6 \\ 14 - 6 \end{bmatrix} \cdot \text{K}$$

Then solve for all five Q_{COLD} values, using element-by-element matrix mathematics. The resulting equation looks like this:

$$q_{COLD} := \overrightarrow{flow \cdot C_p \cdot \Delta T_{COLD}}$$

The arrow over the right side of the equation is the *vectorize* operator (from the Matrix Toolbar). It is used to tell Mathcad to multiply element by element and is required for this problem.

Note: A temperature change of one degree Celsius is equal to a temperature change of one degree on the Kelvin scale—so kelvins can be specified for the ΔT_{COLD} matrix.

4.5 Element-by-Element Matrix Mathematics

The preceding problem considered only the energy acquired by the cold fluid as it passed through the heat exchanger (see Figure 4.). This problem considers the energy transferred from the hot fluid to the cold fluid (the energy passing across the tube of the heat exchanger).

The rate at which energy crosses a heat exchanger tube is predicted with the use of a heat transfer coefficient h. The equation is

$$q_{HX} = hA\Delta T_{LM}$$

where A is the area through which the energy passes (the surface area of the tube) and ΔT_{LM} is the *log mean temperature difference*, defined for the heat exchanger shown as

$$\Delta T_{LM} = \frac{(T_{H\_out} - T_{C\_in}) - (T_{H\_in} - T_{C\_out})}{\ln\left[\frac{(T_{H\_out} - T_{C\_in})}{(T_{H\_in} - T_{C\_out})}\right]} = \frac{\Delta T_{left} - \Delta T_{right}}{\ln\left[\frac{\Delta T_{left}}{\Delta T_{right}}\right]}.$$

Given the following data, use element-by-element matrix mathematics to determine all of the q_{HX} values with one calculation:

Experiment	H	A	T_{C_IN}	T_{C_OUT}	T_{H_IN}	T_{H_OUT}	q_{HX}
	W/m²K	m²	°C	°C	°C	°C	watts
1	300	0.376	6	36	75	54	
2	450	0.376	6	48	73	51	
3	600	0.376	6	52	71	47	
4	730	0.376	6	57	77	40	
5	1200	0.376	6	60	74	35	

Note: The natural logarithm is not defined for matrices, so element-by-element mathematics is required for this problem. Use the *vectorize* operator (from the Matrix Toolbar) over the entire right side of the defining equations to tell Mathcad to use element-by-element mathematics.

4.6 Material Balances on a Gas Absorber

The equations that follow are material balances for CO_2, SO_2, and N_2 around the gas absorber shown in Figure 4.78. Stream S_1 is known to contain 99 mole% MEA and 1 mole% CO_2. The flow rate in S_1 is 100 moles per minute.

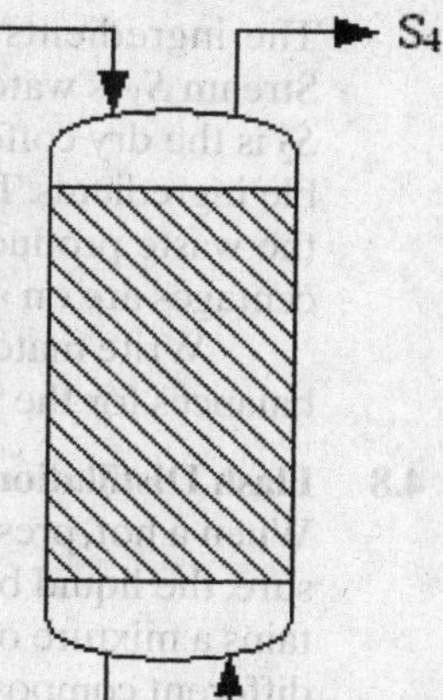

Figure 4.78
Gas absorber schematic.

The compositions used in the material balances are tabulated as follows:

Component	S_1	S_2	S_3	S_4
CO_2	0.01000	0.07522	0.08000	0.00880
SO_2	0	0.01651	0.02000	0.00220
N_2	0	0	0.90000	0.98900
MEA	0.99000	0.90800	0	0

CO_2 Balance: $CO_2 \text{ in } S_1 + CO_2 \text{ in } S_3 = CO_2 \text{ in } S_2 + CO_2 \text{ in } S_4$

$$1 \text{ mole} + 0.08000 \cdot S_3 = 0.07522 \cdot S_2 + 0.00880 \cdot S_4$$

SO_2 Balance: $SO_2 \text{ in } S_1 + SO_2 \text{ in } S_3 = SO_2 \text{ in } S_2 + SO_2 \text{ in } S_4$

$$0 + 0.02000 \cdot S_3 = 0.01651 \cdot S_2 + 0.00220 \cdot S_4$$

N_2 Balance: $N_2 \text{ in } S_1 + N_2 \text{ in } S_3 = N_2 \text{ in } S_2 + N_2 \text{ in } S_4$

$$0 + 0.90000 \cdot S_3 = 0 + 0.98900 \cdot S_4$$

Solve the material balances for the unknown flow rates S_2 through S_4.

4.7 Material Balances on an Extractor

This problem focuses on a low-cost, high-performance chemical extraction unit: a drip coffeemaker, illustrated in Figure 4.79.

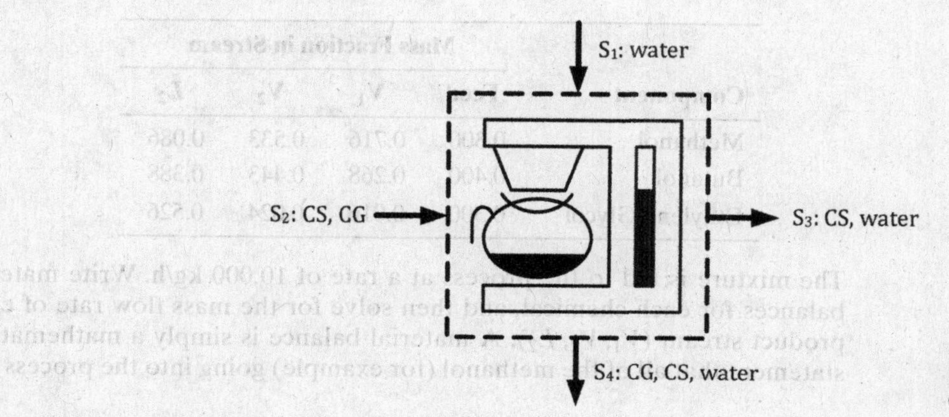

Figure 4.79
Coffeemaker.

The ingredients are water, coffee solubles (CS), and coffee grounds (CG). Stream S_1 is water only, and the coffeemaker is designed to hold 1 liter. Stream S_2 is the dry coffee placed in the filter and contains 99% grounds and 1% soluble ingredients. The product coffee contains 0.4% CS and 99.6% water. Finally, the waste product (S_3) contains 80% CG, 19.6% water, and 0.4% CS. (All percentages are on a volume basis.)

Write material balances on water, CS, and CG, and then solve the material balances for the volumes S_2 through S_4.

4.8 Flash Distillation

When a hot, pressurized liquid is pumped into a tank (flash unit) at a lower pressure, the liquid boils rapidly. This rapid boiling is called a *flash*. If the liquid contains a mixture of chemicals, the vapor and liquid leaving the flash unit will have different compositions, and the flash unit can be used as a separator, called *flash distillation*. The physical principle involved is *vapor–liquid equilibrium*: The vapor and the liquid leaving the flash unit are in equilibrium. This allows the composition of the outlet streams to be determined from the operating temperature and pressure of the flash unit. Multiple flash units can be used together to separate multicomponent mixtures, as illustrated in Figure 4.80.

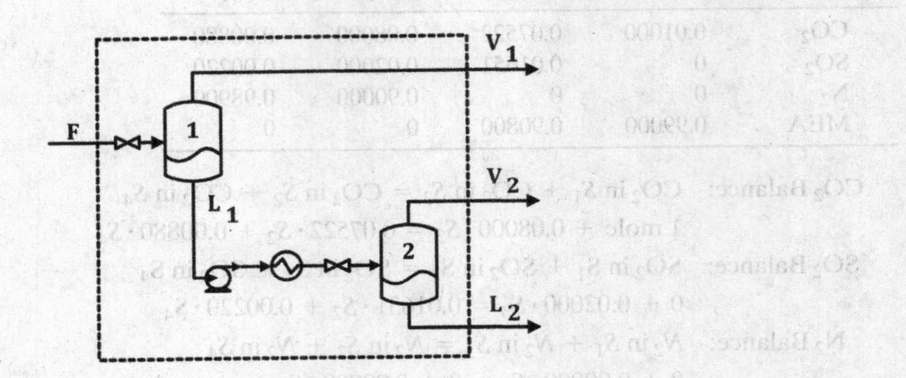

Figure 4.80
Flash distillation schematic.

A mixture of methanol, butanol, and ethylene glycol is fed to a flash unit operating at 165°C and 7 atm. The liquid from the first flash unit is recompressed, reheated, and sent to a second flash unit operating at 105°C and 1 atm. The compositions of the feed stream F and the three product streams are listed in the following table:

	Mass Fraction in Stream			
Component	**Feed**	**V_1**	**V_2**	**L_2**
Methanol	0.300	0.716	0.533	0.086
Butanol	0.400	0.268	0.443	0.388
Ethylene Glycol	0.300	0.016	0.024	0.526

The mixture is fed to the process at a rate of 10,000 kg/h. Write material balances for each chemical, and then solve for the mass flow rate of each product stream (V_1, V_2, L_2). A material balance is simply a mathematical statement that all of the methanol (for example) going into the process has to come out again (assuming a steady state and no chemical reactions). The methanol balance is as follows:

$$\text{methanol in the feed} = \text{methanol in } V_1 + \text{methanol in } V_2 + \text{methanol in } L_2;$$
$$0.300 \cdot (10{,}000\text{kg/h}) = 0.716 \cdot V_1 + 0.533 \cdot V_2 + 0.086 \cdot L_2.$$

4.9 Finding the Median Grade

It is common for instructors to report the average grade on an examination. An alternative way to let students know how their scores compare with the rest of the class is to find the *median* score. If the scores are sorted, the median score is the value in the middle of the data set.

(a) Use the **sort** function to sort the 15 scores (elements 0 to 14) listed at the left, and then display the value of element 7 in the sorted array. Element 7 will hold the median score for the sorted grades.

(b) Check your result by using Mathcad's **median** function on the original vector of grades.

Grades
87
85
43
62
97
88
87
58
67
79
91
82
80
73
58

4.10 Currents in Multiloop Circuits

Multiloop circuits are analyzed using Kirchhoff's voltage law, and Kirchhoff's current law, namely,

At any junction in a circuit, the input current(s) must equal the output current(s).

The latter is simply a statement of conservation of current: All of the electrons entering a junction have to leave again, at least at steady state.

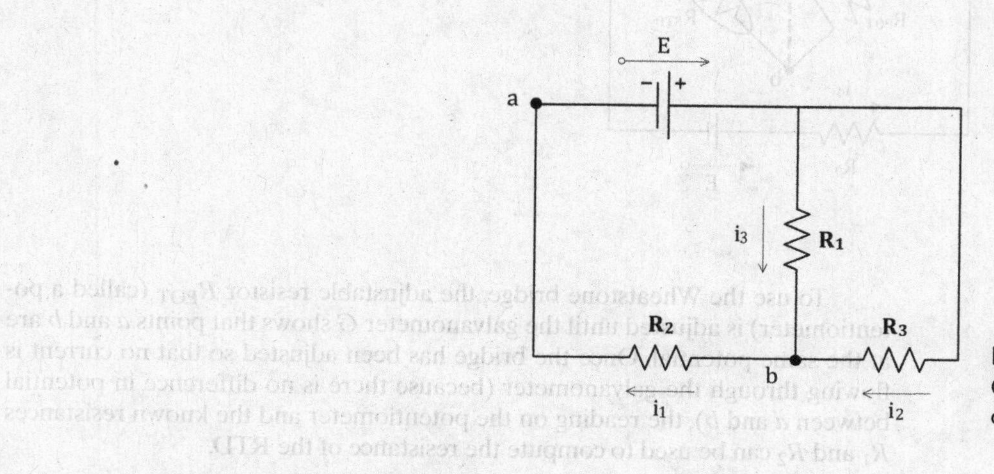

Figure 4.81
Currents in multiloop circuits.

Applying the current law at point *b* in Figure 4.81 yields

$$i_3 + i_2 = i_1.$$

Applying the voltage law to the left loop and the overall loop provides two more equations:

$$E - i_3R_1 - i_1R_2 = 0;$$
$$E - i_2R_3 - i_1R_2 = 0.$$

We thus have three equations in the unknowns i_1, i_2, i_3.

Use the following resistance values, and solve the system of simultaneous equations to determine the current in each portion of the circuit:

$$E = 9 \text{ volts};$$
$$R_1 = 20 \text{ ohms};$$
$$R_2 = 30 \text{ ohms};$$
$$R_3 = 40 \text{ ohms}.$$

4.11 The Wheatstone Bridge

Resistance temperature devices (RTDs) are often used as temperature sensors. As the temperature of the device changes, its resistance changes. If you can determine an RTD's resistance, you can look on a table to find the temperature. A circuit known as a *Wheatstone bridge* can be used to measure unknown resistances. In Figure 4.82, an RTD has been built into the bridge as the unknown resistance.

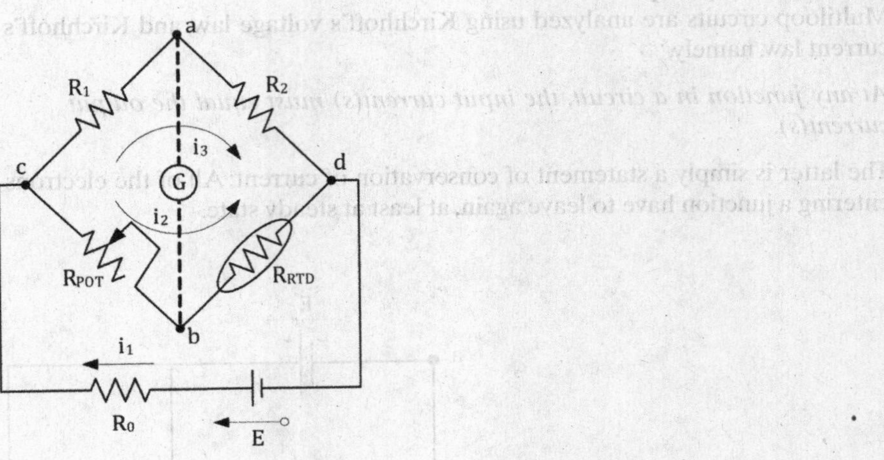

Figure 4.82
Wheatstone bridge.

To use the Wheatstone bridge, the adjustable resistor R_{POT} (called a potentiometer) is adjusted until the galvanometer G shows that points a and b are at the same potential. Once the bridge has been adjusted so that no current is flowing through the galvanometer (because there is no difference in potential between a and b), the reading on the potentiometer and the known resistances R_1 and R_2 can be used to compute the resistance of the RTD.

How Does It Work?

Because resistors R_1 and R_{POT} come together at point c, and we know that the voltages at a and b are the same, there must be the same voltage drop across R_1 and R_{POT} (not the same current). Thus, we have

$$i_3 \cdot R_1 = i_2 \cdot R_{POT}.$$

Similarly, because R_2 and R_{RTD} are connected at point d, we can say that

$$i_3 \cdot R_2 = i_2 \cdot R_{RDT}.$$

If you solve for i_3 in one equation and substitute into the other, you get

$$R_{\text{RDT}} = R_{\text{POT}} \frac{R_2}{R_1}.$$

In this way, if you know R_1 and R_2 and have a reading on the potentiometer, you can calculate R_{RDT}.

(a) Given the following resistances, what is R_{RDT}? Assume that a 9-volt battery is used for E.

$$\begin{aligned} R_0 &= 20 \text{ ohms}; \\ R_1 &= 10 \text{ ohms}; \\ R_2 &= 5 \text{ ohms}; \\ R_{\text{POT}} &\text{ adjusted to 12.3 ohms}. \end{aligned}$$

(b) Use Kirchhoff's current law at either point c or d, and Kirchhoff's voltage law to determine the values of i_1, i_2, and i_3.

4.12 Steady-State Conduction, I

Steady-state conduction in two dimensions is described by Laplace's equation:

$$\frac{\partial^2 T}{\partial x^2} + \frac{\partial^2 T}{\partial y^2} = 0.$$

The partial derivatives in Laplace's equation can be replaced by finite differences to obtain approximate solutions giving the expected temperature at various points in the two-dimensional region. The difference equation is

$$\frac{T_{i-1,j} - 2T_{i,j} + T_{i+1,j}}{(\Delta x)^2} + \frac{T_{i,j-1} - 2T_{i,j} + T_{i,j+1}}{(\Delta y)^2} = 0,$$

where $T_{i,j}$ is the temperature at a point (i, j) in the conducting region, $T_{i-1,j}$ is the temperature at the point to the left of point (i, j), and so forth. If Δx and Δy are equal, these two terms can be combined and the subscripts replaced by more descriptive terms, yielding

$$T_{\text{left}} + T_{\text{above}} + T_{\text{right}} + T_{\text{below}} - 4T_{\text{center}} = 0.$$

Using this equation at each of the nine points in the region shown in Figure 4.83 generates nine equations in nine unknowns that can be solved simultaneously for the temperature at each point.

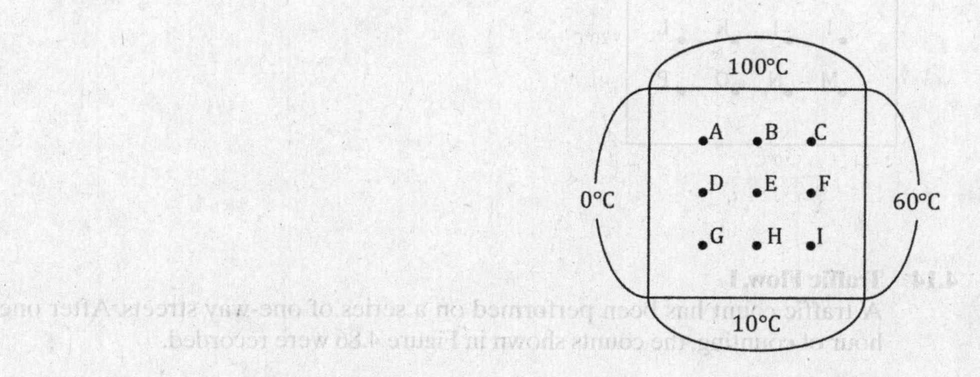

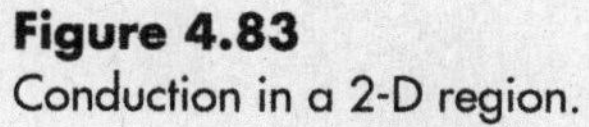
Figure 4.83
Conduction in a 2-D region.

For example, applying the equation at point F (illustrated in Figure 4.84) gives the equation

$$T_E + T_C + 60° + T_I - 4T_F = 0.$$

Develop the system of nine equations, write them in matrix form, and determine the temperature at points *A* through *I*.

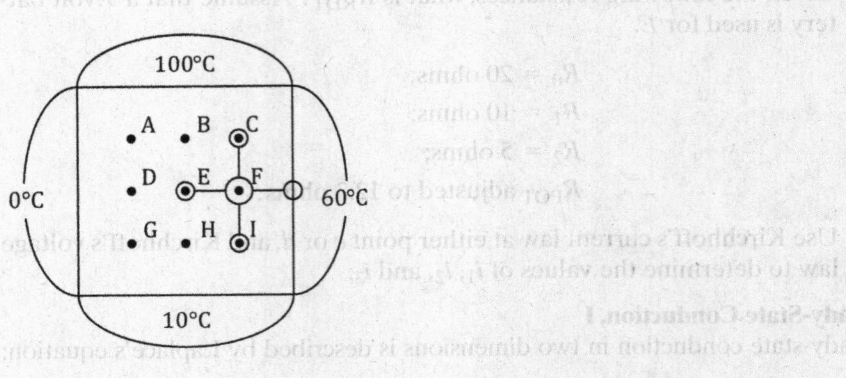

Figure 4.84
Laplace's equation is applied at each grid point.

4.13 Steady-State Conduction, II

In the previous problem, Laplace's equation was applied to steady-state conduction in a square region. Actually, the equation can be applied to any region composed of rectangles or any shape that can be approximated with rectangles. Again, the final version of the equation presented in the last problem does require that the distance between the points be the same in the *x* and *y* directions.

Apply Laplace's equation to the L-shaped region shown in Figure 4.85, and determine the temperatures at points *A* through *P*.

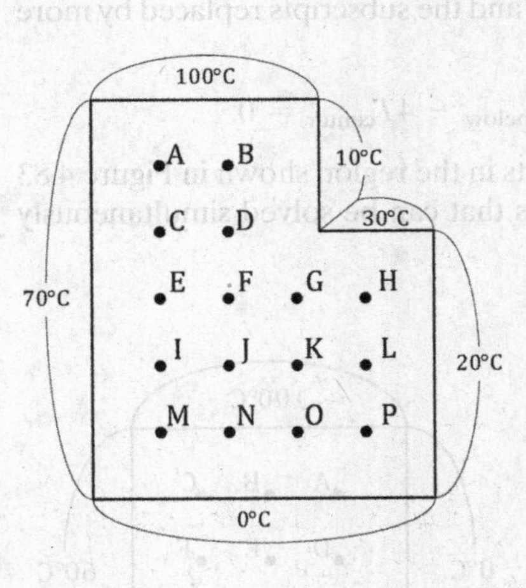

Figure 4.85
Conduction in a non-square 2-D region.

4.14 Traffic Flow, I

A traffic count has been performed on a series of one-way streets. After one hour of counting, the counts shown in Figure 4.86 were recorded.

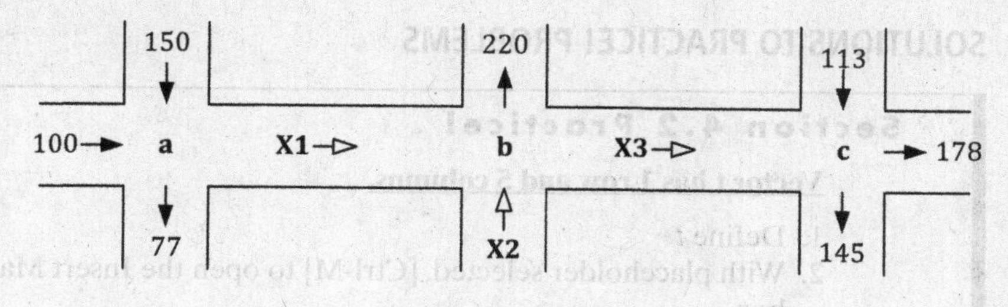

Figure 4.86
Traffic flow, version I.

Since all of the vehicles that enter an intersection must leave again, we can write total vehicle balances on each intersection:

Intersection a:	$100 + 150 = X1 + 77$	or	$X1 = 173;$
Intersection b:	$X1 + X2 = 220 + X3$	or	$X1 + X2 - X3 = 220;$
Intersection c:	$X3 + 113 = 178 + 145$	or	$X3 = 210.$

These balances at the intersections represent three equations in three unknowns and can be solved using matrix methods. (Admittedly, this simple problem hardly needs to be solved using such methods.)

(a) What would the coefficient matrix and right-hand-side vector look like for the preceding system of three equations in three unknowns?

(b) Solve the three equations simultaneously using matrix methods.

4.15 Traffic Flow, II

Since the number of vehicles entering and leaving an intersection must be equal, the four intersections shown in Figure 4.87 allow four vehicle balances to be written that can be solved for four unknown vehicle counts.

(a) Write vehicle balances for each of the four intersections.

(b) Solve for the four unknown traffic counts using matrix methods.

(c) Is there any evidence that people are making U-turns in the intersections?

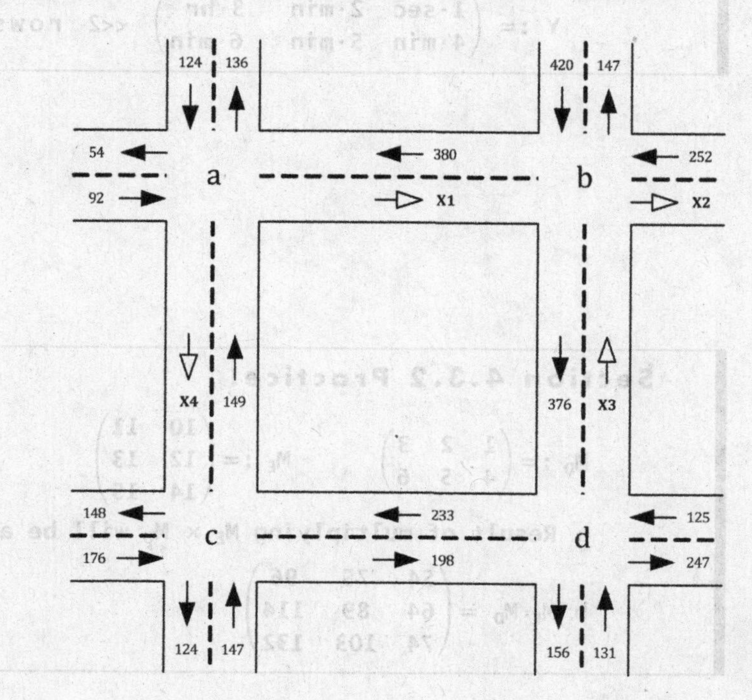

Figure 4.87
Traffic flow, version II.

SOLUTIONS TO PRACTICE! PROBLEMS

Section 4.2 Practice!

Vector t has 1 row and 5 columns.

1. Define *t*
2. With placeholder selected, [Ctrl-M] to open the Insert Matrix dialog box
3. Specify 1 row and 5 columns

t :=

Insert Matrix

Rows: 1 OK

Columns: 5 Insert

Delete

Cancel

4. Click **OK** to create an array
5. Fill in array placeholders

$t := \begin{pmatrix} 2 & 4 & 6 & 8 & 10 \end{pmatrix}$ <<1 row, 5 columns

$W := \begin{pmatrix} 1 & 2 & 3 & 4 \\ 3 & 1 & 5 & 7 \end{pmatrix}$ <<2 rows, 4 columns

$Y := \begin{pmatrix} 1\cdot sec & 2\cdot min & 3\cdot hr \\ 4\cdot min & 5\cdot min & 6\cdot min \end{pmatrix}$ <<2 rows, 3 columns

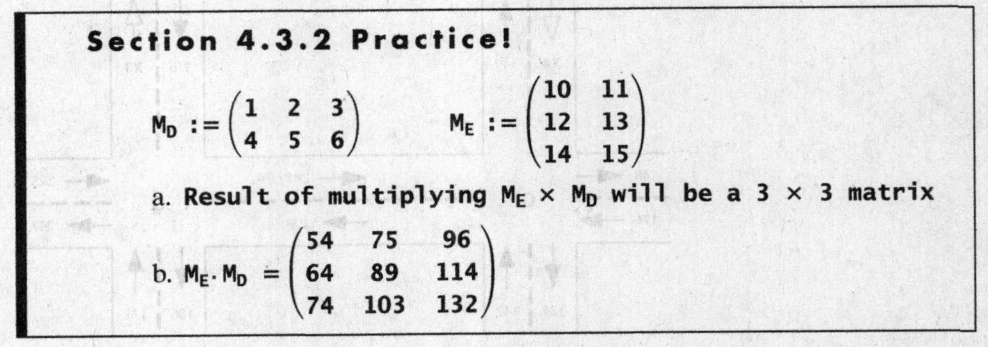

Section 4.3.2 Practice!

$M_D := \begin{pmatrix} 1 & 2 & 3 \\ 4 & 5 & 6 \end{pmatrix}$ $M_E := \begin{pmatrix} 10 & 11 \\ 12 & 13 \\ 14 & 15 \end{pmatrix}$

a. **Result of multiplying $M_E \times M_D$ will be a 3×3 matrix**

b. $M_E \cdot M_D = \begin{pmatrix} 54 & 75 & 96 \\ 64 & 89 & 114 \\ 74 & 103 & 132 \end{pmatrix}$

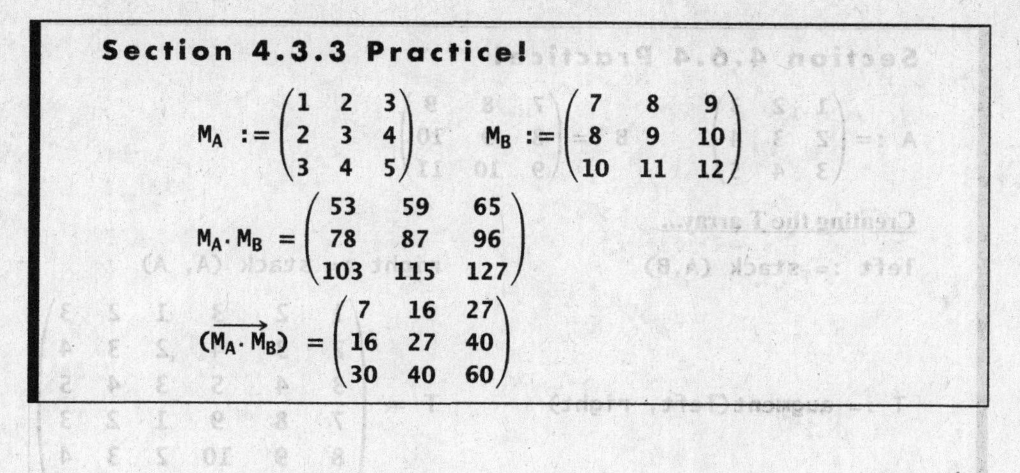

Section 4.3.3 Practice!
M_A := (1 2 3; 2 3 4; 3 4 5)
M_B := (7 8 9; 8 9 10; 10 11 12)
M_A · M_B = (53 59 65; 78 87 96; 103 115 127)
(M_A · M_B) = (7 16 27; 16 27 40; 30 40 60)

Section 4.4 Practice!

a. $C_a := \begin{pmatrix} 2 & 3 & 1 \\ 1 & 4 & 7 \\ 1 & 4 & 7 \end{pmatrix}$ $\quad |C_a| = 0$

b. $C_b := \begin{pmatrix} 2 & 3 & 1 \\ 1 & 4 & 7 \\ 3 & 7 & 8 \end{pmatrix}$ $\quad |C_b| = 8.882 \times 10^{-15}$

^^^ essentially zero

c. $C_c := \begin{pmatrix} 2 & 3 & 1 \\ 1 & 4 & 7 \\ 4 & 1 & 3 \end{pmatrix}$ $\quad |C_c| = 70$

$r_c := \begin{pmatrix} 12 \\ 16 \\ 9 \end{pmatrix}$

$x_c := \text{lsolve}(C_c, r_c)$ $\quad x_c = \begin{pmatrix} 1.214 \\ 3.071 \\ 0.357 \end{pmatrix}$

d. $C_d := \begin{pmatrix} 2 & 3 & 1 \\ 1 & 4 & 7 \\ 7 & 18 & 23 \end{pmatrix}$ $\quad |C_d| = 1.998 \times 10^{-14}$

^^^ essentially zero

Section 4.6.4 Practice!

$$A := \begin{pmatrix} 1 & 2 & 3 \\ 2 & 3 & 4 \\ 3 & 4 & 5 \end{pmatrix} \qquad B := \begin{pmatrix} 7 & 8 & 9 \\ 8 & 9 & 10 \\ 9 & 10 & 11 \end{pmatrix}$$

Creating the T array...

left := stack (A,B)

right := stack (A, A)

T := augment(left, right)

$$T = \begin{pmatrix} 1 & 2 & 3 & 1 & 2 & 3 \\ 2 & 3 & 4 & 2 & 3 & 4 \\ 3 & 4 & 5 & 3 & 4 & 5 \\ 7 & 8 & 9 & 1 & 2 & 3 \\ 8 & 9 & 10 & 2 & 3 & 4 \\ 9 & 10 & 11 & 3 & 4 & 5 \end{pmatrix}$$

Extracting T_{sub} array...

T_{sub} := submatrix (T, 2, 5, 2,4)]

$$T_{sub} = \begin{pmatrix} 5 & 3 & 4 \\ 9 & 1 & 2 \\ 10 & 2 & 3 \\ 11 & 3 & 4 \end{pmatrix}$$

Section 4.7.4 Practice!

1. i := 0..4

   M_i := 2 · i

$$M = \begin{pmatrix} 0 \\ 2 \\ 4 \\ 6 \\ 8 \end{pmatrix}$$

2. i := 0..4

   j := 0..4

   $M_{i,j}$:= 3 + 2 · i = j

$$M = \begin{pmatrix} 3 & 2 & 1 & 0 & -1 \\ 5 & 4 & 3 & 2 & 1 \\ 7 & 6 & 5 & 4 & 3 \\ 9 & 8 & 7 & 6 & 5 \\ 11 & 10 & 9 & 8 & 7 \end{pmatrix}$$

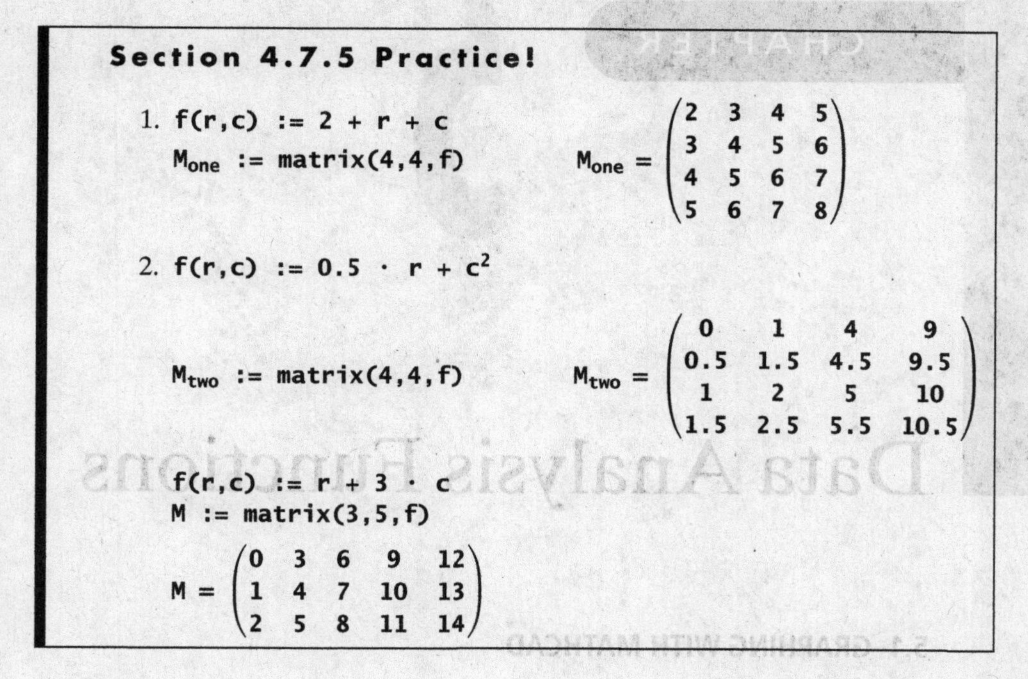

Section 4.7.5 Practice!

1. f(r,c) := 2 + r + c

   M_{one} := matrix(4,4,f) $\qquad M_{one} = \begin{pmatrix} 2 & 3 & 4 & 5 \\ 3 & 4 & 5 & 6 \\ 4 & 5 & 6 & 7 \\ 5 & 6 & 7 & 8 \end{pmatrix}$

2. f(r,c) := 0.5 · r + c^2

   M_{two} := matrix(4,4,f) $\qquad M_{two} = \begin{pmatrix} 0 & 1 & 4 & 9 \\ 0.5 & 1.5 & 4.5 & 9.5 \\ 1 & 2 & 5 & 10 \\ 1.5 & 2.5 & 5.5 & 10.5 \end{pmatrix}$

   f(r,c) := r + 3 · c

   M := matrix(3,5,f)

   $M = \begin{pmatrix} 0 & 3 & 6 & 9 & 12 \\ 1 & 4 & 7 & 10 & 13 \\ 2 & 5 & 8 & 11 & 14 \end{pmatrix}$

ANSWERS TO SELECTED PROBLEMS

Problem 1	b. not defined; matrix must be square to calculate a determinant
	h. 30
Problem 3	b. $x_1 = 1, x_2 = 0, x_3 = 1$
Problem 5	Experiment 1. $\Delta T_{LM} = 43.3°C$, $q_{HX} = 4890$ watts
Problem 8	$V_2 = 3765$ kg/hr
Problem 10	$i_3 = 0.138$ amps
Problem 12	$T_B = 64.9°C$, $T_F = 52.2°C$
Problem 15	$x_3 = 194$ vehicles

CHAPTER 5

Data Analysis Functions

Objectives

After reading this chapter, you will

- be able to create graphs in Mathcad
- know how to use a QuickPlot to see what a function looks like when plotted
- know that Mathcad provides built-in functions for basic statistical calculations
- be able to use Mathcad's linear and spline interpolation functions on data sets
- know two ways to perform linear regression on data sets

5.1 GRAPHING WITH MATHCAD

Engineers spend a lot of time analyzing data. By looking at data from a current design, you can learn what can be changed to make the device, system, or process work better. Mathcad's data analysis functions can provide a lot of help in understanding and evaluating data.

One of the most fundamental steps in analyzing data is visualizing the data, usually in the form of a graph. Mathcad provides a variety of graphing options. In this section we will look at three different ways to create a graph: plotting vector against vector, plotting element against element, and evaluating a function using Mathcad's QuickPlot feature. We will also look at how to plot multiple curves on a graph and how to modify the display characteristics of a graph.

5.1.1 Plotting Vector against Vector

The easiest way to create a graph in Mathcad is simply to plot one vector of values against another. For example, the vector of temperature values shown in Figure 5.1 can be plotted against the vector of time values. This method works well for many data analysis situations because the data are often readily available in vector form.

When you plot one vector against another, the vectors must have the same number of elements. The vectors in Figure 5.1 each have ten elements, so they can be plotted against each other. To create an ***X-Y graph*** do any of the following:

- click on the **X-Y Plot** button on the Graph Toolbar (Figure 5.2), or
- use the keyboard shortcut [Shift-2], or
- use menu options **Insert/Graph/X-Y Plot**.

Mathcad will create an empty graph with placeholders centered on each axis. (See Figure 5.3.)

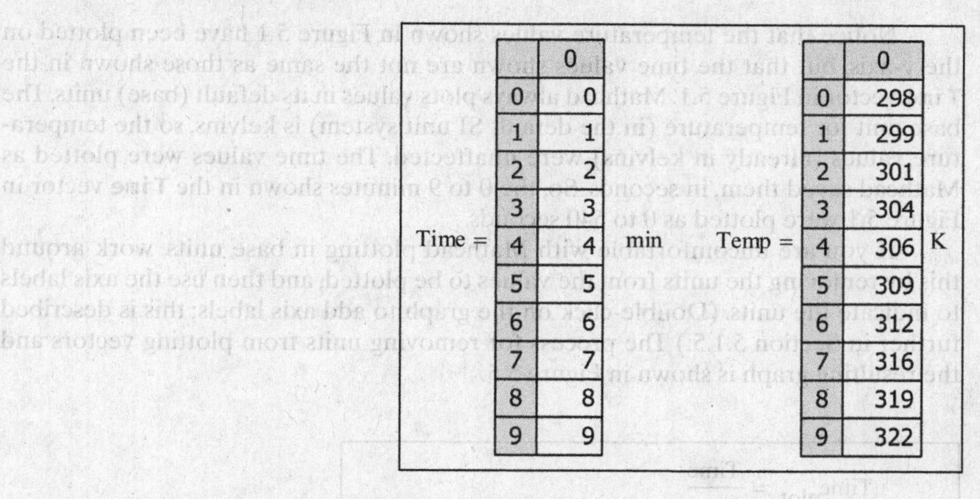

Time =

	0
0	0
1	1
2	2
3	3
4	4
5	5
6	6
7	7
8	8
9	9

min

Temp =

	0
0	298
1	299
2	301
3	304
4	306
5	309
6	312
7	316
8	319
9	322

K

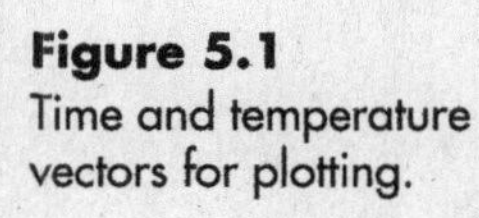

Figure 5.1
Time and temperature vectors for plotting.

Figure 5.2
Mathcad's Graph Toolbar.

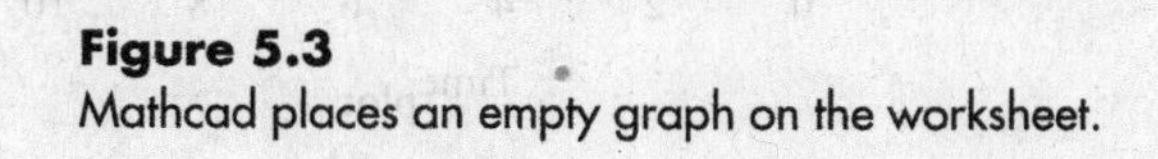

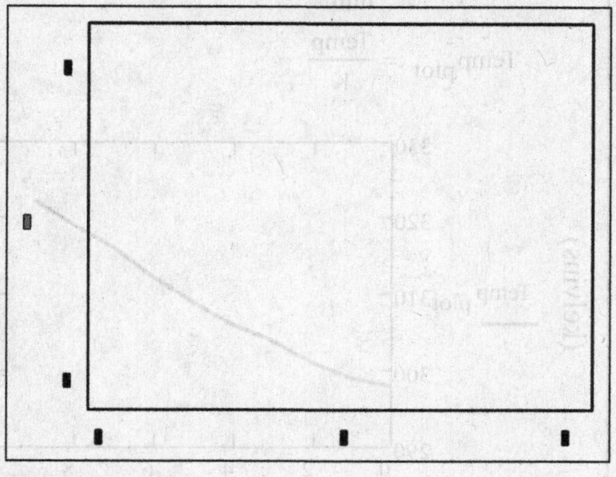

Figure 5.3
Mathcad places an empty graph on the worksheet.

Enter the vector names into the placeholders to complete the graph. The result is shown in Figure 5.4.

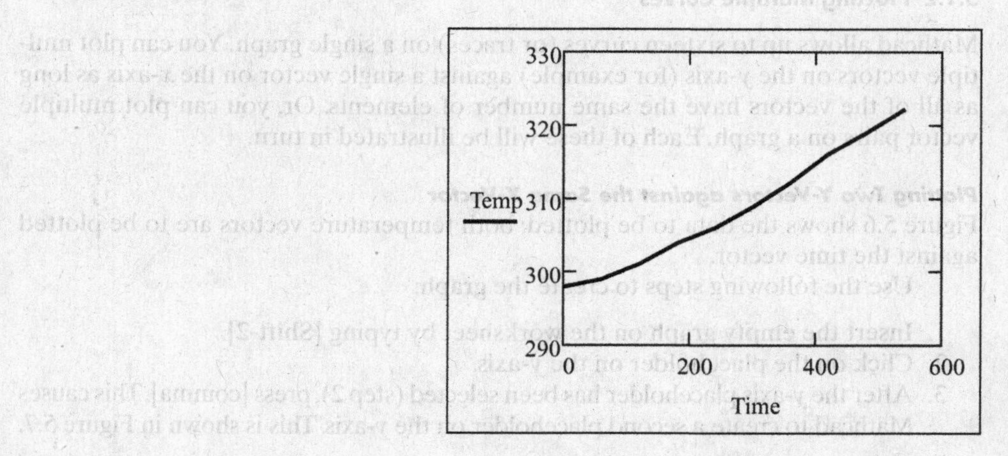

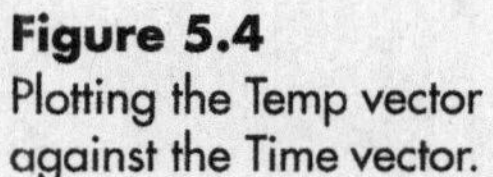

Figure 5.4
Plotting the Temp vector against the Time vector.

Notice that the temperature values shown in Figure 5.1 have been plotted on the *y*-axis, but that the time values shown are not the same as those shown in the `Time` vector in Figure 5.1. Mathcad always plots values in its default (base) units. The base unit for temperature (in the default SI unit system) is kelvins, so the temperature values (already in kelvins) were unaffected. The time values were plotted as Mathcad saved them, in seconds. So, the 0 to 9 minutes shown in the `Time` vector in Figure 5.1 were plotted as 0 to 540 seconds.

If you are uncomfortable with Mathcad plotting in base units, work around this by removing the units from the values to be plotted, and then use the axis labels to indicate the units. (Double-click on the graph to add axis labels; this is described further in Section 5.1.5.) The process for removing units from plotting vectors and the resulting graph is shown in Figure 5.5.

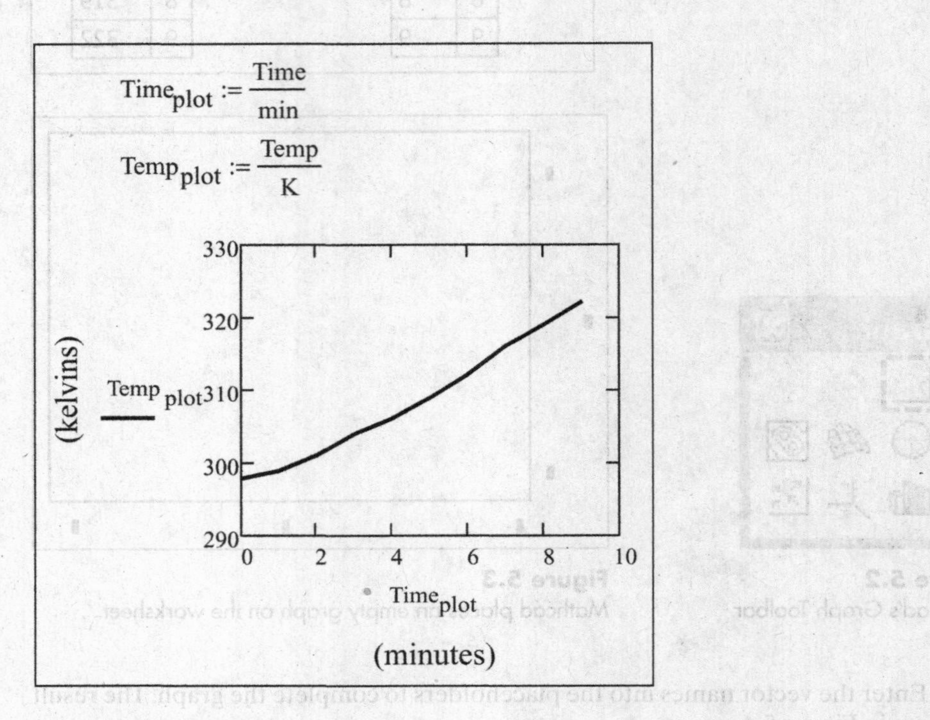

Figure 5.5
Plotting dimensionless values and indicating units with axis labels.

5.1.2 Plotting Multiple Curves

Mathcad allows up to sixteen curves (or traces) on a single graph. You can plot multiple vectors on the *y*-axis (for example) against a single vector on the *x*-axis as long as all of the vectors have the same number of elements. Or, you can plot multiple vector pairs on a graph. Each of these will be illustrated in turn.

Plotting Two Y-Vectors against the Same X-Vector

Figure 5.6 shows the data to be plotted; both temperature vectors are to be plotted against the time vector.

Use the following steps to create the graph:

1. Insert the empty graph on the worksheet by typing [Shift-2].
2. Click on the placeholder on the *y*-axis.
3. After the *y*-axis placeholder has been selected (step 2), press [comma]. This causes Mathcad to create a second placeholder on the *y*-axis. This is shown in Figure 5.7.

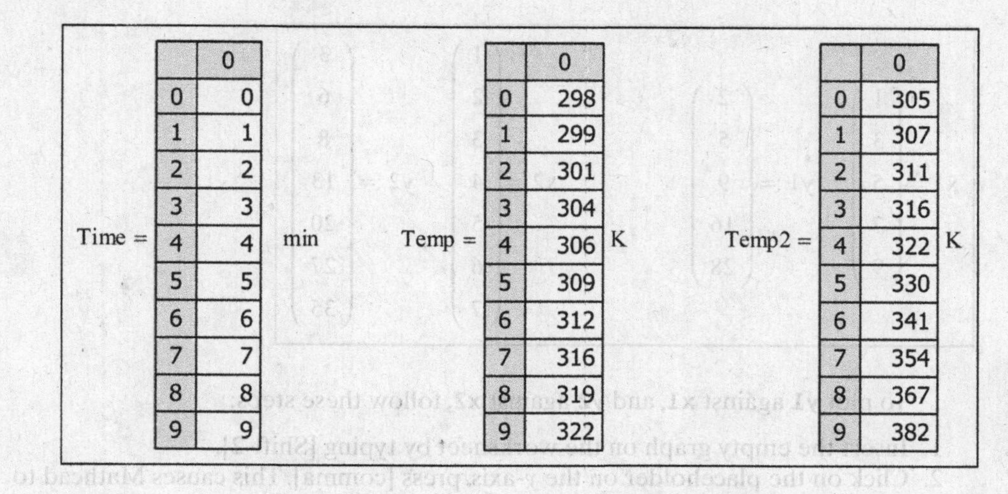

Time =

	0
0	0
1	1
2	2
3	3
4	4
5	5
6	6
7	7
8	8
9	9

min

Temp =

	0
0	298
1	299
2	301
3	304
4	306
5	309
6	312
7	316
8	319
9	322

K

Temp2 =

	0
0	305
1	307
2	311
3	316
4	322
5	330
6	341
7	354
8	367
9	382

K

Figure 5.6
Vectors for plotting two temperature vectors against one time vector.

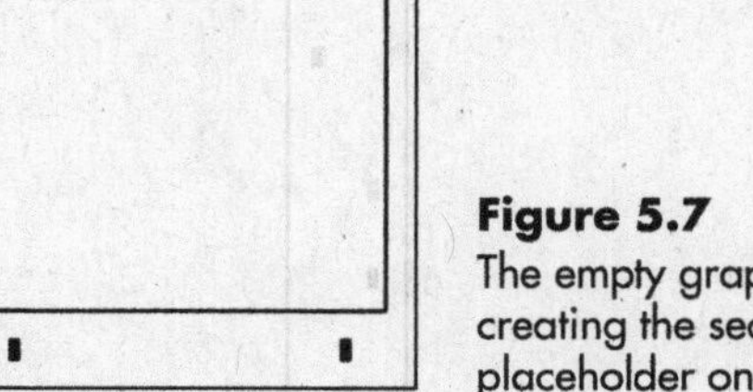

Figure 5.7
The empty graph after creating the second placeholder on the *y*-axis.

4. Enter the vector names Temp and Temp2 in the y-axis placeholders, and Time in the *x*-axis placeholder. The result is shown in Figure 5.8.

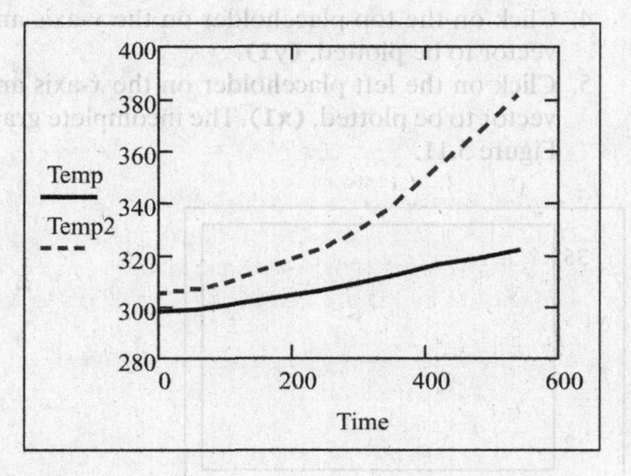

Figure 5.8
The completed graph with two temperature curves.

Plotting Paired X- and Y-Vectors

If the two vectors of *y*-values that you want to plot are of different sizes or correspond to different *x*-values, then you will need to plot paired *x*- and *y*-vectors. Some paired ***sample*** data vectors are illustrated in Figure 5.9.

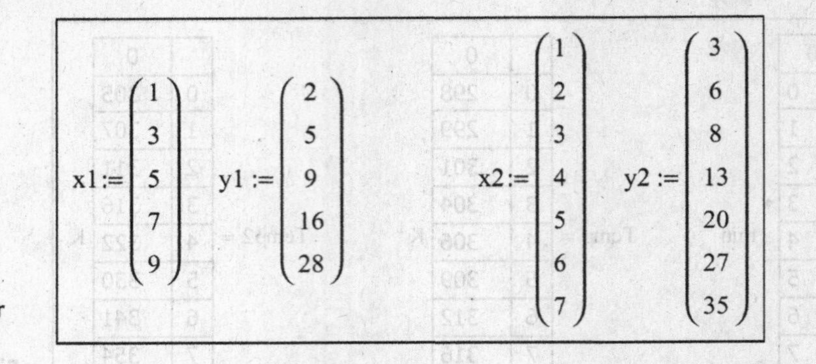

Figure 5.9
Paired *x*- and *y*-vectors for plotting.

To plot **y1** against **x1**, and **y2** against **x2**, follow these steps:

1. Insert the empty graph on the worksheet by typing [Shift-2].
2. Click on the placeholder on the *y*-axis, press [comma]. This causes Mathcad to create a second placeholder on the *y*-axis.
3. Click on the placeholder on the *x*-axis, press [comma]. This causes Mathcad to create a second placeholder on the *x*-axis. The result of the first three steps is shown in Figure 5.10.

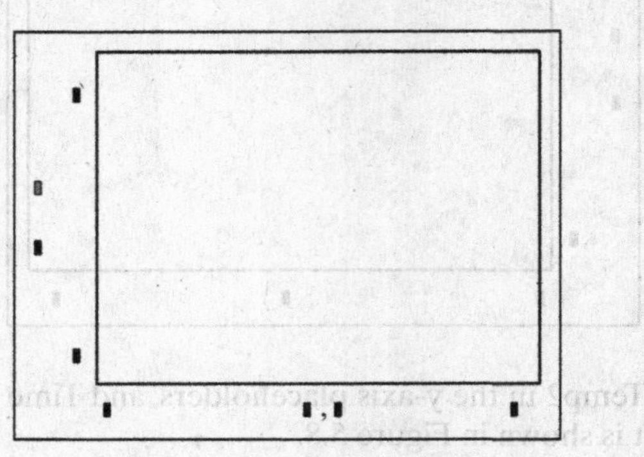

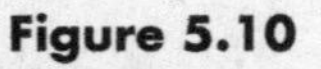

Figure 5.10
The empty graph with two placeholders on each axis.

4. Click on the top placeholder on the *y*-axis and enter the name of the first *y*-vector to be plotted, **(y1)**.
5. Click on the left placeholder on the *x*-axis and enter the name of the first *x*-vector to be plotted, **(x1)**. The incomplete graph after steps 4 and 5 is shown in Figure 5.11.

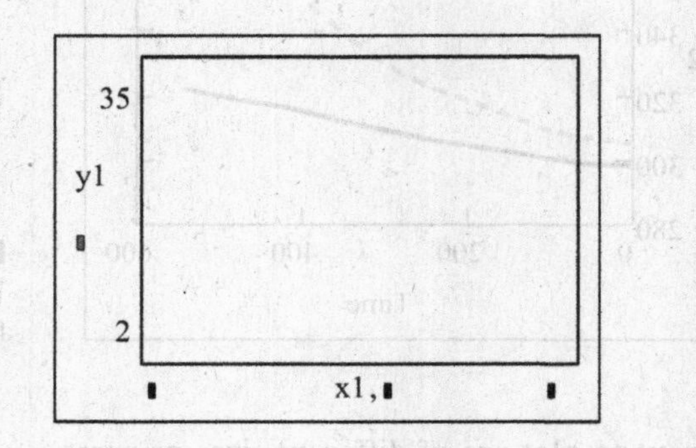

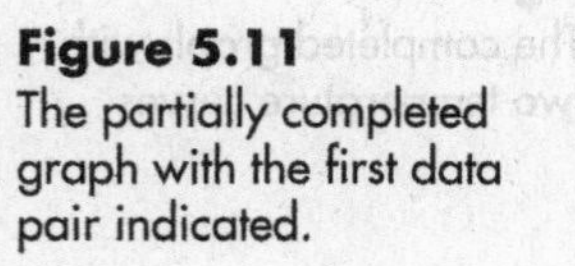

Figure 5.11
The partially completed graph with the first data pair indicated.

6. Click on the lower placeholder on the *y*-axis and enter the name of the second *y*-vector to be plotted, **(y2)**.

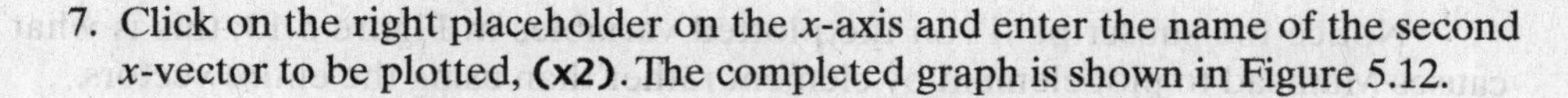

7. Click on the right placeholder on the *x*-axis and enter the name of the second *x*-vector to be plotted, **(x2)**. The completed graph is shown in Figure 5.12.

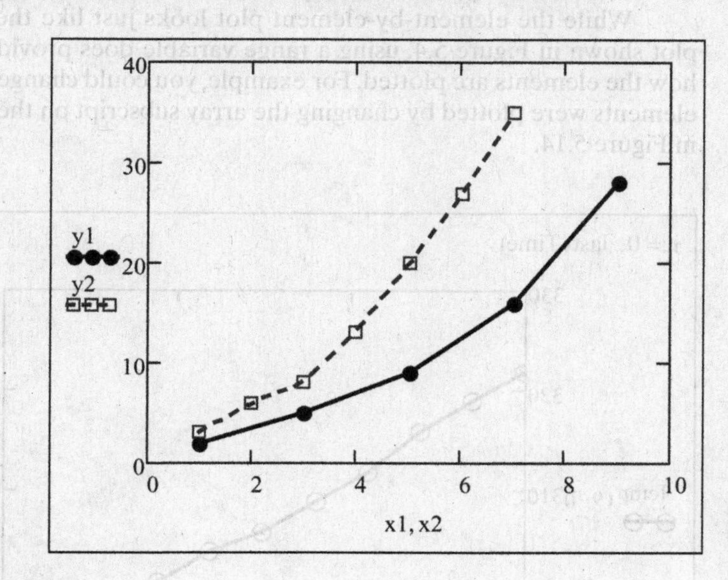

Figure 5.12
The completed graph.

5.1.3 Element-by-Element Plotting

In older versions of Mathcad, the standard method for graphing was to plot each element of one vector against the corresponding elements of the other vector. This method is still available but rarely needed. To plot the temperature and time data (from Figure 5.1) element by element, we must define a range variable containing as many elements as each of the vectors. An easy way to do this is to use the **last(v)** function, which returns the array index of the last element of vector **v**. The use of the **last()** function to define range variable **i** is illustrated in Figure 5.13, along with the element-by-element graph.

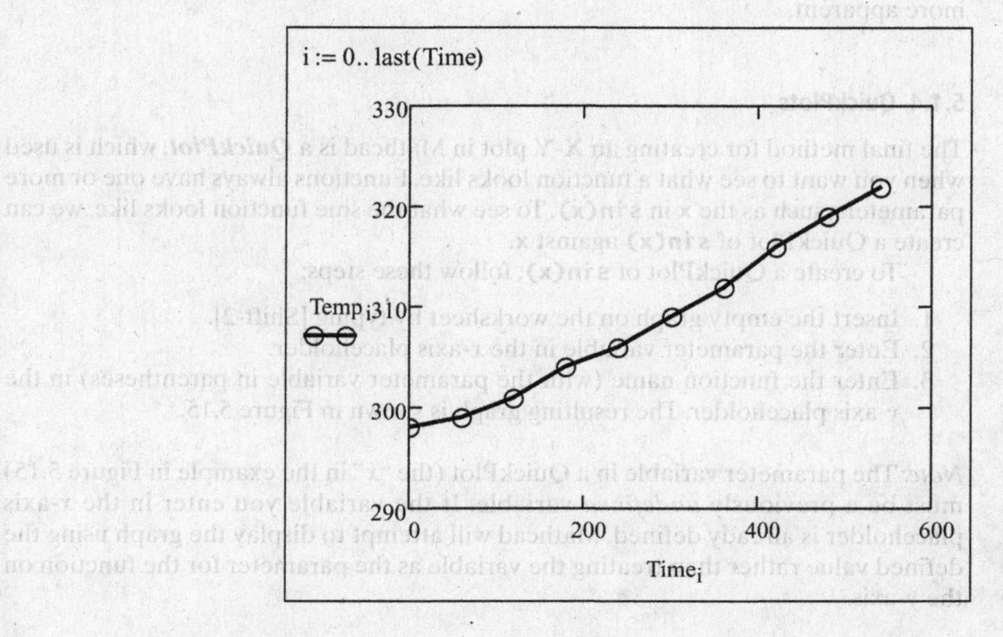

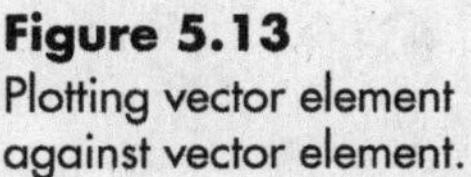

Figure 5.13
Plotting vector element against vector element.

Notice the subscript "i" on the plotted variables in Figure 5.13; this is what causes Mathcad to plot element by element rather than using the entire vectors.

While the element-by-element plot looks just like the vector-against-vector plot shown in Figure 5.4, using a range variable does provide a lot of control over how the elements are plotted. For example, you could change the order in which the elements were plotted by changing the array subscript on the **Temp** vector, as shown in Figure 5.14.

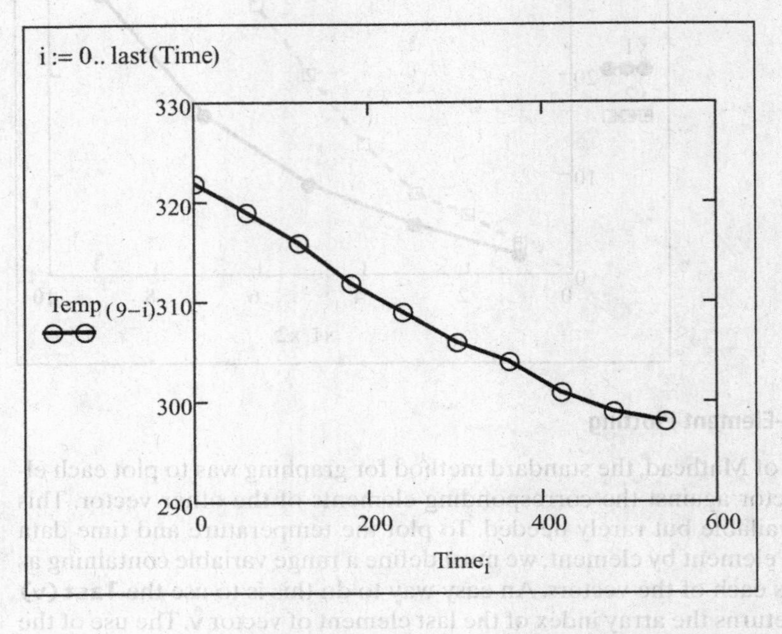

Figure 5.14
Reversing the plotting order by modifying the subscript on the **Temp** vector from **i** to **9−i**.

Note: The parentheses on the "9–i" in the Temp variable's subscript are not required, but they are included in Figure 5.14 to make the change to the graphed variable more apparent.

5.1.4 QuickPlots

The final method for creating an X-Y plot in Mathcad is a ***QuickPlot***, which is used when you want to see what a function looks like. Functions always have one or more parameters, such as the **x** in **sin(x)**. To see what the sine function looks like, we can create a QuickPlot of **sin(x)** against **x**.

To create a QuickPlot of **sin(x)**, follow these steps:

1. Insert the empty graph on the worksheet by typing [Shift-2].
2. Enter the parameter variable in the *x*-axis placeholder.
3. Enter the function name (with the parameter variable in parentheses) in the *y*-axis placeholder. The resulting graph is shown in Figure 5.15.

Note: The parameter variable in a QuickPlot (the "*x*" in the example in Figure 5.15) must be a previously *undefined* variable. If the variable you enter in the *x*-axis placeholder is already defined, Mathcad will attempt to display the graph using the defined value rather than treating the variable as the parameter for the function on the *y*-axis.

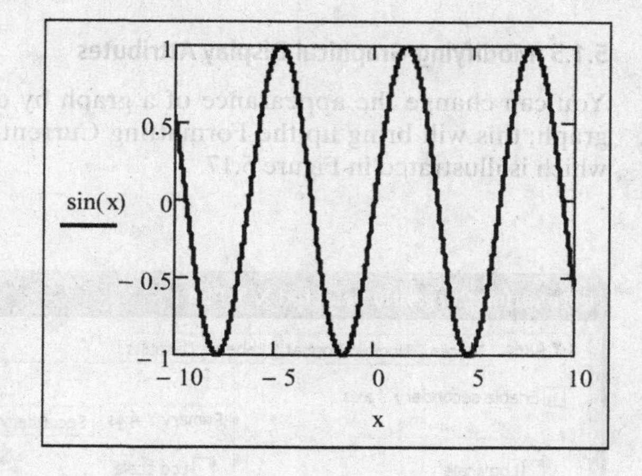

Figure 5.15
A QuickPlot of **sin**(*x*).

By default, Mathcad evaluates the function being plotted over a range of parameter values from -10 to $+10$. (If the function is not defined over the entire range, Mathcad will plot only the portion of the range for which the function is defined.) To plot the function over a different range of *x*-values, simply click on the *x*-axis of the graph and change the displayed axis limits. In Figure 5.16, the axis limits were changed to 0 and 2π to plot one complete sine wave.

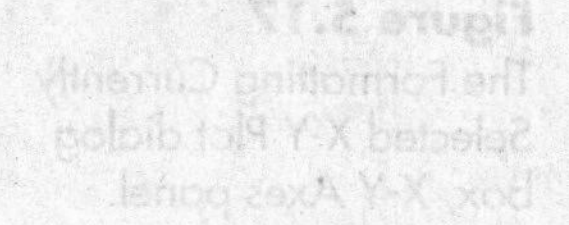

Figure 5.16
A QuickPlot of `sin(x)` with *x* axis limits changed to 0 and 2π.

Practice!

a. Try QuickPlots of other common functions, such as `ln(x)` and `exp(x)`.

b. See what QuickPlots of the following functions look like:

- `if(sin(x)>0, -sin(x), sin(x))`
- `Φ(|mod(x,2)| - 1)`

The last example uses the Heaviside step function, Φ, to generate a square wave. The Heaviside function symbol is available as capital phi on the Greek Toolbar. The absolute value operator is available on the Calculator Toolbar.

5.1.5 Modifying Graphical Display Attributes

You can change the appearance of a graph by double-clicking anywhere on the graph; this will bring up the Formatting Currently Selected X-Y Plot dialog box, which is illustrated in Figure 5.17.

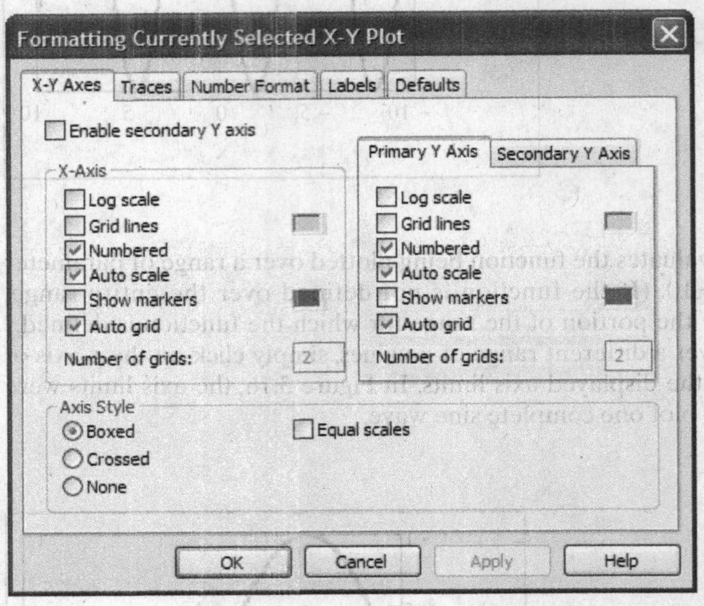

Figure 5.17
The Formatting Currently Selected X-Y Plot dialog box, X-Y Axes panel.

From this dialog box you can access four panels of graphing options:

X-Y Axes Panel (Figure 5.17)
- enable a second *y*-axis (on the right side of the graph)
- select a log axis on either the *x*- or *y*-axis
- add grid lines to either the *x*- or *y*-axis
- change the way the axes are displayed

Traces Panel (Figure 5.18)
- change the way the curves (traces) are displayed
- show or hide the legend

Labels Panel
- add a title to the graph
- add axis labels to the graph

Defaults Panel
- reset the graphing defaults

As shown in Figure 5.18, you can change the way each curve, or trace, on the graph is displayed. The default is to connect the data points with a line but show

Formatting Currently Selected X-Y Plot

X-Y Axes | Traces | Number Format | Labels | Defaults

Legend label	Symbol Frequency	Symbol	Symbol Weight	Line	Line Weight	Col
trace 1	1		1	———	1	
trace 2	1		1	· · · ·	1	
trace 3	1		1	—	1	
trace 4	1		1	– · – ·	1	
trace 5	1		1	———	1	
trace 6	1		1	· · · ·	1	
trace 7	1		1	– –	1	

Hide arguments

Hide legend

Top-left Top-right

Bottom-left Bottom-right

Below

OK Cancel Apply Help

Figure 5.18
The Formatting Currently Selected X-Y Plot dialog box, Traces panel.

no symbol. The columns in the **Traces** panel that are used to change the symbol display are:

- **Symbol**—changes the symbol used to indicate each data point (circles, squares, etc.)
- **Symbol Weight**—changes the size of the symbols used
- **Symbol Frequency**—allows you to show the symbol on every N^{th} symbol, where N is the value entered in the frequency column
- **Color**—changes the color used for both the line and the symbol

The columns that allow you to change the appearance of the line are:

- **Line**—changes the type of line (solid, dashed, etc.)
- **Line Weight**—changes the line thickness
- **Color**—changes the color used for both the line and the symbol

There is one additional column in the Traces panel that allows you to control how the graph is drawn:

- **Type**—selects the way the curve is displayed
  - **lines**—This is the default. The data values are connected with a line (no symbols).
  - **points**—The data values are indicated with symbols (no connecting line).
  - **error**—When two adjacent traces are set to **Type = error** the data values associated with the two traces are used to define the upper and lower limits of error bars.
  - **bar**—The data values are indicated with an outlined (empty) bar.
  - **solidbar**—The data values are indicated with a filled bar.
  - **step**—The line connecting the data points steps from one data level to the next.
  - **stem**—A line is drawn from the x-axis to the symbol for each data point.
  - **draw**—This line type is used when a single pair of vectors contains more than one set of curves. The draw type prevents the "double back" lines that would connect the end of the first curve with the beginning of the second curve.

APPLICATION

Using a Graph to Illustrate a Relationship

It seems reasonable that the number of hours of sunlight that a city receives is related to the average monthly temperature in that city. Data on sunrise and sunset times are available from the U.S. Naval Observatory website at http://aa.usno.navy.mil, and average monthly temperature data (1971–2000) are available from the National Climatic Data Center at http://www.ncdc.noaa.gov. The data shown in Figure 5.19 are for Fairbanks, Alaska.

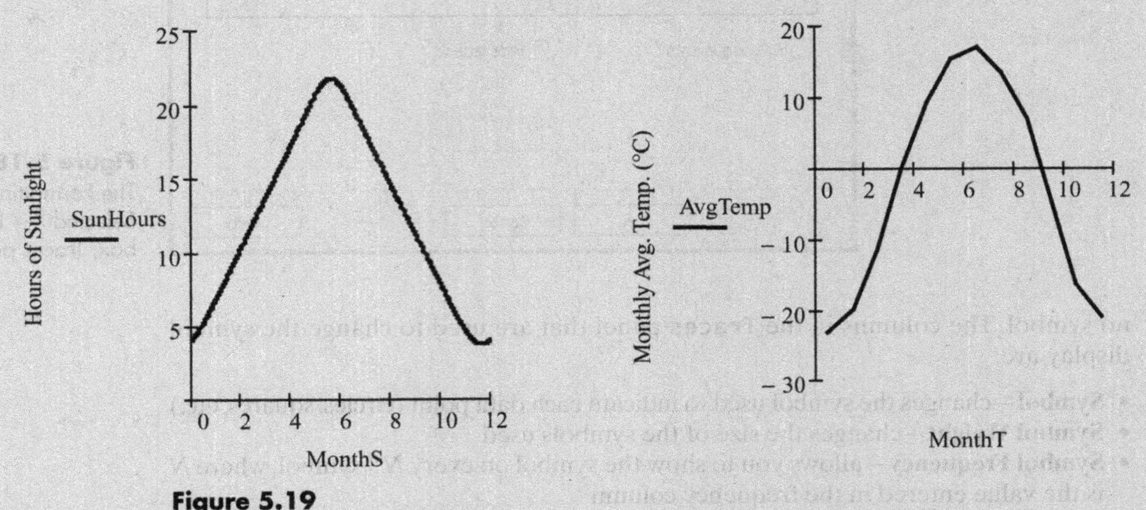

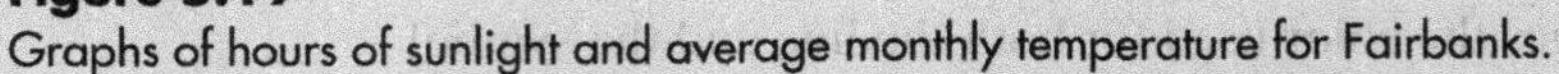

Figure 5.19
Graphs of hours of sunlight and average monthly temperature for Fairbanks.

The relationship between the two graphed quantities is more apparent if hours of sunlight and average monthly temperature are plotted on the same graph, as in Figure 5.20. However, the different units make the curves difficult to compare.

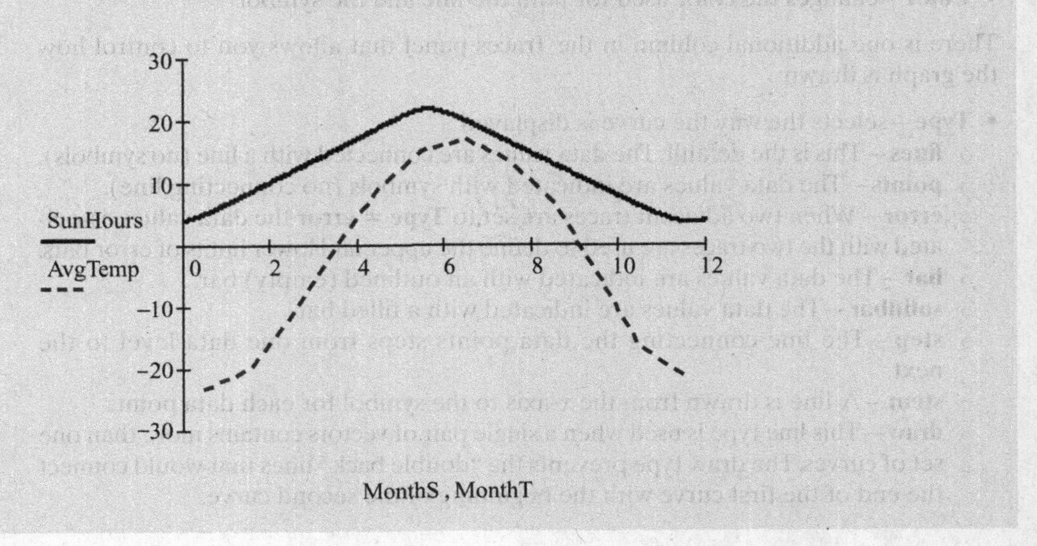

Figure 5.20
First attempt at plotting the two curves on the same graph.

This is a situation where a secondary y-axis can make it easier to compare the two curves. A plot of these two curves using a secondary y-axis is shown in Figure 5.21.

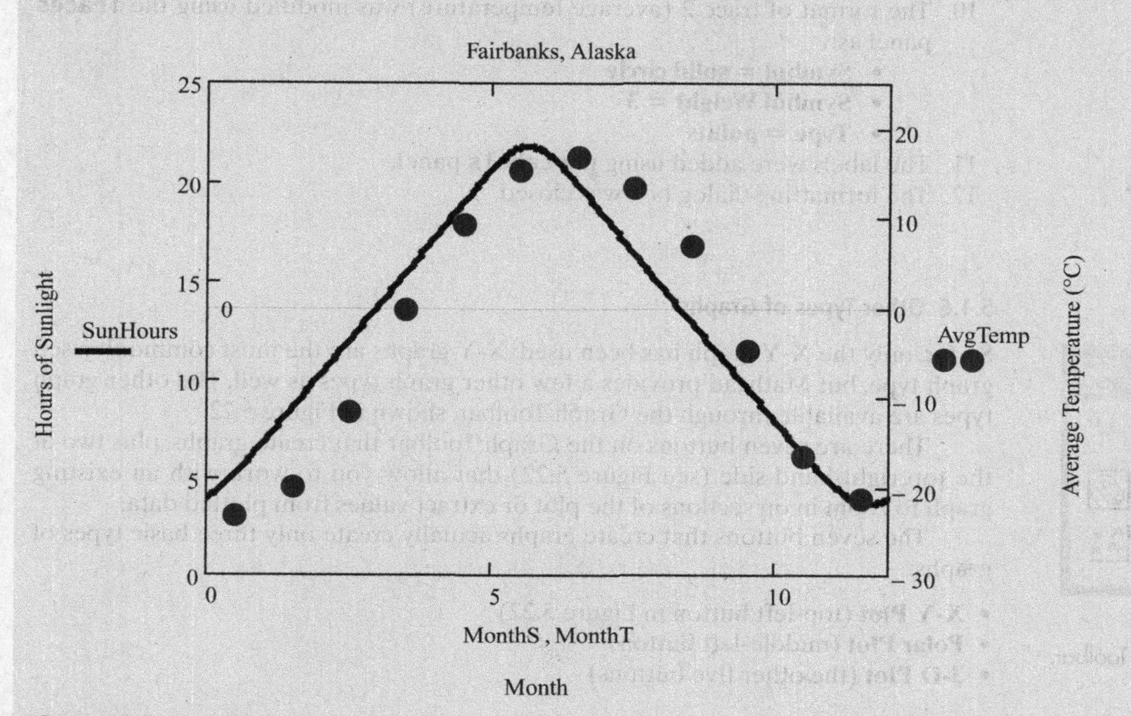

Figure 5.21
Comparing two related quantities using a graph with a secondary y-axis.

By graphing the two data sets on the same plot with a secondary y-axis, it becomes apparent that the temperature curve is shifted slightly to the right of the hours of sunlight curve, suggesting a lag time between the two curves. This in an interesting result that was not immediately apparent when the two curves were plotted separately. (See Figure 5.19.) Using graphs for ***visualization*** can help you learn more from your data.

The graph shown in Figure 5.21 was made as follows:

1. An empty X-Y graph was inserted on the worksheet using [Ctrl 2].
2. Vector **SunHours** was plotted against vector **MonthS** by entering the vector names in the axis placeholders.
3. The formatting dialog box was opened by double-clicking on the graph.
4. The secondary y-axis was enabled using the checkbox on the **X-Y Axes** panel. (See Figure 5.17).
5. The formatting dialog box was closed.
6. A second x-axis placeholder was created by selecting **MonthS** and pressing [comma].
7. **VectorAvgTemp** was plotted against vector **MonthT** by entering the vector names in the axis placeholders.
8. The formatting dialog box was opened (again) by double-clicking on the graph.

9. The format of trace 1 (hours of sunlight) was modified using the **Traces** panel as:
    - **Line Weight = 2**
10. The format of trace 2 (average temperature) was modified using the **Traces** panel as:
    - **Symbol = solid circle**
    - **Symbol Weight = 3**
    - **Type = points**
11. The labels were added using the **Labels** panel.
12. The formatting dialog box was closed.

5.1.6 Other Types of Graphs

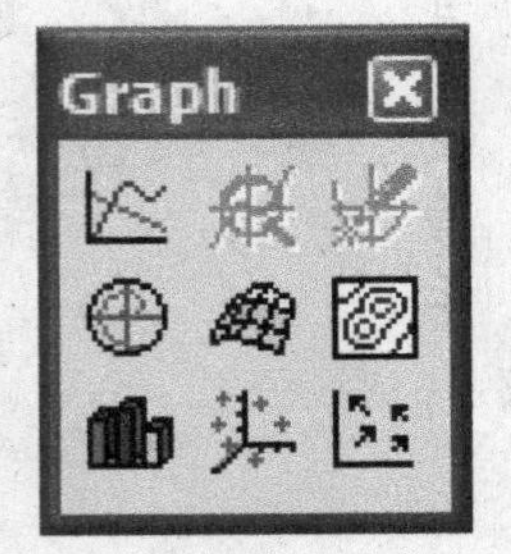

Figure 5.22
Mathcad's Graph Toolbar.

So far, only the X-Y graph has been used; X-Y graphs are the most commonly used graph type, but Mathcad provides a few other graph types as well. The other graph types are available through the Graph Toolbar, shown in Figure 5.22.

There are seven buttons on the Graph Toolbar that create graphs, plus two at the top right-hand side (see Figure 5.22) that allow you to work with an existing graph to zoom in on sections of the plot or extract values from plotted data.

The seven buttons that create graphs actually create only three basic types of graphs:

- **X-Y Plot** (top-left button in Figure 5.22)
- **Polar Plot** (middle-left button)
- **3-D Plot** (the other five buttons)

3-D Plots

The five buttons for 3-D plots all present a 2-D array of values. They just use different graph styles for presenting the same information. To illustrate these plot types, we will show the temperature distribution in the rectangular region shown in Figure 5.23. This might be a metal block with a square opening slightly off center. If water is boiled in the opening and the block is set in ice water, the temperature distribution will be the temperatures in the wall once the system has reached a steady state (thermal equilibrium).

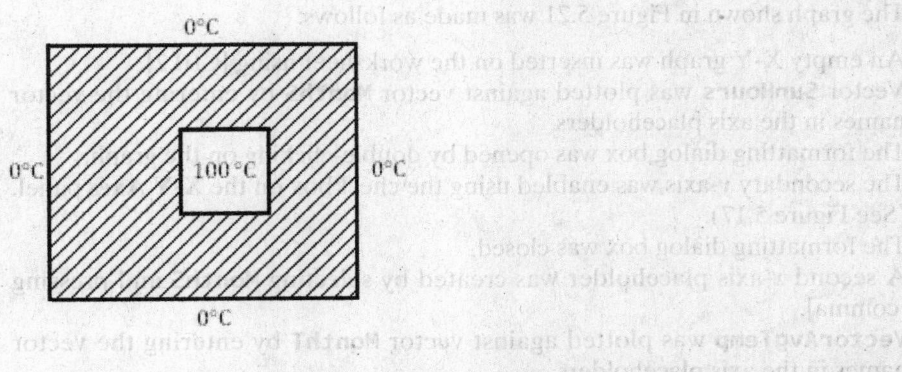

Figure 5.23
The rectangular region used to determine a temperature distribution.

To display these temperatures, a system of grid points is superimposed on the region, as shown in Figure 5.24.

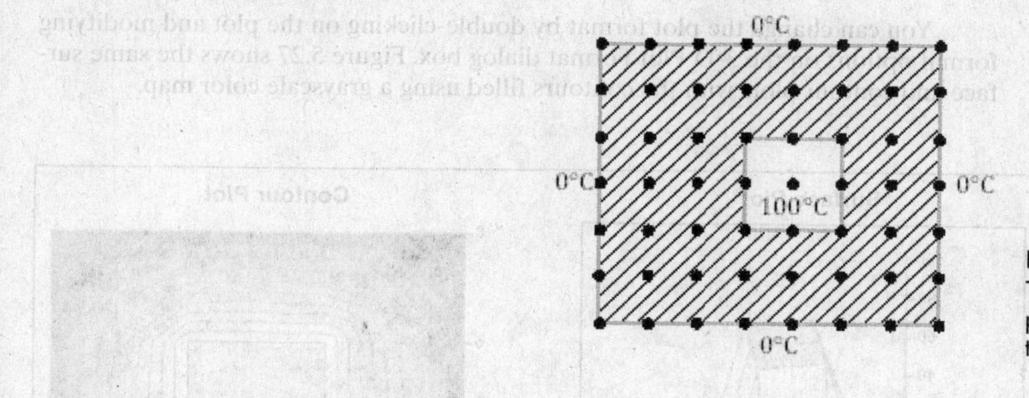

Figure 5.24
Temperatures at each grid point will be displayed on the graph.

The temperature distribution is determined using matrix methods. The values at each grid point are indicated in the **T** matrix, shown in Figure 5.25.

$$T = \begin{pmatrix} 0 & 0 & 0 & 0 & 0 & 0 & 0 & 0 \\ 0 & 12 & 27 & 43 & 46 & 42 & 21 & 0 \\ 0 & 22 & 52 & 100 & 100 & 100 & 42 & 0 \\ 0 & 25 & 57 & 100 & 100 & 100 & 46 & 0 \\ 0 & 22 & 52 & 100 & 100 & 100 & 42 & 0 \\ 0 & 12 & 27 & 43 & 46 & 42 & 21 & 0 \\ 0 & 0 & 0 & 0 & 0 & 0 & 0 & 0 \end{pmatrix}$$

Figure 5.25
The temperatures at each grid point, as matrix **T**.

The 2-D temperature matrix can be plotted using any of the 3-D plots. The surface plot and contour plot are illustrated in Figure 5.26.

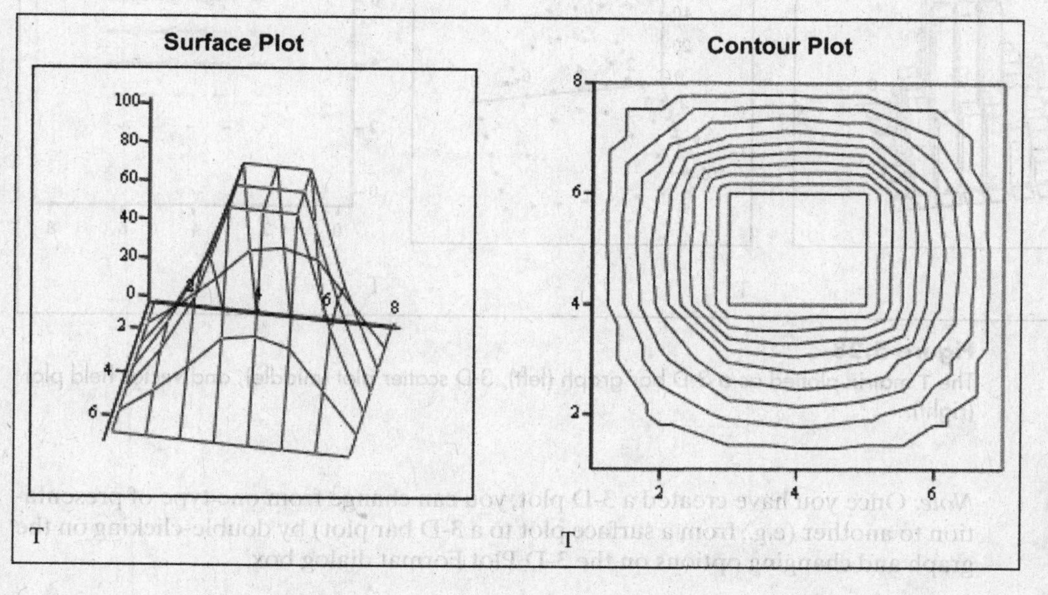

Figure 5.26
Matrix **T** plotted as a surface plot (left) and a contour plot (right).

You can change the plot format by double-clicking on the plot and modifying format options on the 3-D Plot Format dialog box. Figure 5.27 shows the same surface and contour plots with the contours filled using a grayscale color map.

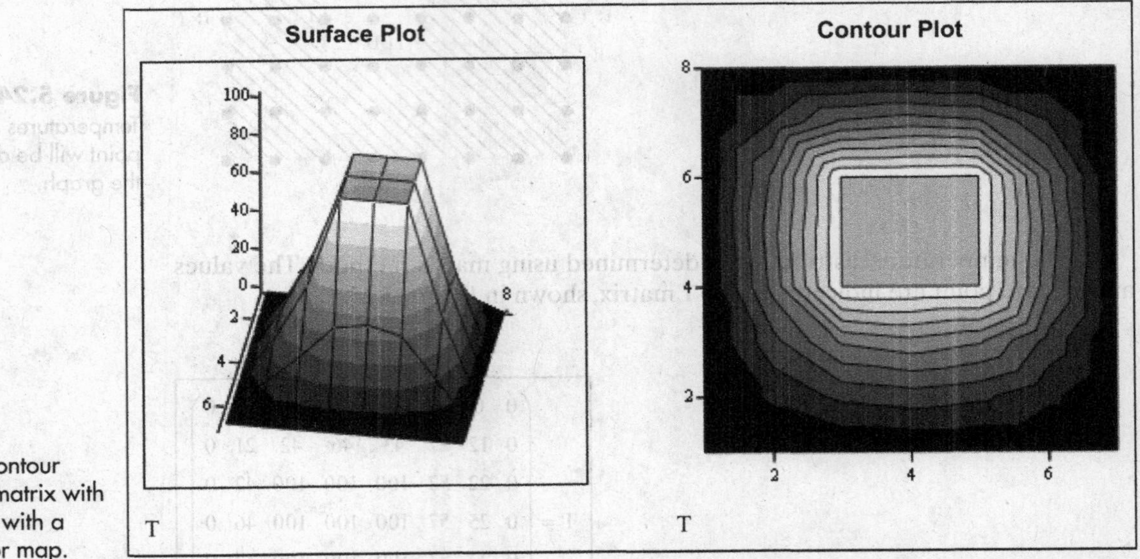

Figure 5.27
Surface and contour plots of the **T** matrix with contours filled with a grayscale color map.

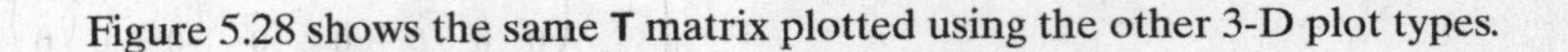

Figure 5.28 shows the same **T** matrix plotted using the other 3-D plot types.

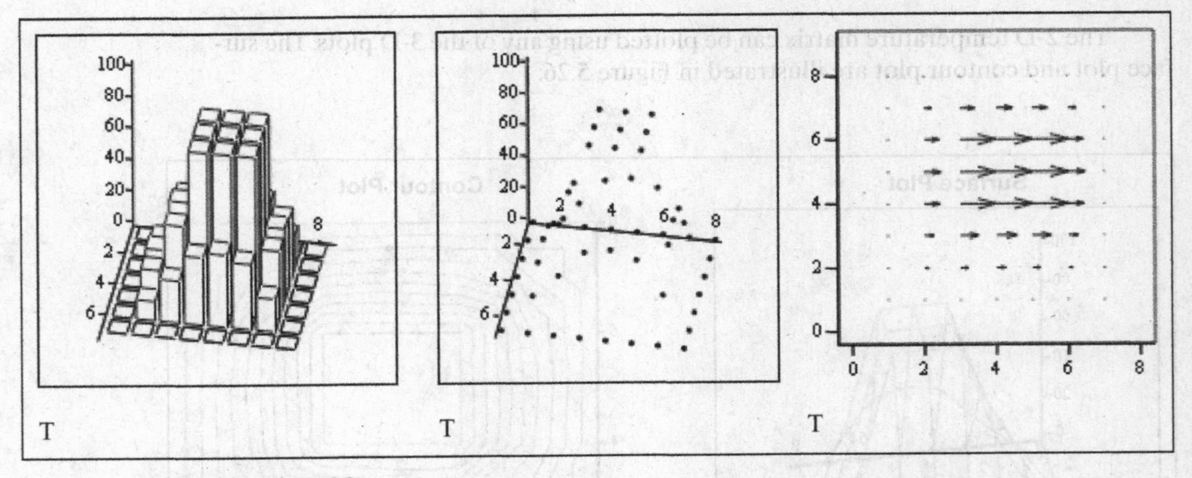

Figure 5.28
The **T** matrix plotted as a 3-D bar graph (left), 3-D scatter plot (middle), and vector field plot (right).

Note: Once you have created a 3-D plot, you can change from one type of presentation to another (e.g., from a surface plot to a 3-D bar plot) by double-clicking on the graph and changing options on the 3-D Plot Format dialog box.

Polar Plots

The polar plot is used to plot functions that vary with angle, such as `sin(θ)` for example. You can use a polar plot as a QuickPlot by entering a dummy variable in the angle placeholder, located at the bottom of the graph, and a function in the function placeholder on the left side of the graph. Figure 5.29 shows a polar plot of `sin(θ)`.

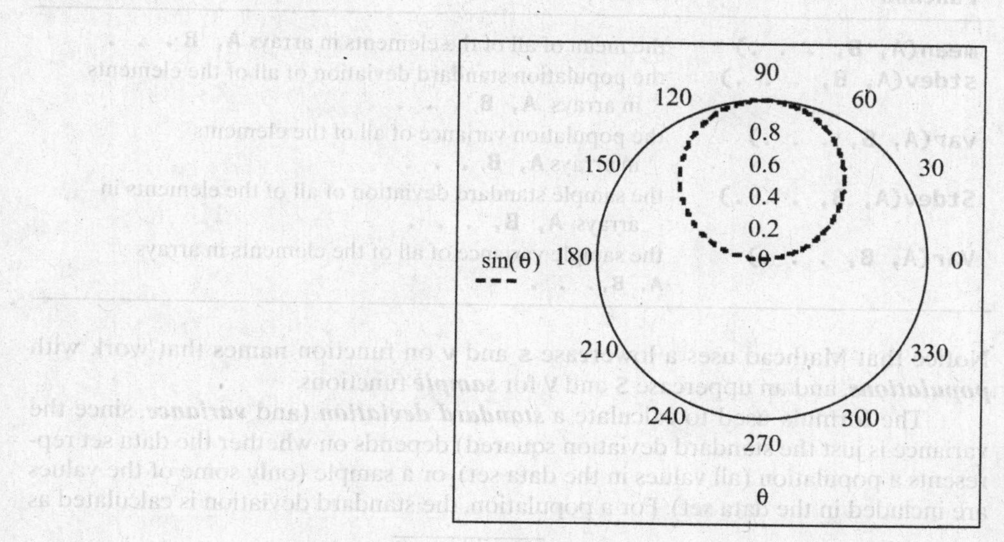

Figure 5.29
A polar QuickPlot of `sin(θ)`.

As usual, you can change the display by double-clicking on the graph and modifying format options on the formatting dialog box.

There is one limitation to using a polar plot as a QuickPlot; the angle is only evaluated for 2π radians (360°). If you need more than one revolution to completely display your function, you will need to use a range variable to define the range of the angle variable. In the example in Figure 5.30, it takes 720° (4π radians) to see the complete curve. This is handled by defining the angle, θ, as a range variable.

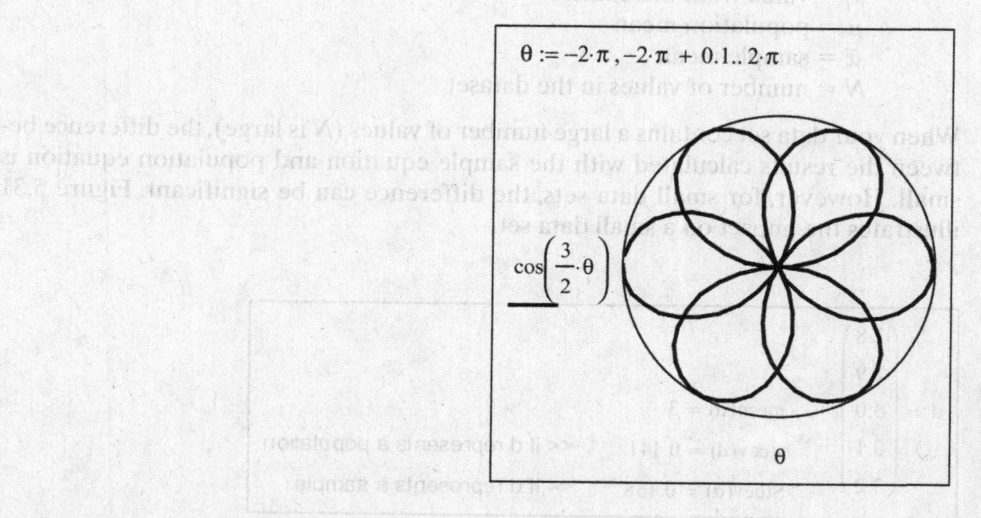

Figure 5.30
Using a range variable to evaluate a function over more than 2π radians.

5.2 STATISTICAL FUNCTIONS

Mathcad provides several functions for commonly used statistical calculations, including

Function	Calculates
`mean(A, B, . . .)`	the mean of all of the elements in arrays **A, B . . .**
`stdev(A, B, . . .)`	the population standard deviation of all of the elements in arrays **A, B, . . .**
`var(A, B, . . .)`	the population variance of all of the elements in arrays **A, B, . . .**
`Stdev(A, B, . . .)`	the sample standard deviation of all of the elements in arrays **A, B, . . .**
`Var(A, B, . . .)`	the sample variance of all of the elements in arrays **A, B,. . .**

Notice that Mathcad uses a lowercase **s** and **v** on function names that work with ***populations***, and an uppercase **S** and **V** for ***sample*** functions.

The formula used to calculate a ***standard deviation*** (and ***variance***, since the variance is just the standard deviation squared) depends on whether the data set represents a population (all values in the data set), or a sample (only some of the values are included in the data set). For a population, the standard deviation is calculated as

$$\sigma = \sqrt{\sum_{i=1}^{N} \frac{(x_i - \mu)^2}{N}}.$$

But for a sample, the equation is

$$s = \sqrt{\sum_{i=1}^{N} \frac{(x_i - \overline{x})^2}{N - 1}},$$

where

σ = population standard deviation
s = sample standard deviation
x_i = value from the dataset
μ = population mean
$\overline{x}$ = sample mean
N = number of values in the dataset

When your data set contains a large number of values (N is large), the difference between the results calculated with the sample equation and population equation is small. However, for small data sets, the difference can be significant. Figure 5.31 illustrates the impact on a small data set.

$$d := \begin{pmatrix} 2.8 \\ 2.9 \\ 3.0 \\ 3.1 \\ 3.2 \end{pmatrix}$$

mean(d) = 3
stdev(d) = 0.141 << if d represents a population
Stdev(d) = 0.158 << if d represents a sample

Figure 5.31
Populations and samples use different standard deviation functions.

EXAMPLE

Samples and Populations

To observe the difference between samples and populations, consider a manufacturing plant that produces windshields.

- Every windshield might be tested for transparency as it leaves the production facility. Since every windshield is being tested, the windshield transparency data set is called a *population*.
- One out of every thousand windshields might be tested for impact strength (i.e., broken), and the results from the tested windshields used to represent the impact strength of all the windshields. Since the impact strength data set does not include all of the windshields, it is called a *sample*.

Any time you use a sample, you need to be careful to try to get a ***representative sample***. The idea behind a representative sample is that the sample results should accurately reflect the population's characteristics. For example, choosing all of the cracked windshields for impact testing would not produce a representative sample. The usual approach to trying to generate a representative sample is choosing a ***random sample***.

The statistical calculation functions work with single or multiple vectors or arrays. In Figure 5.32, two replicate sets of data are first evaluated separately and then combined.

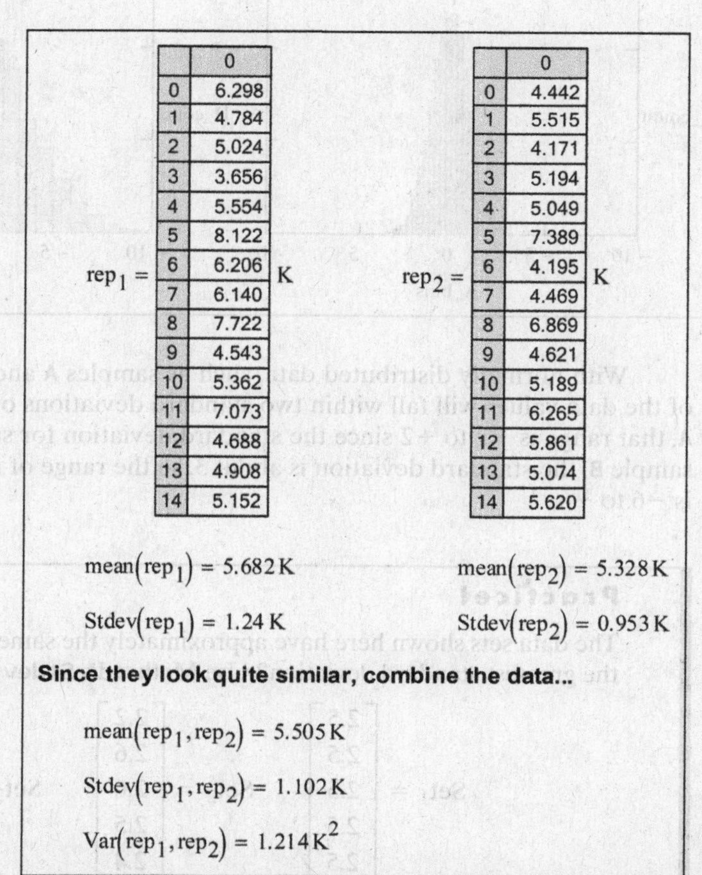

	rep_1 = (K)	rep_2 = (K)
0	6.298	4.442
1	4.784	5.515
2	5.024	4.171
3	3.656	5.194
4	5.554	5.049
5	8.122	7.389
6	6.206	4.195
7	6.140	4.469
8	7.722	6.869
9	4.543	4.621
10	5.362	5.189
11	7.073	6.265
12	4.688	5.861
13	4.908	5.074
14	5.152	5.620

$\text{mean}(rep_1) = 5.682\,\text{K}$ $\text{mean}(rep_2) = 5.328\,\text{K}$

$\text{Stdev}(rep_1) = 1.24\,\text{K}$ $\text{Stdev}(rep_2) = 0.953\,\text{K}$

Since they look quite similar, combine the data...

$\text{mean}(rep_1, rep_2) = 5.505\,\text{K}$

$\text{Stdev}(rep_1, rep_2) = 1.102\,\text{K}$

$\text{Var}(rep_1, rep_2) = 1.214\,\text{K}^2$

Figure 5.32
Comparing replicate data sets.

The example in Figure 5.32 also illustrates that the statistical functions can be used with units.

The standard deviation provides information about the spread of the values in a data set around the ***mean*** value. In the example shown in Figure 5.33, two data sets have been created using the **rnorm** function. These data sets, called samples **A** and **B**, are normally distributed and have a mean of approximately 0, but sample **A** has a standard deviation of about 1, while sample **B** has a standard deviation of about 3. The *x*-axis for the histogram graphs for both sample **A** and sample **B** have the same range (−10 to 10) to illustrate that the data set with the larger standard deviation (sample **B**) has a much broader distribution of values.

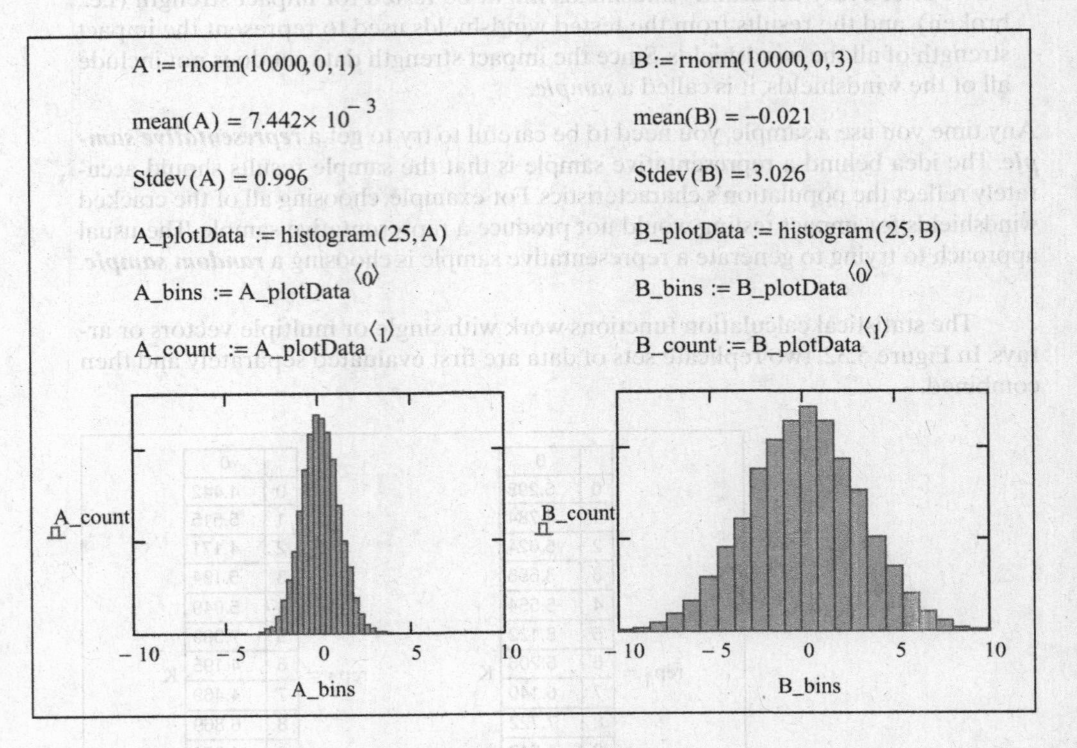

Figure 5.33
Comparing data sets with differing standard deviations.

With normally distributed data, such as samples **A** and **B** in Figure 5.33, 95% of the data values will fall within two standard deviations of the mean. For sample **A**, that range is −2 to +2 since the standard deviation for sample **A** is about 1. For sample **B**, the standard deviation is about 3, so the range of most of the data values is −6 to +6.

Practice!

The data sets shown here have approximately the same mean value. Which has the greatest standard deviation? Use Mathcad's **Stdev** function to find out.

$$Set_1 = \begin{bmatrix} 2.5 \\ 2.5 \\ 2.5 \\ 2.5 \\ 2.5 \end{bmatrix} \quad Set_2 = \begin{bmatrix} 2.2 \\ 2.6 \\ 2.6 \\ 2.5 \\ 2.4 \end{bmatrix} \quad Set_3 = \begin{bmatrix} 1.6 \\ 1.6 \\ 2.8 \\ 2.1 \\ 3.8 \end{bmatrix}$$

APPLICATION

Checking a Distribution

The apparent lack of reliability in some pH readings has caused a dispute between two researchers. Researcher A believes the problem is a faulty pH meter, and the solution is to buy a new meter. Researcher B suspects the problem is a poorly designed experimental system that is causing the acid concentrations to vary, and that the pH meter is accurately measuring the randomly fluctuating concentrations. If this is true, the solution is to rebuild the experimental apparatus.

Researcher B thinks he can use information on the distributions of pH and concentration to determine the source of the problem. Since there is a logarithm relating hydrogen ion concentration to pH,

$$pH = -\log(H^+)$$

the distributions of the two variables will not both be normally distributed. He hypothesizes that if the problem is in the electronics of the meter, the pH meter readings are probably normally distributed. However, if the problem is randomly fluctuating acid concentrations, then the concentrations are probably normally distributed, and the pH readings will not be normally distributed.

Researcher B comes in on a Saturday and runs the system for eight hours after adding enough acid to lower the system pH to 3. The data acquisition system is set to record a pH reading every 10 seconds, so at the end of the experiment, he has a collection of 2,880 pH readings.

To check both distributions, he first rearranges the definition of pH in order to calculate the hydrogen ion concentration

$$H^+ = 10^{-pH}.$$

Then he uses the **histogram** function to graph the distributions of both the pH values and the concentration values. His Mathcad worksheet is shown in Figure 5.34.

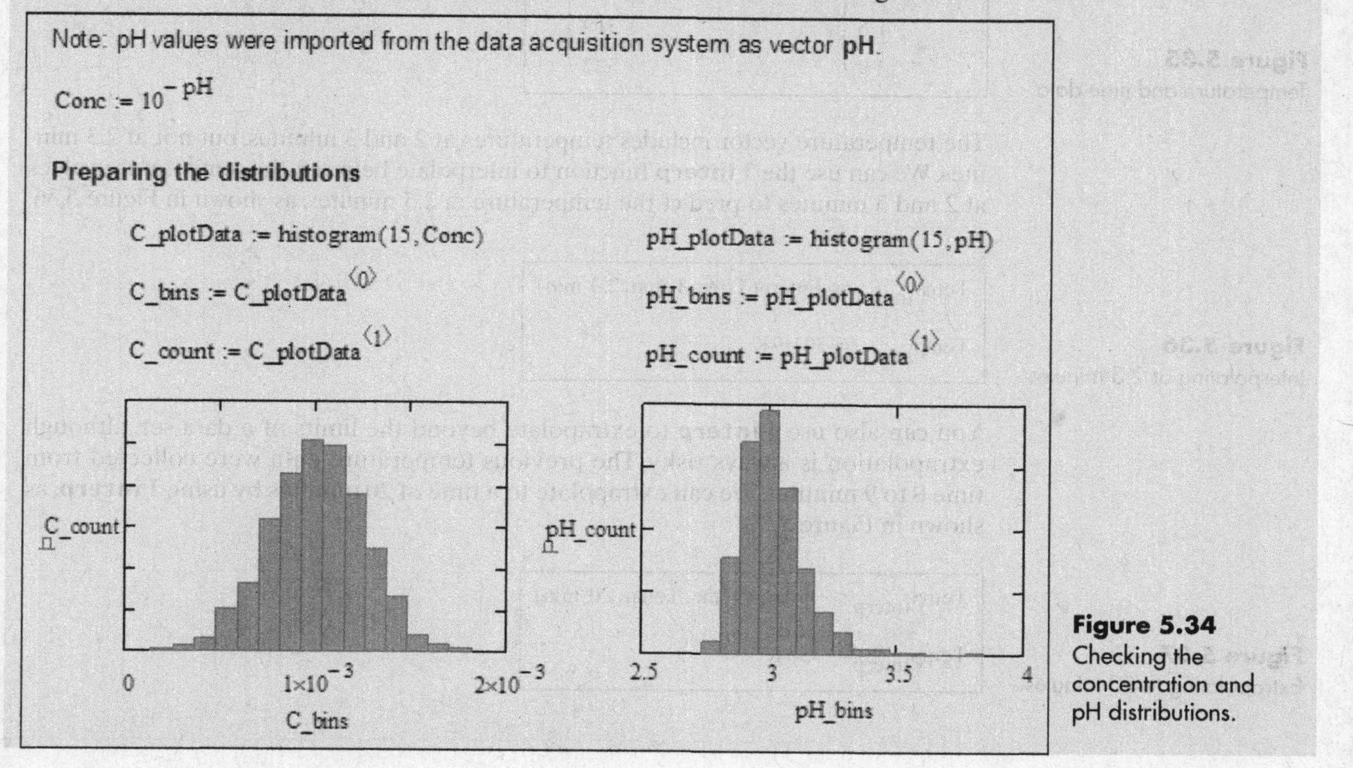

Figure 5.34 Checking the concentration and pH distributions.

The results make it clear that the pH values are not normally distributed, but that the concentration appears to be normally distributed. It looks as though the noise is in the concentration values. Researcher B makes plans to start rebuilding the experiment on Monday.

5.3 INTERPOLATION

If you have a data set—for example, a set of temperature values at various times—and you want to predict a temperature at a new time, you could fit a function to the data and calculate the predicted temperature at the new time. Or you can interpolate between data set values. Mathcad provides a number of functions for linear ***interpolation*** and ***cubic spline interpolation***. The ***linear interpolation*** function, **linterp(vx, vy, x_{new})**, predicts a new **y** value at the new **x** value (specified in the function's argument list) by using a linear interpolation and by using the **x** values in the data set nearest to the new **x** value.

As an example, consider the temperature and time data shown in Figure 5.35.

$$\text{Time} := \begin{pmatrix} 0 \\ 1 \\ 2 \\ 3 \\ 4 \\ 5 \\ 6 \\ 7 \\ 8 \\ 9 \end{pmatrix} \cdot \text{min} \qquad \text{Temp} := \begin{pmatrix} 298 \\ 299 \\ 301 \\ 304 \\ 306 \\ 309 \\ 312 \\ 316 \\ 319 \\ 322 \end{pmatrix} \cdot \text{K}$$

Figure 5.35
Temperature and time data.

The temperature vector includes temperatures at 2 and 3 minutes, but not at 2.3 minutes. We can use the **linterp** function to interpolate between the temperature values at 2 and 3 minutes to predict the temperature at 2.3 minutes, as shown in Figure 5.36.

$$\text{Temp}_{\text{interp}} := \text{linterp}(\text{Time}, \text{Temp}, 2.3 \cdot \text{min})$$

$$\text{Temp}_{\text{interp}} = 301.9\,\text{K}$$

Figure 5.36
Interpolating at 2.3 minutes.

You can also use **linterp** to extrapolate beyond the limits of a data set, although extrapolation is always risky. The previous temperature data were collected from time 0 to 9 minutes. We can extrapolate to a time of 20 minutes by using **linterp**, as shown in Figure 5.37.

$$\text{Temp}_{\text{interp}} := \text{linterp}(\text{Time}, \text{Temp}, 20 \cdot \text{min})$$

$$\text{Temp}_{\text{interp}} = 355\,\text{K}$$

Figure 5.37
Extrapolating to 20 minutes.

Extrapolation will give you a result, but there is no guarantee that it is correct. Since the result is outside the range of the data set, the result is uncertain. For this data set, the heater might have been turned off after 9 minutes, and the temperatures might have started to decrease with time. Because the data set includes only temperatures between 0 and 9 minutes, we cannot know what happened at later times.

An alternative to linear interpolation is *cubic spline interpolation*. This technique puts a curve (a cubic polynomial) through the data points, which results in a curve that passes through each data point with continuous first and second derivatives. Interpolation for values between points can be done using the cubic polynomial. To use cubic spline interpolation in Mathcad, you first need to fit the cubic polynomial to the data by using the **cspline(vx, vy)** function. This function returns a vector of second-derivative values, *vs*, that is then used in the interpolation, which is performed with the **interp(vs, vx, vy, x_{new})** function. This process is illustrated in Figure 5.38.

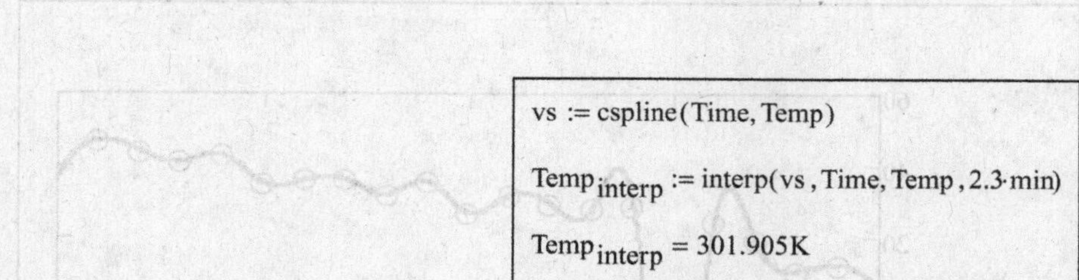

vs := cspline(Time, Temp)

$Temp_{interp}$:= interp(vs, Time, Temp, 2.3·min)

$Temp_{interp}$ = 301.905K

Figure 5.38
Cubic spline interpolation.

You can also extrapolate with the cubic spline, but doing so is risky. Because the cubic spline technique fits three adjacent points with a cubic polynomial, it runs into trouble at each end of the data set since the first and last points do not have another point on each side. Hence, the cubic spline method must do something special for these endpoints. Mathcad provides three methods for handling the endpoints:

- **cspline**—creates a spline curve that is cubic at the endpoints.
- **pspline**—creates a spline curve that is parabolic at the endpoints.
- **lspline**—creates a spline curve that is linear at the endpoints.

These spline functions will all yield the same interpolated results for interior points but can give widely varying results if you extrapolate your data set. This is demonstrated in Figure 5.39.

vs := cspline(Time, Temp)

interp(vs, Time, Temp, 20·min) = 949.695K

vs := pspline(Time, Temp)

interp(vs, Time, Temp, 20·min) = 384.003K

vs := lspline(Time, Temp)

interp(vs, Time, Temp, 20·min) = 232.425K

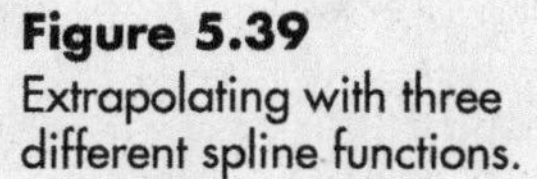

Figure 5.39
Extrapolating with three different spline functions.

CAUTIONS

1. Extrapolation (predicting values outside the range of the data set) should be avoided whenever possible. The temperatures used in the foregoing example were recorded for times ranging from 0 to 9 minutes. Predicting a temperature at 9.2 minutes is safer than predicting one at 90 or 900 minutes, but you are still making an assumption that nothing changed between 9.0 and 9.2 minutes. (The researcher might have turned off the heater, and the temperatures might have started falling.)
2. A cubic spline fitted to noisy data can produce some amazing results. In Figure 5.40, an outlier at $x = 5$ was included in the data set to demonstrate how this single bad point affects the spline curve for adjacent points as well. Interpolation using these curves is a very bad idea. Regression to find the "best fit" curve and then using the regression result to predict new values is a better idea for noisy data.

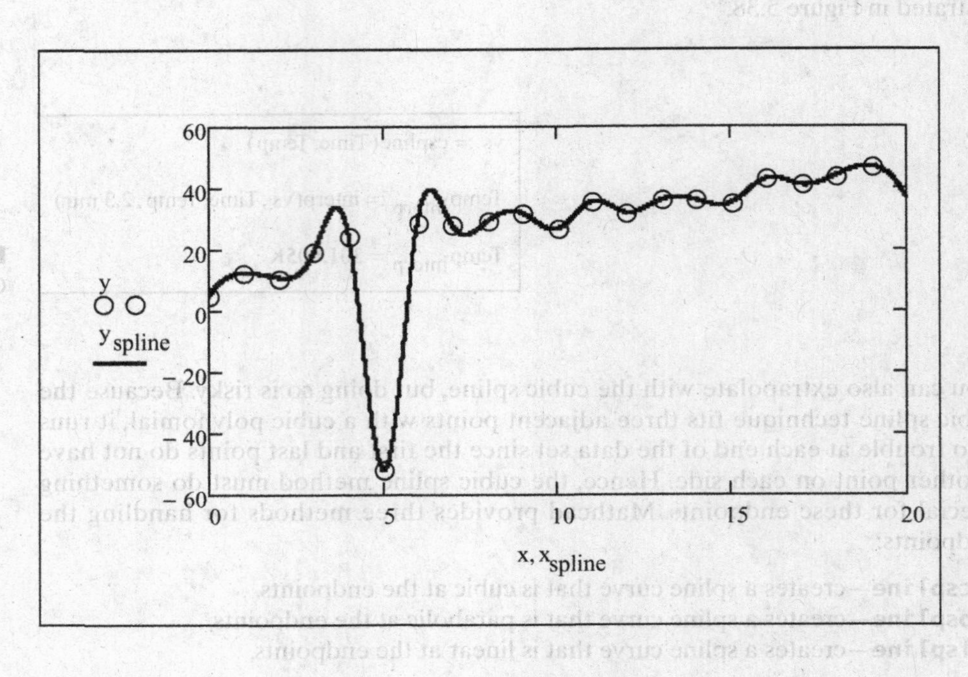

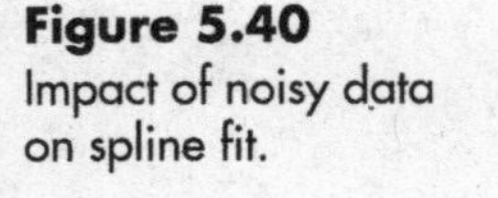

Figure 5.40
Impact of noisy data on spline fit.

Using a QuickPlot to Plot the Spline Curve

If you want to see what the spline curve looks like, Mathcad's QuickPlot feature will allow you to see the curve with little effort. Since a QuickPlot will evaluate a function multiple times over a range of values, simply let Mathcad evaluate the `interp` function many times and display the result. For the temperature–time data presented in Figure 5.35, this means calculating the second-derivative values by using `cspline` (or `pspline` or `lspline`) and then creating the QuickPlot.

In Figure 5.41, the t in the `interp` function and on the x-axis is a dummy variable. Mathcad evaluates the `interp` function for values of t between -10 and 10 (by default) and displays the result. We need to change the limits on the x-axis to coincide with the time values in the data set, 0 to 9 minutes. To change the axis limits, click on the x-axis and then edit the limit values. The result is shown in Figure 5.42. (Note that units have been removed from the *Time* and *Temp* vectors in Figure 5.42.)

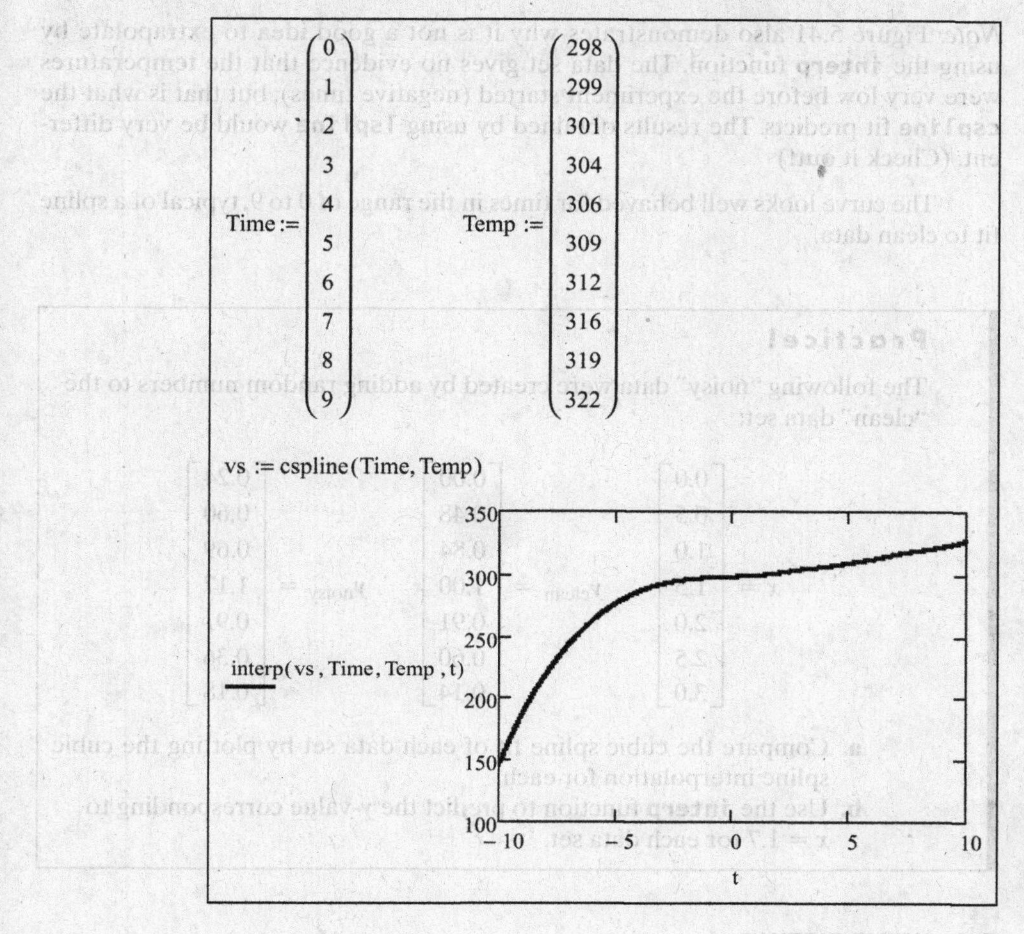

Figure 5.41
By default, a QuickPlot evaluates between *x*-axis values of −10 and 10.

vs := cspline(Time, Temp)

interp(vs, Time, Temp, t)

330 320 310 300 290

0 1 2 3 4 5 6 7 8 9

t

Figure 5.42
The spline fit QuickPlot with corrected *x*-axis limits.

Note: Figure 5.41 also demonstrates why it is not a good idea to extrapolate by using the **interp** function. The data set gives no evidence that the temperatures were very low before the experiment started (negative times), but that is what the **cspline** fit predicts. The results obtained by using **lspline** would be very different. (Check it out!)

The curve looks well behaved for times in the range of 0 to 9, typical of a spline fit to clean data.

Practice!

The following "noisy" data were created by adding random numbers to the "clean" data set:

$$x = \begin{bmatrix} 0.0 \\ 0.5 \\ 1.0 \\ 1.5 \\ 2.0 \\ 2.5 \\ 3.0 \end{bmatrix} \quad y_{\text{clean}} = \begin{bmatrix} 0.00 \\ 0.48 \\ 0.84 \\ 1.00 \\ 0.91 \\ 0.60 \\ 0.14 \end{bmatrix} \quad y_{\text{noisy}} = \begin{bmatrix} 0.24 \\ 0.60 \\ 0.69 \\ 1.17 \\ 0.91 \\ 0.36 \\ 0.18 \end{bmatrix}$$

a. Compare the cubic spline fit of each data set by plotting the cubic spline interpolation for each.
b. Use the **interp** function to predict the y-value corresponding to $x = 1.7$ for each data set.

5.4 CURVE FITTING

Simple Linear Regression

Mathcad provides a number of functions to fit curves to data and to use fitted curves to predict new values. One of the most common curve-fitting applications is ***linear regression.*** Simple linear regression is carried out in Mathcad by using two or three of the following functions: **slope(vx, vy)**, **intercept(vx, vy)**, and **corr(vx, vy)**. These functions respectively return the ***slope, intercept,*** and ***correlation coefficient*** of the best-fit straight line through the data represented by the x- and y-vectors. For example, for the time–temperature data shown in Figure 5.43.

The temperature values predicted by the model equation can be calculated from the slope and intercept values by using a range variable to keep track of the times. Then the quality of the fit can be determined by using the **corr** function. The results are shown in Figure 5.44.

Here, the correlation coefficient was squared to calculate the ***coefficient of determination*** (usually called the R^2 value) for the regression. An R^2 value of 1 implies that the regression line is a perfect fit to the data. The value 0.98866 suggests that the regression line is a pretty good fit to the data, but it is always a good idea to plot the original data together with the regression line. The result is shown in Figure 5.45.

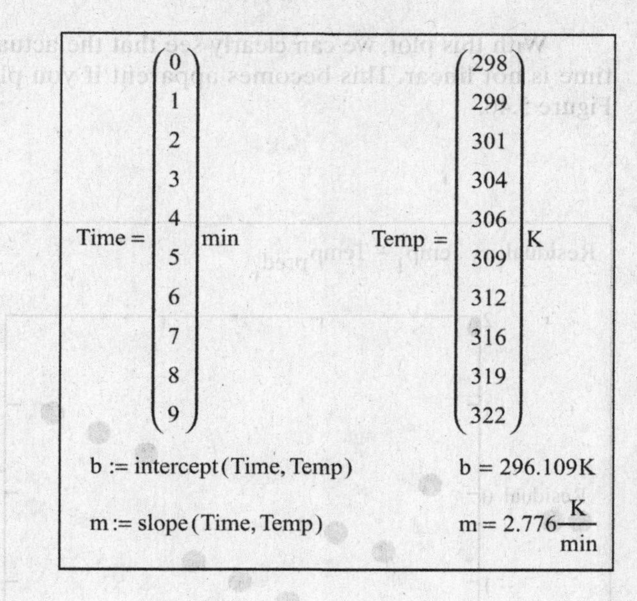

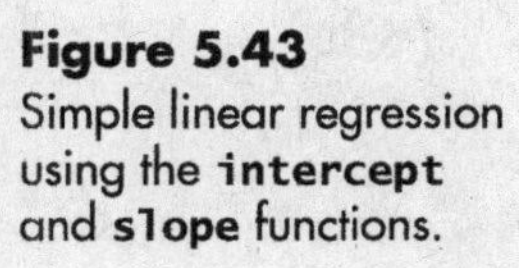

Figure 5.43
Simple linear regression using the **`intercept`** and **`slope`** functions.

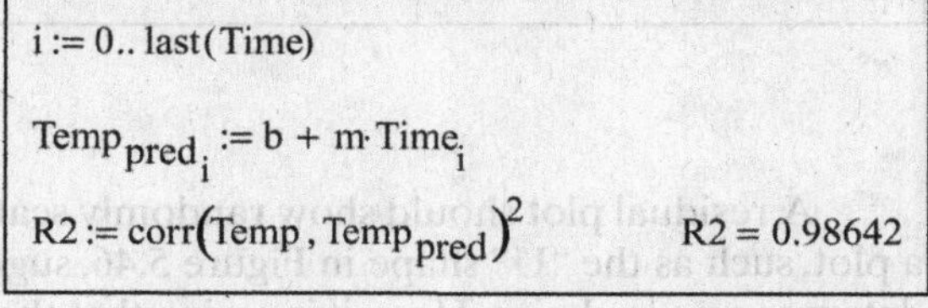

Figure 5.44
Using the slope, *m*, and intercept, *b*, to calculate predicted temperatures, in turn, to find the R^2 value.

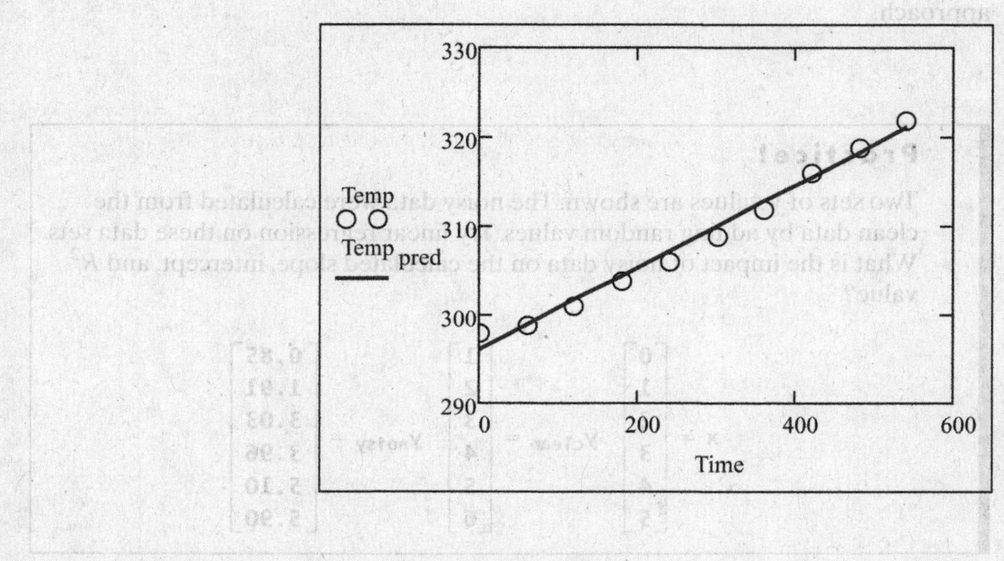

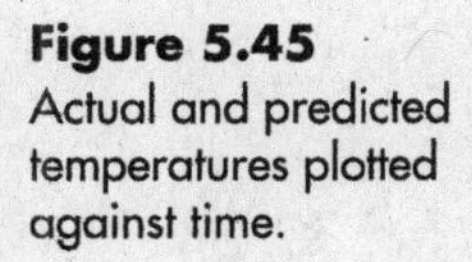

Figure 5.45
Actual and predicted temperatures plotted against time.

With this plot, we can clearly see that the actual curve of temperature versus time is not linear. This becomes apparent if you plot the ***residuals***, as shown in Figure 5.46.

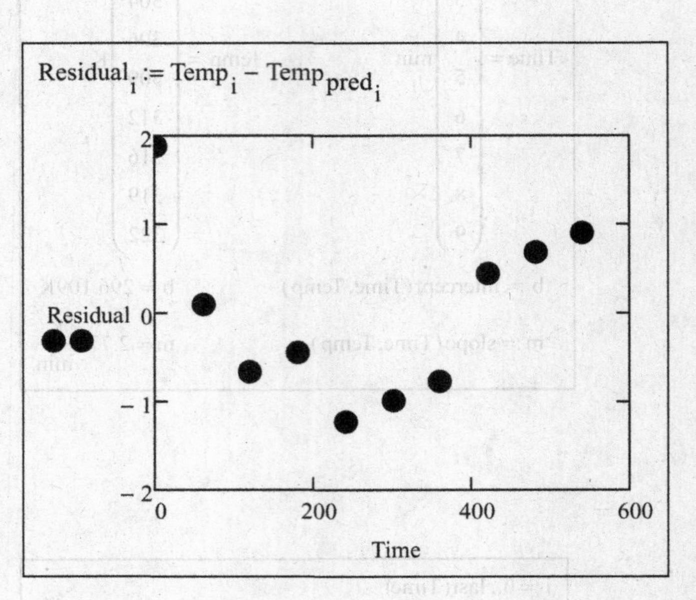

Figure 5.46
The residual plot shows that a straight line is not the best way to fit this data.

A residual plot should show randomly scattered points. Obvious trends in such a plot, such as the "U" shape in Figure 5.46, suggest that the equation used to fit the data was a poor choice. Here, it is saying that the linear equation is not a good choice for fitting these nonlinear data.

The foregoing example was included as a reminder that, while you can perform a simple (straight-line) linear regression on any data set, it is not always a good idea to do so. Nonlinear data require a more sophisticated curve-fitting approach.

Practice!

Two sets of *y* values are shown. The noisy data were calculated from the clean data by adding random values. Try linear regression on these data sets. What is the impact of noisy data on the calculated slope, intercept, and R^2 value?

$$x = \begin{bmatrix} 0 \\ 1 \\ 2 \\ 3 \\ 4 \\ 5 \end{bmatrix} \quad y_{clean} = \begin{bmatrix} 1 \\ 2 \\ 3 \\ 4 \\ 5 \\ 6 \end{bmatrix} \quad y_{noisy} = \begin{bmatrix} 0.85 \\ 1.91 \\ 3.03 \\ 3.96 \\ 5.10 \\ 5.90 \end{bmatrix}$$

Professional Success

Visualize your data and results.

Numbers, such as those in the following data set, are a handy way to store information:

X	Y	X	Y
−1.0000	−0.5478	0.1367	0.0004
−0.9900	−0.4252	0.1543	0.3042
−0.9626	−0.3277	0.2837	0.5407
−0.9577	−0.7865	0.4081	0.0122
−0.9111	−0.2327	0.4242	0.7593
−0.8391	−0.9654	0.5403	0.0304
−0.7597	−0.1030	0.6469	0.0565
−0.7481	−0.9997	0.6603	0.9821
−0.6536	−0.9791	0.7539	0.0999
−0.5477	−0.0319	0.8439	0.9617
−0.5328	−0.8888	0.9074	0.2277
−0.4161	−0.0130	0.9147	0.8781
−0.2921	−0.0043	0.9602	0.7795
−0.2752	−0.5260	0.9887	0.4184
−0.1455	−0.0005	0.9912	0.6536
−0.0044	−0.0088	1.0000	0.5403

For most of us, however, a data set is hard to visualize. A quick glance at the data suggests that the y values get bigger as the x values get bigger. To try to describe these data, linear regression could be performed to calculate a slope and an intercept as shown in Figure 5.47. But by graphing the data, it becomes clear that simple linear regression is not appropriate for this data set.

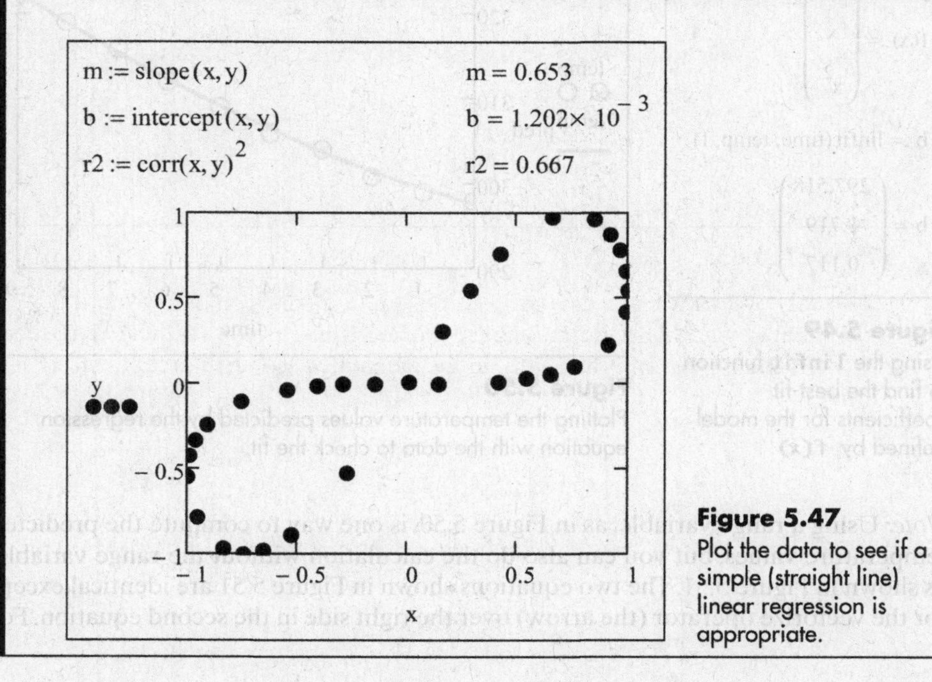

Figure 5.47
Plot the data to see if a simple (straight line) linear regression is appropriate.

Generalized Linear Regression

The `linfit(vx, vy, vf)` function performs a linear regression with an arbitrary linear model. For example, we might try to improve the fit of the regression line to the earlier temperature–time data by using a second-order polynomial, such as

$$\mathbf{Temp_{pred}} = \mathbf{b_0} + \mathbf{b_1 time} + \mathbf{b_2 time^2}$$

The `linfit` function will find the coefficients $\mathbf{b_0}$, $\mathbf{b_1}$, and $\mathbf{b_2}$ that best fit the model to the data. However, the `linfit` function does not handle units, so we first remove the units from the `Time` and `Temp` vectors, as shown in Figure 5.48:

Figure 5.48
Removing units from the `Time` and `Temp` vectors in preparation for using the `linfit` function.

$$time := \frac{Time}{min} \qquad temp := \frac{Temp}{K}$$

Then, we define the linear model. The second-order polynomial has three terms: a constant, a term with *time* raised to the first power, and a term with *time* raised to the second power. We define this functionality in the vector function **f** and then perform the regression by using the `linfit` function. The coefficients computed by `linfit` are stored in the vector **b**. This is illustrated in Figure 5.49.

We can then use the coefficients with the second-order polynomial to compute predicted temperature values at each time, as shown in Figure 5.50.

$$f(x) := \begin{pmatrix} 1 \\ x \\ x^2 \end{pmatrix}$$

$$b := linfit(time, temp, f)$$

$$b = \begin{pmatrix} 297.518 \\ 1.719 \\ 0.117 \end{pmatrix}$$

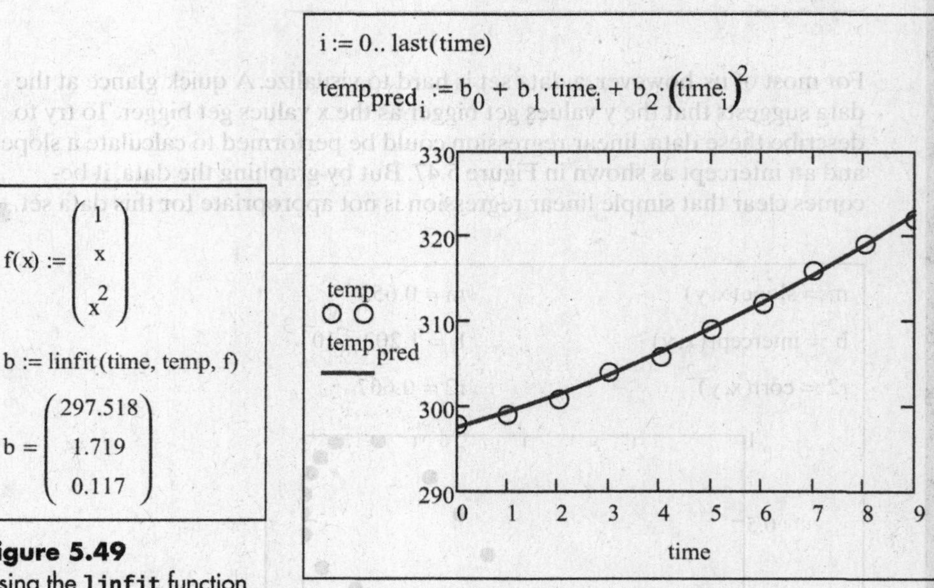

Figure 5.49
Using the `linfit` function to find the best-fit coefficients for the model defined by `f(x)`.

$$i := 0..\ last(time)$$

$$temp_{pred_i} := b_0 + b_1 \cdot time_i + b_2 \cdot (time_i)^2$$

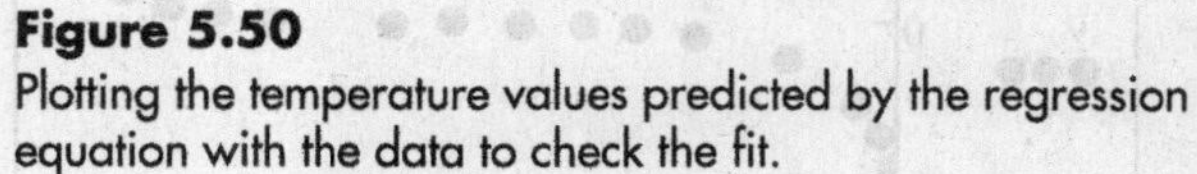

Figure 5.50
Plotting the temperature values predicted by the regression equation with the data to check the fit.

Note: Using a range variable, as in Figure 5.50, is one way to compute the predicted temperature values, but you can also do the calculation without the range variable, as shown in Figure 5.51. The two equations shown in Figure 5.51 are identical except for the vectorize operator (the arrow) over the right side in the second equation. For

$temp_{pred} := b_0 + b_1 \cdot time + b_2 \cdot time^2$

$temp_{pred} := \overrightarrow{\left(b_0 + b_1 \cdot time + b_2 \cdot time^2\right)}$

Figure 5.51
Two ways to calculate the predicted temperature values.

this straightforward calculation the vectorize operator is not needed, but for more complex regression equations it might be required.

After computing predicted values, the graph was produced by plotting the resulting temperature vector against the associated time vector.

As Figure 5.50 shows, the second-order polynomial appears to do a much better job of fitting the data than the linear model does. We can quantify the "goodness of fit" using the **corr** function as shown in Figure 5.52.

$r2 := corr(temp, temp_{pred})$

$r2 = 0.99885781$

Figure 5.52
Calculating the coefficient of correlation, R^2.

The R^2 value is much closer to 1 than the R^2 we obtained by using simple slope–intercept linear regression, indicating that the polynomial is a better fit than the straight line obtained via simple linear regression.

Practice!

Look at the following data and decide whether or not to include an intercept in the regression model:

$$X = \begin{bmatrix} 0.0 \\ 0.5 \\ 1.0 \\ 1.5 \\ 2.0 \\ 2.5 \\ 3.0 \end{bmatrix} \quad Y_{clean} = \begin{bmatrix} 0.00 \\ 0.48 \\ 0.84 \\ 1.00 \\ 0.91 \\ 0.60 \\ 0.14 \end{bmatrix}$$

Now try using the following polynomial models to fit the data with the **linfit** function:

a. $y_p = b_0 + b_1x + b_2x^2$ *or,* $y_p = b_0x + b_1x^2$

b. $y_p = b_0 + b_1x + b_2x^2 + b_3x^3$ *or,* $y_p = b_0x + b_1x^2 + b_2x^3$

Other Linear Models

The models used in the earlier examples, namely,

$$temp_{pred} = b + m \cdot time$$

and

$$temp_{pred} = b_0 + b_1 \cdot time + b_2 \cdot time^2$$

are both linear models (linear in the coefficients, not in **time**). Since **linfit** works with any linear model, you could try fitting an equation such as

$$temp_{pred} = b_0 + b_1 \cdot \sinh(time) + b_2 \cdot \operatorname{atan}(time^2)$$

or

$$\text{temp}_{\text{pred}} = b_0 \cdot \exp(\text{time}^{0.5}) + b_1 \cdot \ln(\text{time}^3)$$

These equations are linear in the coefficients (the **b**'s) and are compatible with `linfit`. The `f(x)` functions for the respective equations would look like this:

$$f(x) := \begin{bmatrix} 1 \\ \sinh(x) \\ \text{atan}(x) \end{bmatrix} \qquad f(x) := \begin{bmatrix} \exp(x^{0.5}) \\ x^3 \end{bmatrix}$$

There is no reason to suspect that either of these last two models would be a good fit to the temperature–time data. In general, you choose a linear model either from some theory that suggests a relationship between your variables or from looking at a plot of the data set.

Practice!

A plot of the data used in the spline fit and polynomial regression "Practice!" boxes has a shape something like half a sine wave. Try fitting the data with a linear model such as

a. $y_p = b_0 + b_1 \sin(x)$

b. $y_p = b_0 + b_1 \cos(x)$

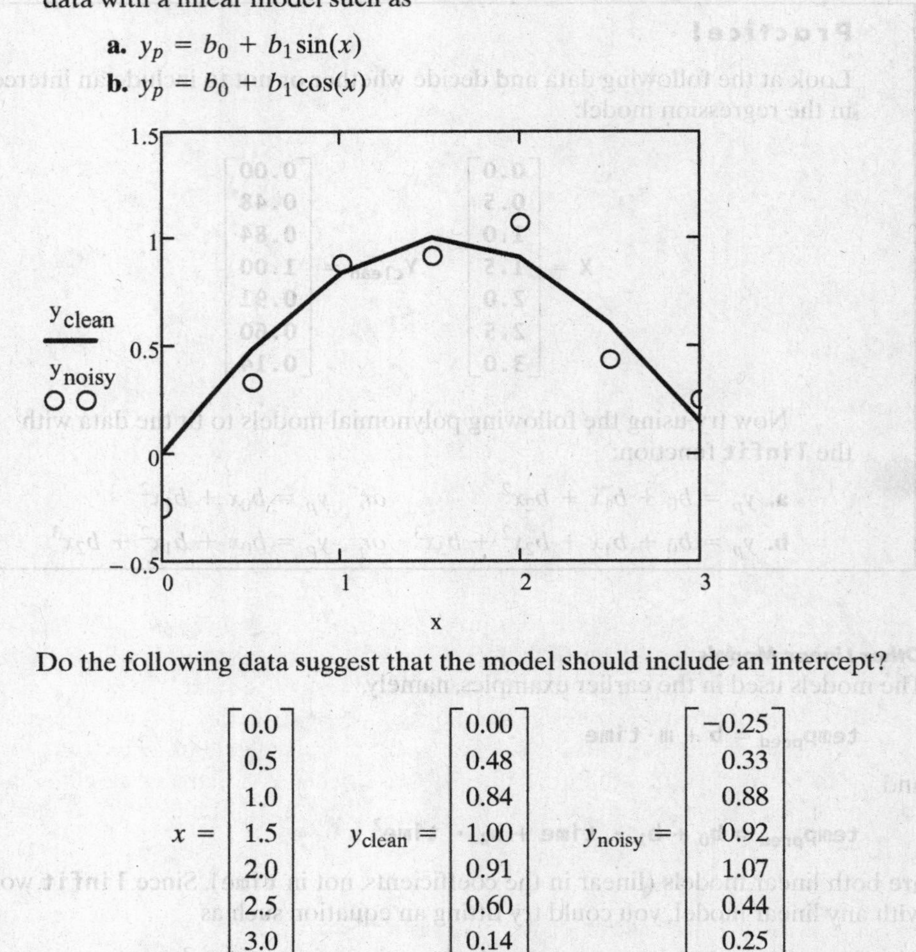

Do the following data suggest that the model should include an intercept?

x	y_{clean}	y_{noisy}
0.0	0.00	−0.25
0.5	0.48	0.33
1.0	0.84	0.88
1.5	1.00	0.92
2.0	0.91	1.07
2.5	0.60	0.44
3.0	0.14	0.25

APPLICATION

Fitting Physical Property Data

Because engineers work on projects that can affect large numbers of people, safety and risk management are a standard practice. Understanding the potential risks associated with a process or product requires a thorough knowledge of the process or product, as well as the ability and willingness to consider what might go wrong.

Silane is an interesting chemical that is being increasingly employed in the manufacture of silicon wafers and chips used in the electronics industry. It is highly flammable and, under the right conditions, will ignite spontaneously upon contact with air. The manufacturers of silane must take special precautions to minimize the risks associated with this gas.

The following table gives the vapor pressure of silane, in pascals, at various temperatures, in kelvins:

Silane Temp. (K)	Vapor Pressure, P_{VAPOR} (Pa)
88.48	21
97.54	167
106.60	640
115.66	2227
124.72	6110
133.79	14144
142.85	30318
151.91	57430
160.97	100078
170.03	163913
179.09	254015
188.15	378009
197.21	541607
206.27	754829
215.33	1027536
224.40	1372259
233.46	1804187
242.52	2341104
251.58	3006269
260.64	3827011
269.70	4838992

Analyzing the risks associated with possible accidents is an important part of a chemical facility's safety program and a big part of some chemical engineers' jobs. There are computer programs to help perform risk analyses, but they require some knowledge of the chemical, physical, and biomedical properties of the chemicals involved. *Risk analysis* programs have been written to use standardized fitting equations, so if you add your own information, it must be in a standard form. The accompanying silane vapor pressure data were calculated by using the ChemCad Physical Properties Database.[1]

[1]ChemCad is a product of Chemstations, Inc., Houston, TX.

STANDARD FITTING EQUATION FOR VAPOR PRESSURE[2]

The standard equation for fitting a curve to vapor pressure p^* is

$$\ln(p^*) = a + \frac{b}{T} + c\ln(T) + dT^*$$

You must specify a and b; a, b, and c; or a, b, c, and d.

To try fitting these data by using the first three terms (a, b, and c), we would use the **linfit** function, with the **f** vector defined as

$$\mathbf{f(x)} := \begin{bmatrix} 1 \\ \frac{1}{x} \\ \ln(x) \end{bmatrix}$$

The coefficients themselves would be computed by **linfit** as shown in Figure 5.53.

Figure 5.53 shows a plot of the original data and the predicted vapor pressures by using the computed coefficients, just to make sure that the process worked and to verify that the last term is not required.

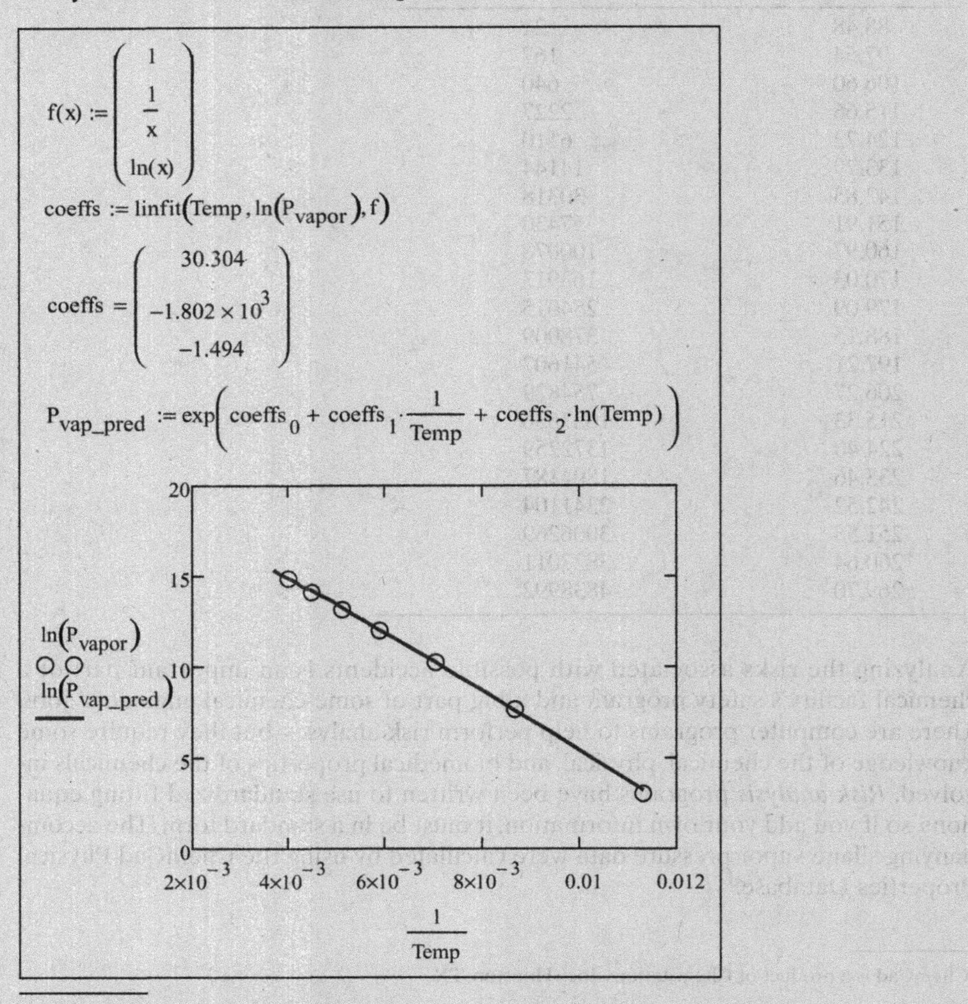

Figure 5.53
Regressing vapor pressure data

[2] *ChemCad User Guide*, version III, p. 11.64.

Specialized Regression Equations

Mathcad provides functions for finding coefficients for a number of commonly used fitting equations. These functions use iterative methods to find the coefficients that best fit the data, so a set of initial guesses for the coefficients might be needed.

As an example, consider fitting the data shown in Figure 5.54 with an ***exponential curve***.

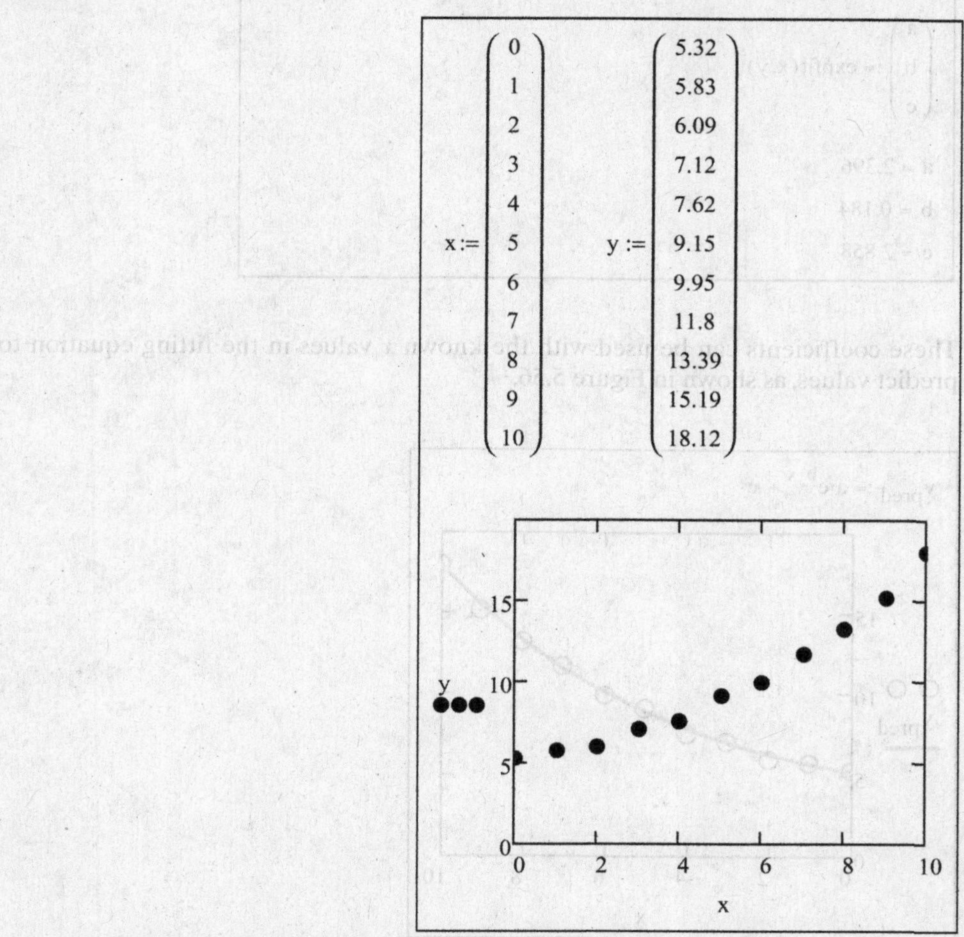

Figure 5.54
Plotting data to determine an appropriate-fitting equation

The data have been plotted, and the y values do seem to increase exponentially (albeit, weakly), so an exponential curve may be a good choice.

The **`expfit`** function finds values for a, b, and c that best fit the following equation to the data:

$$y_{\text{pred}} = ae^{bx} + c$$

Note: In older versions of Mathcad (e.g., before Mathcad 11), in order to use **`expfit`** a vector of initial guesses for **a**, **b**, and **c** had to be provided:

$$\textbf{guesses} := \begin{pmatrix} 1 \\ 1 \\ 1 \end{pmatrix}$$

Then the vector of x values and the vector of y values (and the guesses, if desired) are sent to **`expfit`**. The coefficients of the exponential equation are returned as shown in Figure 5.55.

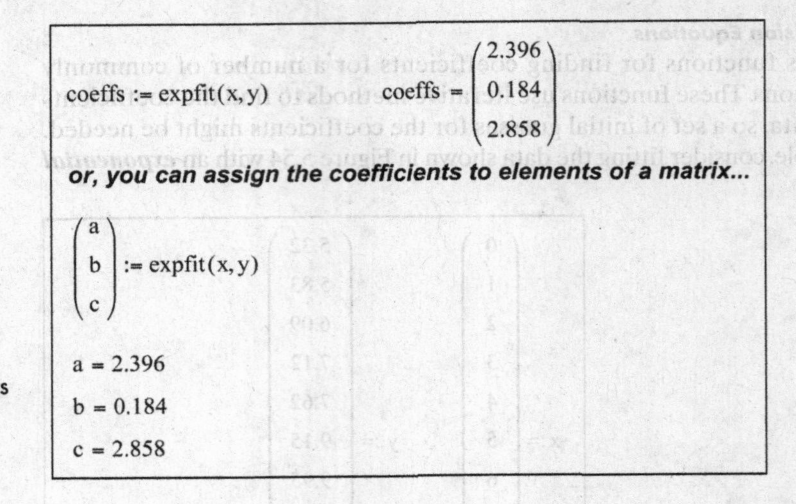

coeffs := expfit(x, y)
coeffs = (2.396, 0.184, 2.858)
or, you can assign the coefficients to elements of a matrix...
(a, b, c) := expfit(x, y)
a = 2.396
b = 0.184
c = 2.858

Figure 5.55
Determining the coefficients for the best-fit exponential curve using the **expfit** function.

These coefficients can be used with the known x values in the fitting equation to predict values, as shown in Figure 5.56.

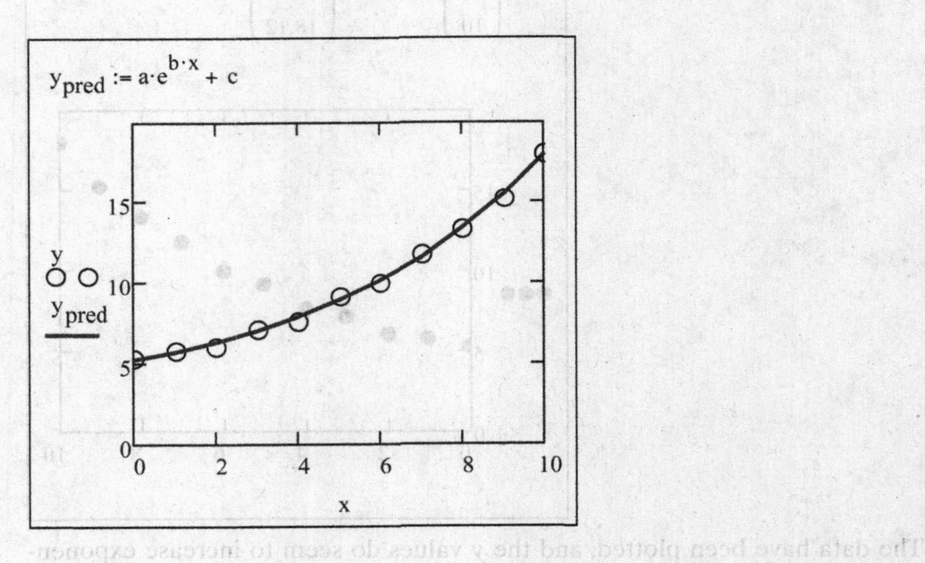

Figure 5.56
Plotting the data values with the predicted values to check the fit.

The following table summarizes the specialized fitting equations available in Mathcad:

	Function	Equation of Curve
Exponential	`expfit(vx, vy, vg)`	$y_{pred} = ae^{bx} + c$
Logistic	`lgsfit(vx, vy, vg)`	$y_{pred} = \frac{a}{(1 + be^{-\alpha})}$
Logarithmic	`logfit(vx, vy, vg)`	$y_{pred} = a\ln(x)^{bx} + c$
Power	`pwrfit(vx, vy, vg)`	$y_{pred} = ax^{b} + c$
Sine	`sinfit(vx, vy, vg)`	$y_{pred} = a\sin(x + b) + c$

Note that the parameters for each function are the same: a vector of x values, a vector of y values, and a vector of initial guesses for the three coefficients (a, b, c). (Guess values are optional with **expfit**.)

APPLICATION

Bode Plots

Engineers are often concerned with how systems or processes respond to changes in applied conditions. It is common to apply a known change to a system input and monitor how the system's output responds to the change. To see how a simple process responds to changes in input conditions, consider a well-mixed tank, as illustrated in Figure 5.57.

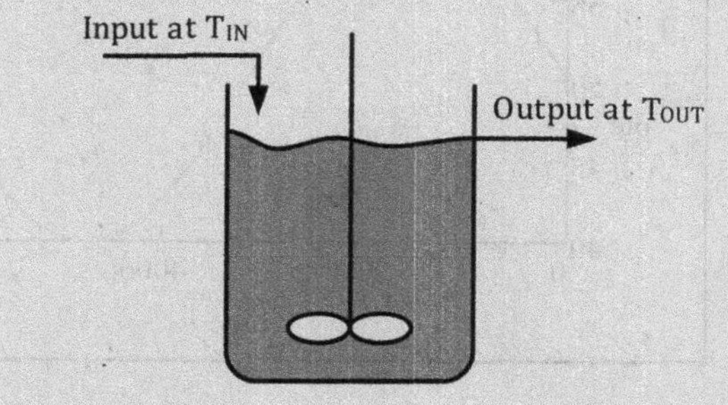

Figure 5.57
Well-mixed tank (first-order process).

If the input temperature suddenly increases, the output temperature will increase as well, but more slowly because of the mass of fluid in the tank. The *process time constant*, t_p, provides information on how fast the system will respond to a change in input temperature. For a well-mixed tank, the time constant is equal to the tank volume, V, divided by the input flow rate, Q.

$$\tau_p = \frac{V}{Q}$$

Engineers sometimes investigate how a process responds to sinusoidal inputs of varying frequency in order to learn about the dynamics and stability of the process. Mathcad can be used to simulate a sinusoidal inlet temperature; this is illustrated in Figure 5.58. Here the tank's inlet temperature oscillates around a mean temperature of 50°.

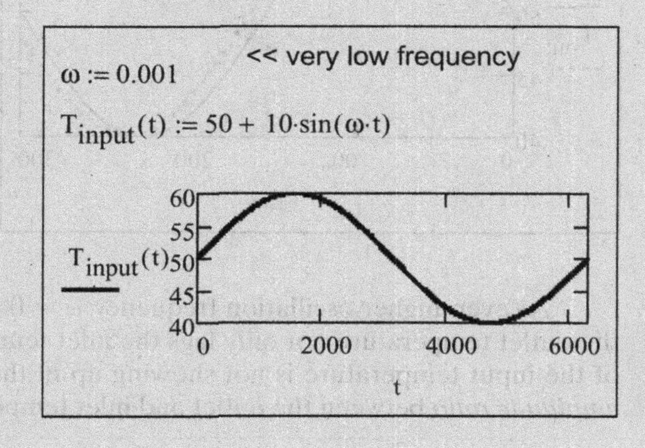

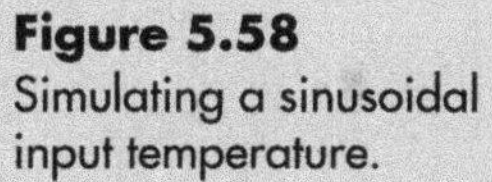

Figure 5.58
Simulating a sinusoidal input temperature.

The tank outlet temperature, T, can be determined by integrating the differential equation

$$\frac{dT}{dt} = \frac{Q}{V}[T_{\text{input}} - T].$$

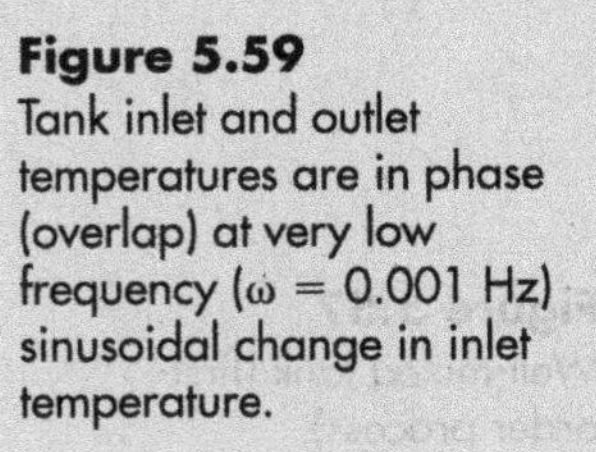

Although it is beyond the scope of this text, Mathcad can be used to integrate this equation to determine how the outlet temperature changes in response to the sinusoidal input temperature. The results only are shown in Figure 5.59.

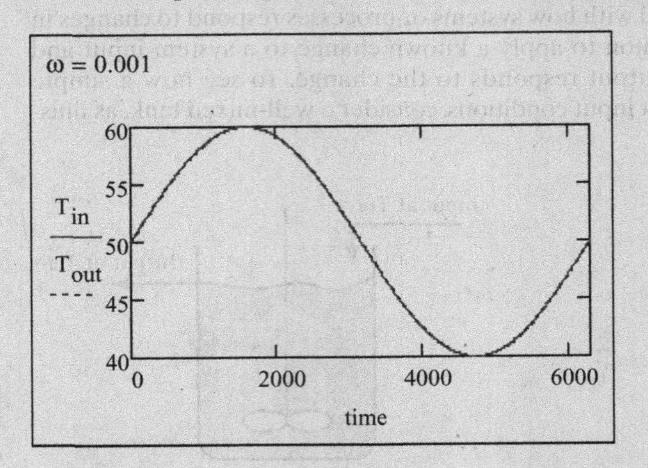

Figure 5.59
Tank inlet and outlet temperatures are in phase (overlap) at very low frequency ($\omega = 0.001$ Hz) sinusoidal change in inlet temperature.

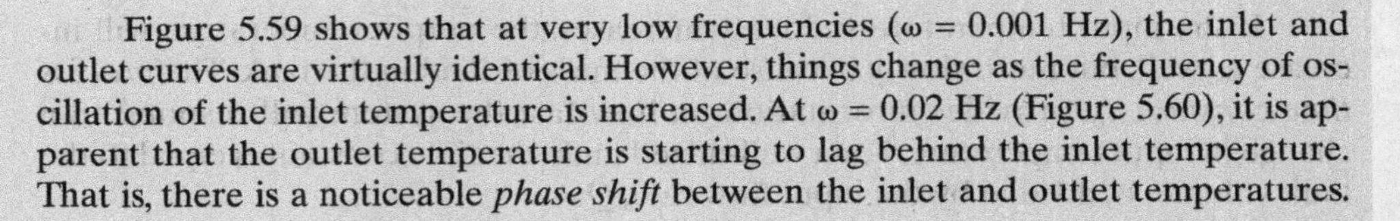

Figure 5.59 shows that at very low frequencies ($\omega = 0.001$ Hz), the inlet and outlet curves are virtually identical. However, things change as the frequency of oscillation of the inlet temperature is increased. At $\omega = 0.02$ Hz (Figure 5.60), it is apparent that the outlet temperature is starting to lag behind the inlet temperature. That is, there is a noticeable *phase shift* between the inlet and outlet temperatures.

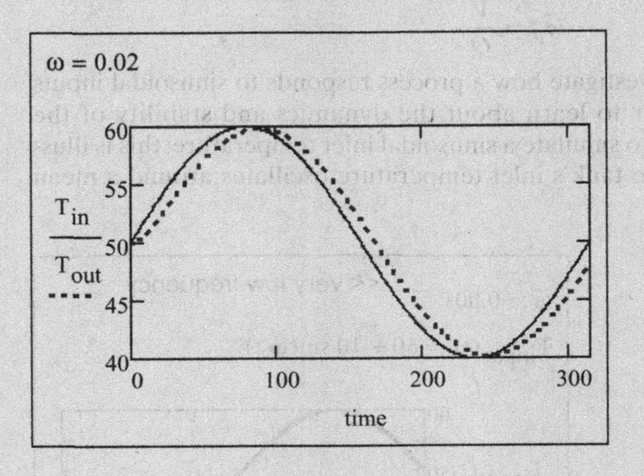

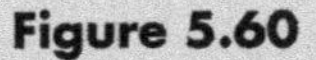

Figure 5.60
Tank response at a slightly higher frequency ($\omega = 0.02$ Hz).

At even higher oscillation frequency, $\omega = 0.1$ Hz, we see (in Figure 5.61) that the outlet temperature not only lags the inlet temperature, but the $\pm 10°$ amplitude of the input temperature is not showing up in the outlet temperature. That is, the *amplitude ratio* between the outlet and inlet temperatures is now less than 1.

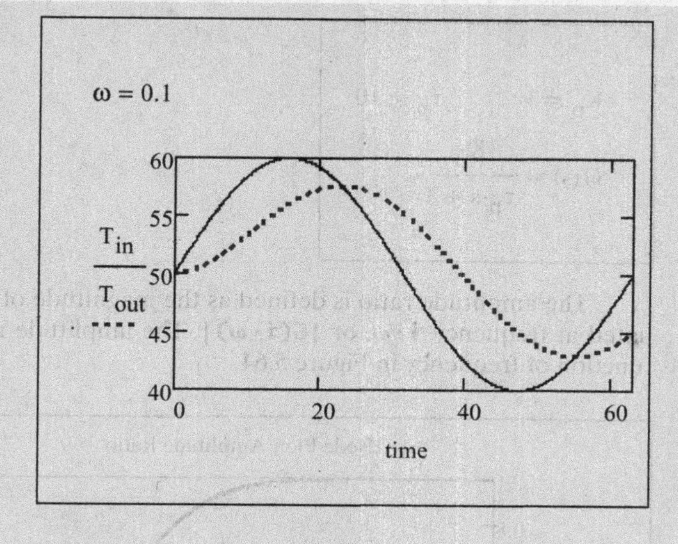

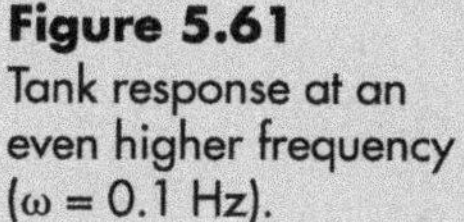

Figure 5.61
Tank response at an even higher frequency (ω = 0.1 Hz).

Although it is possible to investigate phase shifts and amplitude ratios by solving differential equations (as we have done here), it is more common to use Bode Plots. The remainder of this application will show how to construct Bode Plots using Mathcad.

TRANSFER FUNCTIONS AND BODE PLOTS

A *transfer function* is simply a mathematical formula that describes the relationship between the input and output for a process. Transfer functions are often studied in the *Laplace domain*, which means the mathematical model of the system's behavior has been passed through a *Laplace transform*. While Laplace transforms are beyond the scope of this text, Mathcad's symbolic processor will handle them.

A transfer function for a first-order process (such as our well-mixed tank) can be written as

$$G(s) = \frac{K_p}{\tau_p s + 1}$$

where K_p is the process gain ($K_p = 1$ for a well-mixed tank).
t_p is the process time constant, and
s is the Laplace domain variable.

For our tank, the transfer function describes how the outlet temperature responds to a change in the inlet temperature.

Bode Plots are plots of amplitude ratio, A_r, against input frequency, ω, and phase shift, φ, against input frequency, ω. Amplitude ratio and frequency are normally plotted using a logarithm scale. Mathcad has a function, `logspace(min, max, N)` that can be used to automatically create a vector of log-spaced frequencies. The use of the `logspace` function is illustrated in Figure 5.62.

$\omega := \text{logspace}(0.001, 1, 300)$

Figure 5.62
Using the `logspace` function to create a vector of 300 frequency values between 0.001 and 1 Hz.

A transfer function is defined in Mathcad just like any other function, as shown in Figure 5.63.

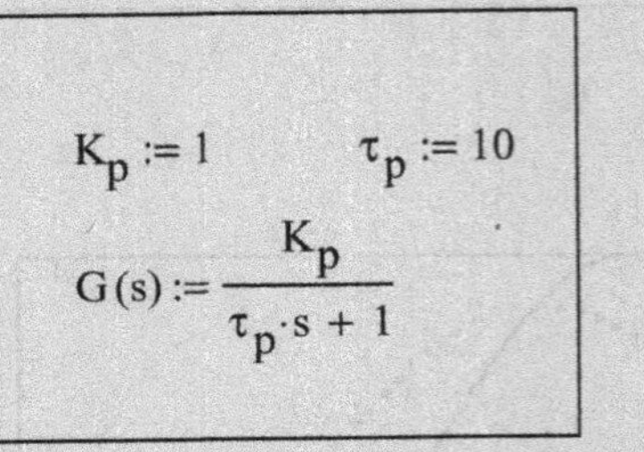

Figure 5.63
Defining a transfer function in Mathcad.

The amplitude ratio is defined as the magnitude of the transfer function evaluated at frequency **i** $\cdot\, \omega$, or $|G(i \cdot \omega)|$. The amplitude ratio has been plotted as a function of frequency in Figure 5.64.

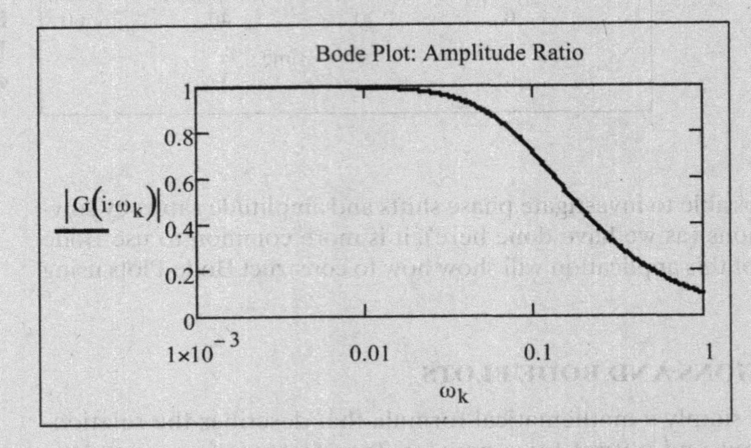

Figure 5.64
The Bode Plot for amplitude ratio.

Notice that at low frequencies (below 0.01 Hz), the amplitude ratio is 1; that is, the inlet and outlet temperatures both oscillate ±10° around the 50° mean. This is exactly the behavior we saw in Figure 5.59. However, at higher frequencies, the amplitude ratio falls off and the outlet temperature oscillates with less than ±10°. At a frequency of 0.1 Hz, the amplitude ratio is about 0.7, as we saw in Figure 5.61.

The phase shift is calculated as the argument of the transfer function, **arg(G(i·ω))**. The phase shift has been plotted as a function of frequency in Figure 5.65.

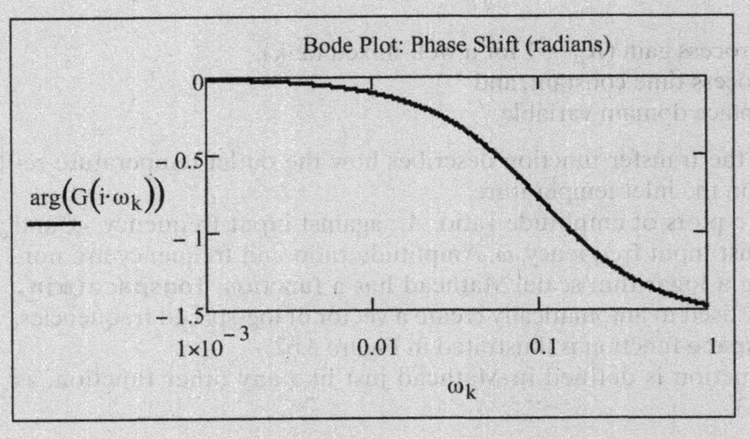

Figure 5.65
The Bode Plot for phase shift.

Figure 5.65 indicates that at very low frequencies (near 0.001 Hz), there is no phase shift and the inlet and outlet temperature curves should coincide; we saw this in Figure 5.59. But even at a frequency of around 0.01 Hz, a small phase shift is apparent, and the outlet temperature begins to lag the inlet temperature. We saw this behavior in Figure 5.60.

Bode Plots are commonly used by various engineering disciplines to investigate how processes or systems respond to input changes, and Mathcad makes it fairly easy to construct Bode Plots.

SUMMARY

In this chapter, you learned to use Mathcad's built-in functions for data analysis. You learned to make X-Y graphs from data, to calculate statistical values from data, to interpolate between data set values, and to fit curves (linear models) to data. You also learned about the risk associated with extrapolating data, especially with nonlinear methods such as cubic splines.

MATHCAD SUMMARY

X-Y Graphs Vector against Vector:

1. Select **X-Y Plot** from the Graph Toolbar.
2. Enter the name of your vector of independent values on the *x*-axis.
3. Enter the name of your vector of dependent values on the *y*-axis.

Element by Element:

1. Declare a range variable starting at 0 and going to **last(*v*)**, where *v* is one of your data set vectors.
2. Select **X-Y Plot** from the Graph Toolbar.
3. Enter the name of your vector of independent values on the *x*-axis – with an index subscript containing the range variable.
4. Enter the name of your vector of dependent values on the *y*-axis – with an index subscript containing the range variable.

Quick Plot:

1. Select **X-Y Plot** from the Graph Toolbar.
2. Enter your function name on the *y*-axis, with a dummy variable as an argument.
3. Enter the dummy variable on the *x*-axis.
4. Adjust the *x*-axis limits to change the displayed range.

Statistical Functions:

mean(A)	Returns the mean (arithmetic average) of the values in **A**.
stdev(A)	Returns the population standard deviation of the values in **A**.
var(A)	Returns the population variance of the values in **A**.
Stdev(A)	Returns the sample standard deviation of the values in **A**.
Var(A)	Returns the sample variance of the values in **A**.

Interpolation:

`linterp(vx, vy, xnew)`	Returns the y value corresponding to $x = x_{new}$ computed by using linear interpolation on the x and y data.
`cspline(vx, vy)`	Returns the vector of second derivatives that specify a spline curve that is cubic at the endpoints.
`pspline(vx, vy)`	Returns the vector of second derivatives that specify a spline curve that is parabolic at the endpoints.
`lspline(vx, vy)`	Returns the vector of second derivatives that specify a spline curve that is linear at the endpoints.
`interp(vs, vx, vy, xnew)`	Uses the vector of second derivatives from any of the spline functions previously listed and returns the y value corresponding to $x = x_{new}$, which is computed by using spline interpolation on the x and y data.

Regression:

`slope(vx, vy)`	Returns the slope of the best-fit (minimum total squared error) straight line through the data in *vx* and *vy*.
`intercept(vx, vy)`	Returns the intercept of the best-fit straight line through the data in *vx* and *vy*.
`corr(vx, vy)`	Returns the coefficient of correlation (usually called R) of the best-fit straight line through the data in *vx* and *vy*. The coefficient of determination, R^2, can be computed from R and is more commonly used.
`linfit(vx, vy, vf)`	Returns the coefficients that best fit the linear model described by *vf* to the data in *vx* and *vy*. The vector *vf* is a vector of functions you provide that describes the linear model you want to use to fit to the data.

Specialized Fitting Functions:

`expfit(vx, vy, vg)`	Fits an exponential curve($y_{pred} = ae^{bx} + c$) to the data in the *vx* and *vy* vectors. A three-element vector of guessed values for the coefficients (a, b, c) may be provided (optional).
`lgsfit(vx, vy, vg)`	Fits a ***logistic curve*** $y_{pred} = a/(1 + be^{-cx})$ to the data in the *vx* and *vy* vectors. A three-element vector of guessed values for the coefficients (a, b, c) must be provided.
`logfit(vx, vy, vg)`	Fits a ***logarithmic curve*** ($y_{pred} = a \ln(x)^b + c$) to the data in the *vx* and *vy* vectors. A three-element vector of guessed values for the coefficients (a, b, c) must be provided.

`pwrfit(vx, vy, vg)` Fits a ***power curve*** ($y_{pred} = ax^b + c$) to the data in the *vx* and *vy* vectors. A three-element vector of guessed values for the coefficients (a, b, c) must be provided.

`sinfit(vx, vy, vg)` Fits a ***sine curve*** ($y_{pred} = a \sin(x + b) + c$) to the data in the *vx* and *vy* vectors. A three-element vector of guessed values for the coefficients (a, b, c) must be provided.

KEY TERMS

Bode Plot
coefficient of determination (R^2)
correlation coefficient (R)
exponential curve
intercept
interpolation
linear interpolation
linear regression
logarithmic curve
logistic curve
mean
population
power curve
QuickPlot
random sample
representative sample
residuals
risk analysis
sample
sine curve
slope
spline interpolation
standard deviation
variance
visualization
X-Y graph

PROBLEMS

5.1 Statistics: Competing Cold Remedies

Two proposed cold remedies — *CS1* (cough syrup) and *CS2* (chicken soup)—were compared in a hospital study. The objective was to determine which treatment enabled patients to recover more quickly. The duration of sickness was defined as the time (days) between when patients reported to the hospital requesting treatment and when they were discharged from the hospital. The results are tabulated as follows:

CS1		CS2	
Patient	**Duration**	**Patient**	**Duration**
1	8.10	1	5.79
2	7.25	2	2.81
3	5.89	3	3.27
4	7.12	4	4.87
5	5.60	5	5.49
6	1.85	6	6.60
7	4.00	7	2.63
8	5.58	8	4.64
9	9.92	9	6.96
10	4.25	10	5.83

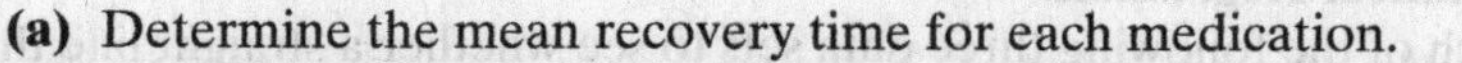

(a) Determine the mean recovery time for each medication.
(b) Which remedy was associated with the fastest recoveries?

5.2 Statistics: Quality Control
A beverage bottling plant uses a sampling protocol for quality control. At random intervals, a bottle is pulled off the line and the volume is measured to make sure that the filling operation is working correctly. The company has several criteria it watches:

- Ideally, the company wants to put an average of 2.03 liters in each bottle. Since the plant must shut down whenever it is found that less than the labeled amount (on average) is being put in each bottle, setting the target amount higher than 2.0 liters reduces the number of times the company has to shut down the process to recalibrate the equipment. If the average fill volume is less than 2.03 liters but greater than 2.00 liters, the equipment will not be shut down immediately, but it will be recalibrated the next time the equipment is down for any reason (i.e., cleaning, breakage, etc.)
- If the average fill volume exceeds 2.06 liters, the equipment must be shut down for recalibration to prevent bottles from overflowing.
- If the standard deviation exceeds 40 milliliters, the equipment must be shut down for recalibration. This is to eliminate customer complaints about incompletely filled bottles, as well as to prevent bottles from overflowing.

2.00	2.09
1.99	2.08
2.00	2.09
1.97	2.07
1.93	1.97
2.03	2.03
2.02	1.99
2.06	2.07
2.16	2.02
2.08	1.99

Given the sample volumes (liters) in the accompanying table, should the equipment be shut down and recalibrated?

5.3 Using X-Y Trace with Graphs
Create a QuickPlot of the function

$$f(x) = 1 - e^{-x}.$$

Use Mathcad's X-Y Trace dialog box to evaluate this function at $x = 1, 2,$ and 3. (Click on the graph, and then use the menu commands **Format/Graph/Trace.**)

5.4 Using QuickPlots to Solve Problems
The equation describing the process of warming a hot tub (see Figure 5.66) by adding hot water is

Figure 5.66
Heating a hot tub.

$$T = T_{IN} - (T_{IN} - T_{START})e^{\frac{-Q}{V}t},$$

where

T	is the temperature of the water in the hot tub,
T_{IN}	is the temperature of the water flowing into the hot tub (130°F).
T_{START}	is the initial temperature of the water in the hot tub (65°F),
Q	is the hot-water flow rate (5 gpm),
V	is the volume of the hot tub (500 gallons), and
t	is the elapsed time since the hot water started flowing.

(a) Use QuickPlot to see how long it will take the hot tub to reach 110°F.
(b) If the hot-water flow rate was increased to 10 gpm, how long would it take for the water temperature in the tub to reach 110°F?

5.5 Simple Linear Regression
Plot each of the following three data sets to see whether a straight line through each set of points seems reasonable:

x	y_1	y_2	y_3
0	2	0.4	10.2
1	5	3.6	4.2
2	8	10.0	12.6
3	11	9.5	11.7
4	14	12.0	28.5
5	17	17.1	42.3
6	20	20.4	73.6
7	23	21.7	112.1

If a straight line is reasonable, use the **`slope`** and **`intercept`** functions to calculate the regression coefficients for the set. (The same x data have been used in each example to minimize typing.)

5.6 Choosing a Fitting Function

Half the challenge of regression analysis is choosing the right fitting function. The purpose of this exercise is to demonstrate the general shape of some typical fitting functions and then to have you choose an appropriate function and fit a curve to some data.

Typical Fitting Functions

Make QuickPlots of the following functions over the indicated range of the independent variable x:

Function	Coefficients	Range
$a + bx$	$a = 5$ $b = 1$	$0 < x < 10$
$a + bx + cx^2$	$a = 5$ $b = 1$ $c = 1$	$0 < x < 10$
$a + bx + cx^2 + dx^3$	$a = -20$ $b = 50$ $c = -10$ $d = 0.6$	$0 < x < 10$
$a + b\ln(x)$	$a = 2$ $b = 1$	$0 < x < 1$
$a + be^x$	$a = 2$ $b = 1$	$0 < x < 3$
$a + be^{-x}$	$a = 2$ $b = 1$	$0 < x < 3$
$a + b/x$	$a = 1$ $b = 1$	$0 < x < 1$

x	y
0.4	38.9
0.9	19.8
1.4	14.2
1.9	10.8
2.4	9.8
2.9	7.3
3.4	7.5
3.9	4.6
4.4	6.6
4.9	6.5
5.4	5.3
5.9	5.0
6.4	3.3
6.9	5.3
7.4	4.9

Selecting and Fitting a Function to Data

Now graph the data shown in the margin table on the previous page, and select an appropriate fitting function. Use **`linfit`** to regress the data with your selected function. Plot the original data and the values predicted by your fitting function on the same graph to visually verify the fit.

5.7 Fitting an Exponential Curve to Data

In Problem 5.4, the exponential function describing the process of heating a hot tub was given. If you have experimental temperature vs. time data taken as the hot tub warms up, Mathcad's **expfit** function could be used to fit an exponential function to the experimental data. The general form of an exponential curve is

$$y_{pred} = a \cdot e^{b \cdot x} + c.$$

Suppose the hot-tub warming data are as follows:

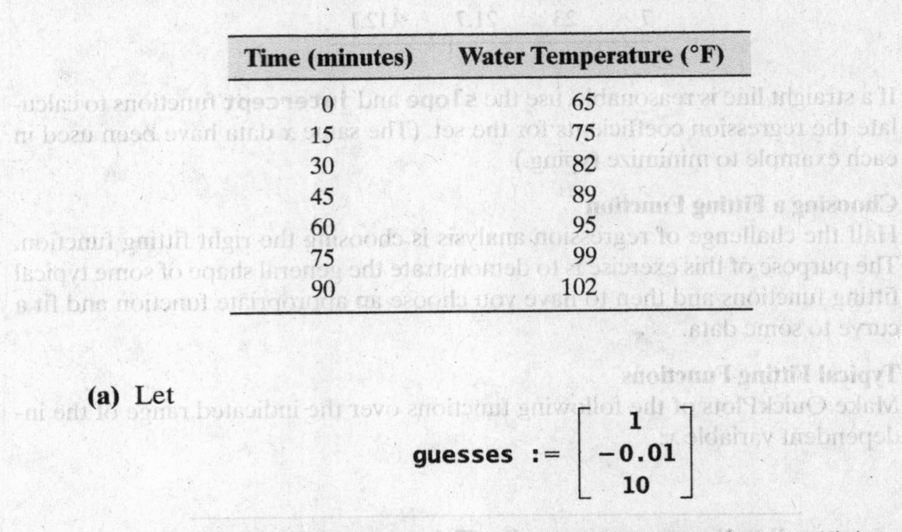

Time (minutes)	Water Temperature (°F)
0	65
15	75
30	82
45	89
60	95
75	99
90	102

(a) Let

$$\text{guesses} := \begin{bmatrix} 1 \\ -0.01 \\ 10 \end{bmatrix}$$

be the initial estimates of the exponential function coefficients. Use **expfit** to find the coefficients of the exponential curve that fits the experimental data.

(b) Plot the experimental values and the exponential curve on the same graph. Does the curve fit the data?

5.8 Checking Thermocouples

Two thermocouples have the same color codes on the wires, and they appear to be identical. But when they were calibrated (illustrated in Figure 5.67), they were found to give somewhat different output voltages. The calibration data are shown in the following table:

Water Temp. (°C)	TC 1 Output (mV)	TC 2 Output (mV)
20.0	1.019	1.098
30.0	1.537	1.920
40.0	2.059	2.526
50.0	2.585	2.816
60.0	3.116	2.842
70.0	3.650	4.129
80.0	4.187	4.266
90.0	4.726	4.340

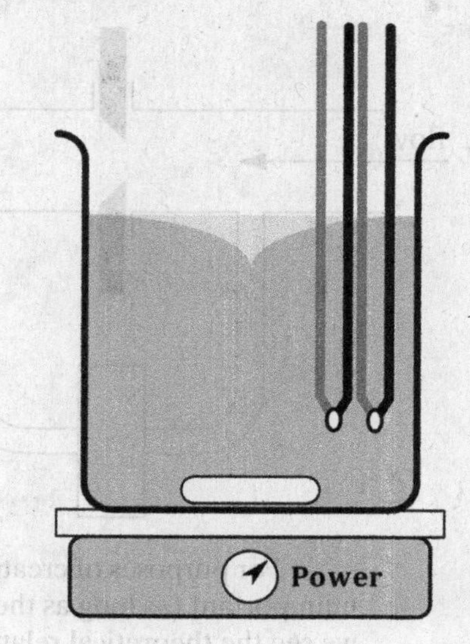

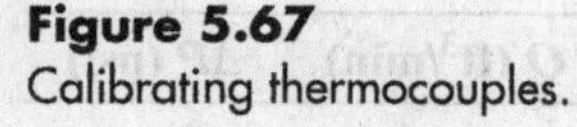
Figure 5.67
Calibrating thermocouples.

Plot the output voltages (y-axis) against the temperature (x-axis) for each thermocouple. Which thermocouple would you want to use in an experiment? Why?

5.9 Thermocouple Calibration

Thermocouples are made by joining two dissimilar metal wires. Contact between the two metals results in a small, but measurable, voltage drop across the junction. This voltage drop changes as the temperature of the junction changes; thus, the thermocouple can be used to measure temperature if you know the relationship between temperature and voltage. Equations for common types of thermocouples are available, or you can simply take a few data points and prepare a calibration curve. This is especially easy for thermocouples because, for small temperature ranges, the relationship between temperature and voltage is nearly linear.

T(°C)	V(mV)
10	0.397
20	0.796
30	1.204
40	1.612
50	2.023
60	2.436
70	2.851
80	3.267
90	3.662

(a) Use the **slope** and **intercept** functions to find the coefficients of a straight line through the data at the left.

Note: The thermocouple voltage changes because the temperature changes—that is, the voltage *depends* on the temperature. For regression, the independent variable (temperature) should always be on the x-axis, and the dependent variable (voltage) should be on the y-axis.

(b) Calculate a predicted voltage at each temperature, and plot the original data and the predicted values together to make sure that your calibration curve actually fits the data.

5.10 Orifice Meter Calibration

Orifice meters (see Figure 5.68) are commonly used to measure flow rates but are highly nonlinear devices. Because of this nonlinearity, special care must be taken when one prepares calibration curves for these meters. The equation relating the flow rate **Q** to the pressure drop across the orifice **ΔP** (the measured variable) is

$$Q = \frac{A_0 C_0}{\sqrt{1-\beta^4}} \sqrt{\frac{2 g_c \Delta P}{\rho}}.$$

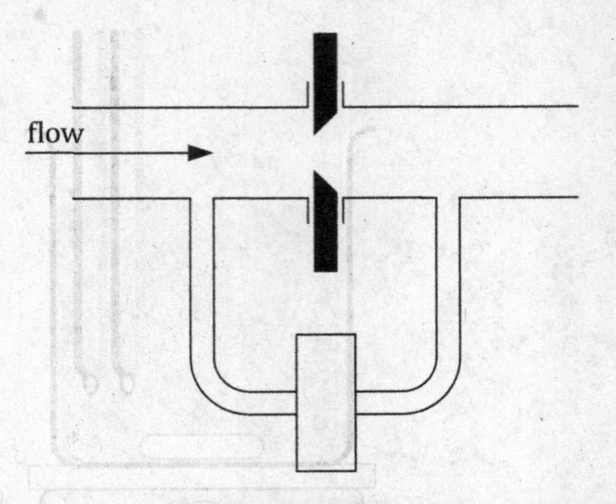

Figure 5.68
Orifice meter.

Q (ft³/min)	ΔP (psi)
3.9	0.13
7.9	0.52
11.8	1.18
15.7	2.09
19.6	3.27
23.6	4.71
27.5	6.41
31.4	8.37
35.3	10.59
39.3	13.08

For purposes of creating a calibration curve, the details of the equation are unimportant (as long as the other terms stay constant). What is necessary is that we see the theoretical relationship between **Q** and **ΔP**, namely,

$$Q \propto \sqrt{\Delta P}.$$

Also, the pressure drop across the orifice plate depends on the flow rate, not the other way around. So $\sqrt{\Delta P}$ should be regressed as the dependent variable (y values) and **Q** as the independent variable (x values).

(a) Using the accompanying data, regress **Q** and $\sqrt{\Delta P}$ to create a calibration curve for the orifice meter.

(b) Calculate and plot predicted values together with the original data to check your results.

Note that if you use the **slope** and **intercept** functions in this problem, you will need to use the *vectorize* operator as well, because taking the square root of an entire matrix is not defined. The vectorize operator tells Mathcad to perform the calculations (taking the square root, for example) on each element of the matrix. The **slope** function might then look like this:

$$m := \text{slope}\left(Q, \overrightarrow{\sqrt{\Delta P}}\right).$$

5.11 Vapor–Liquid Equilibrium

x	y
0.000	0.000
0.022	0.066
0.046	0.133
0.071	0.201
0.098	0.269
0.126	0.335
0.156	0.399
0.189	0.461
0.224	0.519
0.261	0.574
0.302	0.626
0.346	0.674
0.393	0.720
0.445	0.762
0.502	0.802
0.565	0.839
0.634	0.874
0.710	0.907
0.796	0.939
0.891	0.970
1.000	1.000

When a liquid mixture is boiled, the vapor that leaves the vessel is enriched in the more volatile component of the mixture. The vapor and liquid in a boiling vessel are in equilibrium, and vapor–liquid equilibrium (VLE) data are available for many mixtures. The data are usually presented in tabular form, as in the accompanying table, which represents VLE data for mixtures of methanol and *n*-butanol boiling at 12 atm. From the VLE data in the margin table (plotted in Figure 5.69), we see that if a 50:50 liquid mixture of the alcohols is boiled, the vapor will contain about 80% methanol.

Note: The VLE data shown were generated from version 4.0 of Chemcad.[1] The x column represents the mass fraction of methanol in the boiling mixture. (The mass fraction of *n*-butanol in the liquid is calculated as $1 - x$ for any mixture.) The mass fraction of methanol in the vapor leaving the solution is shown in the y column.

[1]Chemcad is a chemical process simulation package produced by Chemstations, Inc., in Houston, Texas, USA.

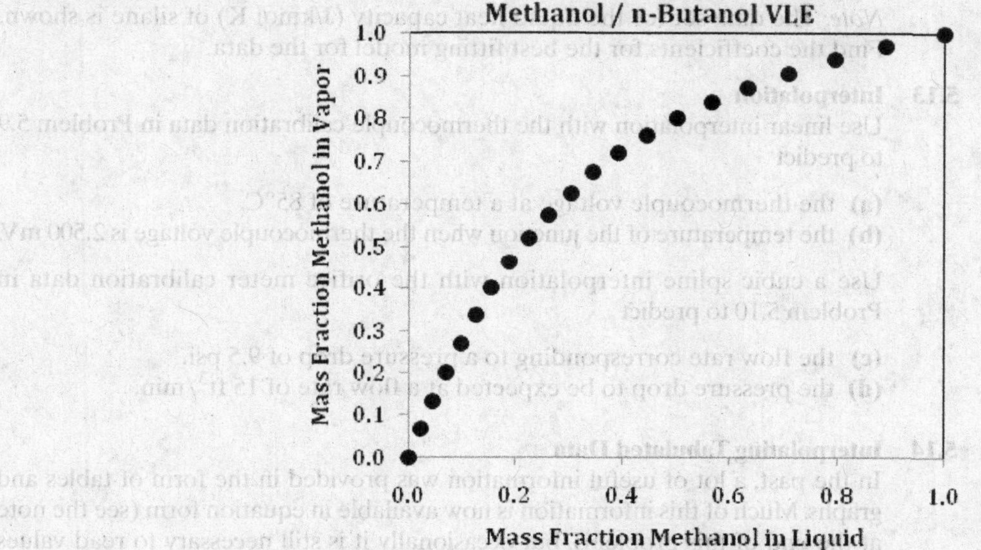

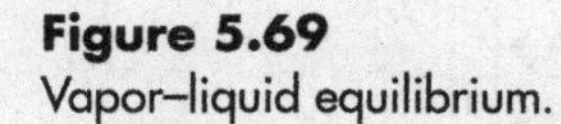
Figure 5.69
Vapor–liquid equilibrium.

VLE data are commonly used in designing distillation columns, but an equation relating vapor mass fraction to liquid mass fraction is a lot handier than tabulated values.

Use the `linfit` function and the tabulated VLE data to obtain an equation relating the mass fraction of methanol in the vapor (y) to the mass fraction of vapor in the liquid (x). Test different linear models (e.g., polynomials) to see which gives the best fit to the experimental data. For your best model, calculate predicted values using the coefficients returned by `linfit`, and plot the predicted values and the original data values on the same graph. Your result should look something like Figure 5.69.

5.12 Fitting Physical Property Data

In an earlier Applications example, experimental data on the vapor pressure of silane were fit to a standard equation. Try fitting a curve to data for silane's liquid heat capacity. The standard fitting equation for this property is

$$C_P = a + bT + cT^2 + dT^3 + eT^4$$

Not all coefficients need to be used (i.e., any order of polynomial is acceptable).

Temp.	C_P	Temp.	C_P
88.48	59874	124.74	61026
92.11	60174	128.37	61231
95.73	60676	131.99	61289
99.36	60156	135.62	61655
102.98	60358	139.24	61160
106.61	60863	142.87	61658
110.24	60604	146.50	62038
113.86	60784	150.12	62085
117.49	61037	153.75	61864
121.11	61215	157.37	61712

Note: The data set for the liquid heat capacity (J/kmol K) of silane is shown. Find the coefficients for the best fitting model for the data.

5.13 Interpolation

Use linear interpolation with the thermocouple calibration data in Problem 5.9 to predict

(a) the thermocouple voltage at a temperature of 85°C.
(b) the temperature of the junction when the thermocouple voltage is 2.500 mV.

Use a cubic spline interpolation with the orifice meter calibration data in Problem 5.10 to predict

(c) the flow rate corresponding to a pressure drop of 9.5 psi.
(d) the pressure drop to be expected at a flow rate of 15 ft^3/min.

5.14 Interpolating Tabulated Data

In the past, a lot of useful information was provided in the form of tables and graphs. Much of this information is now available in equation form (see the note at the end of this problem), but occasionally it is still necessary to read values from tables. Consider the following *compound amount factors* table:

Compound Amount Factors

Year	2%	4%	6%	8%	10%	12%	14%	16%	18%
0	1.000	1.000	1.000	1.000	1.000	1.000	1.000	1.000	1.000
2	1.040	1.082	1.124	1.166	1.210	1.254	1.300	1.346	1.392
4	1.082	1.170	1.262	1.360	1.464	1.574	1.689	1.811	1.939
6	1.126	1.265	1.419	1.587	1.772	1.974	2.195	2.436	2.700
8	1.172	1.369	1.594	1.851	2.144	2.476	2.853	3.278	3.759
10	1.219	1.480	1.791	2.159	2.594	3.106	3.707	4.411	5.234
12	1.268	1.601	2.012	2.518	3.138	3.896	4.818	5.936	7.288
14	1.319	1.732	2.261	2.937	3.797	4.887	6.261	7.988	10.147
16	1.373	1.873	2.540	3.426	4.595	6.130	8.137	10.748	14.129
18	1.428	2.026	2.854	3.996	5.560	7.690	10.575	14.463	19.673
20	1.486	2.191	3.207	4.661	6.727	9.646	13.743	19.461	27.393
22	1.546	2.370	3.604	5.437	8.140	12.100	17.861	26.186	38.142
24	1.608	2.563	4.049	6.341	9.850	15.179	23.212	35.236	53.109

The table can be used to determine the future value of an amount deposited at a given interest rate. One thousand dollars invested at 10% (annual percentage rate) for twenty years would be worth 6.727 × \$1,000 = \$6,727 at the end of the twentieth year. The 6.727 is the compound amount factor for an investment held twenty years at 10% interest.

Interest tables such as the preceding one provided useful information, but it was frequently necessary to interpolate to find the needed value.

(a) Use linear interpolation to find the compound amount factor for 6% interest in year 15.
(b) Use linear interpolation to find the compound amount factor for 11% interest in year 10.

Note: Mathcad provides the **fv** function that can be used to calculate future values directly, eliminating the need for an interest table. The **fv(rate,N,pmt,pv)** function finds the future value of an amount deposited at time zero (**pv**, include a negative sign to indicate out-of-pocket expenses), plus any periodic payments **(pmt)**, invested at a specified interest **rate**, for **N** periods. The example that was worked out earlier in this problem looks like this in Mathcad:

```
F := fv(10%, 20, 0, -1000)
F = 6727
```

5.15 Relating Height to Mass for a Solids Storage Tank

Solids storage tanks are sometimes mounted on *load cells* (large scales, basically) so that the mass of solids in the tank is known, rather than the height of the solids. To make sure they do not overfill the tank, the operators might ask for a way to calculate the height of material in the tank, given the mass reading from the load cells.

The values shown in Figure 5.70 relate height and mass in the tank and are the starting point for this problem. The tank has a conical base section $\theta = 30°$ and a diameter of 12 feet. The apparent density of the solids in the tank is 20 lb/ft^3.

Use a cubic spline to fit a curve to the mass (as **x**) and height (as **y**) data, and then use the **interp** function to predict the height of solids in the tank when the load cells indicate 3200 kg solids.

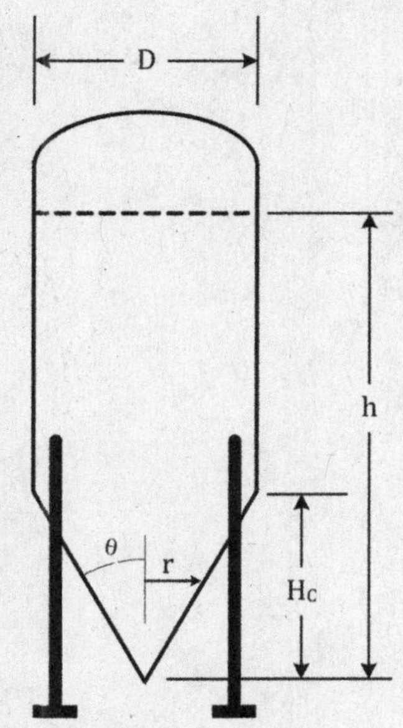

Figure 5.70
Solids storage tank.

Note: The load cells actually measure the mass of the tank and the stored solids, but adjusting the display to read 0 kg before adding any solids effectively causes the load cells to display only the mass of the solids. This is called setting the *tare weight* for the load cells.

h (ft)	*m* (lb)
0	0
2	56
4	447
6	1508
8	3574
10	6981
12	11470
14	16000
16	20520
18	25040
20	29570
22	34090
24	38620

5.16 Methods of Interpolation

In the *x*- and *y*-vectors shown here, the **y** values increase dramatically as **x** increases, suggesting that there might be an exponential relationship between **x** and **y**. The objective of this problem is to try several methods for estimating the value of **y** at **x** = 3 to see which methods work best for this nonlinear data set.

$$x := \begin{pmatrix} 1 \\ 1.1 \\ 1.2 \\ 1.6 \\ 1.9 \\ 2.2 \\ 2.8 \\ 3.6 \end{pmatrix} \qquad y := \begin{pmatrix} 14 \\ 17 \\ 20 \\ 40 \\ 70 \\ 125 \\ 409 \\ 2012 \end{pmatrix}$$

(a) Create an X-Y graph of the *x* and *y* vectors, and use the plot to estimate the value of *y* at *x* = 3.

(b) Use the linear interpolation function **linterp** to estimate the value of *y* at *x* = 3.

(c) Use the cubic spline interpolation to estimate the value of *y* at *x* = 3.

(d) Fit the data with an exponential curve using **expfit**, and use the calculated coefficients to determine the predicted *y* value at *x* = 3

(e) Which of these methods gave the best estimate of *y* at *x* = 3? Which is the worst?

5.17 Choosing Interpolation or Regression

Plot the thermocouple calibration data shown below, and then decide whether it would be best to use regression or cubic spline interpolation to determine the expected voltage when the thermocouple is placed in a 200°C oven. Then use your selected method to find the voltage corresponding to a temperature of 200°C.

Thermocouple Calibration Data (Type K)

Temp. (°C)	Voltage (mV)
20	1.93
60	1.58
100	4.44
140	4.81
180	8.47
220	9.27
260	10.75
300	11.80

5.18 Fitting Thermocouple Data

The relationship between temperature and output voltage for thermocouples is often approximated as

$$T = a \cdot V^b,$$

where a and b are coefficients that are unique values for each type of thermocouple. Determine coefficients **a** and **b** by performing a regression on the thermocouple data tabulated below.

Thermocouple Calibration Data (Type J)

Temp. (°C)	Voltage (mV)
20	1.01
60	3.13
100	5.29
140	7.47
180	9.67
220	11.88
260	14.10
300	16.33

5.19 Fitting Data with a Sine Curve

The hours of sunlight data for Fairbanks, Alaska, shown in Figure 5.19 is cyclic, but far from sinusoidal. As you get closer to the equator, the hours of sunlight curve shows less variation from month to month and is more sinusoidal. Sunrise and sunset data from the U.S. Naval Observatory (http://aa.usno.navy.mil) for Miami, Florida, were used to determine the hours of sunlight data tabulated below and plotted in Figure 5.71. The data have been adjusted to make the shortest day of the year (the winter solstice, December 21) the first day in the data set.

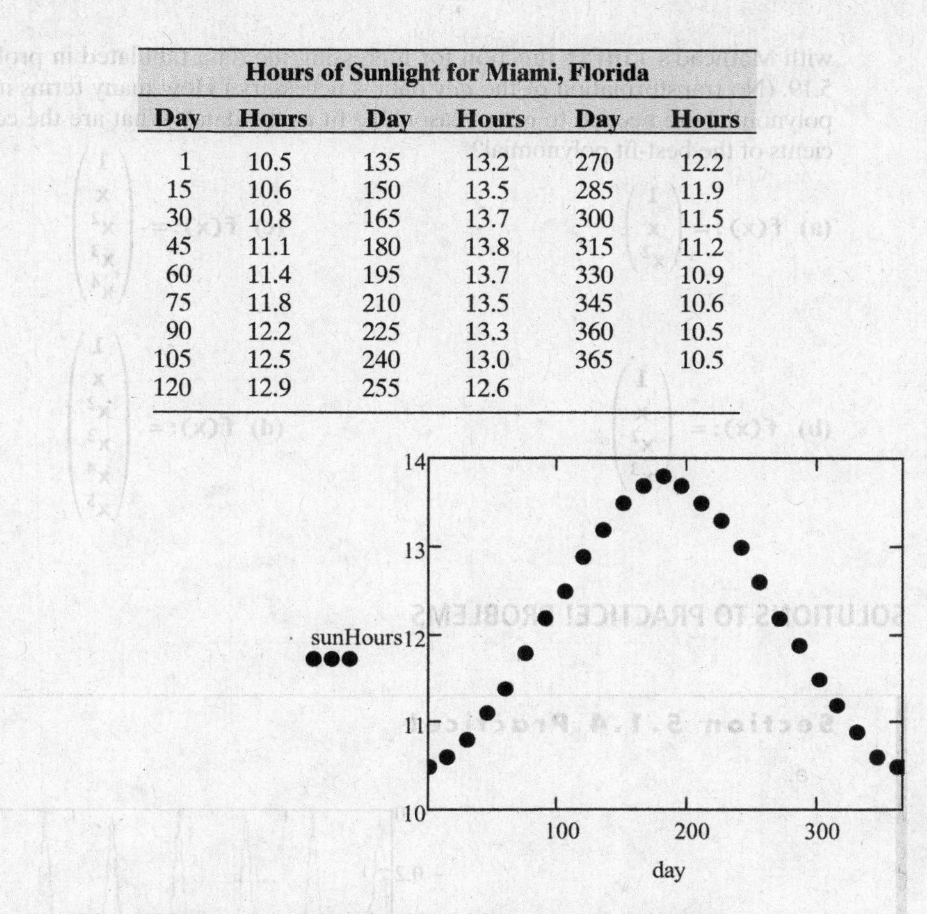

Hours of Sunlight for Miami, Florida

Day	Hours	Day	Hours	Day	Hours
1	10.5	135	13.2	270	12.2
15	10.6	150	13.5	285	11.9
30	10.8	165	13.7	300	11.5
45	11.1	180	13.8	315	11.2
60	11.4	195	13.7	330	10.9
75	11.8	210	13.5	345	10.6
90	12.2	225	13.3	360	10.5
105	12.5	240	13.0	365	10.5
120	12.9	255	12.6		

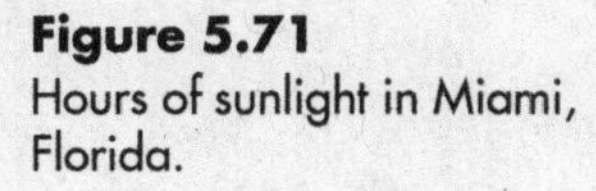

Figure 5.71
Hours of sunlight in Miami, Florida.

For this problem, try to fit the data with a sine curve of the form

$$y_p = a\sin(x + b) + c$$

using Mathcad's **`sinfit(vx, vy, vg)`** function. Since the sine has a period of 2π, you cannot fit the original data that cycles every 365 days. So, transform the **day** data as:

$$x = \text{day} \cdot \frac{2\pi}{365}.$$

and use x in the **`sinfit`** function, not days. The vector of guess values for the coefficients is not optional for the **`sinfit`** function. The following guess value can be used.

$$\mathrm{v}_g = \begin{pmatrix} 1 \\ 1 \\ 1 \end{pmatrix}.$$

Determine the coefficient values that best fit the data, and then plot the predicted values and the hours of sunlight from the original data to check the fit of the sine curve.

5.20 Polynomial Regression
An alternative to fitting a sine curve to the hours of sunlight data in problem 5.19 is to use a polynomial. Try the polynomials defined by the following **f** functions

with Mathcad's `linfit` function for regressing the data tabulated in problem 5.19. (No transformation of the *day* data is necessary.) How many terms in the polynomial are needed to get a reasonable fit to the data? What are the coefficients of the best-fit polynomial?

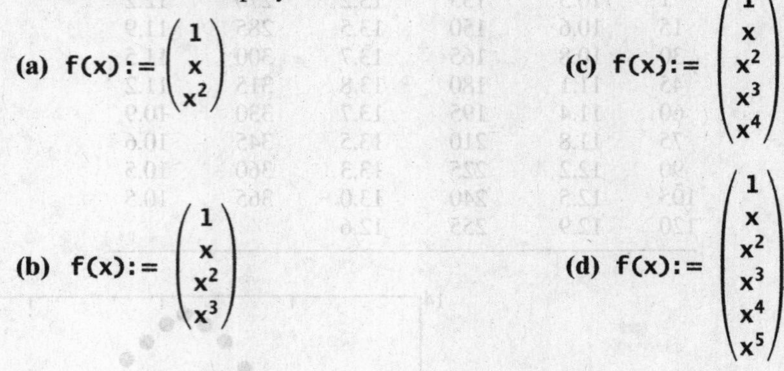

(a) $f(x) := \begin{pmatrix} 1 \\ x \\ x^2 \end{pmatrix}$

(b) $f(x) := \begin{pmatrix} 1 \\ x \\ x^2 \\ x^3 \end{pmatrix}$

(c) $f(x) := \begin{pmatrix} 1 \\ x \\ x^2 \\ x^3 \\ x^4 \end{pmatrix}$

(d) $f(x) := \begin{pmatrix} 1 \\ x \\ x^2 \\ x^3 \\ x^4 \\ x^5 \end{pmatrix}$

SOLUTIONS TO PRACTICE! PROBLEMS

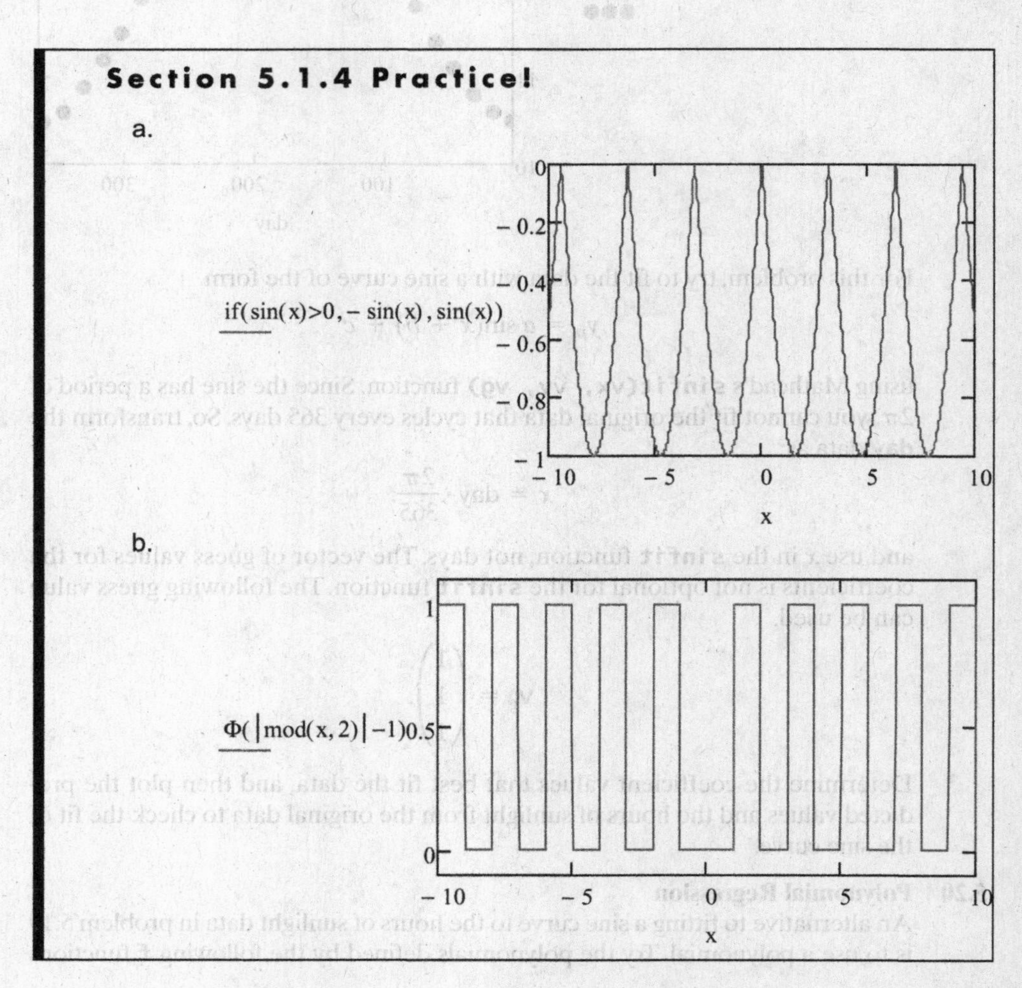

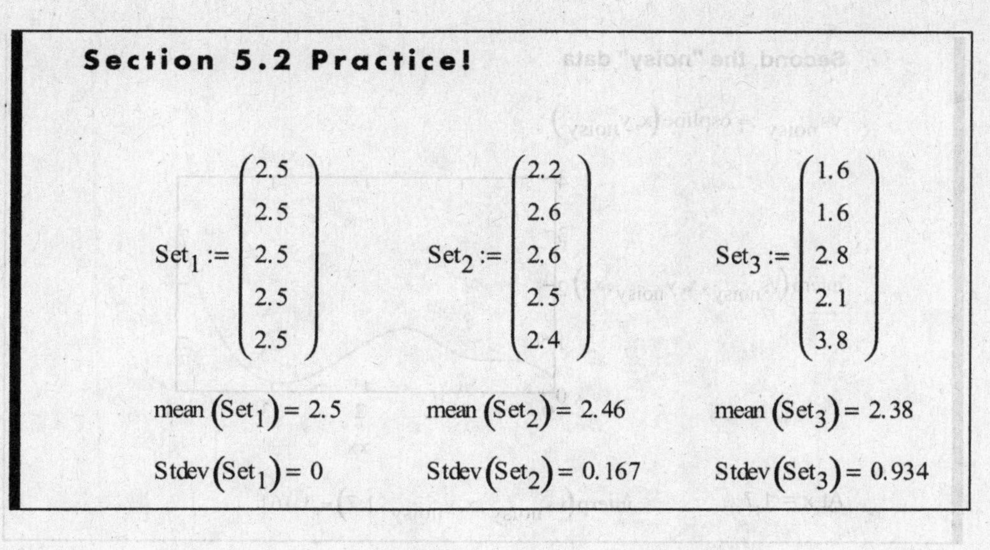
Section 5.2 Practice!
Set1 := (2.5, 2.5, 2.5, 2.5, 2.5)
Set2 := (2.2, 2.6, 2.6, 2.5, 2.4)
Set3 := (1.6, 1.6, 2.8, 2.1, 3.8)
mean(Set1) = 2.5
mean(Set2) = 2.46
mean(Set3) = 2.38
Stdev(Set1) = 0
Stdev(Set2) = 0.167
Stdev(Set3) = 0.934

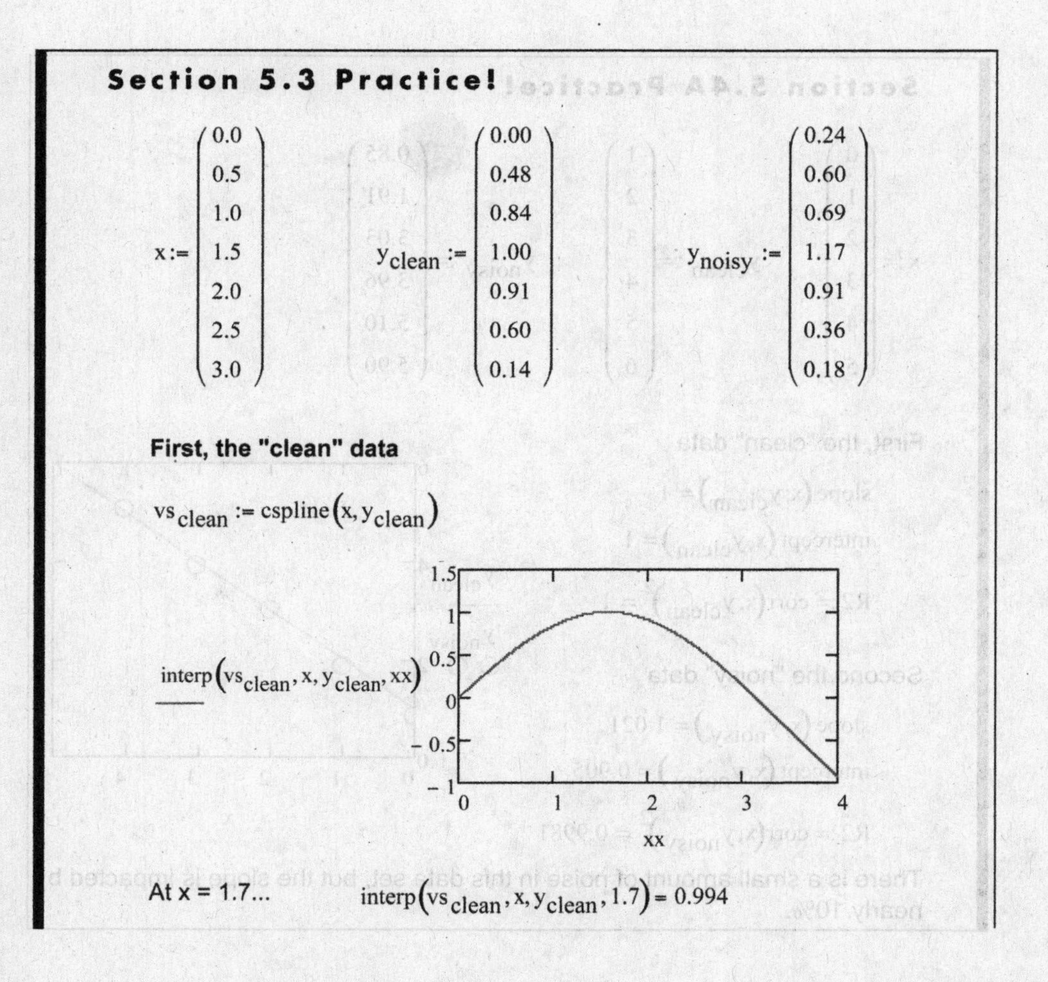
Section 5.3 Practice!
x := (0.0, 0.5, 1.0, 1.5, 2.0, 2.5, 3.0)
yclean := (0.00, 0.48, 0.84, 1.00, 0.91, 0.60, 0.14)
ynoisy := (0.24, 0.60, 0.69, 1.17, 0.91, 0.36, 0.18)
First, the "clean" data
vsclean := cspline(x, yclean)
interp(vsclean, x, yclean, xx)
1.5
1
0.5
0
− 0.5
− 1
0
1
2
3
4
xx
At x = 1.7...
interp(vsclean, x, yclean, 1.7) = 0.994

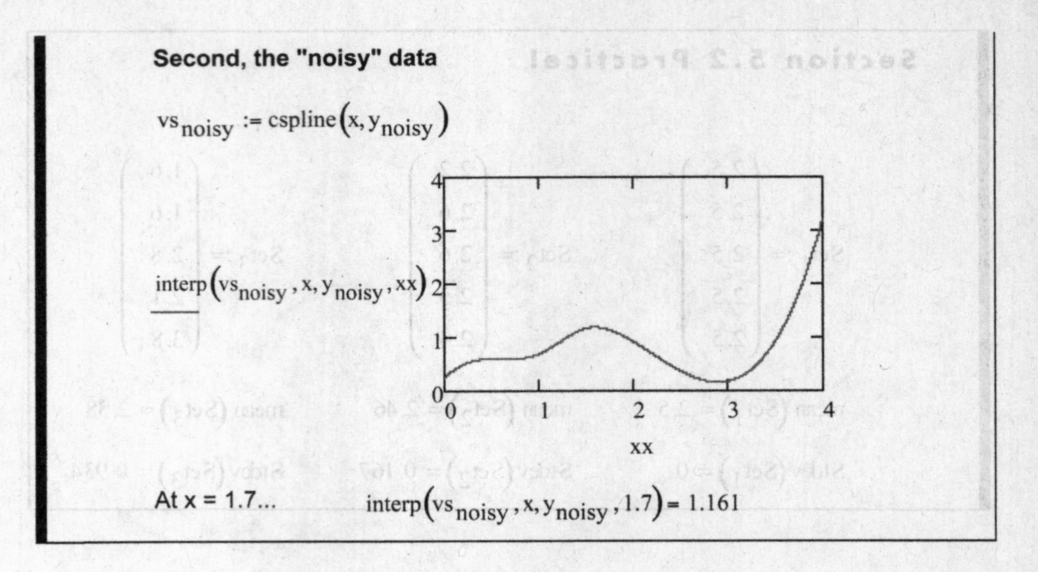
Second, the "noisy" data
vs_noisy := cspline(x, y_noisy)
interp(vs_noisy, x, y_noisy, xx)
xx
At x = 1.7...
interp(vs_noisy, x, y_noisy, 1.7) = 1.161

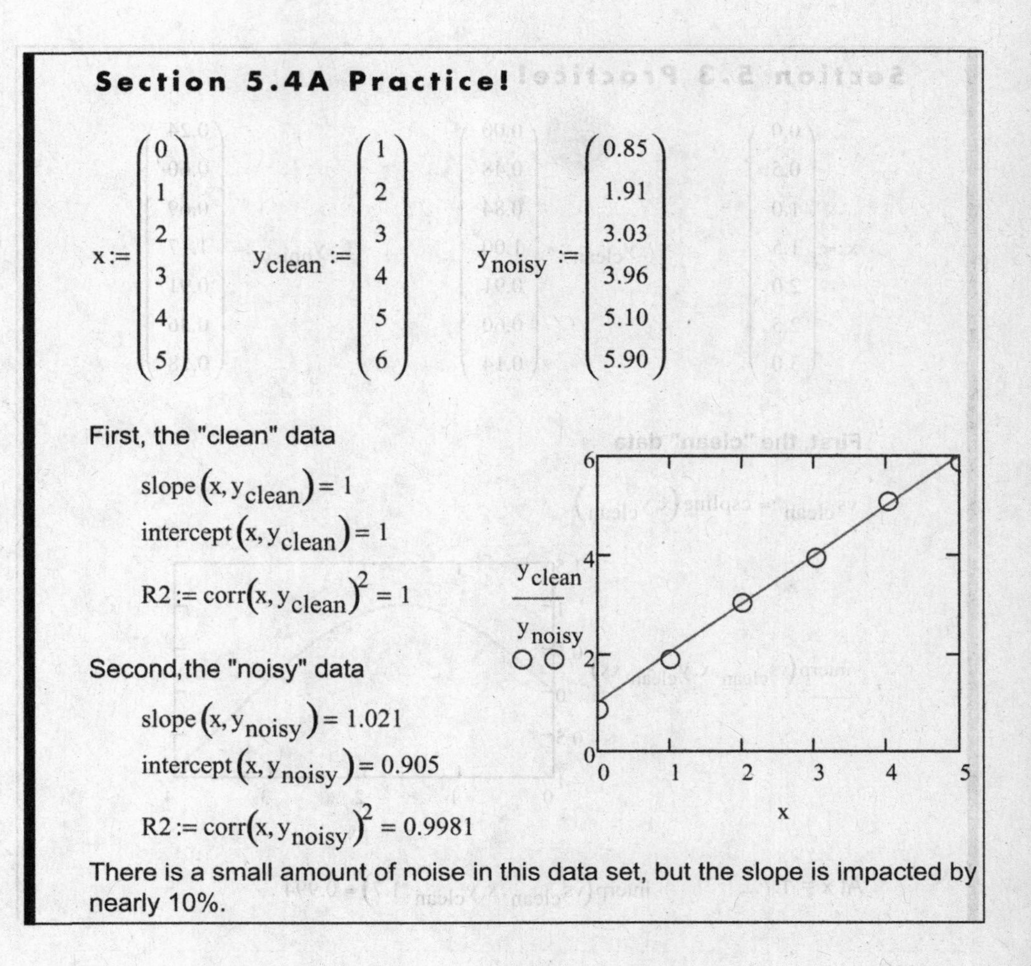
Section 5.4A Practice!
x := (0 1 2 3 4 5)
y_clean := (1 2 3 4 5 6)
y_noisy := (0.85 1.91 3.03 3.96 5.10 5.90)
First, the "clean" data
slope(x, y_clean) = 1
intercept(x, y_clean) = 1
R2 := corr(x, y_clean)^2 = 1
Second, the "noisy" data
slope(x, y_noisy) = 1.021
intercept(x, y_noisy) = 0.905
R2 := corr(x, y_noisy)^2 = 0.9981
y_clean
y_noisy
x
There is a small amount of noise in this data set, but the slope is impacted by nearly 10%.

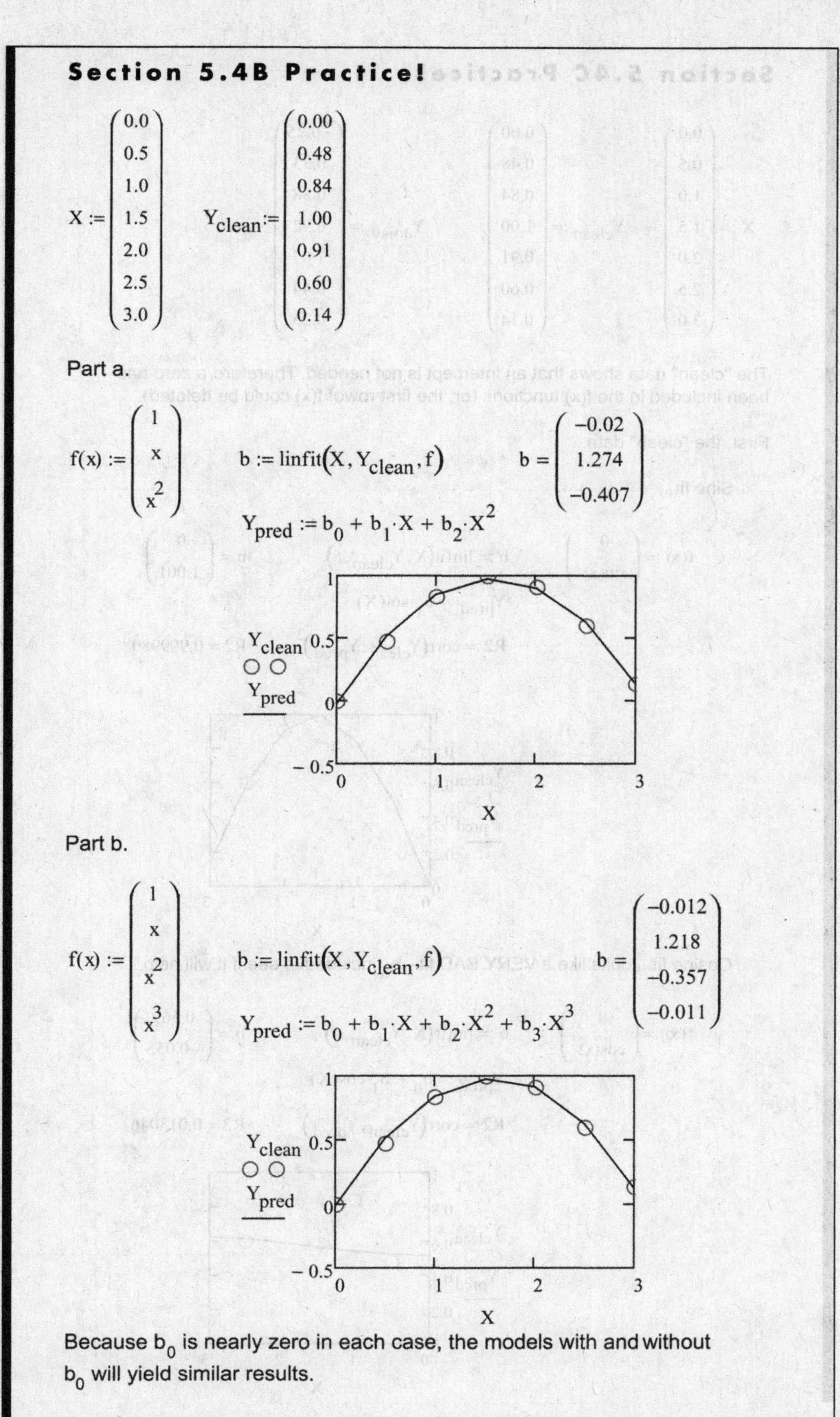

Section 5.4B Practice!

$$X := \begin{pmatrix} 0.0 \\ 0.5 \\ 1.0 \\ 1.5 \\ 2.0 \\ 2.5 \\ 3.0 \end{pmatrix} \qquad Y_{clean} := \begin{pmatrix} 0.00 \\ 0.48 \\ 0.84 \\ 1.00 \\ 0.91 \\ 0.60 \\ 0.14 \end{pmatrix}$$

Part a.

$$f(x) := \begin{pmatrix} 1 \\ x \\ x^2 \end{pmatrix} \qquad b := \text{linfit}(X, Y_{clean}, f) \qquad b = \begin{pmatrix} -0.02 \\ 1.274 \\ -0.407 \end{pmatrix}$$

$$Y_{pred} := b_0 + b_1 \cdot X + b_2 \cdot X^2$$

Part b.

$$f(x) := \begin{pmatrix} 1 \\ x \\ x^2 \\ x^3 \end{pmatrix} \qquad b := \text{linfit}(X, Y_{clean}, f) \qquad b = \begin{pmatrix} -0.012 \\ 1.218 \\ -0.357 \\ -0.011 \end{pmatrix}$$

$$Y_{pred} := b_0 + b_1 \cdot X + b_2 \cdot X^2 + b_3 \cdot X^3$$

Because b_0 is nearly zero in each case, the models with and without b_0 will yield similar results.

Section 5.4C Practice!

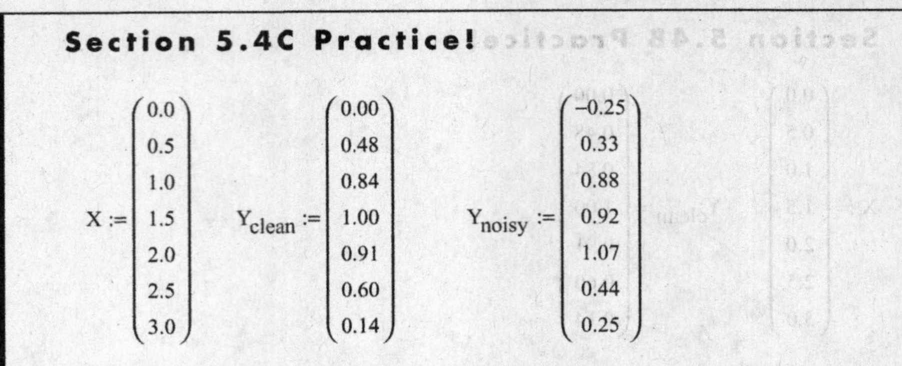

$$X := \begin{pmatrix} 0.0 \\ 0.5 \\ 1.0 \\ 1.5 \\ 2.0 \\ 2.5 \\ 3.0 \end{pmatrix} \quad Y_{clean} := \begin{pmatrix} 0.00 \\ 0.48 \\ 0.84 \\ 1.00 \\ 0.91 \\ 0.60 \\ 0.14 \end{pmatrix} \quad Y_{noisy} := \begin{pmatrix} -0.25 \\ 0.33 \\ 0.88 \\ 0.92 \\ 1.07 \\ 0.44 \\ 0.25 \end{pmatrix}$$

The "clean" data shows that an intercept is not needed. Therefore, a zero has been included in the f(x) functions (or, the first rowof f(x) could be deleted).

First, the "clean" data...

Sine fit...

$$f(x) := \begin{pmatrix} 0 \\ \sin(x) \end{pmatrix} \qquad b := \text{linfit}(X, Y_{clean}, f) \qquad b = \begin{pmatrix} 0 \\ 1.001 \end{pmatrix}$$

$$Y_{pred} := b_1 \cdot \sin(X)$$

$$R2 := \text{corr}(Y_{clean}, Y_{pred})^2 \qquad R2 = 0.999989$$

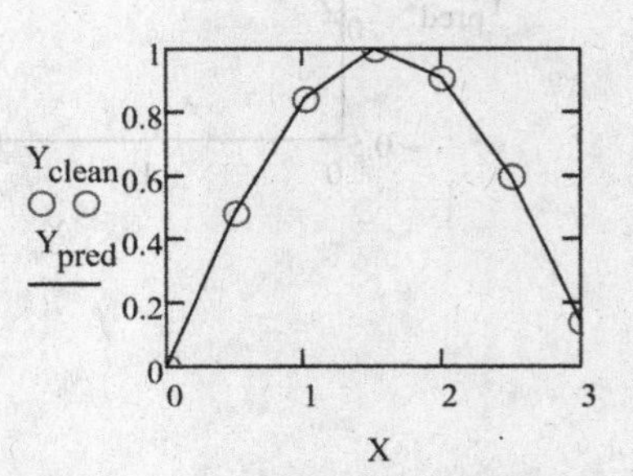

Cosine fit...looks like a VERY BAD fit - b_0 included to see if it will help...

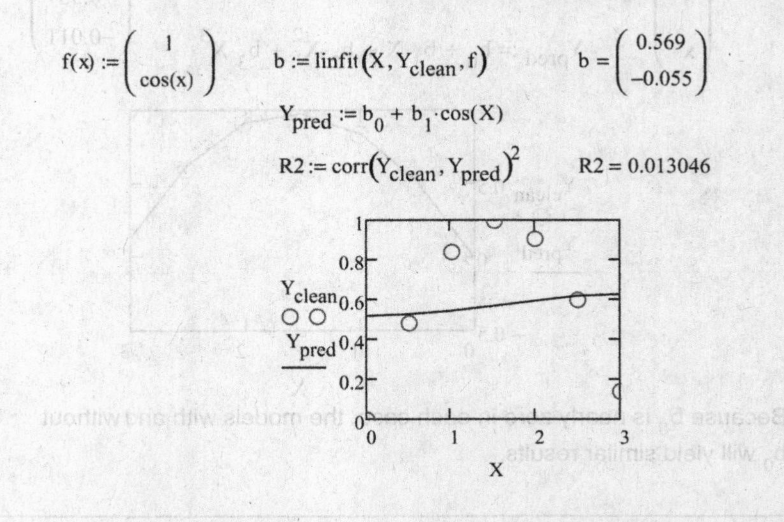

$$f(x) := \begin{pmatrix} 1 \\ \cos(x) \end{pmatrix} \qquad b := \text{linfit}(X, Y_{clean}, f) \qquad b = \begin{pmatrix} 0.569 \\ -0.055 \end{pmatrix}$$

$$Y_{pred} := b_0 + b_1 \cdot \cos(X)$$

$$R2 := \text{corr}(Y_{clean}, Y_{pred})^2 \qquad R2 = 0.013046$$

Second, the "noisy" data...

Sine fit...

$$f(x) := \begin{pmatrix} 0 \\ \sin(x) \end{pmatrix} \qquad b := \text{linfit}\left(X, Y_{noisy}, f\right) \qquad b = \begin{pmatrix} 0 \\ 0.984 \end{pmatrix}$$

$$Y_{pred} := b_1 \cdot \sin(X)$$

$$R2 := \text{corr}\left(Y_{noisy}, Y_{pred}\right)^2 \qquad R2 = 0.908391$$

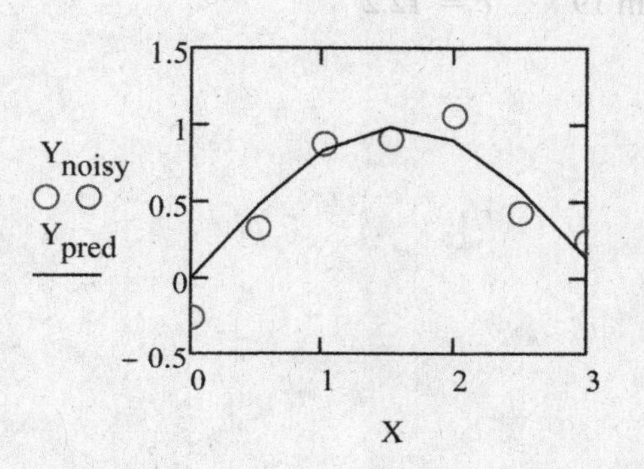

Cosine fit...looks like a VERY BAD fit - b_0 included to see if it will help...

$$f(x) := \begin{pmatrix} 1 \\ \cos(x) \end{pmatrix} \qquad b := \text{linfit}\left(X, Y_{noisy}, f\right) \qquad b = \begin{pmatrix} 0.526 \\ -0.159 \end{pmatrix}$$

$$Y_{pred} := b_0 + b_1 \cdot \cos(X)$$

$$R2 := \text{corr}\left(Y_{noisy}, Y_{pred}\right)^2 \qquad R2 = 0.074654$$

1.5
1
0.5
0
– 0.5
0
1
2
3
X
Y noisy
Y pred

ANSWERS TO SELECTED PROBLEMS

Problem 1	*CS1* mean recovery time = 5.96 days
Problem 4	b. 60 min
Problem 5	a. slope = 3, intercept = 2
Problem 9	a. slope = 0.04mV/°C
Problem 10	a. intercept = $-1.5 \times 10^{-3}\text{psi}^{0.5}$
Problem 13	b. 61.5°C
Problem 16	b. $y(x = 3) = 810$ using linear interpolation
Problem 19	$c = 12.2$

CHAPTER 6

Programming in Mathcad

Objectives

After reading this chapter, you will

- know what Mathcad programs are, and when they are useful
- know how to write a program in Mathcad
- understand how to access program keywords through the Programming Toolbar
- be able to read and write programming flowcharts
- understand the seven basic elements of all programming languages, and understand how each is implemented in Mathcad

6.1 MATHCAD PROGRAMS

A Mathcad ***program*** is essentially a *multistep function*. ***Program regions*** are not complicated, but they can be very handy to make complex functions easier to write, and easier for others to comprehend. For example, Figure 6.1 shows a **thermostat(T)** function that sets a **heater** variable to 1 (on) if the temperature is below 23°C, or −1 (off) if the temperature is above 25°C. If the temperature is between 23°C and 25°C, then **heater** is set to 0, implying that no action should be taken.

Defining the function...

$$thermostat(T) := if(T < 23, 1, if(T > 25, -1, 0))$$

Using the function...

heater := thermostat(22)	heater = 1
heater := thermostat(26)	heater = −1
heater := thermostat(24)	heater = 0

Figure 6.1
The **thermostat** function, version 1.

First, the function is written *without* using a program as shown in Figure 6.1. In Figure 6.2, we rewrite the function using a Mathcad program region (to be explained later).

Note: The **RV** in the **thermostat** program stands for ***return value.*** All functions have return values, and the multistep functions that Mathcad calls programs also have return values. By default, whatever value is assigned on the last assignment line of the program is the value returned. In the **thermostat** program, that will always be variable **RV**. You can return a different result by using the **return** program statement, which will be described later.

Defining the function...

```
thermostat(T) :=  | RV ← 0
                  | RV ← 1  if T < 23
                  | RV ← −1  if T > 25
```

Using the function...

```
heater := thermostat(22)        heater = 1
heater := thermostat(26)        heater = −1
heater := thermostat(24)        heater = 0
```

Figure 6.2
The **thermostat** function, version 2.

In both cases, the **heater** variable is assigned the same values, but most people would find

```
thermostat(T) :=  | RV ← 0
                  | RV ← 1  if T<23
                  | RV ← −1  if T>25
```

easier to understand than

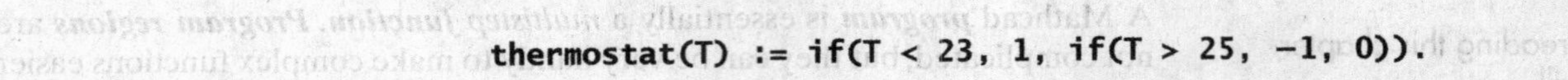

```
thermostat(T) := if(T < 23, 1, if(T > 25, -1, 0)).
```

The version shown in Figure 6.3 might be even easier for someone to understand.

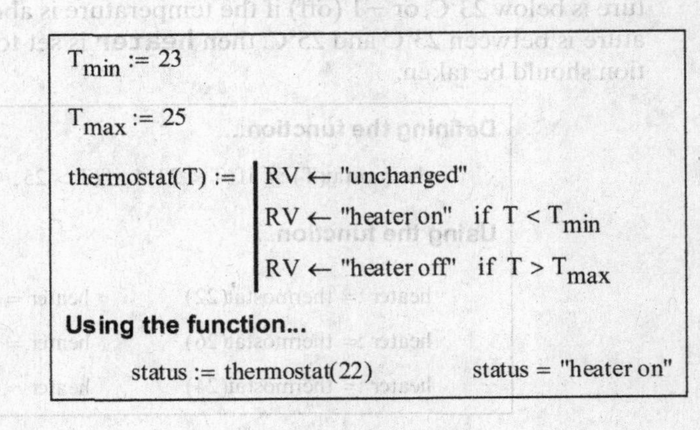

T_{min} := 23
T_{max} := 25
thermostat(T) := | RV ← "unchanged"
| RV ← "heater on" if T < T_{min}
| RV ← "heater off" if T > T_{max}
Using the function...
status := thermostat(22)
status = "heater on"

Figure 6.3
The **thermostat** function, version 3.

6.2 WRITING A SIMPLE PROGRAM

Much of this chapter will be spent presenting the basic elements of programming, but first we will develop a very simple program, just to provide an overview of how to write a program in Mathcad. Our program will calculate the area of a circle given the diameter.

Since Mathcad programs are essentially multistep functions, we begin writing a program by defining a function. This includes three pieces:

- function name (which will become the program name)
- parameter (argument) list
- assignment (define as equal to) ***operator***

The ***program name*** will be used to refer to the program whenever it is needed in the rest of the worksheet. In Figure 6.4, the program is called A_{circle}.

$$A_{circle}(diameter) := \blacksquare$$

Figure 6.4
Writing a Mathcad program, step 1: Program name and parameter list.

The ***parameter list*** (or, ***argument list***) is the list of all the variable information that must be known before the program can do its job. For example, in order to solve for the area of a circle, we need to know the diameter (and the value of π, but π is a predefined variable in Mathcad).

To create a program, simply add two or more program lines in the placeholder on the right side of the assignment operator. To do this, use the **Add Line** button on the ***Programming Toolbar*** (Figure 6.5), or the [ ]] key (i.e., press the right-bracket key). In the example in Figure 6.6, two placeholders were added.

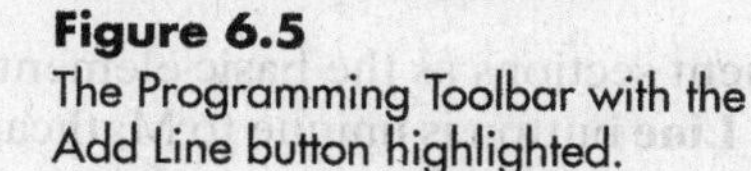

Figure 6.5
The Programming Toolbar with the Add Line button highlighted.

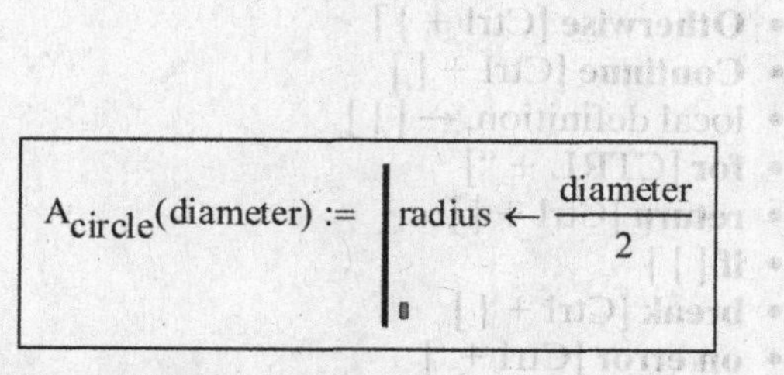

Figure 6.6
Writing a Mathcad program, step 2: Calculating radius from diameter.

Next, click on the top placeholder and add the formula for calculating a radius from a diameter, as shown in Figure 6.6.

We have created a new variable, **radius**, and assigned it a value. The left-arrow symbol, ←, is the ***local definition*** operator, which is only used in a program region. It is available either from the Programming Toolbar or by pressing the left-brace key, [{].

Note: The "local" in local definition operator means that the variable that is being defined and assigned a value (**radius**, in this example) will exist only in the program region. It can be used on any line in the program region, but the rest of the worksheet will not know that the local variable has been created and will not be able to use it. All ***local variables*** (variables defined within a program region) disappear from the computer's memory once the program stops running.

Finally, use the **radius** to calculate the **area** of the circle, as shown in Figure 6.7. By default, the value calculated (or set) on the last line of the program (**area** in this example) is the value returned to the worksheet. So, when the A_{circle} function is used (i.e., when the program is run (see Figure 6.8)), the value of variable **D** (with units) is passed into the program through the A_{circle} parameter list. Then the local variable, **radius**, is calculated (program line 1), and used (line 2) to calculate the **area** (local variable) that is passed back out of the program and assigned to the worksheet's variable **A**.

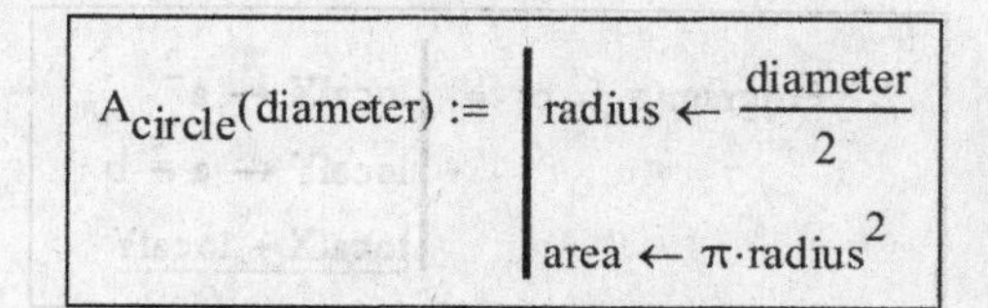

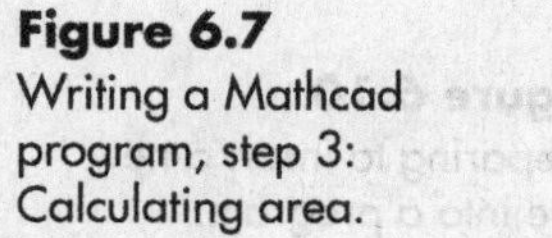

Figure 6.7
Writing a Mathcad program, step 3: Calculating area.

$D := 3 \cdot cm$

$A := A_{circle}(D) \qquad A = 7.069\,cm^2$

Figure 6.8
Using the Mathcad program to find the area of a circle.

6.3 THE PROGRAMMING TOOLBAR

The Programming Toolbar (Figure 6.9) provides access to the keywords used in Mathcad programs. The toolbar contains the following options (with shortcut keys shown next to each item):

- **Add Line** [ ]]
- **While** [Ctrl +]]
- **Otherwise** [Ctrl + }]
- **Continue** [Ctrl + [ ]
- local definition, ← [{]
- **for** [CTRL + “]
- **return** [Ctrl + |]
- **if** [}]
- **break** [Ctrl + {]
- **on error** [Ctrl + ‘]

Most of these options will be discussed in subsequent sections as the basic elements of programming are presented. However, the **Add Line** button is unique to Mathcad and is presented here.

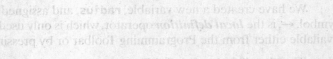

Figure 6.9
The Programming Toolbar.

6.3.1 Add Line

The **Add Line** button is used to create placeholders for each of the lines in a program. You can create as many lines as you will need before filling in any of the placeholders, or you can insert or append lines as needed.

- To insert a placeholder for a new line ahead of an existing line, select the existing line and get the insertion line (vertical blue bar) at the left side (see Figure 6.10).

$$MyProgram(a,b,c) := \left|\begin{array}{l} localX \leftarrow a^2 \\ localY \leftarrow a + b \\ localX + localY \end{array}\right.$$

Figure 6.10
Preparing to insert a new line into a program.

Then, click **Add Line** (see Figure 6.11).

```
MyProgram(a,b,c) := | localX ← a^2
                    | localY ← a + b
                    | ▮
                    | localX + localY
```

Figure 6.11
The new line appears as an empty placeholder.

- To add a line to the end of the program, get the insertion line at the right side of the last program line, as shown in Figure 6.12.

```
MyProgram(a,b,c) := | localX ← a^2
                    | localY ← a + b
                    | localX + localY|
```

Figure 6.12
Preparing to add a new line to the end of a program.

Then, click **Add Line** (see Figure 6.13).

```
MyProgram(a,b,c) := | localX ← a^2
                    | localY ← a + b
                    | localX + localY
                    | ▮
```

Figure 6.13
The new line at the end of the program.

6.4 PROGRAM FLOWCHARTS

A ***flowchart*** is a visual depiction of a program's operation. It is designed to show, step by step, what a program does. Typically, a flowchart is created before the program is written. The flowchart is used by the programmer to assist in developing the program and by others to help understand how the program works.

Some ***standard programming symbols*** used in computer flowcharts are listed in Table 6.1.

Table 6.1 Flowcharting Symbols

Symbol	Name	Usage
(rounded rectangle)	**Terminator**	Indicates the start or end of a program.
(rectangle)	**Operation**	Indicates a computation step.
(parallelogram)	**Data**	Indicates an input or output step.
(diamond)	**Decision**	Indicates a decision point in a program.
(circle)	**Connector**	Indicates that the flowchart continues in another location.

These symbols are connected by arrows to indicate how the steps are connected and the order in which the steps occur.

Flowcharting Example
As an example of a simple flowchart, consider the **thermostat** program illustrated in Figure 6.2. The program's flowchart is shown in Figure 6.14.

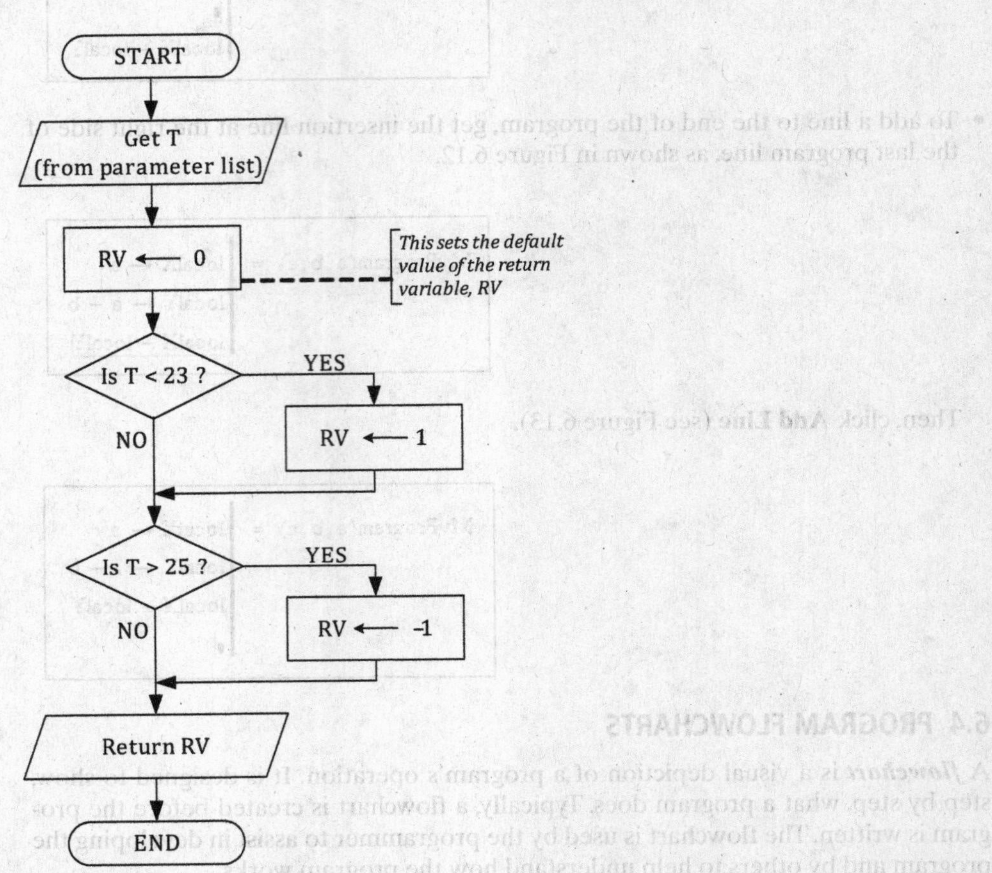

Figure 6.14
Flowchart for the **thermostat** program.

The flowchart indicates that the program starts (when Mathcad encounters the program region while evaluating the worksheet), and the value of **T** is passed into the program from the parameter list. Then, the first line of the program is an operation step: **RV ← 0**. This step ensures that the return variable, **RV**, has a value no matter what **T** value is passed into the program. This is called assigning a ***default value*** to **RV**.

The next line in the program is a ***decision step***, indicated by the diamond symbol on the flowchart. The **if T < 23** portion of the second program line (Figure 6.2) is the ***condition*** that is checked, and the left side of the line, **RV ← 1**, is the operation that is performed if the condition is found to be true.

The last line of the program is another decision step. The condition **if T > 25** is tested, and if the condition is found to be true, the operation **RV ← −1** is performed.

Just before the program ends, the value of *RV* is returned through **thermostat** so that it is available to the worksheet.

You might have noticed that the **thermostat** program tests **T** twice. Even if it has already been determined that **T < 23**, it still checks to see if **T > 25**. This is not a particularly efficient way to write a program. The last steps of a better version are shown in Figure 6.15.

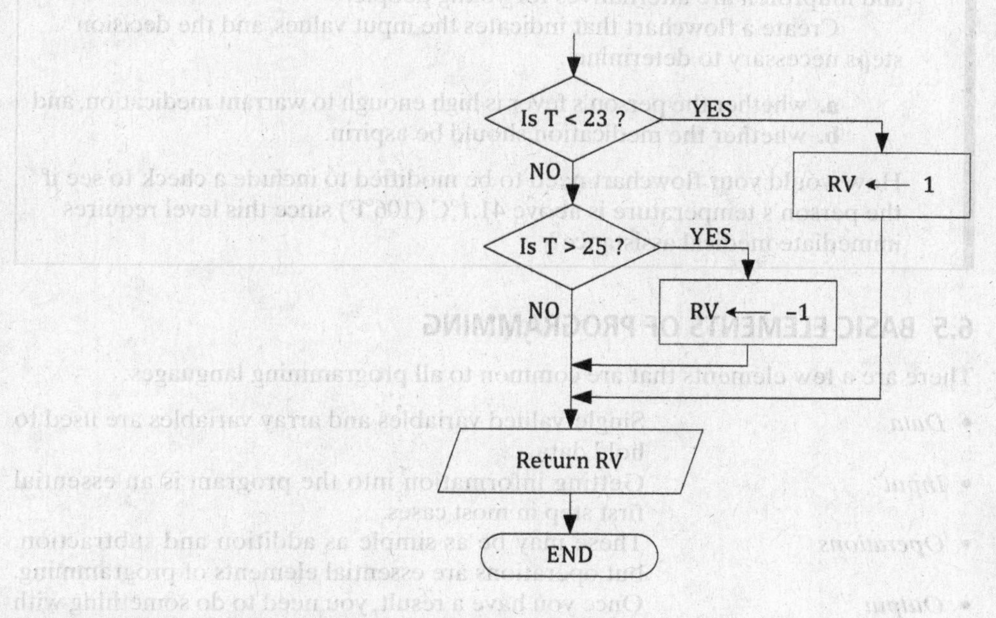

Figure 6.15
An improved **thermostat** program.

This illustrates two of the reasons for using flowcharts:

- to help identify inefficient programming
- to indicate how the program should be written

Practice!

Flowcharts may be used to depict virtually any multistep process, and they are frequently used to illustrate a decision tree, or process leading to a particular decision. This Practice! exercise is about creating a flowchart to show a common decision process: determining if someone has a fever and should take some medicine, and if that medicine should be aspirin.

Disclaimer: There is no universally accepted criterion for deciding whether someone's body temperature is high enough to require medication. The values given here are sometimes used but certainly are no replacement for good medical advice:

- For babies less than a year old, if their temperature is over 38.3°C (101°F), they should receive medication.
- For children less than twelve years old, if their temperature is over 38.9°C (102°F), they should receive medication.
- For people over twelve years old, if their temperature is over 38.3°C (101°F), they should receive medication.

Oral or ear temperatures are assumed in all cases. The threshold is slightly higher for children because their body temperatures fluctuate more than baby or adult temperatures.

Aspirin is a good fever reducer, but you should not give it to children under twelve years old (some say under nineteen years) because of the risk of a rare but serious illness known as Reye's syndrome. Acetaminophen and ibuprofen are alternatives for young people.

Create a flowchart that indicates the input values, and the decision steps necessary to determine

a. whether the person's fever is high enough to warrant medication, and
b. whether the medication should be aspirin.

How would your flowchart need to be modified to include a check to see if the person's temperature is above 41.1°C (106°F) since this level requires immediate medical assistance?

6.5 BASIC ELEMENTS OF PROGRAMMING

There are a few elements that are common to all programming languages:

- *Data* Single-valued variables and array variables are used to hold data.
- *Input* Getting information into the program is an essential first step in most cases.
- *Operations* These may be as simple as addition and subtraction, but operations are essential elements of programming.
- *Output* Once you have a result, you need to do something with it. This usually means assigning the result to a variable, saving it to a file, or displaying the result on the screen.
- *Conditional Execution* The ability to have a program decide how to respond to a situation is a very important and powerful aspect of programming.
- *Loops* Loop structures make repetitive calculations easy to perform in a program.
- *Functions* The ability to create reusable code elements that (typically) perform a single task is considered an integral part of a modern programming language.

Mathcad provides each of these elements. Access to many of these elements is provided by the Programming Toolbar.

6.5.1 Data

What a program does is manipulate ***data*** stored in variables. The data values can

- come from variables on the worksheet,
- come from parameter values passed into the program through the program's parameter list (argument list),
- be read from an external file (covered in the Input section), or
- be computed by calculations within the program.

The variables can hold a single value or an array of values.

Using Worksheet Variables in Programs

Mathcad programs work with variables in nearly the same way they are used in a Mathcad worksheet. In fact, variables assigned values on the worksheet may be used in your program as long as the variables are assigned values before the program is run. In a typical Mathcad worksheet, this means that the variables must be defined above the program.

Note: Just because you can use worksheet variables in a Mathcad program does not mean it is necessarily a good idea. One of the reasons for writing programs is to be able to copy and use the program in other places (i.e., in other worksheets). If your program depends on certain variables being predefined in the worksheet before the program runs, then every worksheet in which the program is used would have to include the same variable definitions. A preferred approach is to pass all of the information required by the program into the program via the parameter list. This will be discussed further in the next section.

Variables defined within a program can be used only within the program. In programming, variables that are only defined within a program unit are called *local variables*. The local definition symbol in a Mathcad program is a left arrow, ←. It is available either from the programming toolbar or by typing [{] (left brace key).

Consider the following example (Figure 6.16), in which variable **A** is defined on the worksheet, above the program region, and is used in the program named **addTwo**. The program name is used both to identify the program and to access the program's return variable, as shown in the last line of Figure 6.16.

Figure 6.16
Using a worksheet variable in a program.

The variable **B** is defined within the program (on the right side of the vertical line) and is a local variable. It has a value inside the program, but if you type [B] [=] on the worksheet (outside of the program region) to try to see the value of **B**, Mathcad will assume you are trying to define a new variable since **B** has not been defined on the worksheet. This is illustrated in Figure 6.17.

$$A := 12 \cdot cm^2$$

$$addTwo := \left| \begin{array}{l} B \leftarrow A + 2 \cdot cm^2 \\ B \end{array} \right.$$

$$B := \blacksquare$$

Figure 6.17
Local variable **B** is not defined outside of the program.

The value assigned to variable **B** inside the program was passed out of the program through the **addTwo** program name. To see the program's result, use **addTwo** as shown in Figure 6.16.

You should also be aware that Mathcad keeps worksheet variable definitions separate from variables defined locally within a program. In the example in Figure 6.18, variable **A** has been defined on the worksheet, and as a local variable inside the program.

Since Mathcad keeps the two types of variables separated, this is permissible, but it would be confusing to anyone trying to read your program, so it is not a very good way to write a program.

$$A := 12 \cdot cm^2$$

$$addTwo := \left| \begin{array}{l} A \leftarrow A + 2 \cdot cm^2 \\ A \end{array} \right.$$

$$addTwo = 14\ cm^2$$

$$A = 12\ cm^2$$

Figure 6.18
This program uses **A** from the worksheet to assign a value to local variable **A** (not a good programming practice).

Notice that in the last line of Figure 6.18, after running the program, worksheet variable *A* still has a value of 12 cm^2. When local variable *A* was defined inside the program, it did not change the value assigned to variable *A* on the worksheet.

To summarize:

- Variables defined on the worksheet above the program region may be used in the program, but their values will not be permanently changed by the program.
- Variables defined within a program can only be used within that program and will disappear when the program ends. (However, their value may be returned from the program before it stops running.)

Passing Values through a Parameter List

It is a good programming practice to pass all of the information required by a function into the function through the parameter list. This allows the function to be self-contained and ready to be used anywhere. This wisdom about functions also applies to Mathcad programs. If the programs are self-contained, they can be copied and reused wherever they are needed.

While you are writing a program, the variables you include in a parameter list and then use in the program definition are simply there to indicate how the various parameter values should be manipulated by the program. That is, while you are defining the program, the variables in the parameter list are ***dummy variables***. You can use any variables you want, but well-named variables will make your program easier to understand.

The two program definitions in Figure 6.19 are functionally equivalent, although the first one is preferred because the variable names have more meaning.

The flowchart for the **CylinderArea** program is shown in Figure 6.20.

Also, it does not matter if the variables used in the program definition already have their own definitions. While the program is being defined, Mathcad is not calculating anything, just defining the program. In the following example (see Figure 6.21), *D* and *L* are defined before the program is defined. (That doesn't hurt anything, but the values of *D* and *L* are not used while the program is being defined.)

When you include variables in a parameter list, only the *values* assigned to those variables are actually passed into the program. In programming terms, this is called ***passing arguments by value***. The standard programming alternative is ***passing arguments by address***, but this is not available in Mathcad. What this means is that you cannot permanently change the value of a variable passed through the parameter list. You can change its value within the program, but once the program

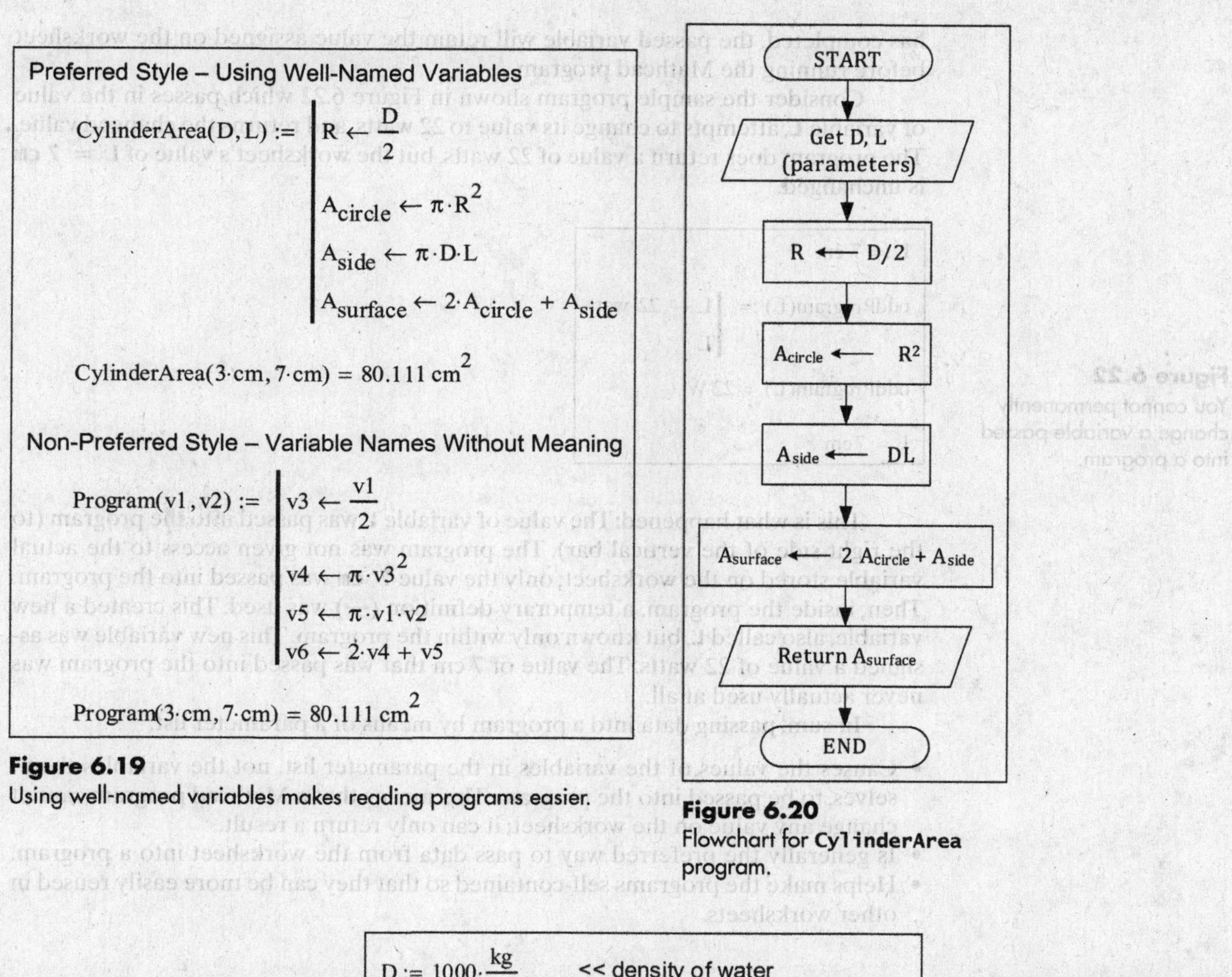

Preferred Style – Using Well-Named Variables
CylinderArea(D, L) := R ← D/2
A_circle ← π·R^2
A_side ← π·D·L
A_surface ← 2·A_circle + A_side
CylinderArea(3·cm, 7·cm) = 80.111 cm^2
Non-Preferred Style – Variable Names Without Meaning
Program(v1, v2) := v3 ← v1/2
v4 ← π·v3^2
v5 ← π·v1·v2
v6 ← 2·v4 + v5
Program(3·cm, 7·cm) = 80.111 cm^2
START
Get D, L
(parameters)
R ← D/2
A_circle ← R^2
A_side ← DL
A_surface ← 2 A_circle + A_side
Return A_surface
END

Figure 6.19
Using well-named variables makes reading programs easier.

Figure 6.20
Flowchart for `CylinderArea` program.

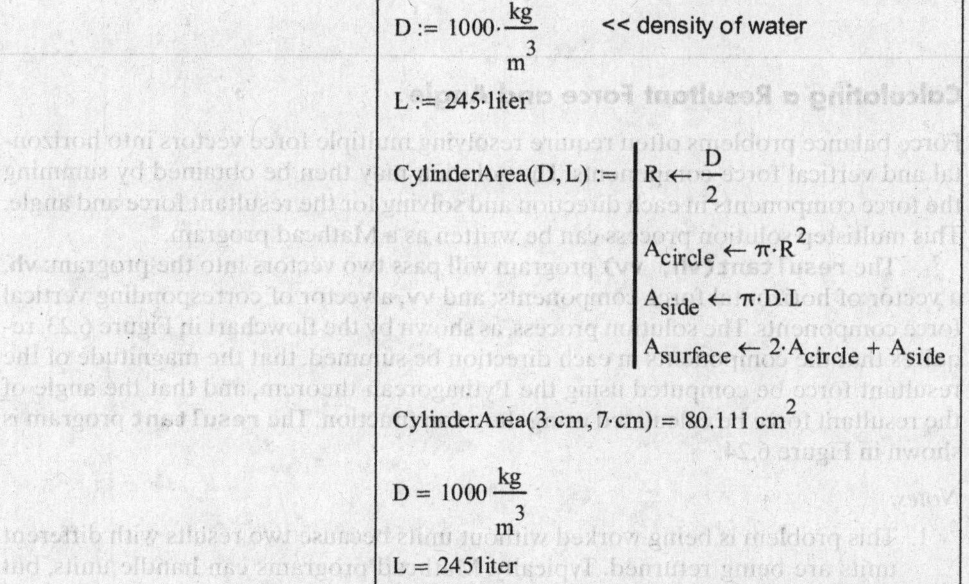

Figure 6.21
Variables used in program definitions are dummy variables and do not overwrite worksheet definitions.

has completed, the passed variable will retain the value assigned on the worksheet before running the Mathcad program.

Consider the sample program shown in Figure 6.22 which passes in the value of variable **L**, attempts to change its value to 22 watts, and returns the changed value. The program does return a value of 22 watts, but the worksheet's value of **L = 7 cm** is unchanged.

```
L := 7·cm

oddProgram(L) := | L ← 22·watt
                 | L

oddProgram(L) = 22 W

L = 7cm
```

Figure 6.22
You cannot permanently change a variable passed into a program.

This is what happened: The value of variable **L** was passed into the program (to the right side of the vertical bar). The program was not given access to the actual variable stored on the worksheet; only the value **7 cm** was passed into the program. Then, inside the program, a temporary definition (←) was used. This created a new variable, also called **L**, but known only within the program. This new variable was assigned a value of 22 watts. The value of 7 cm that was passed into the program was never actually used at all.

In sum, passing data into a program by means of a parameter list:

- Causes the values of the variables in the parameter list, not the variables themselves, to be passed into the program. This means that a Mathcad program cannot change any value on the worksheet; it can only return a result.
- Is generally the preferred way to pass data from the worksheet into a program.
- Helps make the programs self-contained so that they can be more easily reused in other worksheets.

APPLICATION

Calculating a Resultant Force and Angle

Force balance problems often require resolving multiple force vectors into horizontal and vertical force components. The solution may then be obtained by summing the force components in each direction and solving for the resultant force and angle. This multistep solution process can be written as a Mathcad program.

The **resultant(vh, vv)** program will pass two vectors into the program: **vh**, a vector of horizontal force components; and **vv**, a vector of corresponding vertical force components. The solution process, as shown by the flowchart in Figure 6.23, requires that the components in each direction be summed, that the magnitude of the resultant force be computed using the Pythagorean theorem, and that the angle of the resultant force be calculated using the **atan** function. The **resultant** program is shown in Figure 6.24.

Notes:

1. This problem is being worked without units because two results with different units are being returned. Typically, Mathcad programs can handle units, but

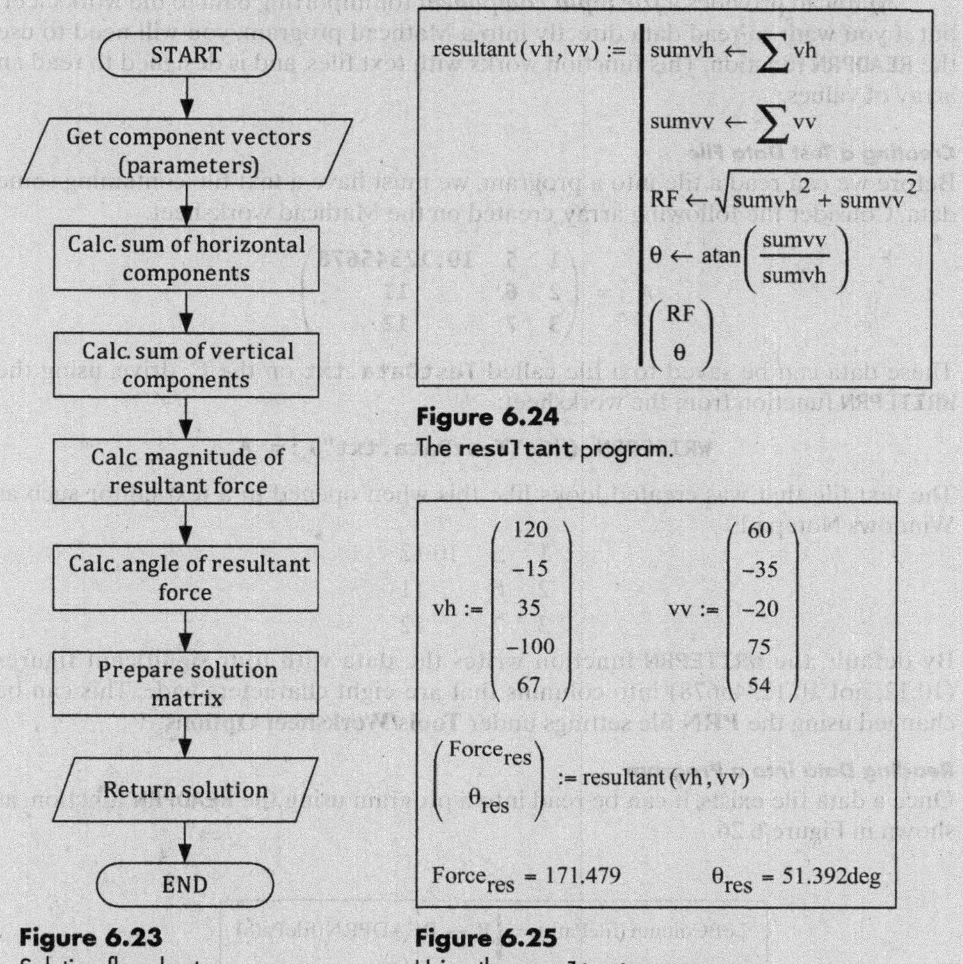

Figure 6.23
Solution flowchart.

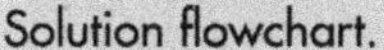

Figure 6.24
The **resultant** program.

Figure 6.25
Using the **resultant** program.

when multiple results are returned as a vector, each result must have the same units.

2. The summation operators used in the first two lines of the program are Mathcad's vector sum operators from the Matrix toolbar. These operators provide a quick way to calculate the sum of the values in a vector.

When the resultant program is run, the magnitude and direction of the resultant vector are returned. The angle is in radians (Mathcad's default), but here it has been converted to degrees using the **deg** unit. (See Figure 6.25.)

6.5.2 Input

There are a variety of ***input sources*** available on computers: keyboard, disk drives, USB drive, a mouse, microphone, and more. Because Mathcad programs are housed within a worksheet, there are basically two available input sources: data available on the worksheet itself, and data in files. The use of worksheet data was covered in the previous section, so this section deals only with reading data files.

Mathcad provides a ***file input component*** for importing data to the worksheet, but if you want to read data directly into a Mathcad program, you will need to use the **READPRN** function. This function works with text files, and is designed to read an array of values.

Creating a Test Data File

Before we can read a file into a program, we must have a text file containing some data. Consider the following array, created on the Mathcad worksheet:

$$\mathbf{A} := \begin{pmatrix} 1 & 5 & 10.12345678 \\ 2 & 6 & 11 \\ 3 & 7 & 12 \end{pmatrix}$$

These data can be saved to a file called `TestData.txt` on the C: drive, using the **WRITEPRN** function from the worksheet:

```
WRITEPRN ("C:\TestData.txt") := A
```

The text file that was created looks like this when opened in a text editor such as Windows Notepad:

```
1   5   10.12
2   6      11
3   7      12
```

By default, the **WRITEPRN** function writes the data with four significant figures (10.12, not 10.12345678) into columns that are eight characters wide. This can be changed using the **PRN** file settings under **Tools/Worksheet Options**.

Reading Data into a Program

Once a data file exists, it can be read into a program using the **READPRN** function, as shown in Figure 6.26.

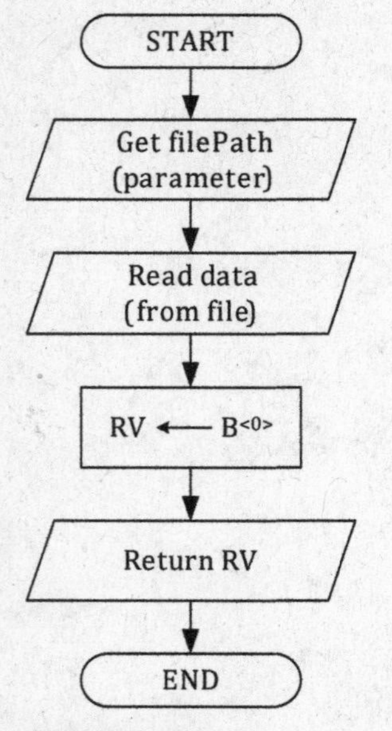

Figure 6.27
Flowchart of `LeftColumn` program.

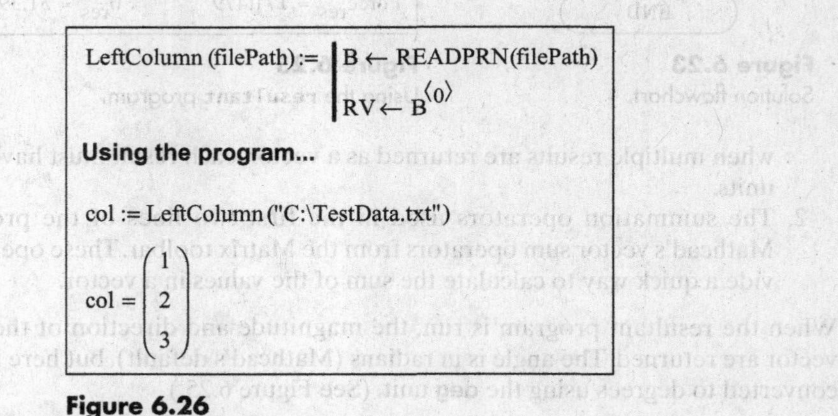

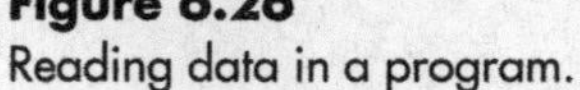

Figure 6.26
Reading data in a program.

In the `LeftColumn` program (see flowchart in Figure 6.27), the file name and drive location are passed into the program through variable `filePath`. The file is read and the data is assigned to local variable **B**. On the last line of the program, the left column of array **B** is selected using the column operator and assigned to variable **RV** to be returned by the program. When the program is run, the `TestData.txt` file is read, and the left column is returned.

Note: **RV** was being used in the previous program to indicate where the program's return value is set, but the use of that particular variable name (or any variable at all) is not required. The program shown in Figure 6.28 returns exactly the same result.

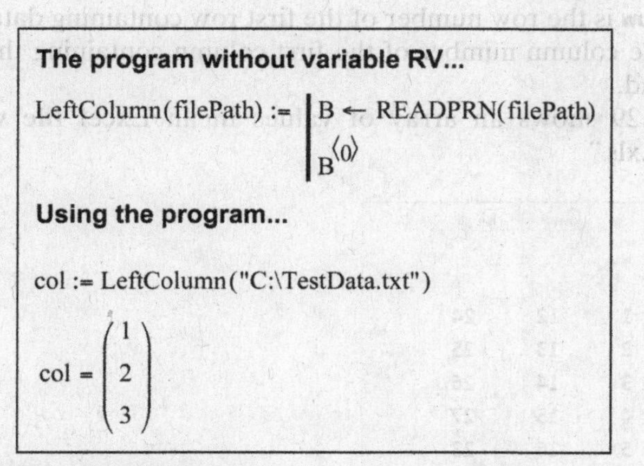

Figure 6.28
The explicit return variable, **RV**, is not required.

An alternative to using the **READPRN** function to read data from a text file is to use the **READFILE** function. **READFILE** can read delimited files (just like **READPRN**), fixed column width files, and Excel files. The most likely reason to use **READFILE** is to read data from an Excel file into a program.

Reading Excel Files

When used to read data from an Excel file, the **READFILE** function requires either two or four parameters. One option is to supply just the Excel path name and the file type, as

```
Readfile(filePath, "Excel")
```

When only the path name and file type are supplied, Mathcad will attempt to read all contiguous data from the first sheet of the Excel workbook. Empty cells are filled with *NAN* (not a number) by default. If you use only two parameters to read an Excel file, your Excel data should be in a cell range that starts in cell *A1*.

Notes:

- Be sure to save the Excel file before reading values into Mathcad; this ensures that Mathcad is reading the current values in the spreadsheet.
- Mathcad 15 can read data from files saved in the Excel 2007 default format; Mathcad 14 cannot. You must save your Excel file in Excel 2003 format for Mathcad 14.
- With Mathcad 15 you can use the **READEXCEL** function to access Excel 2007 files. To read all values on the first worksheet, simply send the file path as the function's argument. To read a specific range of cells (including cells on various worksheets) send both the file path and the cell range (as a text string) to the function. Both usages are illustrated below:

```
READEXCEL("C:\ExcelFile.xlsx")
READEXCEL("C:\ExcelFile.xlsx", "Sheet2!B2:E5")
```

If you want Mathcad to read a cell range that does not start in cell *A1*, use the four-parameter version, as

```
READFILE(filePath, "Excel", startRow, startCol)
```

where **startRow** is the row number of the first row containing data to be read, and **startCol** is the column number of the first column containing the data you want Mathcad to read.

Figure 6.29 shows an array of values in an Excel file with path name "C:\ExcelData.xls."

	A	B	C	D	E
1					
2					
3		1	12	24	
4		2	13	25	
5		3	14	26	
6		4	15	27	
7		5	16	28	
8					

Figure 6.29
Data in Excel file C:\ExcelData.xls.

Since the Excel data array does not start in cell *A1*, we must use the four-parameter version of **READFILE** function call to read the Excel data. First, we will use **READFILE** outside of a program, on the worksheet. This is illustrated in Figure 6.30.

rowStart := 3

colStart := 2

G := READFILE("C:\ExcelData.xls", "Excel", rowStart, colStart)

$$G = \begin{pmatrix} 1 & 12 & 24 \\ 2 & 13 & 25 \\ 3 & 14 & 26 \\ 4 & 15 & 27 \\ 5 & 16 & 28 \end{pmatrix}$$

Figure 6.30
Using **READFILE** on the worksheet.

In Figure 6.30, the **rowStart** and **colStart** variables were used to tell Mathcad to start reading Excel data at cell *B3*, and Mathcad read the entire array.

If you want to read only part of the array, you have to tell Mathcad where to start and stop. This is done by passing **READFILE** a two-element vector indicating the starting and stopping rows, and a two-element vector indicating the starting and stopping columns. For example, to read only the values in the cell range *C4:D6* (row 4, col 3 to row 6, col 4), we would use **READFILE** as illustrated in Figure 6.31.

Since it is easier to get data from an Excel file into a Mathcad worksheet using the ***Data File Import component***, the **READFILE** function tends to be used only in programs since the components cannot be used inside programs. An example of using the **READFILE** function within a program is shown in Figure 6.32.

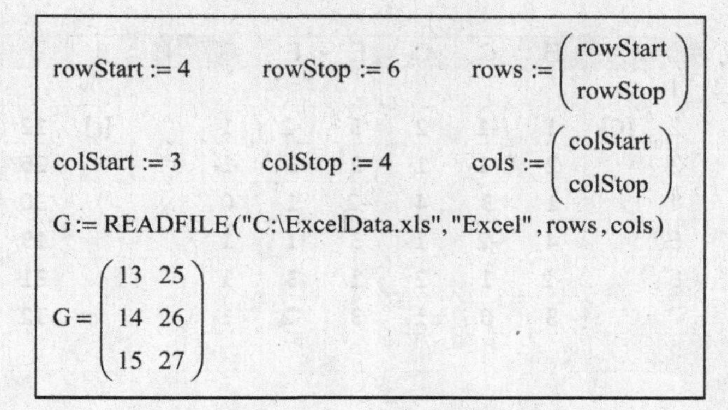

Figure 6.31
Reading only part of an Excel array with **READFILE**.

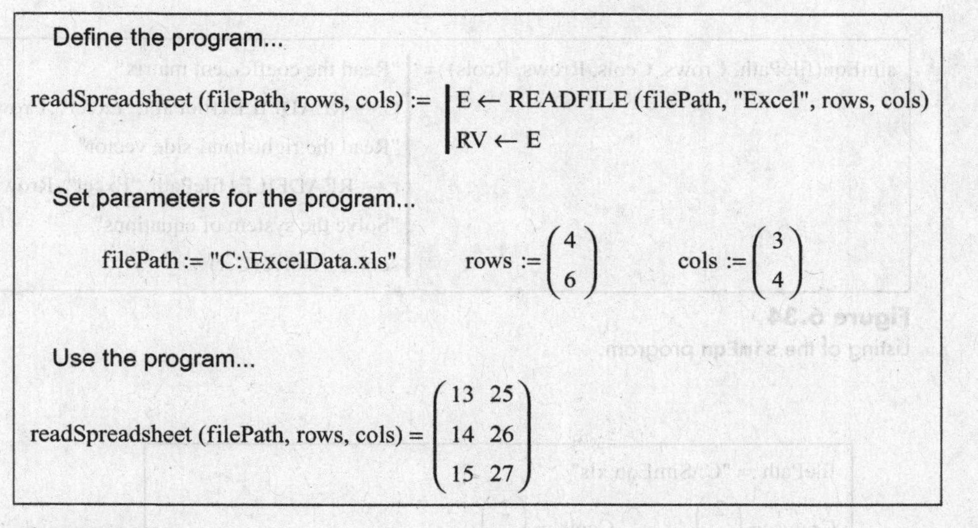

Figure 6.32
A program that reads data from an Excel file.

A program that actually does something with the data from an Excel file is shown in the following Application.

APPLICATION

Using Excel Data

The Excel data shown in Figure 6.33 represent the coefficient matrix and right-hand-side vector for a set of simultaneous linear equations. The file has been saved as C:\SimEqn.xls. The coefficient matrix is in cells *B2:G7* (row 2, col 2 to row 7, col 7), while the right-hand-side vector is in cells *J2:J7* (row 2, col 10 to row 7, col 10).

We can write a program that will read the Excel file and solve the system of linear equations using Mathcad's **lsolve** function. The program listing is shown in Figure 6.34.

The comments (in quotes) in the program are designed to help readers understand the program but are not necessary for the program to operate correctly. The usage of the program is illustrated in Figure 6.35. The solution to the set of simultaneous linear equations is shown as the **x**-vector in Figure 6.35.

	A	B	C	D	E	F	G	H	I	J	K
1											
2	[C]	1	-1	2	3	-2	1		[r]	12	
3		2	-1	1	2	3	1			26	
4		1	3	4	-2	1	0			30	
5		4	-2	1	3	1	1			19	
6		-2	1	2	1	3	1			31	
7		3	0	1	3	-2	1			12	
8											

Figure 6.33
Excel data for simultaneous equations problem.

```
simEqn(filePath, Crows, Ccols, Rrows, Rcols) := | "Read the coeffecient matrix"
                                                | C ← READFILE(filePath, "Excel", Crows, Ccols)
                                                | "Read the right-hand-side vector"
                                                | r ← READFILE(filePath, "Excel", Rrows, Rcols)
                                                | "Solve the system of equations"
                                                | soln ← lsolve(C, r)
```

Figure 6.34
Listing of the **simEqn** program.

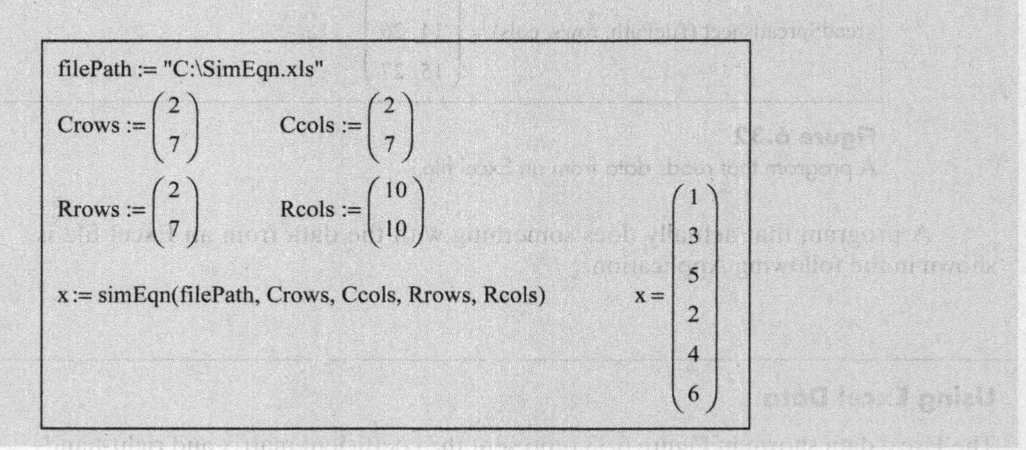

filePath := "C:\SimEqn.xls"
Crows := (2 7)
Ccols := (2 7)
Rrows := (2 7)
Rcols := (10 10)
x := simEqn(filePath, Crows, Ccols, Rrows, Rcols)
x = (1 3 5 2 4 6)

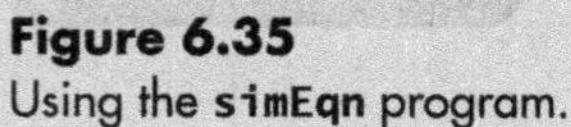

Figure 6.35
Using the **simEqn** program.

6.5.3 Operations

Most of the Mathcad's standard operators can be used in programs. These operators are listed in the table that follows. The exceptions are the "define as equal to" operators, := and ≡, which may not be used inside a program. Only the local definition operator, ←, may be used inside of a program.

Standard Math Operators

Symbol	Name	Shortcut Key
+	Addition	+
−	Subtraction	−
*	Multiplication	[Shift-8]
/	Division	/
e^x	Exponential	
1/x	Inverse	
x^y	Raise to a Power	[^], or [Shift-6]
n!	Factorial	!
|x|	Absolute Value	
$\sqrt{\ }$	Square Root	\
$\sqrt[n]{\ }$	N^{th} Root	[Ctrl-\]

Matrix Operators

Symbol	Name	Shortcut Key
+	Addition	+
−	Subtraction	−
*	Multiplication	[Shift-8]
/	Division	/
A_n	Array Element	[
M^{-1}	Matrix Inverse	
|M|	Determinant	|
→	Vectorize	[Ctrl- −]
$M^{<x>}$	Matrix Column	[Ctrl-6]
M^T	Matrix Transpose	[Ctrl-1]
$M_1 \cdot M_2$	Dot Product	*, or [Shift-8]
$M_1 \times M_2$	Cross Product	[Ctrl-8]

Operator Precedence Rules

Mathcad evaluates expressions from left to right starting at the assignment operator following standard ***operator precedence*** rules.

Operator Precedence

Precedence	Operator	Name
First	^	Exponentiation
Second	*, /	Multiplication, Division
Third	+ , −	Addition, Subtraction

For example, you might see the following equation in a Mathcad worksheet:

```
C := A · B + E · F
```

You would need to know that Mathcad multiplies before it adds (operator precedence) in order to understand that the equation would be evaluated as

```
C := (A · B) + (E · F)
```

It is a good idea to include the parentheses to make the order of evaluation clear.

Local Definition Operator, ←

Variables may be defined and assigned values inside a program, but such variables can be used only inside the program (they are of ***local scope***) and lose their values when the program terminates (they are ***temporary variables***). The local definition operator, ←, is used to assign values to these local variables. You can either use the Programming Toolbar to enter the local definition operator or press the left-brace key, [{]. The local definition operator is the only way to assign a value to a variable inside a program; the "define as equal to" operator, :=, cannot be used inside a program (except, of course, right after the program name).

APPLICATION

Alternative Fuel Calculations

The search is on to find the fuels of the future. Potential alternative fuels include nuclear, solar, wind, and tidal power. Although there are substantial obstacles to overcome if some of these sources are to be effectively utilized on a large scale, there also are significant problems to solve if we continue to rely on predominantly fossil fuels. Securing the world's energy supply is a major challenge facing the next generation of engineers.

According to information provided by the Energy Information Administration, a branch of the U.S. Department of Energy, the people of the world use over 150 quadrillion British thermal units (BTUs) (150×10^{15} BTU $= 158 \times 10^{15}$ kJ) of energy from petroleum products each year. To reduce the use of petroleum, some of that energy would need to come from alternative sources such as solar or wind energy. This problem considers what it would take to replace 10% of the current world petroleum usage using today's solar and wind technologies.

SOLAR POWER

Current commercial photovoltaic cells can produce, at peak light intensity, about 250 watts per square meter of panel surface. Assuming that the panels will receive peak light intensity for eight hours per day, how many square meters of solar panels would be required to replace 10% of the world's petroleum usage? Figure 6.36 shows the **solar** program used to calculate the answer.

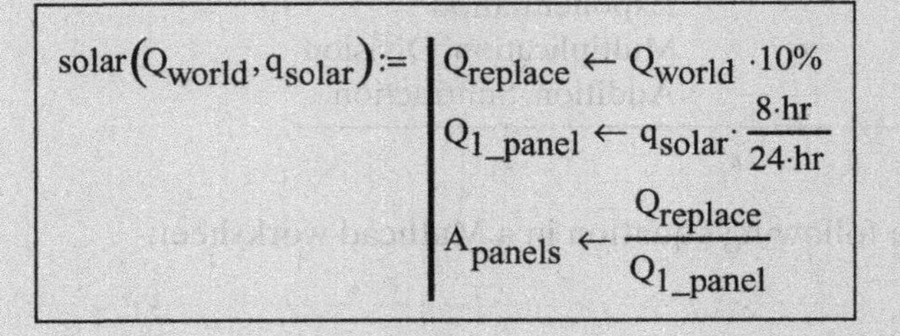

$solar(Q_{world}, q_{solar}) :=$
$Q_{replace} \leftarrow Q_{world} \cdot 10\%$
$Q_{1_panel} \leftarrow q_{solar} \cdot \frac{8 \cdot hr}{24 \cdot hr}$
$A_{panels} \leftarrow \frac{Q_{replace}}{Q_{1_panel}}$

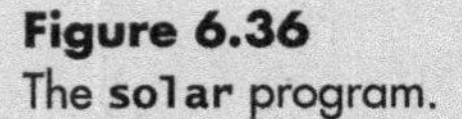

Figure 6.36
The **solar** program.

The required area is about 2300 square miles (6000 km^2) (see Figure 6.37).

$kJ := 1000 \cdot joule$

$Q_{world} := 150 \cdot 10^{15} \cdot \frac{BTU}{yr}$

$Q_{world} = 5.015 \times 10^{9}\ kW$

$q_{solar} := 250 \cdot \frac{watt}{m^2}$

$Area_{needed} := solar(Q_{world}, q_{solar})$

$Area_{needed} = 6.018 \times 10^{9}\ m^2$

$Area_{needed} = 2.324 \times 10^{3}\ mi^2$

Figure 6.37
Using the **solar** program.

WIND ENERGY

As of 2010, a company in Norway is developing a wind turbine capable of generating 10 megawatts of power. How many of these would be required to replace 10% of the world's petroleum usage? Assume that favorable winds keep the turbines running 75% of the time. To arrive at the answer, we use the **wind** program shown in Figure 6.38.

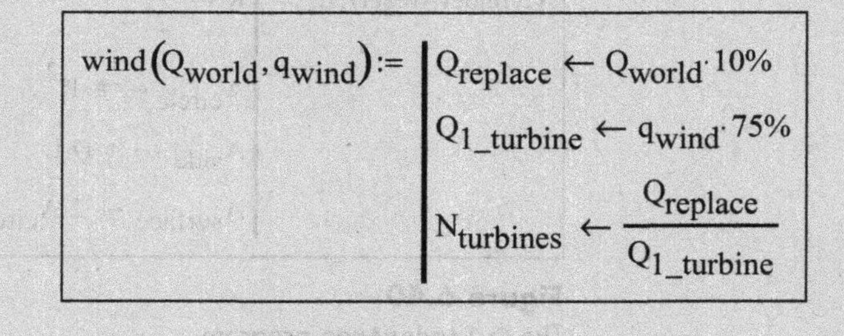

Figure 6.38
The **wind** program.

As shown in Figure 6.39, it would take 67,000 of these wind turbines to replace 10% of the world's petroleum usage.

$Q_{world} := 150 \cdot 10^{15} \cdot \frac{BTU}{yr}$

$Q_{world} = 5.015 \times 10^{9}\ kW$

$q_{wind} := 10 \cdot MW$

$N_{needed} := wind(Q_{world}, q_{wind})$

$N_{needed} = 66867$

Figure 6.39
Using the **wind** program.

Note: Neither of these examples is intended to suggest that it is futile to try to develop alternative energy sources. They were selected to illustrate that current technologies will require dramatic improvements if they are to succeed. The good news is that those improvements are happening fast, but we have a long way to go. By developing alternative energy, technologies will provide lots of opportunities for good engineering work in the years to come!

6.5.4 Output

There are two ways to send information calculated within a program back out of the program:

- return the value or values from the program back to the worksheet, or
- save the value or values to a file using **WRITEPRN(filePath, outputVariable)**.

You cannot assign values to a worksheet variable from within a program.

The usual way to get information out of a program is to return values to the worksheet, and that method will be illustrated in this section.

By default, the last value assigned in a program is the program's return value. This is usually the assignment on the last line of the program, such as $A_{surface}$ as shown in Figure 6.40.

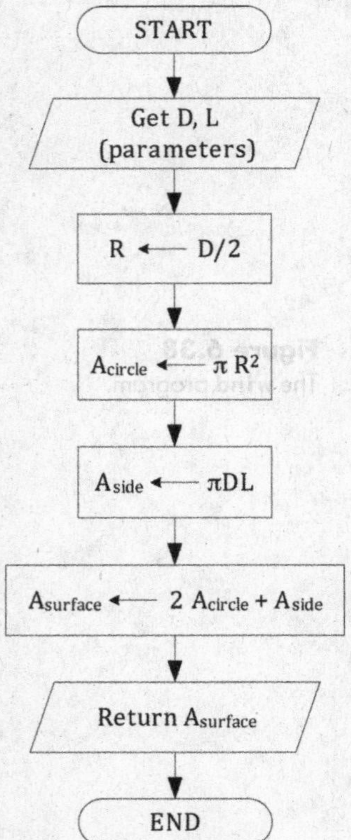

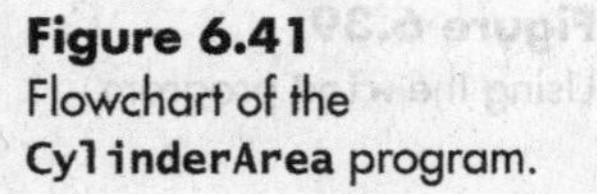
Figure 6.41
Flowchart of the **CylinderArea** program.

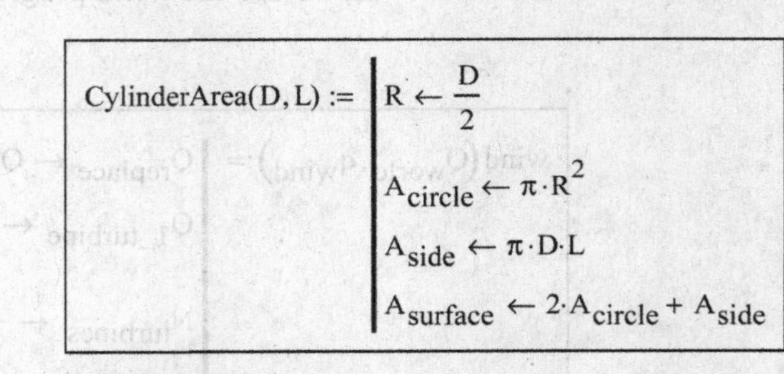

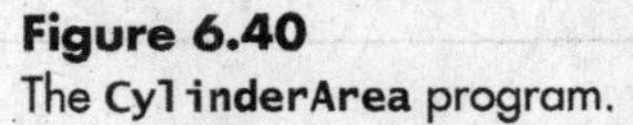
Figure 6.40
The **CylinderArea** program.

In the **CylinderArea** program shown in Figure 6.40 (diagrammed in Figure 6.41), $A_{surface}$ is the last assignment statement, so that is the value returned by the program.

Figure 6.42 shows how to modify the **CylinderArea** program to send the calculated surface area to the file C:\SurfaceData.txt. The calculated surface area is still returned from the program as well as being written to the file.

The only time the value of the variable on the last line might not be the return value if the variable on the last line is never assigned a value. An example of this is the **thermostat** program introduced earlier:

```
thermostat(T) := | RV ← 0
                 | RV ← 1  if T < 23
                 | RV ← -1  if T > 25
```

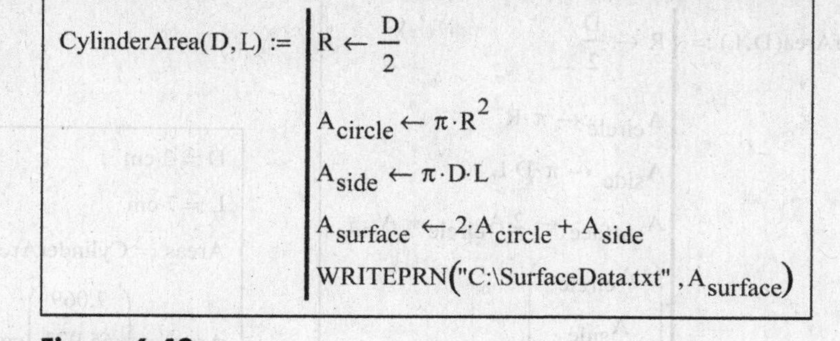

Figure 6.42
The **CylinderArea** program with file output.

The assignment of **−1** to **RV** on the last line happens only if **T > 25**. Otherwise, no assignment is made on the last line of the program. A slight modification of this program might make this more apparent:

```
thermostat(T) := | A ← 0
                 | B ← 1  if T< 23
                 | C ← −1  if T>25
```

Now, **C** is only assigned a value if **T > 25**. If T is not greater than 25, then **C** is never assigned a value at all and this program's return value is held in variable **B** (if **T < 23**) or **A**.

So, a program's return value is the value calculated or assigned on the last line of the program, unless no value is calculated or assigned on the last line of the program. Specifically, the return value is the last value assigned to a variable.

```
demoReturn := | x ← 3
              | y ← 5
              | return  x
              | z ← 7

demoReturn = 3
```

Figure 6.43
Using the **return** statement.

Return Statement

You can use a ***return statement*** to override the default and specify a different value as the value to be returned by the **demoReturn** program shown in Figure 6.43 and diagrammed in Figure 6.44.

In this program, the **return x** statement says to return the value of **x**, which is **3**. Without the **return** statement, the value on the last line (the value of **z**, or **7**) would be returned.

A **return** statement can be used to set the return value from anywhere in the program, which can be useful when there are numerous ***loops*** and/or **if** ***statements***.

Note: You must use the Programming Toolbar (or the keyboard shortcut, [Ctrl-Shift |]) to enter a **return** statement into a function; typing the word "return" does not work.

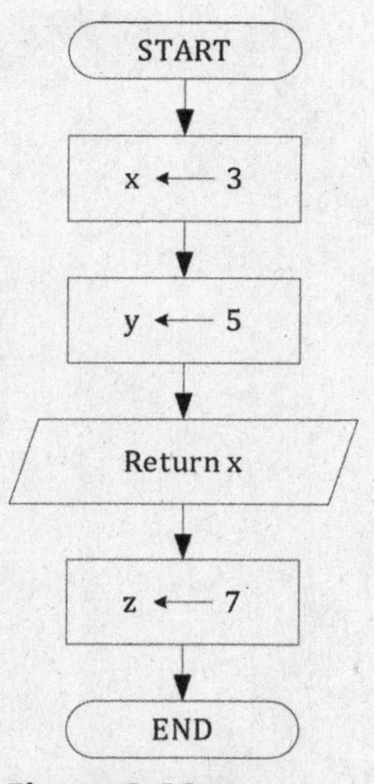

Figure 6.44
Flowchart of **demoReturn** program.

Returning Multiple Values

You can return multiple values from a program by returning them as an array. For example, if you wanted to return each of the computed areas from the **CylinderArea** program, you would add an array definition on the (new) last line of the program, as indicated in Figure 6.45. When the program is used, all three computed areas will be returned as a vector (see Figure 6.46). Figure 6.47 is a flowchart of the modified **CylinderArea** program.

If you want the results available individually after the program is run, assign the returned values to elements of a vector by placing a vector of variables on the left side of an assignment operator, **:=**. This is illustrated in Figure 6.48.

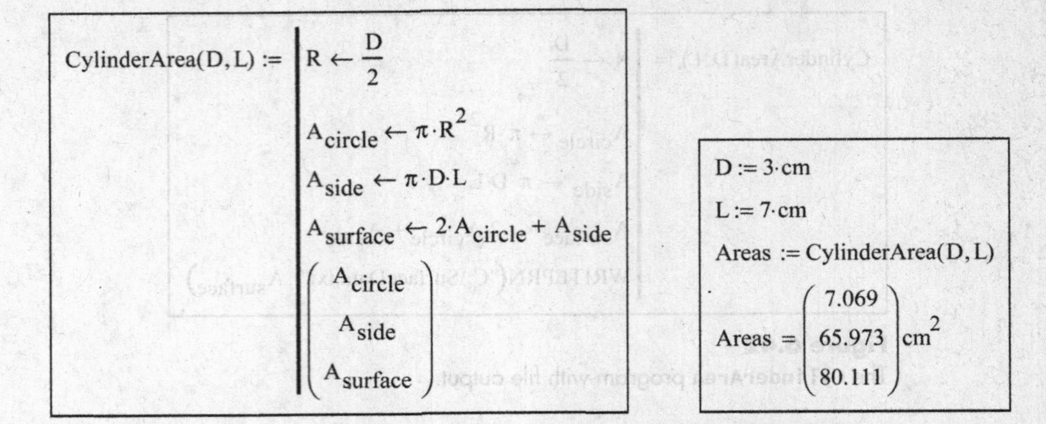

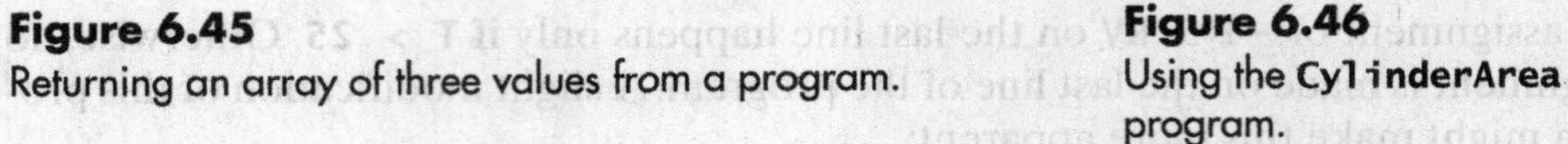

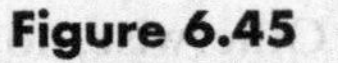

Figure 6.45
Returning an array of three values from a program.

Figure 6.46
Using the **CylinderArea** program.

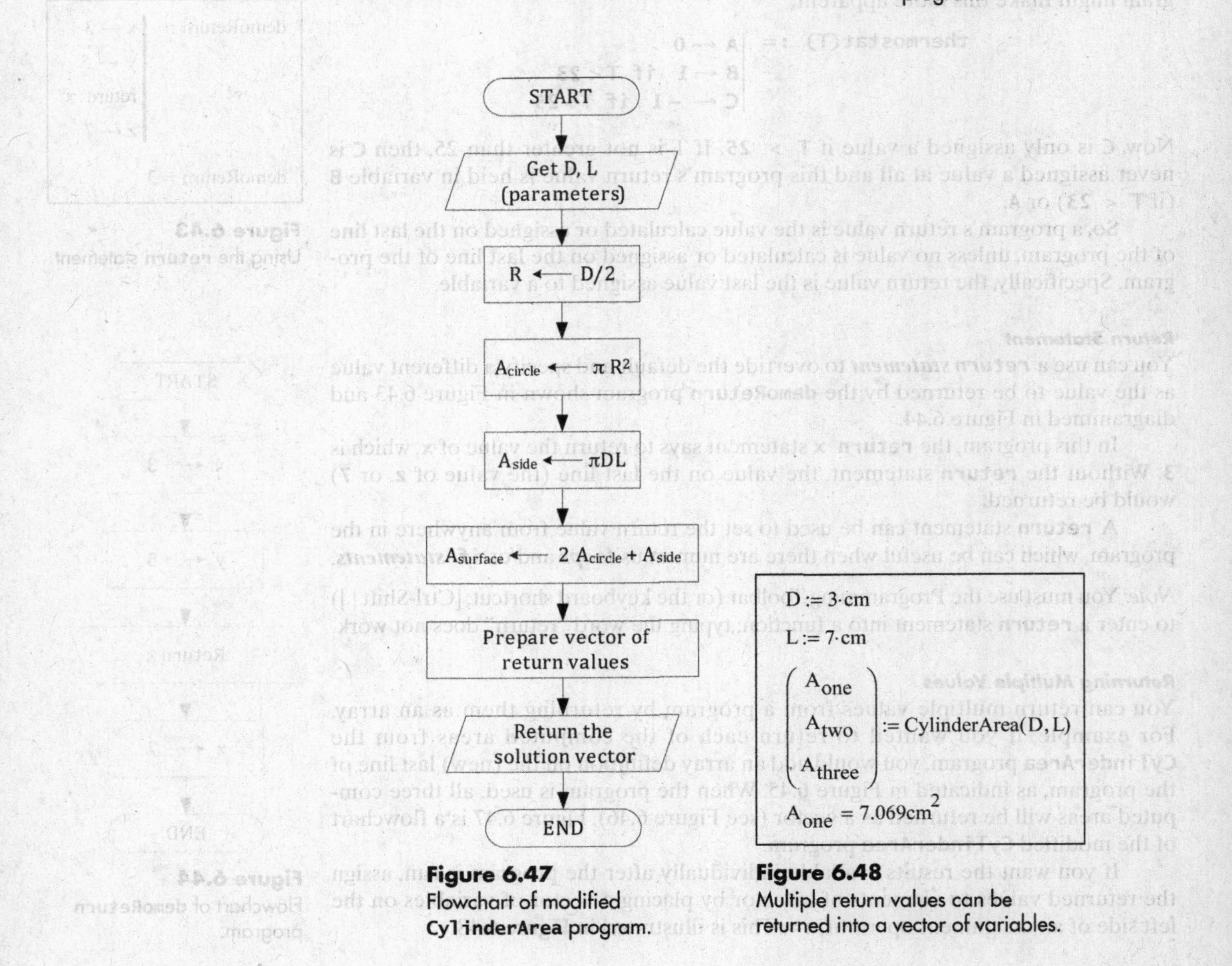

Figure 6.47
Flowchart of modified **CylinderArea** program.

Figure 6.48
Multiple return values can be returned into a vector of variables.

EXAMPLE

Linear Regression of a Data Set

What follows is an example of using a Mathcad program to perform a series of ***operations*** on a data set and return the results as a vector. This program will perform a linear regression (using Mathcad's built-in functions) on x and y values stored in two vectors, and return the slope, intercept, and R^2 values. First, the data vectors are defined as shown in Figure 6.49.

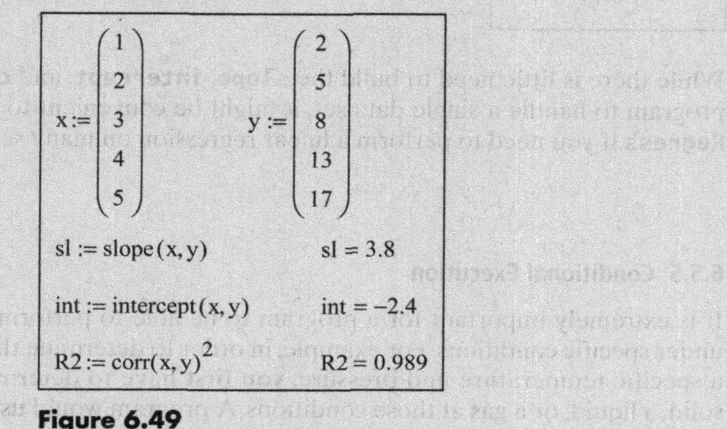

Figure 6.49
Calculating slope and intercept without using a program.

We could calculate the slope, intercept, and R^2 value using three of Mathcad's built-in functions as shown in Figure 6.49, or alternatively, we can combine these three steps into a program that calculates all three results at the same time, as shown in Figure 6.50. Figure 6.51 is a flowchart of the **Regress** program.

```
Regress(x, y) := | sl ← slope(x, y)
                 | int ← intercept(x, y)
                 | R2 ← corr(x, y)^2
                 | (sl, int, R2)
```

Figure 6.50
A program to calculate slope and intercept.

Note: The vector on the fourth line of the program shown in Figure 6.50 was created as follows:

1. **Add Line** was used to create a placeholder in the fourth line of the program.
2. After clicking on the placeholder in the fourth row of the program, a vector of placeholders with three rows and one column was created using the Insert Matrix dialog box, [CTRL-M].
3. The three placeholders in the vector were filled with the program's local variable names: **sl**, **int**, and **R2**.

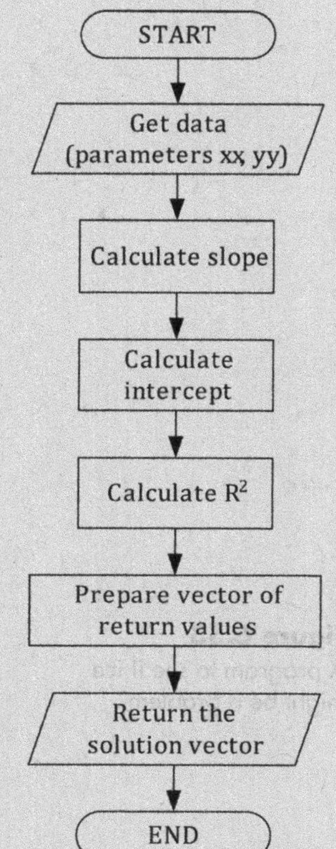

Figure 6.51
Flowchart of the **Regress** program.

The three computed results are returned as a vector containing three values as shown in Figure 6.52.

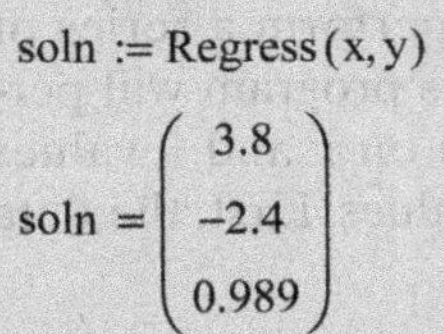

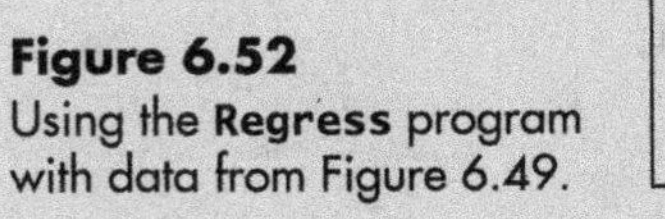

Figure 6.52
Using the **Regress** program with data from Figure 6.49.

While there is little need to build the **slope**, **intercept**, and **corr** functions into a program to handle a single data set, it might be convenient to have a program like **Regress** if you need to perform a linear regression on many sets of data.

6.5.5 Conditional Execution

It is extremely important for a program to be able to perform certain calculations under specific conditions. For example, in order to determine the density of water at a specific temperature and pressure, you first have to determine if the water is a solid, a liquid, or a gas at those conditions. A program would use ***conditional execution*** statements to select the appropriate equation for density.

If Statement

The classic conditional execution statement is the **if** statement. An **if** statement is used to select from two options, depending on the result of a calculated (logical) condition. In the example in Figure 6.53, the temperature is checked to see if freezing is a concern.

```
checkForIce(Temp) := | RV ← "No Problem"
                     | RV ← "Look Out for Ice!"  if  Temp < 273.15K
```

Using the program...

```
checkForIce(280·K) = "No Problem"
checkForIce(250·K) = "Look Out for Ice!"
checkForIce(28°F) = "Look Out for Ice!"
checkForIce(34°F) = "No Problem"
```

Figure 6.53
A program to see if ice might be a problem.

This example also illustrates that you can use units and text strings with Mathcad programs. Figure 6.54 is a flowchart of the **checkForIce** program.

To add an **if** statement to a program, click on the placeholder where you want the statement to be placed, and either click on **if** on the Programming Toolbar or press the right-brace key, [}].

Note: You cannot simply type "if" into the placeholder.

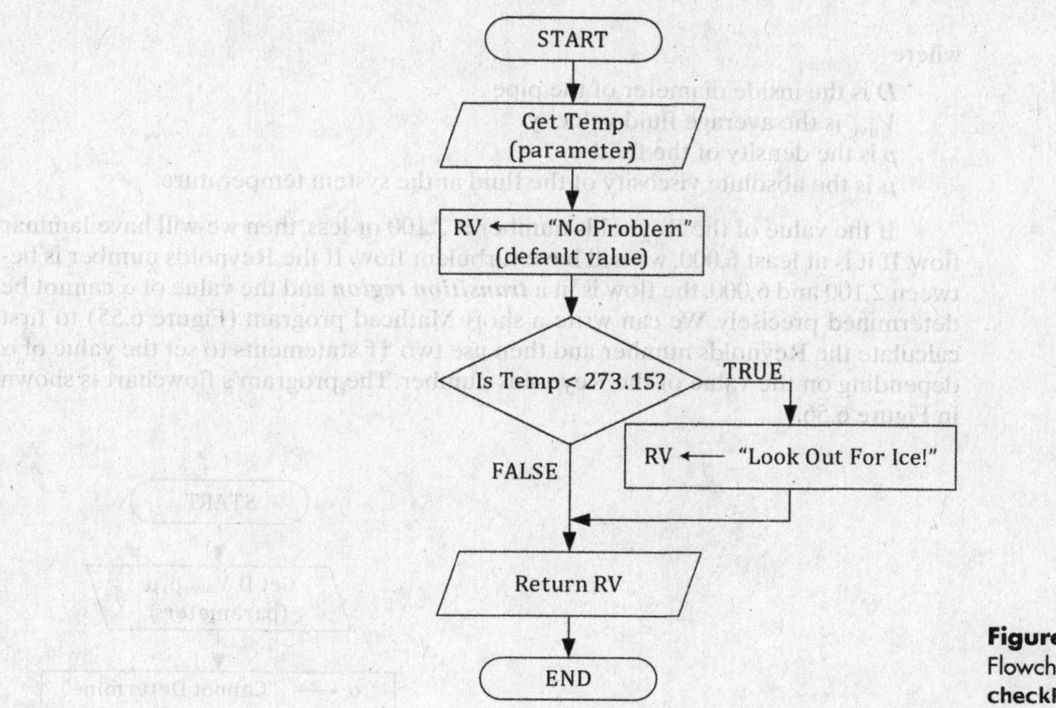

Figure 6.54
Flowchart of the **checkForIce** program.

When the **if** statement is inserted, there will be a placeholder on both sides of the **if**:

```
checkForIce(Temp) := | RV ← "No Problem"
                     | ▮ if ▮
```

The placeholder on the right is for the condition that will be evaluated; it must evaluate to true or false. The placeholder on the left will contain the program code that should execute if the condition evaluates to true. In the example in Figure 6.53, we use the left placeholder to assign the string **"Look Out For Ice!"** to the return variable, **RV**, if the condition in the right placeholder, **Temp < 273.15 K**, evaluates to true.

APPLICATION

Determining the Correct Kinetic Energy Correction Factor, α, for a Particular Flow

The ***mechanical energy balance*** is an equation that is commonly used by engineers for determining the pump power required to move a fluid through a piping system. One of the terms in the equation accounts for the change in ***kinetic energy*** of the fluid, and includes a ***kinetic energy correction factor***, α. The value of α is 2 for fully developed ***laminar flow***, and approximately 1.05 for fully developed ***turbulent flow***. In order to determine if the flow is laminar or turbulent, we must calculate the ***Reynolds number***, defined as

$$\text{Re} = \frac{DV_{\text{avg}}\rho}{\mu}.$$

where

D is the inside diameter of the pipe
V_{avg} is the average fluid velocity
ρ is the density of the fluid
μ is the absolute viscosity of the fluid at the system temperature

If the value of the Reynolds number is 2,100 or less, then we will have laminar flow. If it is at least 6,000, we will have turbulent flow. If the Reynolds number is between 2,100 and 6,000, the flow is in a ***transition region*** and the value of α cannot be determined precisely. We can write a short Mathcad program (Figure 6.55) to first calculate the Reynolds number and then use two `if` statements to set the value of α depending on the value of the Reynolds number. The program's flowchart is shown in Figure 6.56.

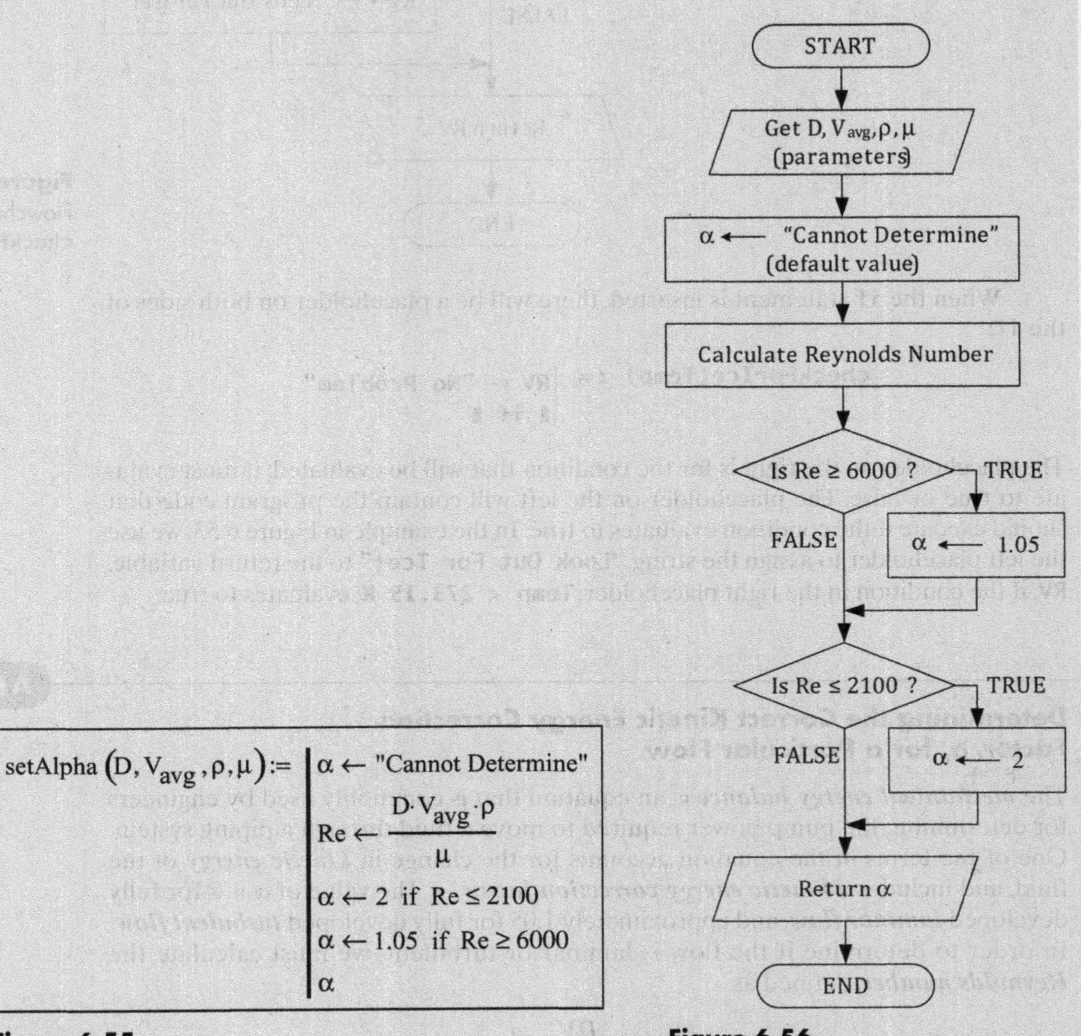

Figure 6.55
A program to determine the correct value for α.

Figure 6.56
Flowchart of the `setAlpha` program.

Notice that the value of α was initially assigned the text string *"Cannot Determine"*. This is the default case; if both of the `if` statements evaluate to false, the returned α value will be the warning text string. Next, the Reynolds number is determined. Then in program line 3, an `if` statement is used to see if the Reynolds number is less than or equal to 2,100. If it is, then the flow is laminar and α is given a value of 2. Line 4 then checks to see if the flow is turbulent and, if so, sets $\alpha = 1.05$. The final line indicates that the value of α is returned by the program since, by default, the value on the last line of the program is returned. (The last line is not actually needed, but it does ensure that someone reading the program knows which variable is returned.)

In Figure 6.57, the flow of a fluid with a density of 950 kg/m^3 and a viscosity of 0.012 poise in a 2-inch pipe at an average velocity of 3 m/s was found to be turbulent, so $\alpha = 1.05$. (The Reynolds number is 120,600.)

High Flow...

$$\text{setAlpha}\left(2\cdot\text{in}, 3\cdot\frac{\text{m}}{\text{s}}, 950\cdot\frac{\text{kg}}{\text{m}^3}, 0.012\,\text{poise}\right) = 1.05$$

Low Flow...

$$\text{setAlpha}\left(2\cdot\text{in}, 0.1\cdot\frac{\text{m}}{\text{s}}, 950\cdot\frac{\text{kg}}{\text{m}^3}, 0.012\,\text{poise}\right) = \text{"Cannot Determine"}$$

Figure 6.57
Using the `setAlpha` program.

If the velocity is decreased to 0.1 m/s as in the "Low Flow" section of Figure 6.57, then the Reynolds number falls below 6,000 and the program indicates this by sending back the warning string.

Otherwise Statement

The ***otherwise statement*** is used in conjunction with an `if` statement when you want the program to do something when the condition in the `if` statement evaluates to false. For example, we might rewrite the **checkForIce** program (Figure 6.58) to test for temperatures below freezing but provide an **otherwise** to set the text string when freezing is not a concern.

```
checkForIce(Temp) := | RV ← "Look Out for Ice!"   if Temp < 273.15K
                     | RV ← "No Problem"   otherwise
```

Using the program...

checkForIce(250·K) = "Look Out for Ice!"

checkForIce(4 °C) = "No Problem"

Figure 6.58
Modified **checkForIce** program using the **otherwise** statement.

This version is functionally equivalent to the earlier version, but it might be more easily read by someone unfamiliar with programming. In the earlier version, the return value was set to **"No Problem"** and was then overwritten by **"Look Out**

`For Ice!"` if the temperature was below freezing. In this version of the program, there is no overwriting; the return value is set to one text string or the other, based on the result of the `if` statement. (See flowchart in Figure 6.59.)

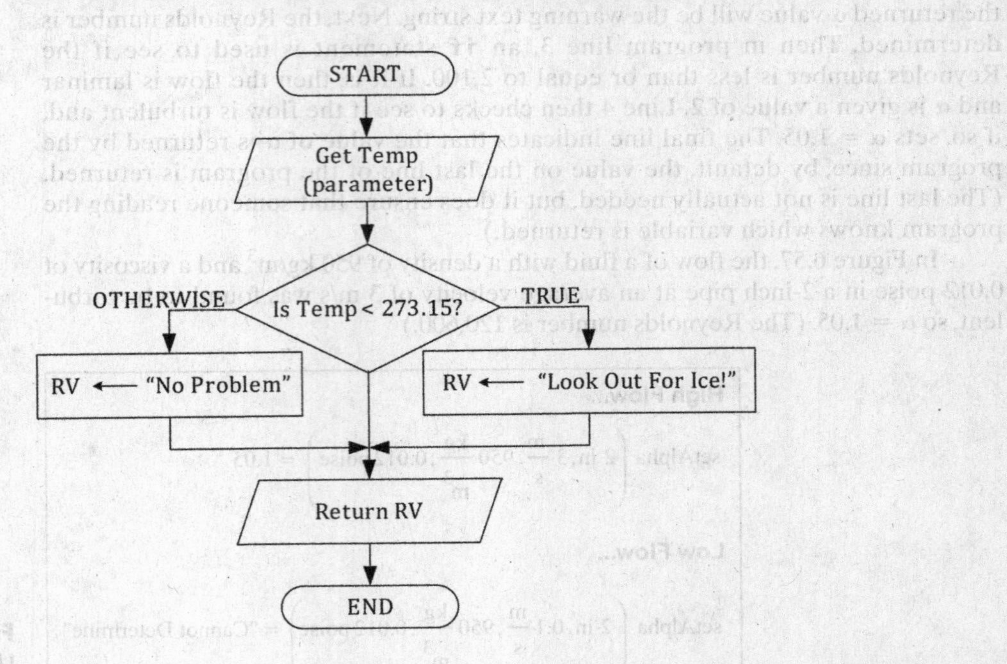

Figure 6.59
Flowchart of modified **checkForIce** program.

Note: The **otherwise** statement must be inserted using the Programming Toolbar or by pressing [Ctrl + }]. You cannot simply type "otherwise" into the placeholder.

On Error Statement

The ***on error statement*** is used for error trapping, which provides an alternative calculation path when certain values are known to cause errors in a program. When you include an **on error** statement on a program line, there is a placeholder on both sides of the statement.

- The placeholder on the right is for the normal calculation, assuming that there is no error.
- The placeholder on the left is for the alternative calculation in case an error occurs.

For example, the **inverseValue(*x*)** program shown in Figure 6.60 (diagrammed in Figure 6.61) will fail if the value of x is zero. However, we can trap this error and return ∞ (evaluates to 1×10^{307}) instead.

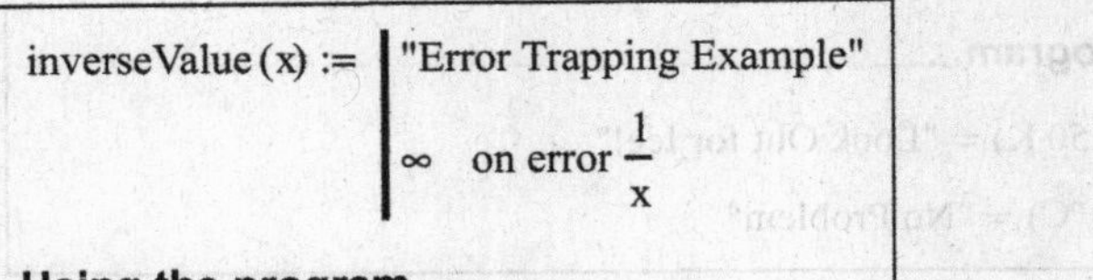

Using the program...

inverseValue (4) = 0.25

inverseValue (0) = 1×10^{307}

Figure 6.60
Using the **on error** statement.

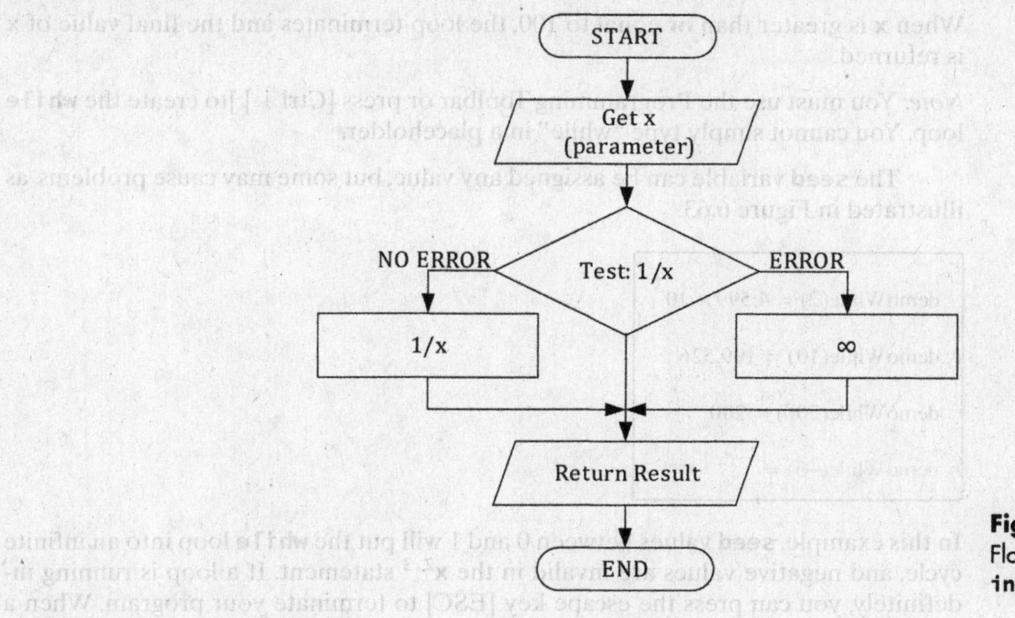

Figure 6.61
Flowchart of the **inverseValue** program.

Note: The **on error** statement must be inserted using the Programming Toolbar or by pressing [Ctrl-']. You cannot simply type "on error" into the placeholder.

6.5.6 Loops

Loop structures are used to cause programs to perform calculations over and over again. There are several instances when these repetitive calculations are desirable:

- If you want to repeat a series of calculations for each value in a data set or matrix.
- If you want to perform an iterative (guess-and-check) calculation until the guessed and calculated values are within a preset tolerance.
- If you want to move through the rows of data in an array until you find a value that meets a particular criterion.

There are two loop structures supported in Mathcad programs: ***while loops*** and ***for loops***.

While Loops

A ***while loop*** is a control structure that causes an action to be repeated (*iteration*) as long as (i.e., while) a condition is true. As soon as the condition is false, the iteration stops. In the **demoWhile** program shown in Figure 6.62, the loop continues to operate as long as the value of local variable **x** is less than 100:

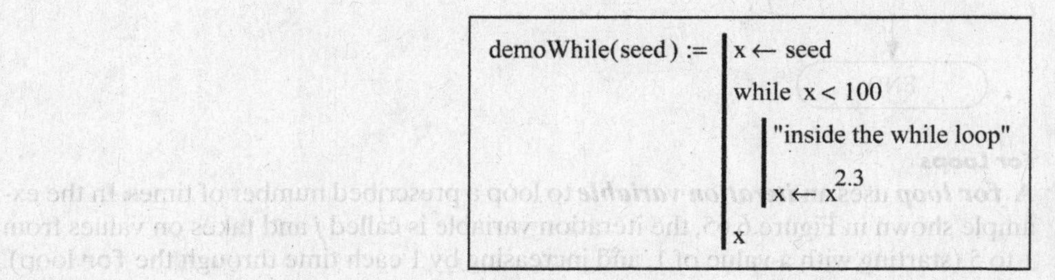

Figure 6.62
Demonstrating the **while** loop.

When **x** is greater than or equal to 100, the loop terminates and the final value of **x** is returned.

Note: You must use the Programming Toolbar or press [Ctrl +]]to create the **while** loop. You cannot simply type "while" in a placeholder.

The **seed** variable can be assigned any value, but some may cause problems, as illustrated in Figure 6.63.

demoWhile(2) = 4.599×10^3

demoWhile(10) = 199.526

demoWhile(200) = 200

demoWhile(−3) =

Figure 6.63
Trying various seed values with the **demoWhile** program.

In this example, **seed** values between 0 and 1 will put the **while** loop into an infinite cycle, and negative values are invalid in the $\mathbf{x^{2.3}}$ statement. If a loop is running indefinitely, you can press the escape key [ESC] to terminate your program. When a program statement contains an error that prevents it from running, Mathcad indicates that the program failed by showing the equation in red.

Figure 6.64 is a flowchart of the **demoWhile** program.

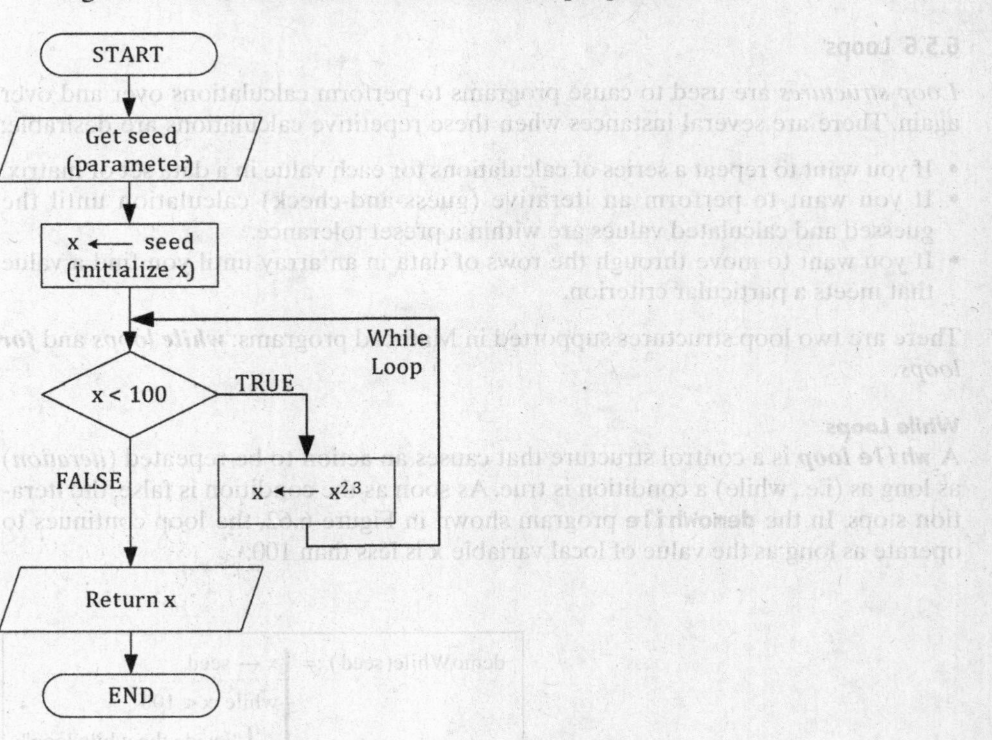

Figure 6.64
Flowchart of the **demoWhile** program.

For Loops

A ***for loop*** uses an ***iteration variable*** to loop a prescribed number of times. In the example shown in Figure 6.65, the iteration variable is called *j* and takes on values from 1 to 5 (starting with a value of 1, and increasing by 1 each time through the **for** loop).

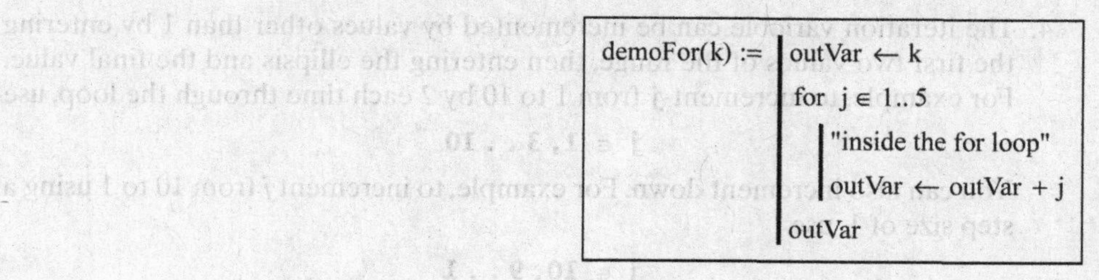

Figure 6.65
Demonstrating the **for** loop.

Notes:

1. You must use the Programming Toolbar or press [Ctrl + "] to create the **for** loop. You cannot simply type "for" in a placeholder.
2. The ∈ symbol is read "in the range of" and is a reminder that the iteration variable is a ***range variable***, and thus is defined as a range variable. The range 1 . . 5 was entered as [1] [;] [5], where the semicolon was used to enter the ellipsis, ". .".
3. In the flowchart for this program (Figure 6.66), the steps in which *j* is initialized and incremented are shown with dotted-line borders. This is meant as a reminder that initialization and incrementation of the index variable, *j*, is automatically handled by the **for** loop.

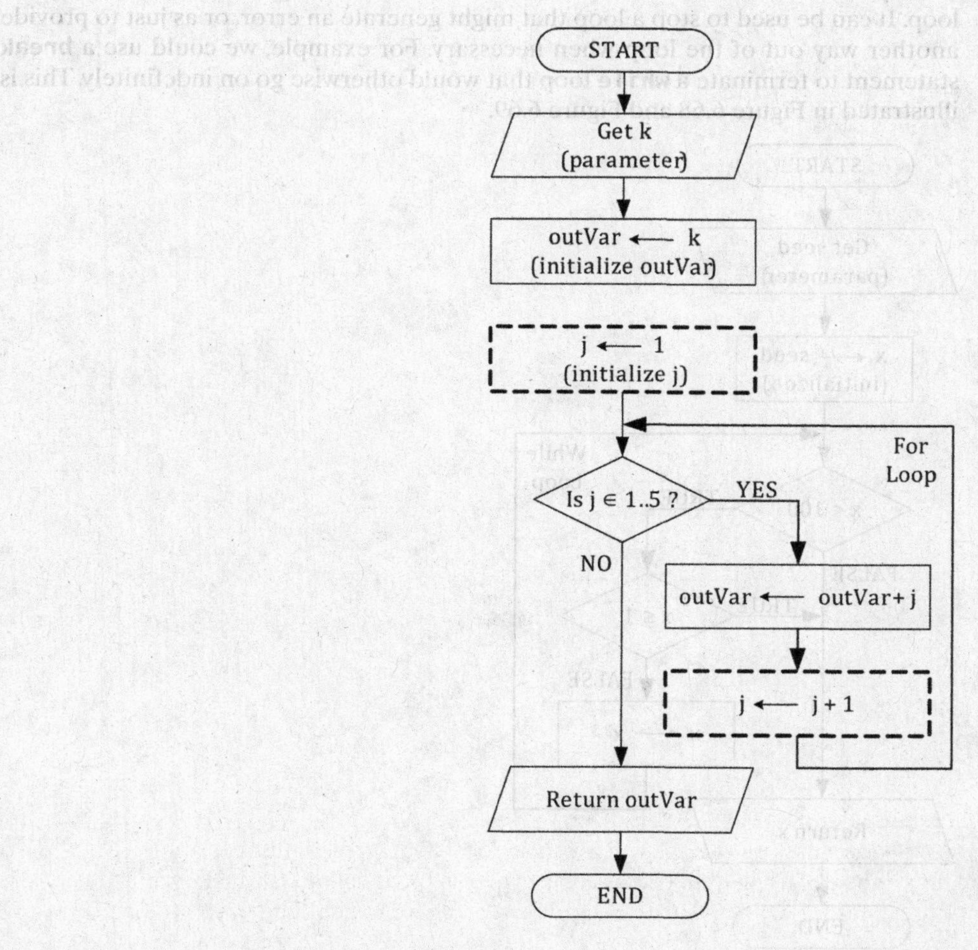

Figure 6.66
Flowchart of the **demoFor** program.

4. The iteration variable can be incremented by values other than 1 by entering the first two values of the range, then entering the ellipsis and the final value. For example, to increment **j** from 1 to 10 by 2 each time through the loop, use

   j ∈ 1, 3 . . 10

   You can also increment down. For example, to increment *j* from 10 to 1 using a step size of 1, use

   j ∈ 10, 9 . . 1

   When the **demoFor** program is run with a **k** value of 0, it should return a value of 0 + 1 + 2 + 3 + 4 + 5 = 15. This is shown in Figure 6.67.

Using the demoFor() program...

demoFor(0) = 15

demoFor(5) = 20

Figure 6.67
Using the **demoFor** program.

Figure 6.67 also shows that when **k** = 5, the return value is 20.

Break Statement

The ***break statement*** is used to halt execution of a **for** or **while** loop. It is used when you want some condition other than the normal loop termination to stop the loop. It can be used to stop a loop that might generate an error, or as just to provide another way out of the loop when necessary. For example, we could use a **break** statement to terminate a **while** loop that would otherwise go on indefinitely. This is illustrated in Figure 6.68 and Figure 6.69.

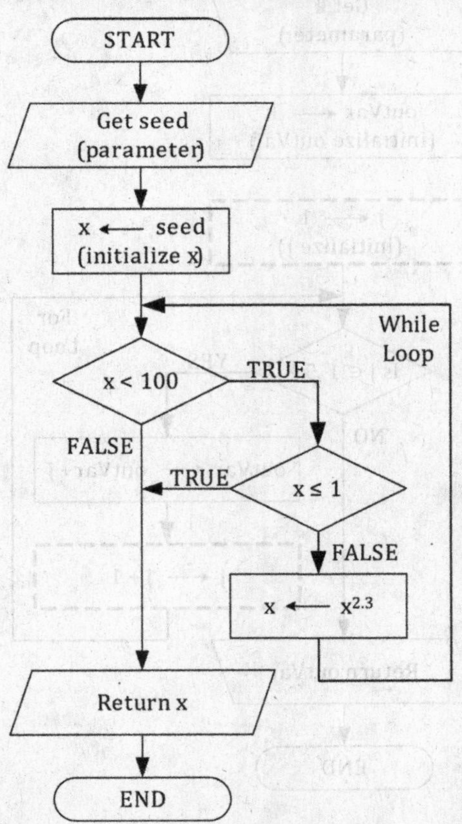

Figure 6.68
Flowchart of a **break** statement in a **while** loop.

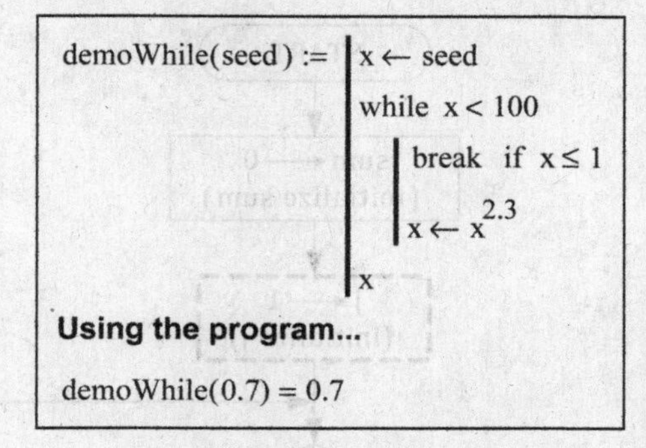

Figure 6.69
Modifying the **demoWhile** program to **break** out of the loop if **seed** is too small.

Note: You must use the Programming Toolbar or press [Ctrl + {] to insert the **break** statement. You cannot simply type "break" in a placeholder.

Continue Statement

A ***continue statement*** is used with ***nested loops*** (loops within other loops) to halt execution of the loop containing the **continue** statement, and continuing the program with the next iteration of the next outer loop. For example, the program shown in Figure 6.70 (diagrammed in Figure 6.71), without a **continue** statement, has three nested for loops using iteration variables **i**, **j**, and **k**.

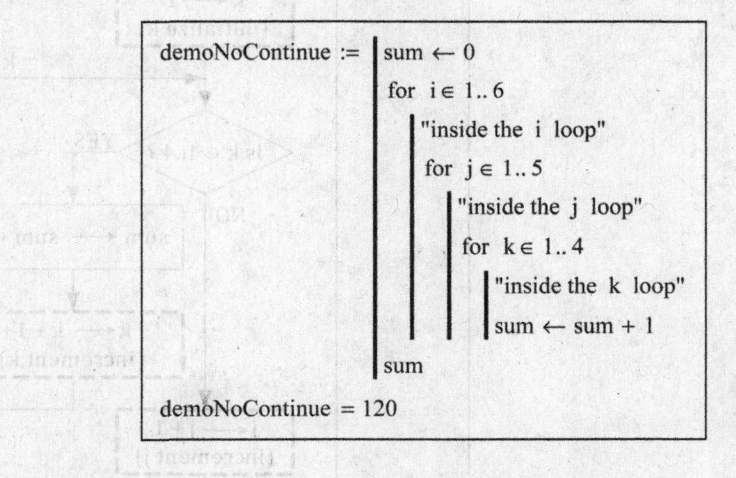

Figure 6.70
A three-loop summing program without a **continue** statement.

The computed sum is 120, because the **i** loop cycled 6 times, the **j** loop cycled 6 × 5 times, and the **k** loop cycled 6 × 5 × 4 = 120 times, and the **sum** calculation was inside the **k** loop.

Now, in Figure 6.72, we add a **continue** statement that halts the **j** loop if **j > 3**:

This time, **sum** has a value of 72, because the **i** loop cycled 6 times, the **j** loop cycled 6 × 3 times, and the **k** loop cycled 6 × 3 × 4 = 72 times.

Notes:

1. You must use the Programming Toolbar or press [Ctrl + [ ] to insert the **continue** statement. You cannot simply type "continue" in a placeholder.
2. The **continue** statement is typically used with an **if** statement. You must enter the **if** statement first to create the placeholder for the **continue** statement.
3. In the flowchart for this program (Figure 6.73), the initialization and incrementation steps that are automatically handled by the **for** loop have been omitted.

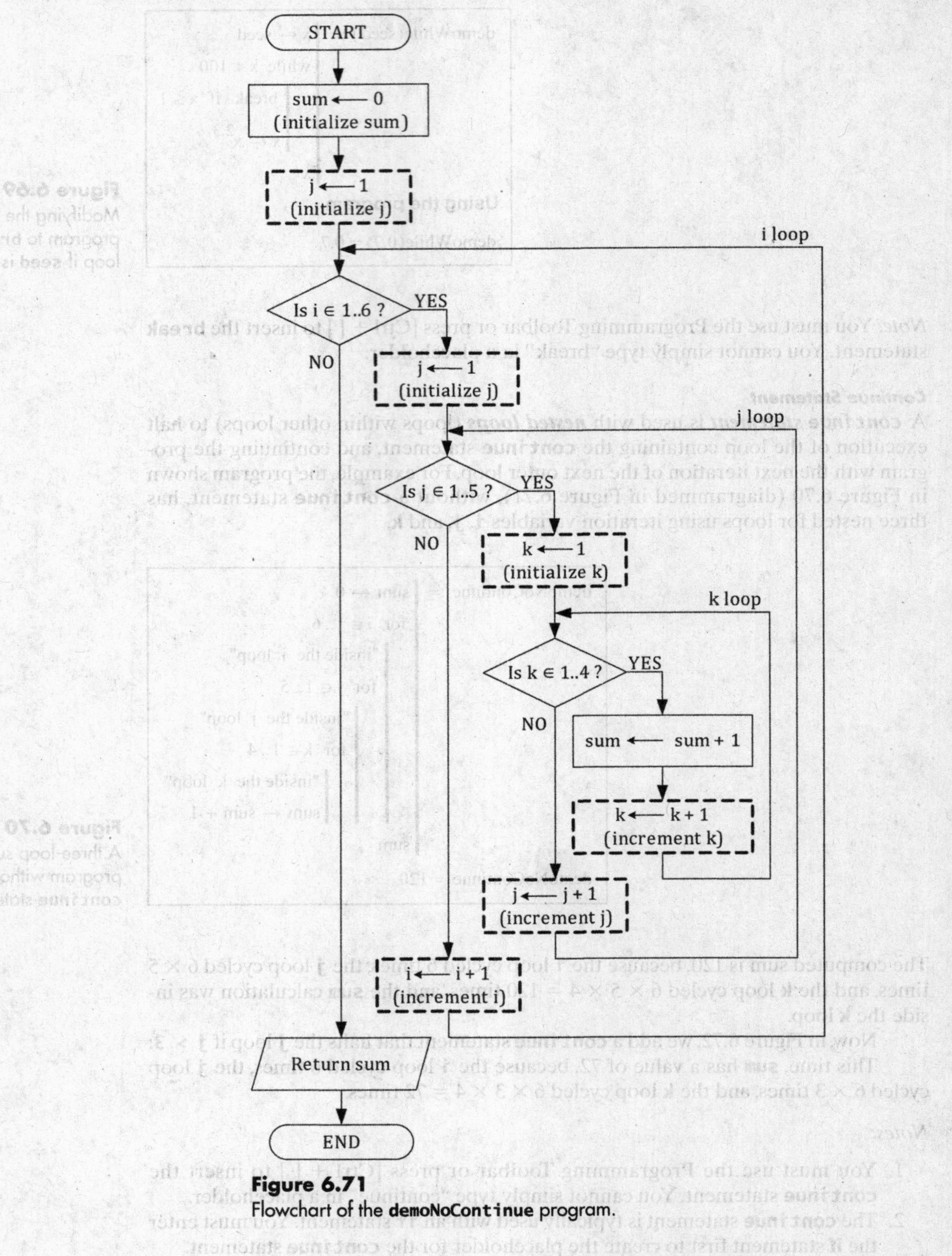

Figure 6.71
Flowchart of the **demoNoContinue** program.

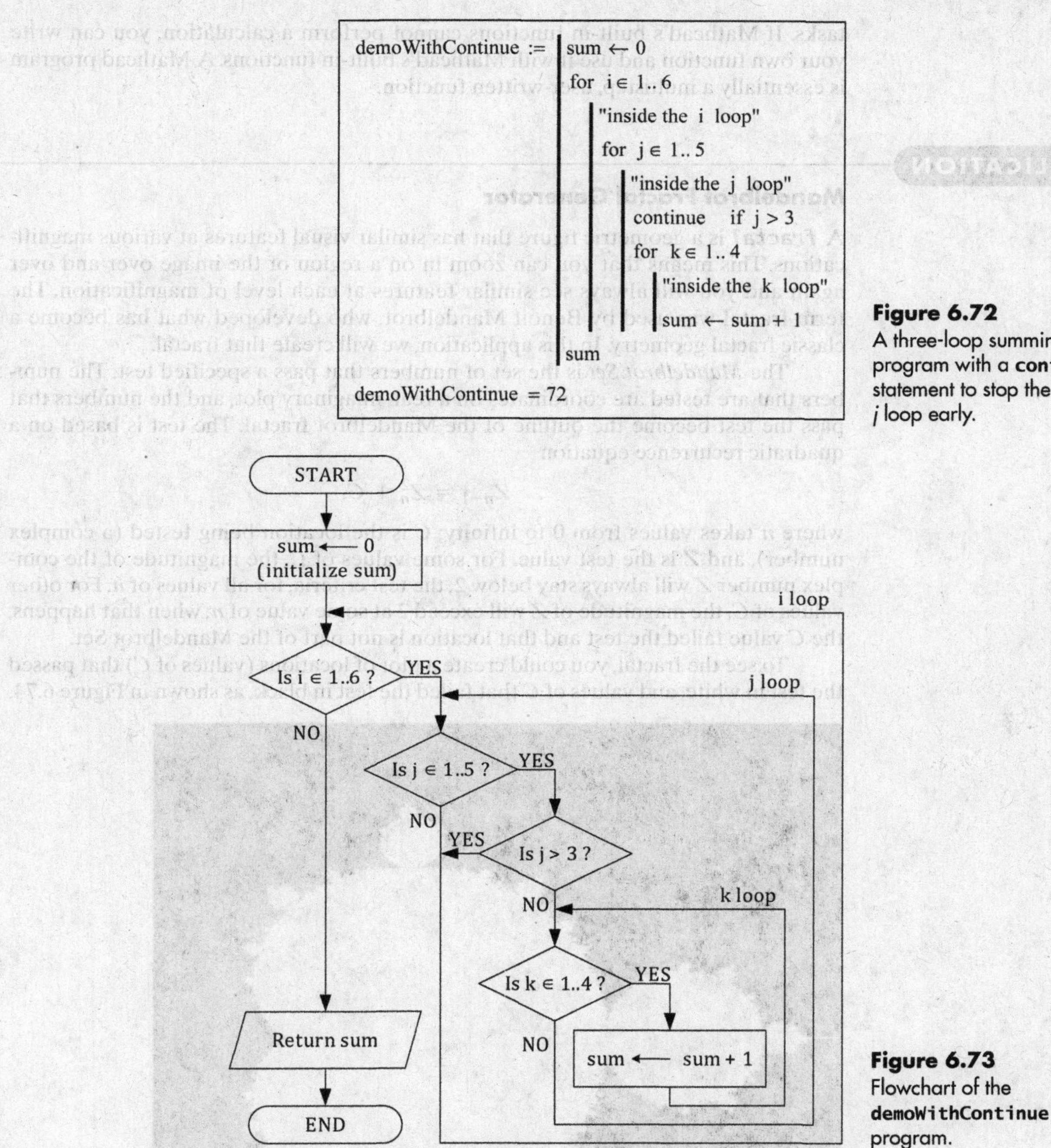

Figure 6.72
A three-loop summing program with a **continue** statement to stop the *j* loop early.

Figure 6.73
Flowchart of the **demoWithContinue** program.

6.5.7 Functions

Functions are an indispensable part of modern programming because they allow a program to be broken down into pieces, each of which ideally handles a single task (i.e., performs a single function). The programmer can then call upon the functions as needed to complete a more complex calculation.

Functions are an indispensable part of Mathcad as well. Mathcad provides built-in functions that can be used as needed to complete a lot of computational tasks. If Mathcad's built-in functions cannot perform a calculation, you can write your own function and use it with Mathcad's built-in functions. A Mathcad program is essentially a multistep, user-written function.

APPLICATION

Mandelbrot Fractal Generator

A ***fractal*** is a geometric figure that has similar visual features at various magnifications. This means that you can zoom in on a region of the image over and over again and you will always see similar features at each level of magnification. The term fractal was used by Benoit Mandelbrot, who developed what has become a classic fractal geometry. In this application, we will create that fractal.

The *Mandelbrot Set* is the set of numbers that pass a specified test. The numbers that are tested are coordinates on a real–imaginary plot, and the numbers that pass the test become the outline of the Mandelbrot fractal. The test is based on a quadratic recurrence equation

$$Z_{n+1} = Z_n^2 + C,$$

where n takes values from 0 to infinity; C is the location being tested (a complex number), and Z is the test value. For some values of C, the magnitude of the complex number Z will always stay below 2, the test criteria, for all values of n. For other values of C, the magnitude of Z will exceed 2 at some value of n; when that happens, the C value failed the test and that location is not part of the Mandelbrot Set.

To see the fractal, you could create a plot of locations (values of C) that passed the test in white, and values of C that failed the test in black, as shown in Figure 6.74.

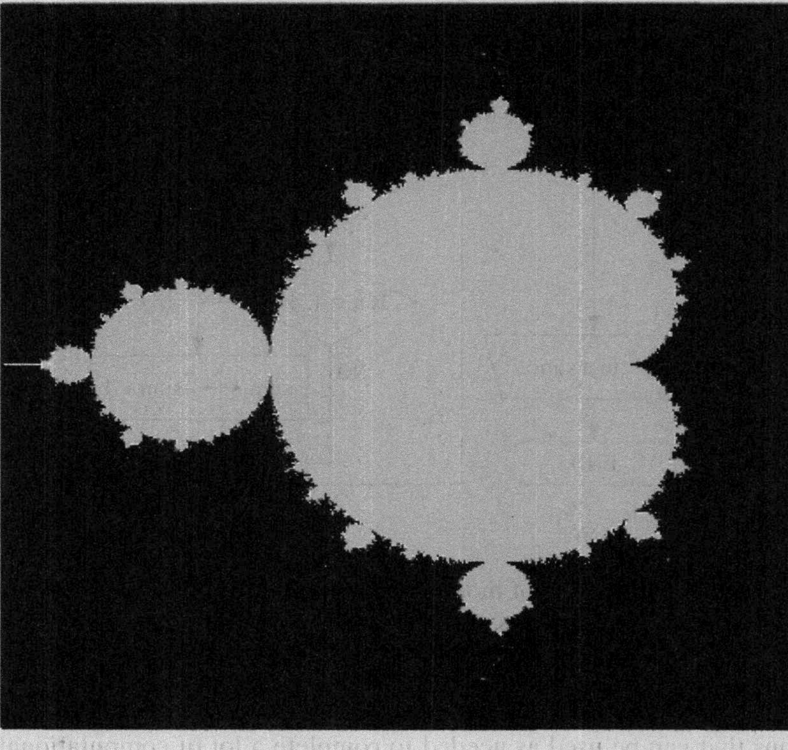

Figure 6.74
The Mandelbrot Set (stored in matrix ZZ). Values in the set that passed the test are in white.

However, you get more dramatic results by plotting the value of n when the test fails. Since it is not possible to vary n from 0 to infinity, we set an arbitrary upper bound, N_{Qual}, and deem that if Z is still less than 2 after N_{Qual} iterations, then the location passes the test and is part of the set. (Set N_{Qual} low for faster calculations, or high for better quality images.)

To generate the Mandelbrot Set, we will write two multistep functions (Mathcad programs):

- **`mandTest`**—tests a location (a value of **`C`**) to determine whether or not it is a member of the Mandelbrot Set, and
- **`mandSet`**—loops through the entire image region and calls **`mandTest`** to test each location.

THE TESTING FUNCTION, MANDTEST

The flowchart for the function that tests a location to see if it is in the Mandelbrot Set is illustrated in Figure 6.75. Note that **`i`** was used as the iteration loop index rather than **`n`**.

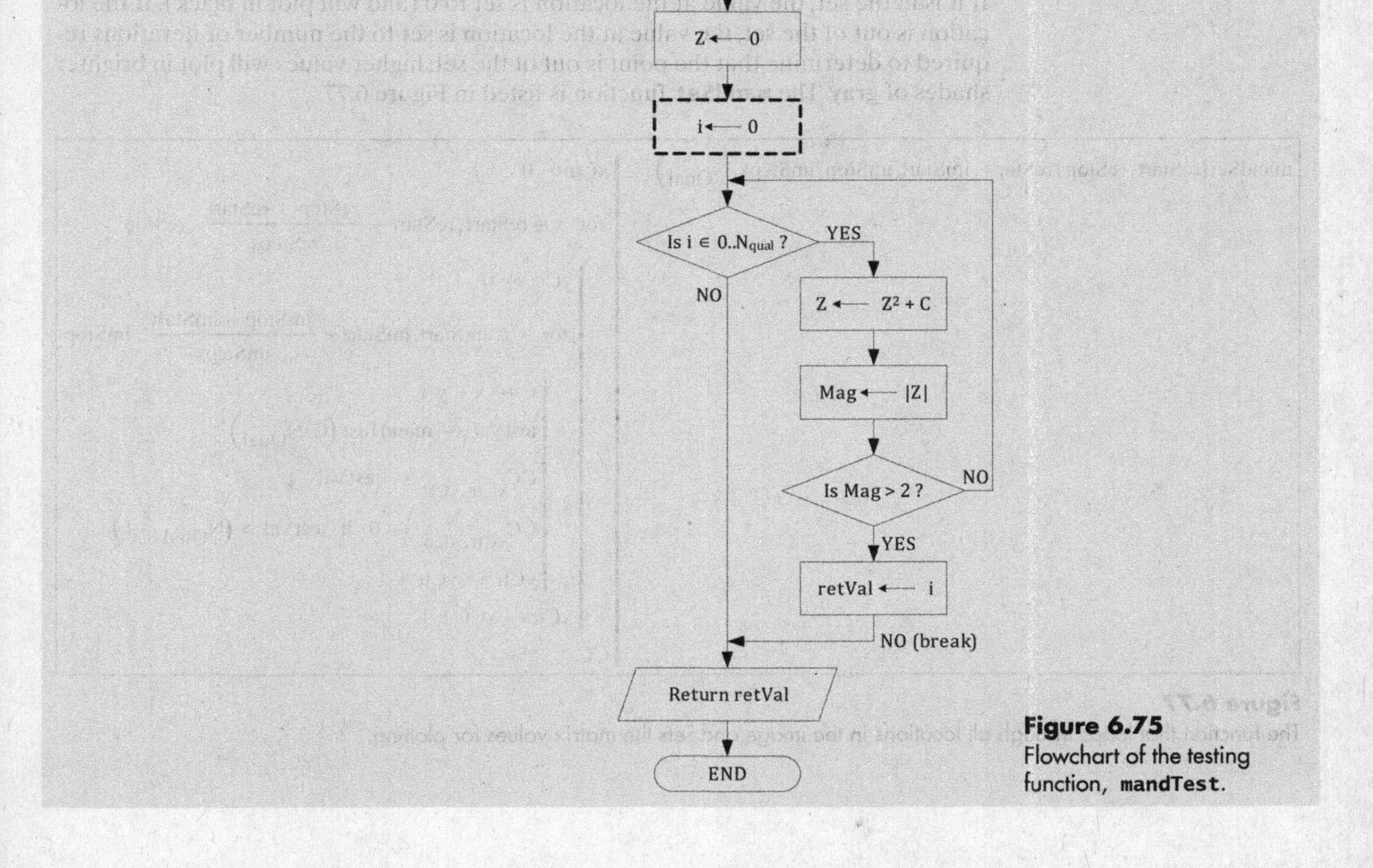

Figure 6.75
Flowchart of the testing function, **`mandTest`**.

The **mandTest(C, N_{Qual})** function is listed in Figure 6.76. If the function returns **retVal = 0**, then the tested **C** value is a member of the set. If the **C** value is not part of the set, the function returns the number of iterations required to determine that the value is not part of the set.

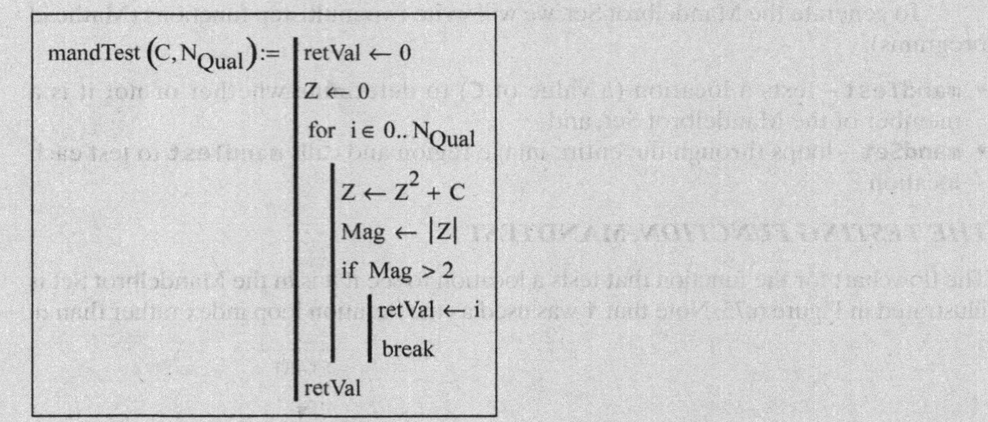

```
mandTest(C, N_Qual) := | retVal ← 0
                       | Z ← 0
                       | for i ∈ 0.. N_Qual
                       |   | Z ← Z^2 + C
                       |   | Mag ← |Z|
                       |   | if Mag > 2
                       |   |   | retVal ← i
                       |   |   | break
                       | retVal
```

Figure 6.76
The function that tests a C value to see if it belongs in the Mandelbrot Set.

THE LOOPING FUNCTION, MANDSET

The other multistep function loops through the entire domain of the image and calls **mandSet** for every location to see if that location is in or out of the Mandelbrot Set. If it is in the set, the value at the location is set to 0 (and will plot in black). If the location is out of the set, the value at the location is set to the number of iterations required to determine that the point is out of the set; higher values will plot in brighter shades of gray. The **mandSet** function is listed in Figure 6.77.

```
mandSet(reStart, reStop, reSteps, imStart, imStop, imSteps, N_Qual) :=
  | xCtr ← 0
  | for x ∈ reStart, reStart + (reStop − reStart)/reSteps .. reStop
  |   | yCtr ← 0
  |   | for y ∈ imStart, imStart + (imStop − imStart)/imSteps .. imStop
  |   |   | C ← x + y·i
  |   |   | testVal ← mandTest(C, N_Qual)
  |   |   | CC_(xCtr, yCtr) ← testVal
  |   |   | CC_(xCtr, yCtr) ← 0 if testVal > (N_Qual − 1)
  |   |   | yCtr ← yCtr + 1
  |   | xCtr ← xCtr + 1
  | CC
```

Figure 6.77
The function that loops through all locations in the image and sets the matrix values for plotting.

The parameters in the **mandSet** function include beginning and ending values in the **x** direction (**reStart, reStop**), the number of steps to include in the **x** direction (**reSteps**), beginning and ending values in the *y* direction (**imStart, imStop**), the number of steps to include in the **y** direction (**imSteps**), and the number of iterations to try before assuming the location is a part of the Mandelbrot Set, N_{Qual}.

To create the matrix for plotting (called *CC* here), use the **mandSet** function as shown in Figure 6.78.

$reStart := -1.5 \qquad reStop := 0.7 \qquad reSteps := 500$

$imStart := -1.2 \qquad imStop := 1.2 \qquad imSteps := 500 \qquad N_{Qual} := 100$

$CC := mandSet(reStart, reStop, reSteps, imStart, imStop, imSteps, N_{Qual})$

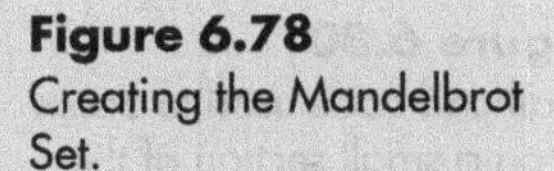

Figure 6.78
Creating the Mandelbrot Set.

Matrix **CC** contains the information required to display the Mandelbrot Set, which is shown in Figure 6.79. Matrix **CC** has been plotted using a ***contour plot*** with a ***grayscale color map***. The brighter spots are the locations that required a large number of iterations before failing the test. The interior black region is composed of locations that did not fail the test and are part of the Mandelbrot Set. The exterior black region is made up of locations that failed the test with very few iterations.

Figure 6.79
The Mandelbrot fractal.

You are supposed to be able to zoom in on any region of a fractal and still see the same type of features. Figure 6.80 shows the Mathcad code used to zoom in on a small part of Figure 6.79 (in this example, we zoom in on the base of the circle on top of the main cardioid). The resulting plot is shown in Figure 6.81. Notice that the value of N_{Qual} was increased; this provides better definition of edges as you zoom in.

```
reStart := -0.4     reStop := -0.1     reSteps := 500

imStart := 0.6      imStop := 0.7      imSteps := 500

N_Qual := 300

CC := mandSet(reStart, reStop, reSteps, imStart, imStop, imSteps, N_Qual)
```

Figure 6.80
Mathcad code used to zoom in on a small section of the image shown in Figure 6.79.

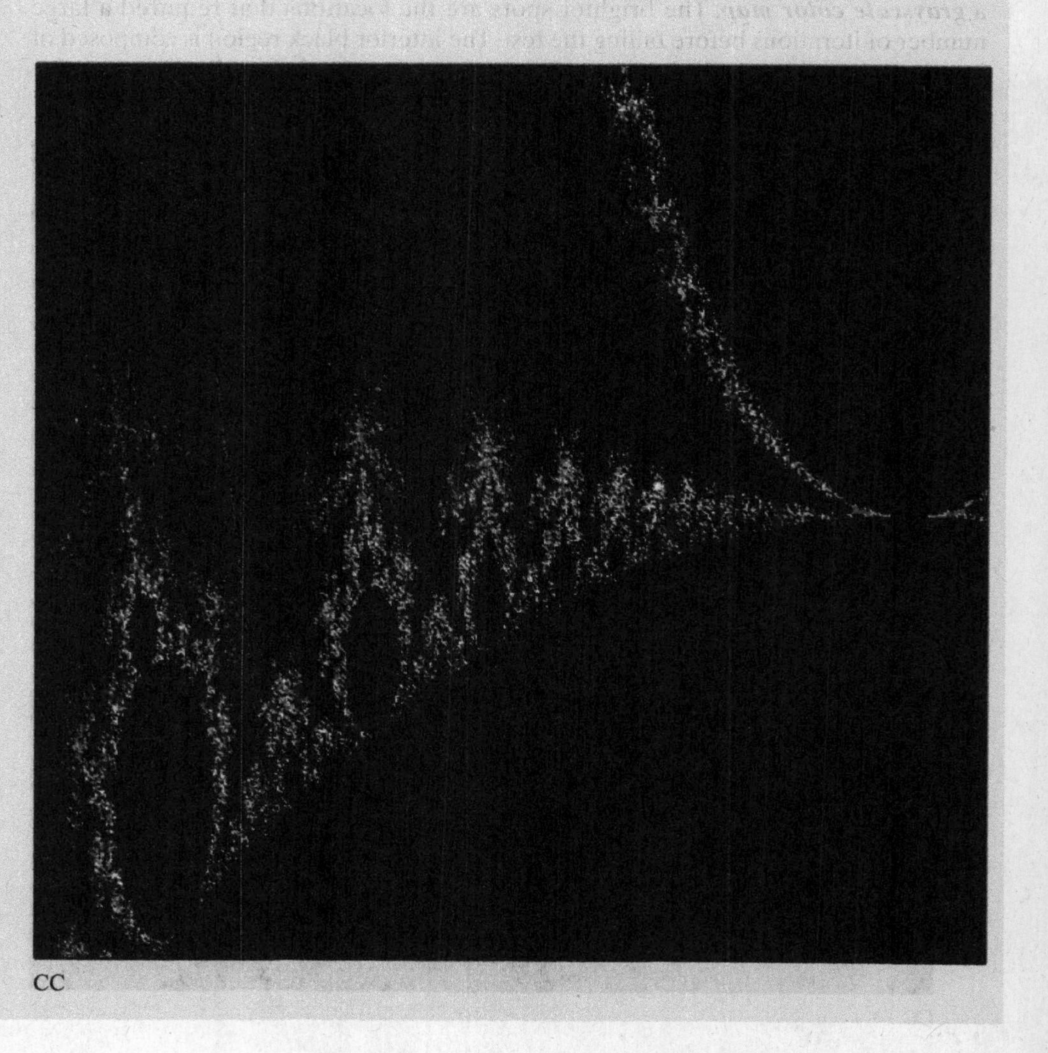

CC

Figure 6.81
Zooming in on the Mandelbrot fractal.

Notice that as you zoom in on a small section of the image, the same features are evident, which is characteristic of a fractal.

The worksheet used to create these images is available on the text's website (www.chbe.montana.edu/mathcad) so that you can download it to your own computer and look at other portions of the image if you would like. Other color maps can also be used to create some striking images.

SUMMARY

Creating a Mathcad Program

Program Name	Identifies the program and is used as a function name when the program is used on the worksheet.
Parameter List	Used to pass values into a program.
Assignment Operator, :=	Connects the program name and parameter list to the program lines, which are indicated by heavy vertical lines.
Add Line	Used to insert lines into the program; the keyboard shortcut is [ ]] (right-bracket key).

Basic Elements of Programming

Data	Data values are stored in Mathcad variables—either as single values or as arrays.
Input	Mathcad programs can receive input from • the worksheet, • the parameter list, or • data files (text files) using the **READPRN** function.
Operations	Nearly all of Mathcad's standard operators may be used in programs, including the standard math operators and matrix operators listed in the tables that follow. Variables defined within a Mathcad program must be defined with the local definition operator, ←. These variables are defined only within the program region (local scope).
Output	Results from Mathcad programs must be returned from the program. By default, the value on the last assignment statement in the program is the return value. You can override the default by using the **return** statement (from the Programming Toolbar) in your program. Multiple values may be returned from Mathcad programs as an array.
Conditional Execution	Mathcad's **if** statement is used for conditional execution—to allow the flow of the program to follow different paths depending on the value of the condition following the **if** statement.

The `if` statement is found on the Programming Toolbar, or it may be entered using [}] (right brace).

Loops — Mathcad programs support two types of loops:

- **`for`** loops—used when you want the loop to execute a specified number of times.
- **`while`** loops—used when you want the looping to continue as long as a specified condition is true.

You can use the **`break`** statement to terminate a loop or a **`continue`** statement to cause the program flow to jump out of an inside loop, but continue in another loop. Both **`break`** [Ctrl + {] and **`continue`** [Ctrl + [ ] are available on the Programming Toolbar.

Functions — Mathcad programs are multistep functions and may include other functions, including Mathcad's built-in functions.

Standard Math Operators

Symbol	Name	Shortcut Key
+	Addition	+
−	Subtraction	−
*	Multiplication	[Shift-8]
/	Division	/
e^x	Exponential	
$1/x$	Inverse	
x^y	Raise to a Power	[^], or [Shift-6]
$n!$	Factorial	!
$\lvert x \rvert$	Absolute Value	
$\sqrt{}$	Square Root	\
$\sqrt[n]{}$	N^{th} Root	[Ctrl-\]

Matrix Operators

Symbol	Name	Shortcut Key
+	Addition	+
−	Subtraction	−
*	Multiplication	[Shift-8]
/	Division	/
A_n	Array Element	[
M^{-1}	Matrix Inverse	
$\lvert M \rvert$	Determinant	|
$\rightarrow$	Vectorize	[Ctrl- −]
$M^{<x>}$	Matrix Column	[Ctrl-6]
M^T	Matrix Transpose	[Ctrl-1]
$M_1 \cdot M_2$	Dot Product	*, or [Shift-8]
$M_1 \times M_2$	Cross Product	[Ctrl-8]

Operator Precedence Rules

Precedence	Operator	Name
First	^	Exponentiation
Second	*, /	Multiplication, Division
Third	+ , −	Addition, Subtraction

It is a good idea to include the parentheses to make the order of evaluation clear.

Flowchart Symbols

Symbol	Name	Usage
(rounded rectangle)	**Terminator**	Indicates the start or end of a program
(rectangle)	**Operation**	Indicates a computation step
(parallelogram)	**Data**	Indicates an input or output step
(diamond)	**Decision**	Indicates a decision point in a program
(circle)	**Connector**	Indicates that the flowchart continues in another location

KEY TERMS

argument list
break statement
condition
conditional execution
continue statement
contour plot
data
data file input component
decision step
default value
dummy variable
file input component
flowchart
for loop
fractal
function
grayscale color map
if statement
input
input source
iteration variable
kinetic energy
kinetic energy correction factor
laminar flow
local definition
local scope
local variables
loops
mechanical energy balance
on error statement
operations
operator
operator precedence
otherwise statement
output
parameter list
passing arguments by address
passing arguments by value
program
program name
program region
Programming Toolbar
range variable
return statement
return value
Reynolds number
standard programming symbol
temporary variables
transition region
turbulent flow
while loop

PROBLEMS

6.1 Checking for Fever

For a Practice! Exercise in this chapter you were asked to create a flowchart illustrating the decision tree for determining whether a person's fever requires medication, and whether the medication should be aspirin. Now, write a Mathcad program that receives the patient's oral temperature and age through the parameter list and returns two text strings, one for each of the following decisions:

Decision 1: Should the patient receive medication?

- "patient should receive medicine" or "patient does not require medicine"

Decision 2: Can the medicine be aspirin?

- "no medicine required" or "aspirin is OK" or "do not give aspirin"

6.2 Calculating Grades

(a) Write a flowchart for a program that receives a numerical score (0–100) through the parameter list and then determines a letter grade based on the following information:

Score $\geq$ 90	A
80 $\geq$ Score $<$ 90	B
70 $\geq$ Score $<$ 80	C
60 $\geq$ Score $<$ 70	D
Score $<$ 60	E

(b) Write a Mathcad program that will receive the numerical score through the parameter list and return the appropriate letter grade.

6.3 Calculating Average and Median Scores

(a) Write a program that receives a vector of scores through the parameter list, then uses Mathcad's `mean()` and `median()` functions to compute the average and median scores.

(b) Test your program with the following values:

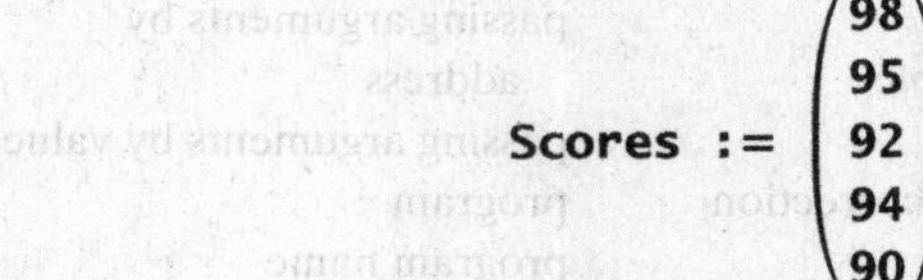

$$\text{Scores} := \begin{pmatrix} 98 \\ 95 \\ 92 \\ 94 \\ 90 \end{pmatrix} \qquad \begin{aligned} \text{mean(Scores)} &= 93.8 \\ \text{median(Scores)} &= 94 \end{aligned}$$

(c) Use your program to determine the average and median scores for the following data set:

$$\text{ExamScores} := \begin{pmatrix} 58 \\ 92 \\ 45 \\ 84 \\ 93 \\ 60 \\ 91 \\ 55 \\ 97 \end{pmatrix}$$

6.4 Resolution of a Force Vector

A common calculation in physics and engineering mechanics is the determination of the horizontal and vertical components of a force acting at an arbitrary angle. For example, Figure 6.82 shows a 250-N force acting at 150° (from three o'clock, which is called zero degrees) has a horizontal component of −216.5 N ($F \cos(150°)$) and a vertical component of 125 N ($F \sin(150°)$):

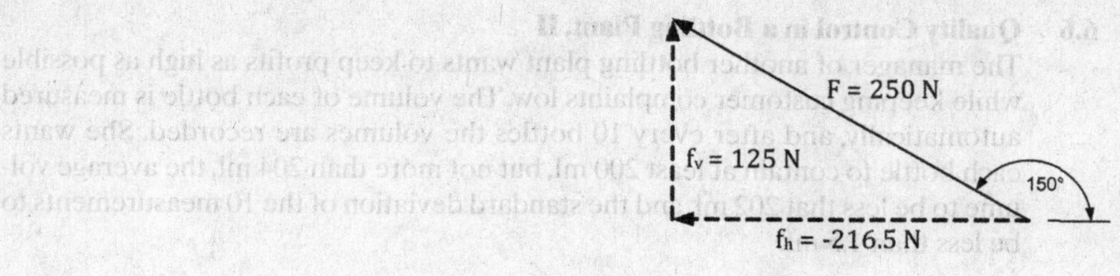

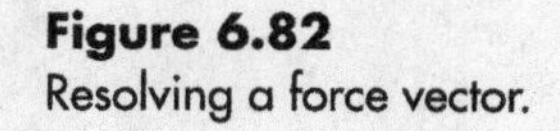
Figure 6.82
Resolving a force vector.

(a) Write a Mathcad program that receives the magnitude and direction (angle) of a force, and returns both the horizontal and vertical components.
(b) Test your program using the 250-N force example.
(c) Use your program to determine the horizontal and vertical components of the following forces:
1. 250 N at 60°.
2. 1,200 N at 220°.
3. 840 lb_f at 5 radians.

6.5 Quality Control in a Bottling Plant, I
The manager of a bottling plant wants to keep profits as high as possible while keeping customer complaints low. The volume of each bottle is measured automatically, and after every 10 bottles the volumes are recorded.

(a) Write a Mathcad program that will receive a vector of 10 volume values, and then use Mathcad's **mean** and **stdev** functions to calculate the average and standard deviation of the volume values. Have your program return both calculated values.
(b) Test your program with the following values:

$$\text{bottles} := \begin{pmatrix} 201.5 \\ 202.3 \\ 203.4 \\ 202.1 \\ 200.5 \\ 203.1 \\ 201.1 \\ 202.0 \\ 201.4 \\ 201.1 \end{pmatrix} \qquad \begin{aligned} \text{mean(bottles)} &= 201.85 \\ \text{stdev(bottles)} &= 0.867 \end{aligned}$$

(c) Use your program to determine the average and standard deviation of the following volume values:

$$\text{largeBottles} := \begin{pmatrix} 503 \\ 497 \\ 512 \\ 502 \\ 517 \\ 505 \\ 499 \\ 501 \\ 511 \\ 482 \end{pmatrix}$$

6.6 Quality Control in a Bottling Plant, II

The manager of another bottling plant wants to keep profits as high as possible while keeping customer complaints low. The volume of each bottle is measured automatically, and after every 10 bottles the volumes are recorded. She wants each bottle to contain at least 200 ml, but not more than 204 ml, the average volume to be less that 202 ml, and the standard deviation of the 10 measurements to be less than 1.5 ml.

(a) Write a flowchart of a program that will
- receive a vector of 10 volume values,
- use a **for** loop to check the individual volumes of each bottle,
- use Mathcad functions to determine the average volume and standard deviation for the 10 values, and
- return a text string that indicates whether the bottles were or were not filled correctly. If they were not, have the program indicate (with a text string) which of the criteria is/are not being met.

(b) Write a Mathcad program that implements your flowchart. Test the program with the following values:

$$\mathbf{bottles} := \begin{pmatrix} 201.5 \\ 202.3 \\ 203.4 \\ 202.1 \\ 200.5 \\ 203.1 \\ 201.1 \\ 202.0 \\ 201.4 \\ 201.1 \end{pmatrix} \qquad \begin{array}{l} \mathrm{mean(bottles)} = 201.85 \\ \mathrm{stdev(bottles)} = 0.867 \end{array}$$

(c) Create a new set of test data that will allow you to demonstrate that your **for** loop is working correctly to check the volume of each bottle.

(d) Use your program to see if the bottles represented by the following values were filled correctly:

$$\mathbf{moreBottles} := \begin{pmatrix} 201.3 \\ 201.7 \\ 203.1 \\ 201.8 \\ 203.5 \\ 203.3 \\ 202.1 \\ 201.0 \\ 201.6 \\ 202.1 \end{pmatrix}$$

(e) How would your program have to be modified if the sample size was increased from 10 bottles to 100?

6.7 Using a for Loop to Count Occurrences, I

Write a program that uses a **for** loop to count the number of exam scores in the 80s.

$$\text{Scores} := \begin{pmatrix} 85 \\ 82 \\ 93 \\ 90 \\ 57 \\ 84 \\ 91 \\ 88 \\ 79 \\ 82 \end{pmatrix}$$

Note: Mathcad provides the `last(v)` function to return the index value of the last element in vector, v. This is an easy way to set the limit of the **for** loop (illustrated in Figure 6.83).

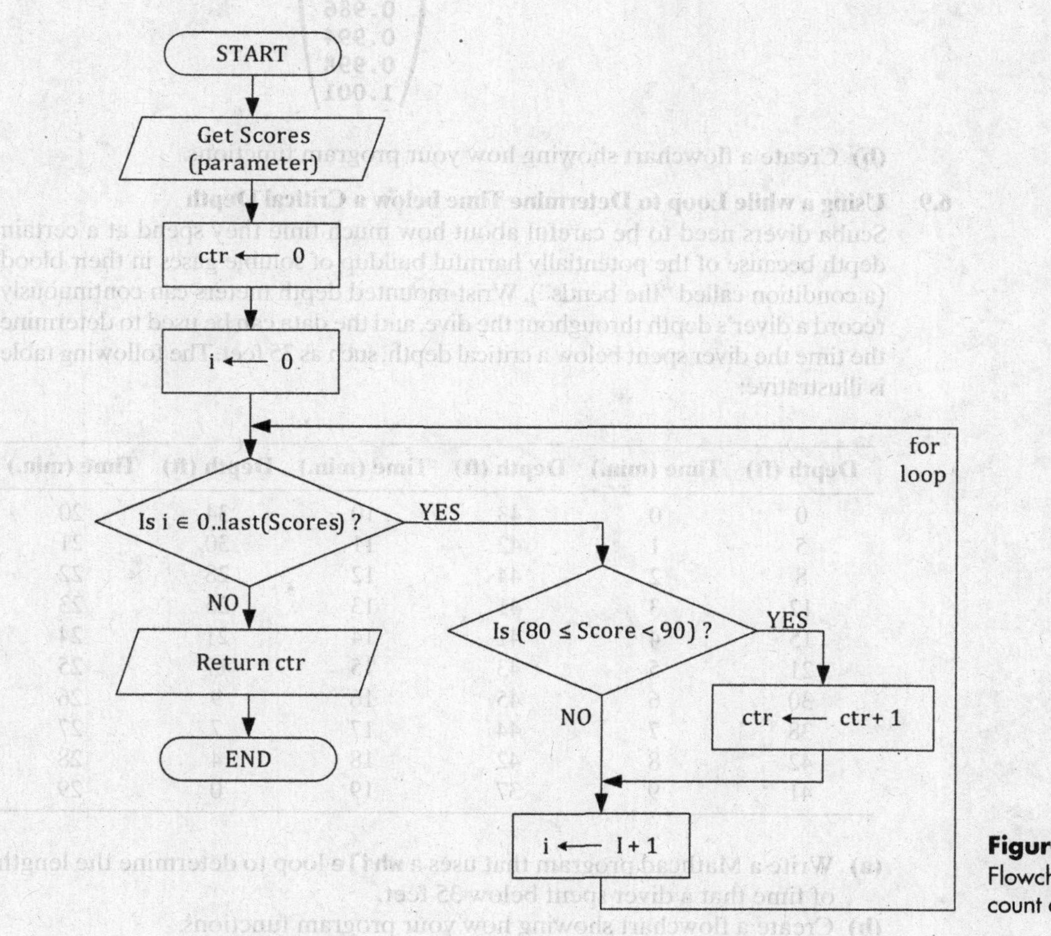

Figure 6.83
Flowchart of a program to count occurrences.

6.8 Using a for Loop to Count Occurrences, II

One approach to quality control is to improve product quality by reducing product variability. For example, an extruded plastic part might have a width that varies somewhat. The parts will fit better with the rest of the system if the variability in width is tightly controlled.

(a) Write a program that uses a **for** loop to count the number of parts that are outside of the width specification range between 0.995 cm and 1.004 cm:

$$\mathbf{width} := \begin{pmatrix} 1.011 \\ 0.997 \\ 0.999 \\ 0.993 \\ 1.004 \\ 0.998 \\ 1.005 \\ 0.999 \\ 1.000 \\ 0.998 \\ 0.989 \\ 0.990 \\ 0.986 \\ 0.994 \\ 0.998 \\ 1.001 \end{pmatrix}$$

(b) Create a flowchart showing how your program functions.

6.9 Using a while Loop to Determine Time below a Critical Depth

Scuba divers need to be careful about how much time they spend at a certain depth because of the potentially harmful buildup of soluble gases in their blood (a condition called "the bends"). Wrist-mounted depth meters can continuously record a diver's depth throughout the dive, and the data can be used to determine the time the diver spent below a critical depth, such as 35 feet. The following table is illustrative:

Depth (ft)	Time (min.)	Depth (ft)	Time (min.)	Depth (ft)	Time (min.)
0	0	43	10	34	20
5	1	42	11	30	21
8	2	44	12	28	22
12	3	41	13	25	23
15	4	41	14	21	24
21	5	43	15	17	25
30	6	45	16	9	26
38	7	44	17	7	27
42	8	42	18	4	28
41	9	37	19	0	29

(a) Write a Mathcad program that uses a **while** loop to determine the length of time that a diver spent below 35 feet.

(b) Create a flowchart showing how your program functions.

6.10 Direct Substitution Iterative Method

One simple iterative solution technique is a method called *direct substitution*, shown in the flowchart. The method requires that the function you are trying to solve be written with an unknown on both sides of the equation, and the unknown alone on the left side of the equation. For example, consider the following equation with two obvious roots ($x = 4, x = 7$):

$$(x - 4)(x - 7) = 0.$$

This can be written as a polynomial:

$$x^2 - 11x + 28 = 0.$$

We can rearrange this equation so that there is an x by itself on the left side:

$$x = \frac{x^2 + 28}{11}.$$

To solve this equation by the direct substitution method, you use your guess value (call it x_G) on the right side of the equation, and solve for a calculated value (x_C) on the left:

$$x_C = \frac{x_G^2 + 28}{11}.$$

The **F(x$_G$)** that appears in the flowchart (see Figure 6.84) is the right side of this equation and is used to compute **x$_C$**:

$$F(X_G) = \frac{x_G^2 + 28}{11}.$$

This function must be defined on your Mathcad worksheet before the direct substitution program is used.

(a) Write a **dirSub(x$_G$)** program based on the flowchart. For now, skip the error-checking section to keep the program a little simpler.

(b) Use the test function **F(x$_G$)** shown previously to test your program. Try various initial-guess values, **x$_G$**. Can you obtain either of the two solutions ($x = 4, x = 7$)?

(c) The direct substitution method cannot find all solutions, and it should fail to find one of the solutions to the test function (unless you enter the solution as the guess value). When the program fails to find a solution, it is said to diverge from, rather than converge on, the solution. How does Mathcad indicate that there is a divergence problem?

(d) Now add the error-checking section of the flowchart to your program. With the error-checking code in place, how does the program respond if the method diverges?

(e) Try a polynomial without an obvious solution, $0.3x^2 - 9x + 24 = 0$. Rearrange the equation as required for the direct substitution method, and use your program to search for roots. (You can use Mathcad's **polyroots** function to check your answer.)

6.11 Mean and Standard Deviation from a Data Vector

Calculating the mean and standard deviation of a vector of values is a very common task. Write a Mathcad program called **SampleStats(v)** that receives a vector of values as a parameter, then uses Mathcad functions to compute the

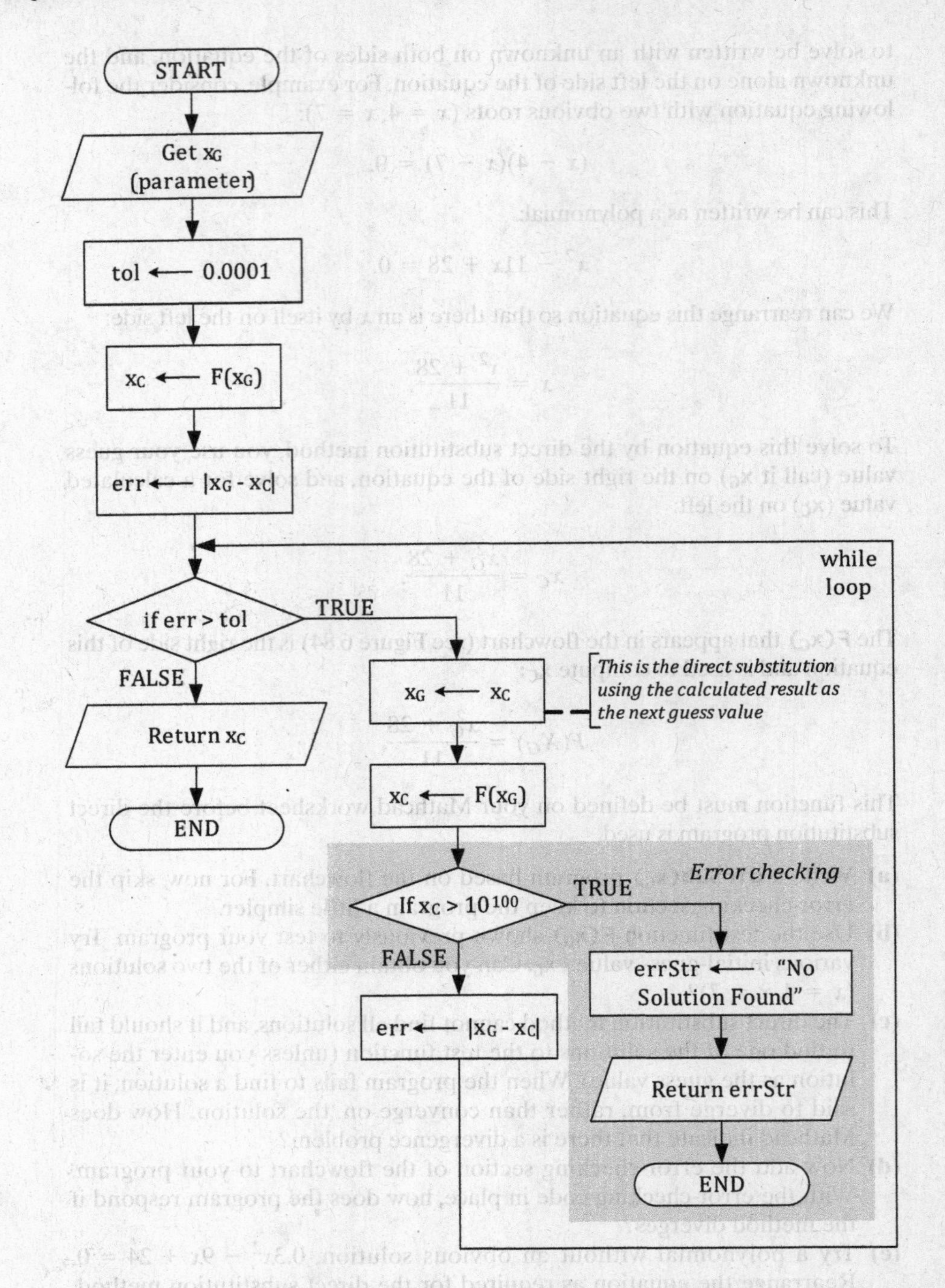

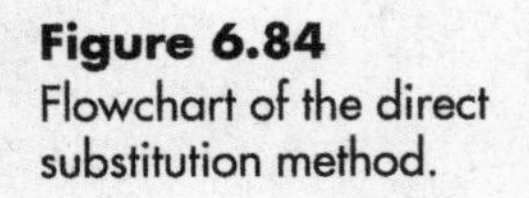
Figure 6.84
Flowchart of the direct substitution method.

mean and sample standard deviation, and returns both values as elements of a results vector. Sample data is provided to allow you to test your program.

$$\mathbf{set_1} := \begin{pmatrix} 1.1 \\ 1.3 \\ 1.9 \\ 1.2 \\ 1.4 \end{pmatrix} \quad \mathbf{SampleStats(set_1)} = \begin{pmatrix} 1.38 \\ 0.311 \end{pmatrix}$$

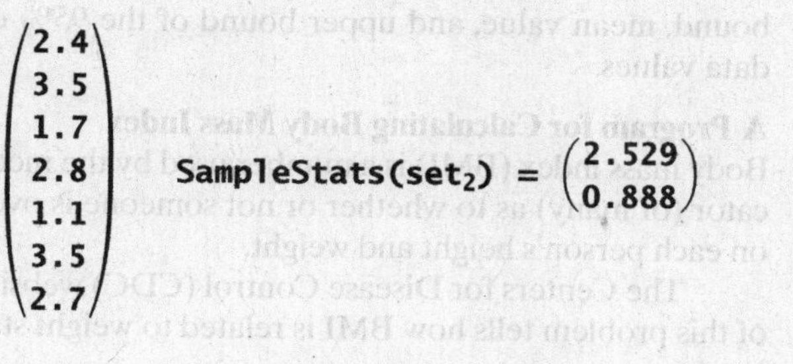

6.12 Confidence Interval

Figure 6.85 shows a Mathcad worksheet for calculating the confidence interval for a set of data values. Use the information in Figure 6.85 to write a function called `ConfInt(v)` that receives a vector of data values and returns the lower

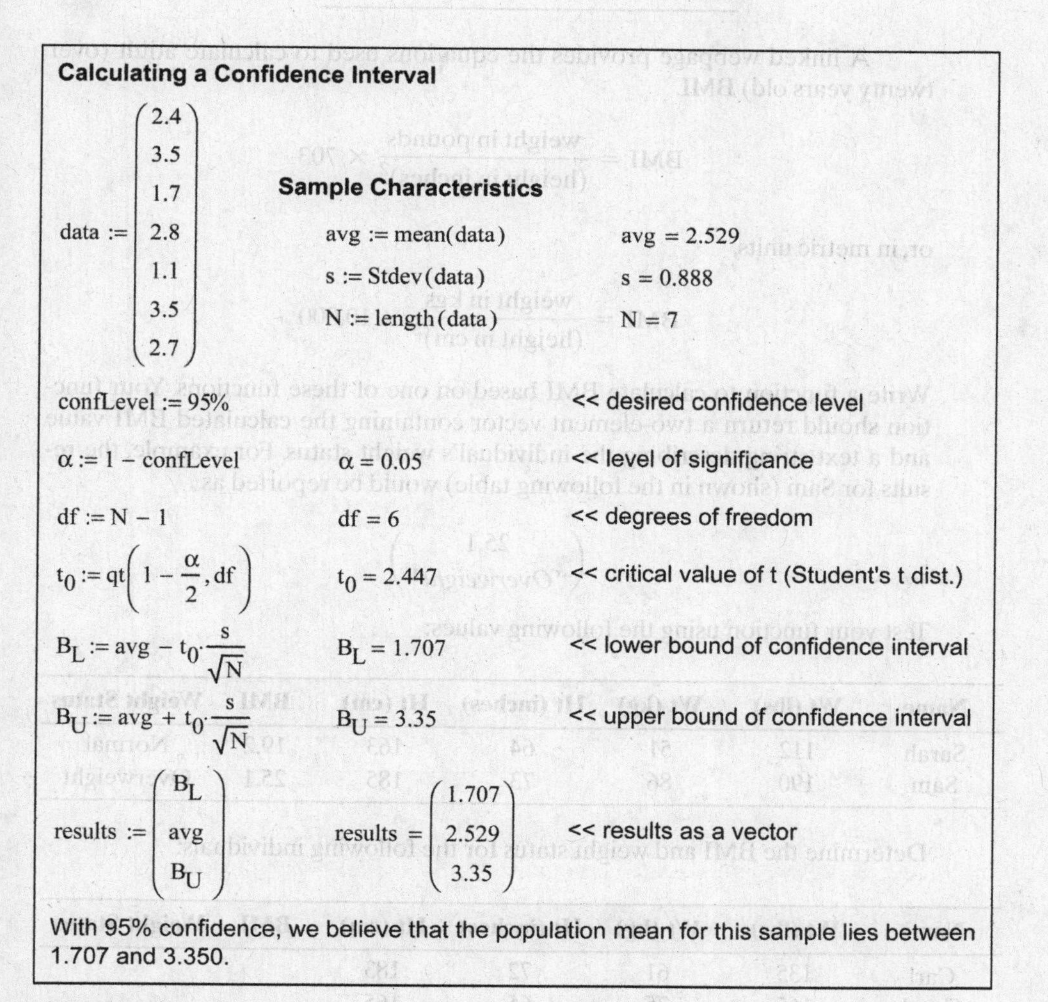

Calculating a Confidence Interval
data := (2.4, 3.5, 1.7, 2.8, 1.1, 3.5, 2.7)
Sample Characteristics
avg := mean(data)
avg = 2.529
s := Stdev(data)
s = 0.888
N := length(data)
N = 7
confLevel := 95%
<< desired confidence level
α := 1 − confLevel
α = 0.05
<< level of significance
df := N − 1
df = 6
<< degrees of freedom
$t_0 := qt\left(1 - \frac{\alpha}{2}, df\right)$
$t_0 = 2.447$
<< critical value of t (Student's t dist.)
$B_L := avg - t_0 \cdot \frac{s}{\sqrt{N}}$
$B_L = 1.707$
<< lower bound of confidence interval
$B_U := avg + t_0 \cdot \frac{s}{\sqrt{N}}$
$B_U = 3.35$
<< upper bound of confidence interval
results := (B_L, avg, B_U)
results = (1.707, 2.529, 3.35)
<< results as a vector
With 95% confidence, we believe that the population mean for this sample lies between 1.707 and 3.350.

Figure 6.85
Mathcad worksheet for computing a 95% confidence interval.

bound, mean value, and upper bound of the 95% confidence interval for the data values.

6.13 A Program for Calculating Body Mass Index
Body mass index (BMI) is a number used by the medical profession as one indicator (of many) as to whether or not someone is overweight. The BMI is based on each person's height and weight.

The Centers for Disease Control (CDC) website referenced at the bottom of this problem tells how BMI is related to weight status.

Adult BMI and Weight Status

BMI	Status
<18.5	Underweight
18.5–24.9	Normal Weight
25.0–29.9	Overweight
30.0 and above	Obese

A linked webpage provides the equations used to calculate adult (over twenty years old) BMI.

$$\text{BMI} = \frac{\text{weight in pounds}}{(\text{height in inches})^2} \times 703$$

or, in metric units,

$$\text{BMI} = \frac{\text{weight in kgs}}{(\text{height in cm})^2} \times 10{,}000$$

Write a function to calculate BMI based on one of these functions. Your function should return a two-element vector containing the calculated BMI value and a text string describing the individual's weight status. For example, the results for Sam (shown in the following table) would be reported as

$$\begin{pmatrix} 25.1 \\ \text{“}Overweight\text{”} \end{pmatrix}$$

Test your function using the following values:

Name	Wt (lbs)	Wt (kg)	Ht (inches)	Ht (cm)	BMI	Weight Status
Sarah	112	51	64	163	19.2	Normal
Sam	190	86	73	185	25.1	Overweight

Determine the BMI and weight status for the following individuals:

Name	Wt (lbs)	Wt (kg)	Ht (inches)	Ht (cm)	BMI	Weight Status
Carl	135	61	72	183		
Carol	165	75	65	165		

References:

- BMI and Weight Status: http://www.cdc.gov/nccdphp/dnpa/bmi/bmi-means.htm
- BMI Equations: http://www.cdc.gov/nccdphp/dnpa/bmi/bmi-adult-formula.htm

6.14 Drop Time and Final Velocity

Ignoring air resistance, the equations relating the position of a falling body to time, velocity, and acceleration are:

position $x(t) = x_0 + v_0 \cdot t + \frac{1}{2} a \cdot t^2$

velocity $v(t) = v_0 + a \cdot t$

If a body is dropped from a tower of known height, **H**, ($\mathbf{x_0} = 0$) with no initial velocity ($\mathbf{v_0} = 0$) and the only acceleration acting on the body is gravity ($\mathbf{a} = 9.8$) m/s^2); it is possible to calculate the duration of the fall from the position equation, and (using that result) the final velocity from the velocity equation.

(a) Write a program that receives the tower height as a parameter, then calculates the duration of fall and final velocity, and returns the values as elements of a vector.

(b) Use your function to find the duration of fall and final velocity when a marble is dropped from a tower 20 meters high.

Note: Since each element of a vector must have the same dimensions, this problem will have to be worked without units in order to return the fall duration and final velocity in the same vector.

6.15 Time Required to Heat a Hot Tub

The time required to change the water temperature in a hot tub from $\mathbf{T_1}$ to $\mathbf{T_2}$ is given by the equation

$$\Delta t = \frac{\rho \cdot V \cdot C_p \cdot (T_2 - T_1)}{Q}$$

where

ρ is the water density, 1000 kg/m^3.
V is the tub volume.
$V = \pi R^2 H$, where R is the radius of the (circular) tub and H is the water depth.
C_p is the heat capacity of water, 4187 J/kg K.
Q is the power of the hot tub's heater, usually expressed in kW (kilowatts).

(a) Write a program called `hotTub(D, H, Q, T1, T2)` that has parameters **D** (tub diameter), **H** (water depth), **Q**, $\mathbf{T_1}$, and $\mathbf{T_2}$; and calculates the tub radius, the tub volume, and the time needed to warm the water from $\mathbf{T_1}$ to $\mathbf{T_2}$. Have the program return only the required heating time. Use units for this problem so that Mathcad can help out with the unit conversions.

(b) Use your program to predict how long (in hours) it will take to warm a hot tub (**D** = 2 m, **H** = 1 m) from 25°C to 40°C with a 7-kW watt heater.

SOLUTIONS TO PRACTICE! PROBLEMS

Section 6.4, part 1

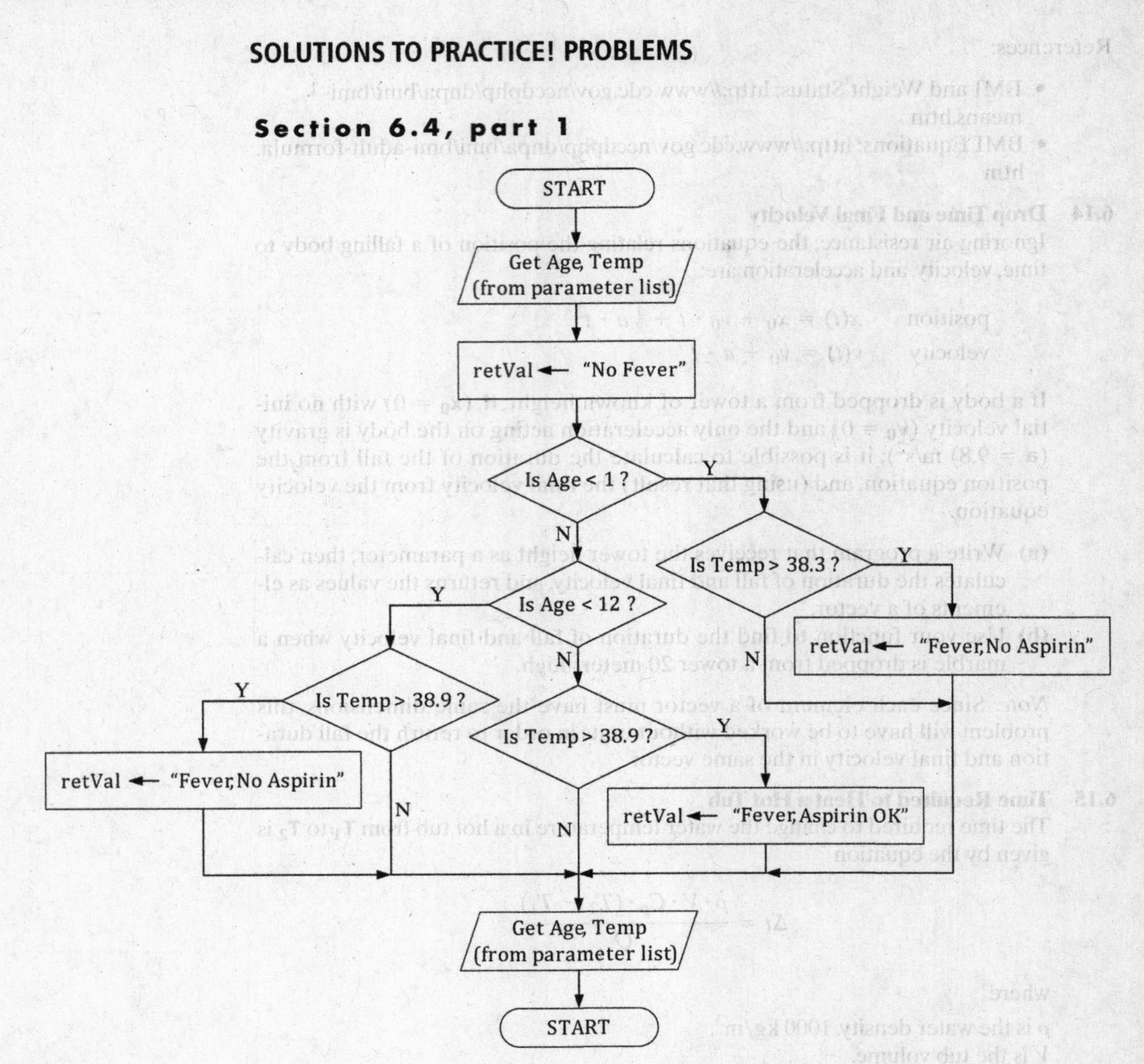

Section 6.4, part 2

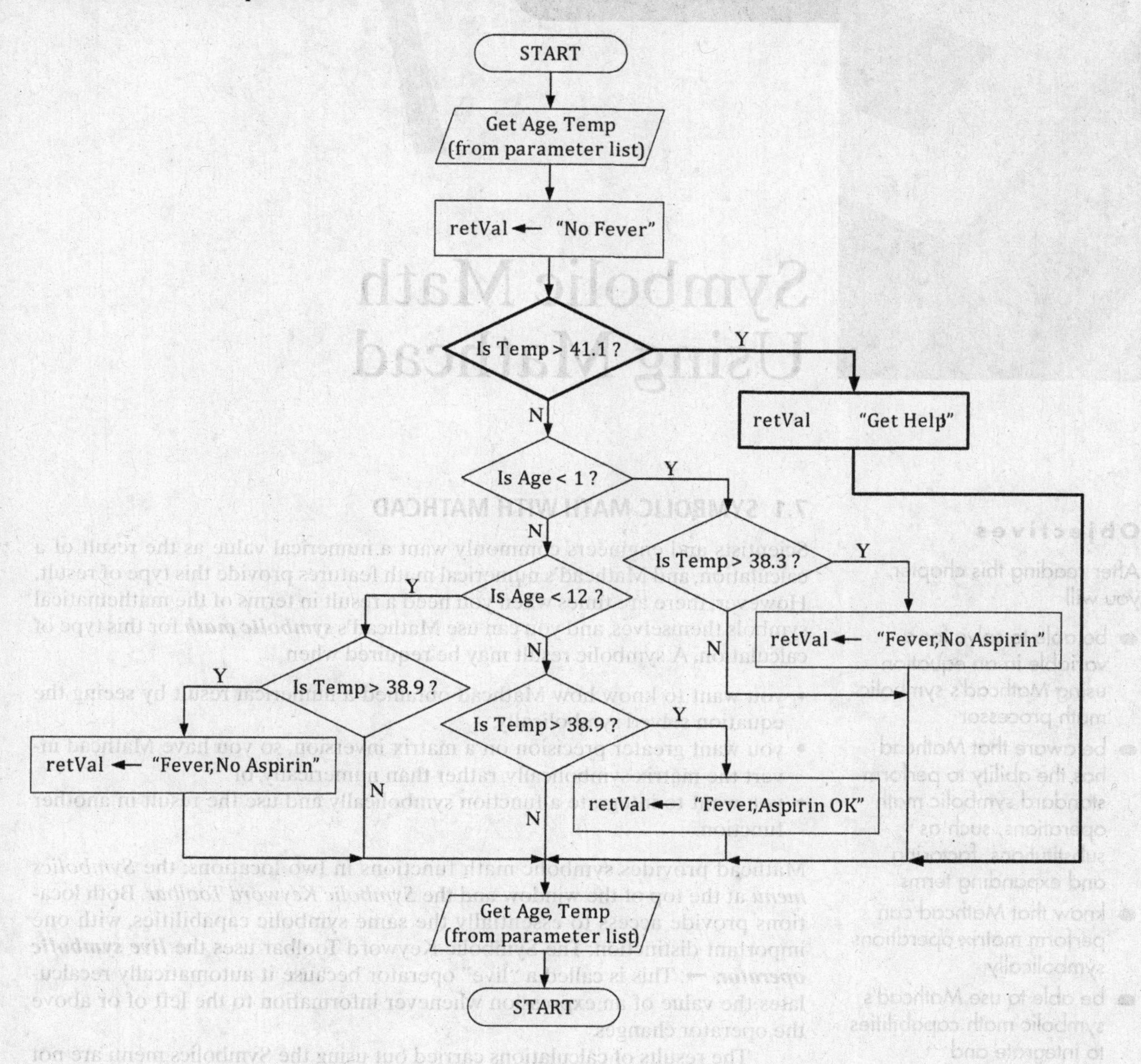

ANSWERS TO SELECTED PROBLEMS

Problem 4	c. ii. $F_V = -771$ N, $F_H = -919$ N
Problem 9	13 minutes below 35 feet
Problem 10	$x = 2.96$ (one of two roots)
Problem 14	b. duration = 2 seconds
Problem 15	b. 7.8 hours

CHAPTER 7

Symbolic Math Using Mathcad

Objectives

After reading this chapter, you will

- be able to solve for a variable in an equation using Mathcad's symbolic math processor
- be aware that Mathcad has the ability to perform standard symbolic math operations, such as substitutions, factoring, and expanding terms
- know that Mathcad can perform matrix operations symbolically
- be able to use Mathcad's symbolic math capabilities to integrate and differentiate functions

7.1 SYMBOLIC MATH WITH MATHCAD

Scientists and engineers commonly want a numerical value as the result of a calculation, and Mathcad's numerical math features provide this type of result. However, there are times when you need a result in terms of the mathematical symbols themselves, and you can use Mathcad's ***symbolic math*** for this type of calculation. A symbolic result may be required when

- you want to know how Mathcad obtained a numerical result by seeing the equation solved symbolically,
- you want greater precision on a matrix inversion, so you have Mathcad invert the matrix symbolically, rather than numerically, or
- you want to integrate a function symbolically and use the result in another function.

Mathcad provides symbolic math functions in two locations: the *Symbolics menu* at the top of the window and the *Symbolic Keyword Toolbar*. Both locations provide access to essentially the same symbolic capabilities, with one important distinction: The Symbolic Keyword Toolbar uses the ***live symbolic operator,*** →. This is called a "live" operator because it automatically recalculates the value of an expression whenever information to the left of or above the operator changes.

The results of calculations carried out using the Symbolics menu are not live. That is, once a symbolic operation has been performed using commands from the Symbolics menu, the result will not be automatically updated, even if the input data change. Thus, to update a result using the Symbolics menu, you must repeat the calculation.

The Symbolics menu is a bit more straightforward than the Symbolic Keyword Toolbar for solving for a particular variable and for factoring an expression. The Symbolic Keyword Toolbar is simpler for substitution. Both approaches will be demonstrated.

Symbolics Menu

The Symbolics menu is on the menu bar at the top of the Mathcad window. One of the first considerations in using the symbolic commands is how you want the results to be displayed. You can set this feature by using the Evaluation Style dialog box (Figure 7.2) from the Symbolics menu (shown in Figure 7.1). The Evaluation Style dialog box controls whether the results of a symbolic operation are presented to the right of the original expression or below the original expression. When results are placed below the original expression, the new equation regions can start running into existing regions. This can be avoided by having Mathcad insert blank lines before displaying the results. By default, Mathcad places the results below the original expression and adds blank lines to avoid overwriting other equation regions.

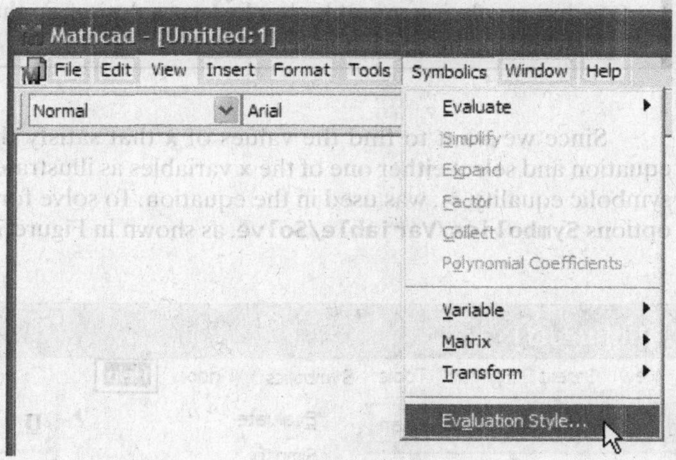

Figure 7.1
The Symbolics menu.

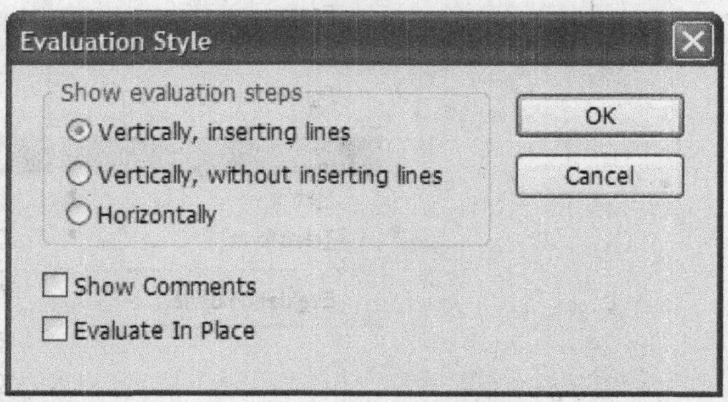

Figure 7.2
Evaluation Style dialog box.

Note: If you select **Evaluate In Place**, your original expression will be replaced by the computed result.

As an example of using the Symbolics menu, we will find the solutions of the equation

$$(x - 3) \cdot (x - 4) = 0.$$

This equation has two obvious solutions ($x = 3, x = 4$), so it is easy to see whether Mathcad is finding the correct ones.

> **Professional Success**
>
> *Keep test expressions as simple as possible.*
>
> When you are choosing mathematical expressions to test the features of a software package or to validate your own functions, try to come up with a test that
>
> - is just complex enough to demonstrate that the function is (or is not) working correctly, and
> - has an obvious, or at least a known, solution, so that it is readily apparent whether the test has succeeded or failed.

Since we want to find the values of **x** that satisfy the equation, click on the equation and select either one of the **x** variables as illustrated in Figure 7.3. Note that symbolic equality, **=**, was used in the equation. To solve for the variable **x**, use menu options **Symbolics/Variable/Solve**, as shown in Figure 7.3.

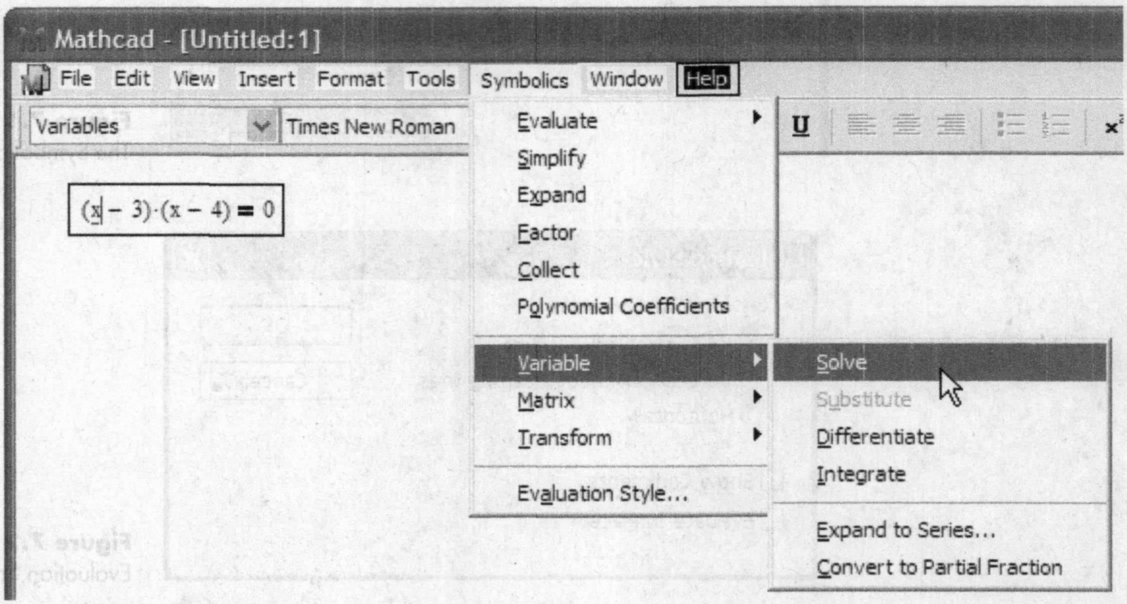

Figure 7.3
Using the Symbolics menu to solve for a variable.

The solutions—the two values of x that satisfy the equation—are presented as a two-element vector (Figure 7.4).

Symbolic operations are often performed on expressions, rather than complete equations. If you solve the expression **(x − 3) · (x − 4)** for **x**, Mathcad will set the expression to zero before finding the solutions. This is illustrated in Figure 7.5.

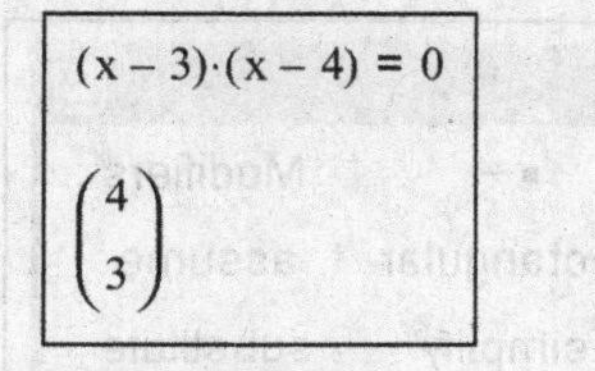

Figure 7.4
The solutions returned as a two-element vector.

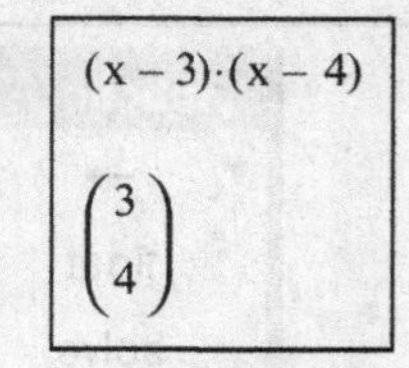

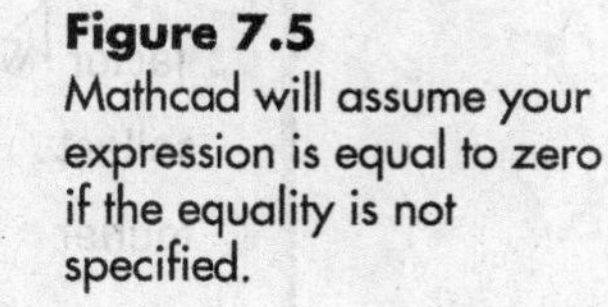

Figure 7.5
Mathcad will assume your expression is equal to zero if the equality is not specified.

Symbolic Keyword Toolbar

The Symbolic Keyword Toolbar is available from the Math Toolbar; click on the button showing a mortarboard icon (indicated in Figure 7.6) to open the Symbolic Keyword Toolbar, shown in Figure 7.7.

To use the features on this toolbar, first enter an expression, and then select the expression and click one of the buttons on the toolbar to perform a symbolic operation on the expression. For instance, from the preceding example, we know that the equation

$$(x - 3) \cdot (x - 4) = 0$$

has two solutions: **x = 3** and **x = 4**. Mathcad can also find those solutions using the **solve** button on the Symbolic Keyword Toolbar. This will be demonstrated in the next section.

Figure 7.6
The icon for Symbolic Keyword Toolbar looks like a mortarboard, or graduation cap.

Symbolic

→	▪ →	Modifiers
float	rectangular	assume
solve	simplify	substitute
factor	expand	coeffs
collect	series	parfrac
fourier	laplace	ztrans
invfourier	invlaplace	invztrans
M^T →	M^{-1} →	|M| →
explicit	combine	confrac
rewrite		

Figure 7.7
The Symbolic Keyword Toolbar.

7.2 SOLVING AN EQUATION SYMBOLICALLY

If you enter an incomplete equation, such as the left-hand side of the previous equation, Mathcad will set the expression equal to zero when the solve operation is performed. To solve the equation in this manner, enter the expression to be solved and select it. Then click on the **solve** button on the Symbolic Keyword Toolbar, as illustrated in Figure 7.8.

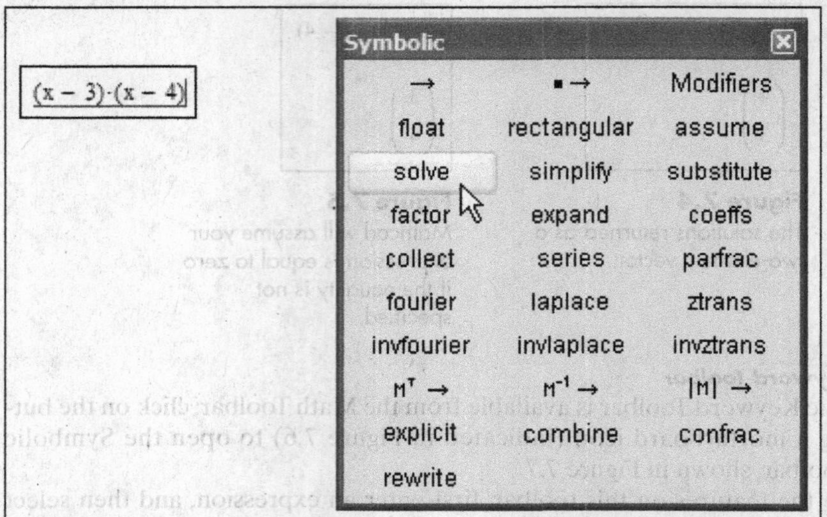

$(x-3)\cdot(x-4)$

Figure 7.8
Solving an expression using the **solve** keyword.

When using the **solve** keyword Mathcad will attempt to automatically determine which variable to solve for (for example, if there is only one variable, Mathcad will solve for that variable). This is illustrated in Figure 7.9.

$$(x-3)\cdot(x-4)\ \text{solve}\ \rightarrow \begin{pmatrix} 3 \\ 4 \end{pmatrix}$$

Figure 7.9
If Mathcad can detect the variable in the expression, the solution will be determined as soon as the **solve** keyword is used.

In this case, since there are two solutions, the solutions are returned as a two-element vector as shown in Figure 7.9.

When there are multiple variables in the expression, you must specify the variable that you want Mathcad to solve for. To do so, type a comma after the **solve** keyword. Mathcad will display a placeholder after the **solve** keyword. Enter the variable to be solved for in the placeholder, as shown in Figure 7.10.

$$(y-1)\cdot(z-2)\ \text{solve}, z\ \rightarrow 2$$
$$(y-1)\cdot(z-2)\ \text{solve}, y\ \rightarrow 1$$

Figure 7.10
After selecting the **solve** keyword from the Symbolics Toolbar, indicate which variable to solve for by entering the variable name in the placeholder.

Note: When using the Symbolic Keyword Toolbar, you tell Mathcad which variable to solve for by typing the variable name into the placeholder (if there are multiple variables in the expression). When using the Symbolics menu to solve for **x**, you tell Mathcad to solve for **x** by first selecting one of the x variables in the expression and then choosing **Solve** from the menu.

Practice!

Use Mathcad to solve the following equations. (You can leave off the zero when entering the expressions into Mathcad.)

a. $(x-1)\cdot(x+1)=0.$
b. $x^2-1=0.$
c. $(y-2)\cdot(y+3)=0.$
d. $3z^2-2z+8=0.$
e. $\dfrac{x+4}{x^2+6x-2}=0.$

7.3 MANIPULATING EQUATIONS

A number of algebraic manipulations, such as factoring out a common variable, are frequently used when one is working with algebraic expressions, and Mathcad implements these through the Symbolic Keyword Toolbar. These routine manipulations include the following:

- **Expanding** a collection of variables (e.g., after factoring).
- **Factoring** a common variable out of a complex expression.
- **Substituting** one variable or expression for another variable.
- **Simplifying** a complex expression.
- **Collecting terms** on a designated variable.

A ***partial-fraction expansion*** in which a complex expression is expanded into an equivalent expression consisting of products of fractions is a somewhat more complex operation, but one that can be helpful in finding solutions. Although this procedure will not work on all fractional expressions, it can be very useful in certain circumstances.

Examples of each of the preceding manipulations are presented in the next sections.

7.3.1 Expand

Expanding the expression `(x - 3)·(x - 4)` yields a polynomial in **x**. To expand an expression using the Symbolics menu, select the expression, and then choose **`Symbolics/Expand`**. This is illustrated in Figure 7.11.

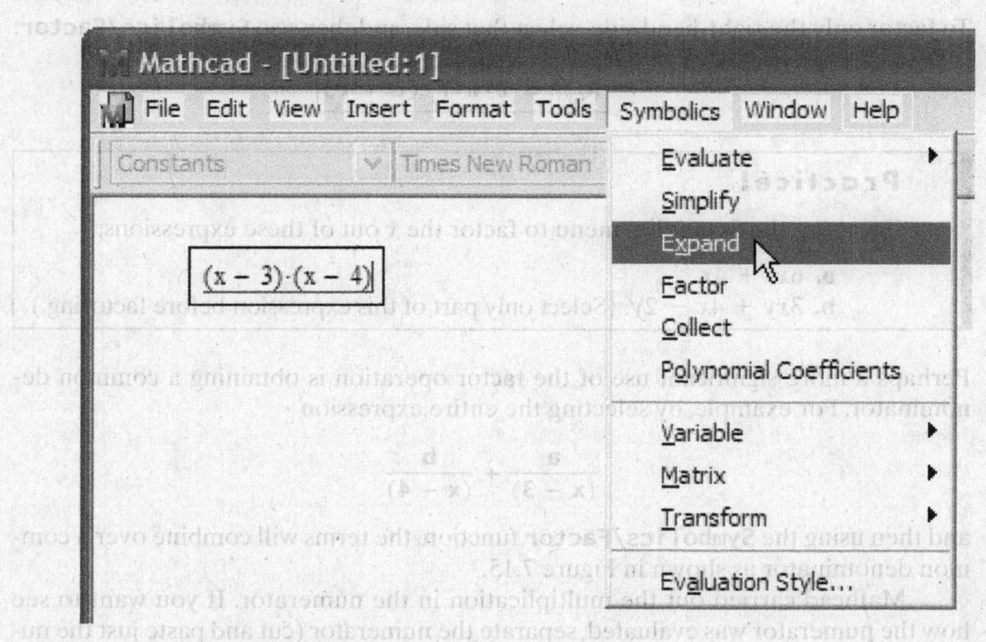

Figure 7.11
The **Expand** menu option on the Symbolics menu.

After expansion, the result is placed below the original expression as shown in Figure 7.12.

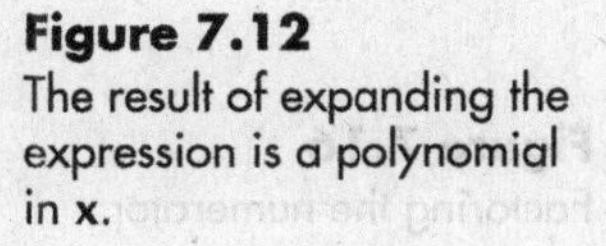

Figure 7.12
The result of expanding the expression is a polynomial in **x**.

Alternatively, select the expression, and then press the **expand** button on the Symbolic Keyword Toolbar. When the **expand** keyword is used, Mathcad will attempt to determine the variable to use in the expansion. If the variable can be automatically detected, the expansion will be completed as soon as the **expand** keyword is used. This is illustrated in Figure 7.13.

$$(x-3)\cdot(x-4)\ \textbf{expand} \rightarrow x^2 - 7\cdot x + 12$$

Figure 7.13
Using the **expand** keyword on the Symbolics Toolbar when Mathcad can determine the variable.

To explicitly enter the variable to expand on, press [,] after the **expand** keyword to create a placeholder, then enter the variable in the placeholder.

7.3.2 Factor

The *factor* operation pulls a similar quantity out of a multiterm expression. For example, $2\pi r$ appears in both terms in the expression for the surface area of a cylinder, as shown in Figure 7.14. To factor this expression, select the terms to be factored (the entire expression in this case), and then use **Symbolics/Factor**. The result is placed below the original expression as shown in Figure 7.14.

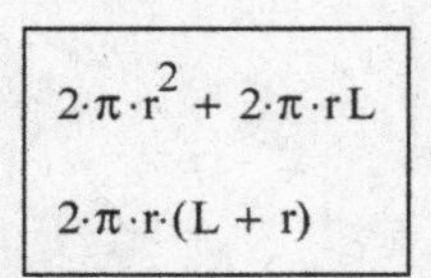

Figure 7.14
Factoring an expression using menu options **Symbolics/Factor**.

You can also factor only a portion of an expression. For example, we could factor only the right-hand side of the following equation:

$$A_{cyl} = 2\cdot\pi\cdot r^2 + 2\cdot\pi r\cdot L$$

To factor only the right-hand side, select that side, and then use **Symbolics/Factor**:

$$A_{cyl} = 2\cdot\pi\cdot r\cdot(L + r)$$

> **Practice!**
>
> Try using the Symbolics menu to factor the x out of these expressions:
>
> **a.** $6x^2 + 4x$.
> **b.** $3xy + 4x - 2y$. (Select only part of this expression before factoring.)

Perhaps a more significant use of the factor operation is obtaining a common denominator. For example, by selecting the entire expression

$$\frac{a}{(x-3)} + \frac{b}{(x-4)}$$

and then using the **Symbolics/Factor** function, the terms will combine over a common denominator as shown in Figure 7.15.

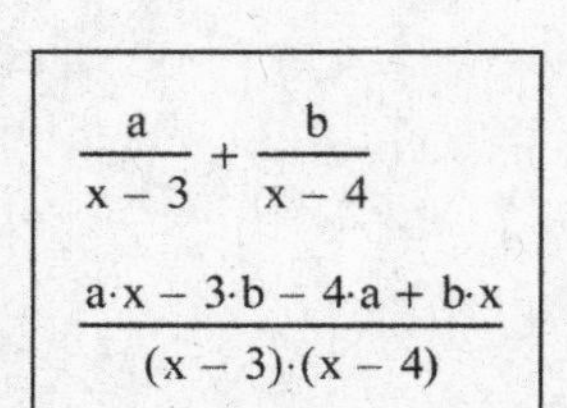

Figure 7.15
Using menu options **Symbolics/Factor** to find a common denominator.

Mathcad carried out the multiplication in the numerator. If you want to see how the numerator was evaluated, separate the numerator (cut and paste just the numerator onto the worksheet) and **collect** variables on a and b. This can be done in two steps, or in a single operation as shown in Figure 7.16. When you click the

$$a\cdot x - 3\cdot b - 4\cdot a + b\cdot x \ \left|\begin{matrix}\text{collect}, a \\ \text{collect}, b\end{matrix}\right. \rightarrow (x-3)\cdot b + a\cdot(x-4)$$

Figure 7.16
Factoring the numerator.

Collect button on the Symbolic Toolbar twice, Mathcad shows the two operations stacked to indicate the order in which they will be evaluated. (You need to use the [,] character after the **collect** keyword to specify the variable to collect on.)

7.3.3 Substitute

The ***substitute*** operation replaces a variable by another expression. Because substitution is a bit simpler using the Symbolic Keyword Toolbar buttons, that approach will be shown first.

Substitution Using the Symbolic Keyword Toolbar

A simple replacement, such as replacing all the **x**'s in an expression with **y**'s, is easily carried out using the **substitute** button on the Symbolic Keyword Toolbar as shown in Figure 7.17.

$$(x-3)\cdot(x-4) \text{ substitute}, x = y \rightarrow (y-3)\cdot(y-4)$$

Figure 7.17
Using the **substitute** keyword to change a variable name.

But you can also replace a variable with a more complicated expression. For example, suppose you wanted to replace the **x**'s with an exponential expression, such as $e^{-t/\tau}$. The **substitute** button can handle this; simply enter the complete exponential expression into the placeholder in the substitute command, as illustrated in Figure 7.18.

$$(x-3)\cdot(x-4) \text{ substitute}, x = e^{-\frac{t}{\tau}} \rightarrow e^{-\frac{2\cdot t}{\tau}} - 7\cdot e^{-\frac{t}{\tau}} + 12$$

Figure 7.18
Using the **substitute** keyword to replace a variable with an expression.

Note that Mathcad not only performed the substitution, but it also expanded the result.

Substitution Using the Symbolics Menu

If you want to use the Symbolics menu to carry out a substitution, there are two things to keep in mind:

1. The new expression (the expression that will be substituted into the existing expression) must be copied to the Windows clipboard before performing the substitution.
2. You must select the variable to be replaced before performing the substitution.

To repeat the last example, the $e^{-t/\tau}$ would be entered into the Mathcad worksheet, selected, and copied to the Windows clipboard, using **Edit/Copy** (or [Ctrl-C]). Then one of the **x** variables in **(x - 3)·(x - 4)** would be selected. (This tells Mathcad to replace all of the **x**'s in the expression with the contents of the Windows clipboard.) Finally, the substitution is performed using **Symbolics/Variable/**

Substitute, and the results are placed below the original expression. The final Mathcad worksheet now looks like Figure 7.19 (with some added comments):

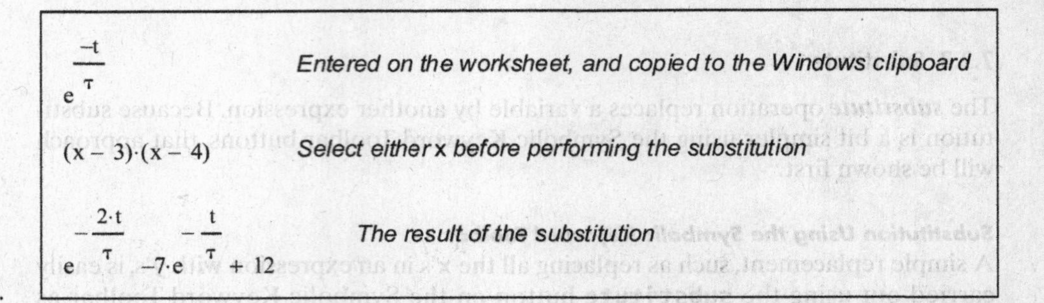

Entered on the worksheet, and copied to the Windows clipboard
(x − 3)·(x − 4)
Select either x before performing the substitution
The result of the substitution

Figure 7.19
Using menu options **Symbolics/Variable/Substitute** to replace a variable with an expression.

7.3.4 Simplify

According to the Mathcad help files, the ***simplify*** menu command "performs arithmetic, cancels common factors, uses basic trigonometric and inverse function identities, and simplifies square roots and powers." If we try to simplify **(x − 3) · (x − 4)** using menu options **Symbolics/Simplify**, the expression is returned unchanged: Mathcad thinks that **(*x* − 3) · (*x* − 4)** is as simple as this expression gets. In order to demonstrate the simplify operation, we need to complicate the example a little. Consider this modification:

$$(x - \sqrt{9}) \cdot (x - 2^2).$$

If we try to simplify this expression using **Symbolics/Simplify**, the original expression is returned as shown in Figure 7.20.

The simplify operation simplified the square root (selecting the positive root) and power.

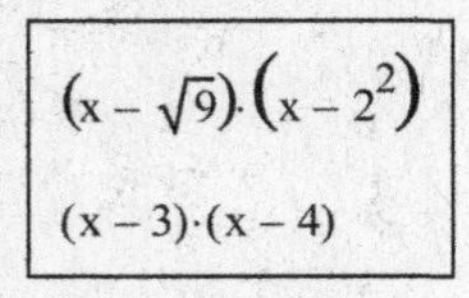

Figure 7.20
Using menu options **Symbolics/Simplify** to simplify an expression.

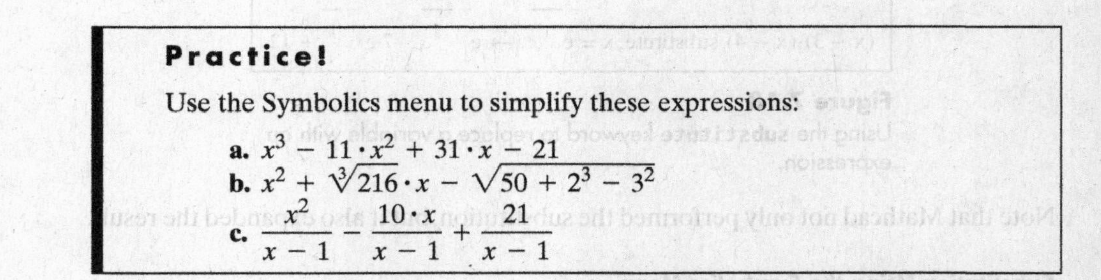

Practice!

Use the Symbolics menu to simplify these expressions:

a. $x^3 - 11 \cdot x^2 + 31 \cdot x - 21$

b. $x^2 + \sqrt[3]{216} \cdot x - \sqrt{50 + 2^3 - 3^2}$

c. $\dfrac{x^2}{x-1} - \dfrac{10 \cdot x}{x-1} + \dfrac{21}{x-1}$

7.3.5 Collect

The ***collect*** operation is used to rewrite a set of summed terms as a polynomial in the selected variable (if it is possible to do so). For example, the **expand** command, operating on the **x** in **(x − 3) · (x − 4)** returned a polynomial, so we know that expression can be written as a polynomial in **x**. The **collect** operation should also return that polynomial, and in Figure 7.21 we observe that it does.

$(x-3)\cdot(x-4)$ collect $\rightarrow x^2 - 7\cdot x + 12$

Figure 7.21
Using the **collect** keyword to return a polynomial.

7.3.6 Partial-Fraction Expansion

A partial-fraction expansion on a variable is a method for expanding a complex expression into a sum of (we hope) simpler expressions with denominators containing only linear and quadratic terms and no functions of the variable in the numerator.

As a first example, consider the expression

$$\frac{a}{x-3}+\frac{b}{x-4}$$

We can modify the expression to obtain a common denominator, as

$$\frac{a\cdot x-3\cdot b-4\cdot a+b\cdot x}{(x-3)\cdot(x-4)}$$

A partial-fraction expansion, based upon *x*, on this expression returns the original function, as seen in Figure 7.22.

$$\frac{a\cdot x-3\cdot b-4\cdot a+b\cdot x}{(x-3)\cdot(x-4)}\ \text{parfrac}, x\ \rightarrow\ \frac{a}{x-3}+\frac{b}{x-4}$$

Figure 7.22
Using the **parfrac** keyword to perform a partial fraction expansion.

This can be accomplished by selecting an **x** (any of them) and then choosing **Symbolics/Variable/Convert to Partial Fraction** from the Symbolics menu, or, as shown in Figure 7.22, by using the **parfrac** keyword on the Symbolics Toolbar.

A slightly more complex example is to take the ratio of two polynomials and expand it by using partial fractions. The process is the same as described in the preceding example:

1. Select the variable upon which you want to expand (**z** in this example).
2. Choose **Symbolics/Variable/Convert to Partial Fraction** from the Symbolics menu.

$$\frac{9\cdot z^2-40\cdot z+23}{z^3-7\cdot z^2+7\cdot z+15}$$

$$\frac{3}{z+1}+\frac{-2}{z-3}+\frac{4}{z\ \ 5}$$

Figure 7.23
Using menu options **Symbolic/Variable/Convert to Partial Function**.

The result is shown in Figure 7.23.

An expanded example involving two variables illustrates how the partial fraction expansion depends on the variable selected. The starting expression, shown in Figure 7.24, now involves both **z** and **y**.

$$\frac{6\cdot z^2-20\cdot z\cdot y+15\cdot z-35\cdot y+6}{z^3-5\cdot z^2\cdot y+4\cdot z^2-20\cdot z\cdot y+3\cdot z-15\cdot y}\ \text{parfrac}, z\ \rightarrow$$

$$\frac{\dfrac{\dfrac{25\cdot y+15}{10\cdot y+6}-20\cdot y+\dfrac{(5\cdot y+2)\cdot(25\cdot y+15)}{10\cdot y+6}-9}{5\cdot y+1}}{z-5\cdot y}\quad\frac{\dfrac{25\cdot y+15}{10\cdot y+6}+6\quad\dfrac{25\cdot y+15}{10\cdot y+6}-20\cdot y+\dfrac{(5\cdot y+2)\cdot(25\cdot y+15)}{10\cdot y+6}-9}{(z+1)\cdot(5\cdot y+1)}+\frac{25\cdot y+15}{(z+3)\cdot(10\cdot y+6)}$$

$$\frac{6\cdot z^2-20\cdot z\cdot y+15\cdot z-35\cdot y+6}{z^3-5\cdot z^2\cdot y+4\cdot z^2-20\cdot z\cdot y+3\cdot z-15\cdot y}\ \text{parfrac}, y\ \rightarrow\ \frac{20\cdot z+35}{5\cdot z^2+20\cdot z+15}+\frac{15\cdot z+6\cdot z^2-\dfrac{(20\cdot z+35)\cdot\left(z^3+4\cdot z^2+3\cdot z\right)}{5\cdot z^2+20\cdot z+15}+6}{(z-5\cdot y)\cdot(z+1)\cdot(z+3)}$$

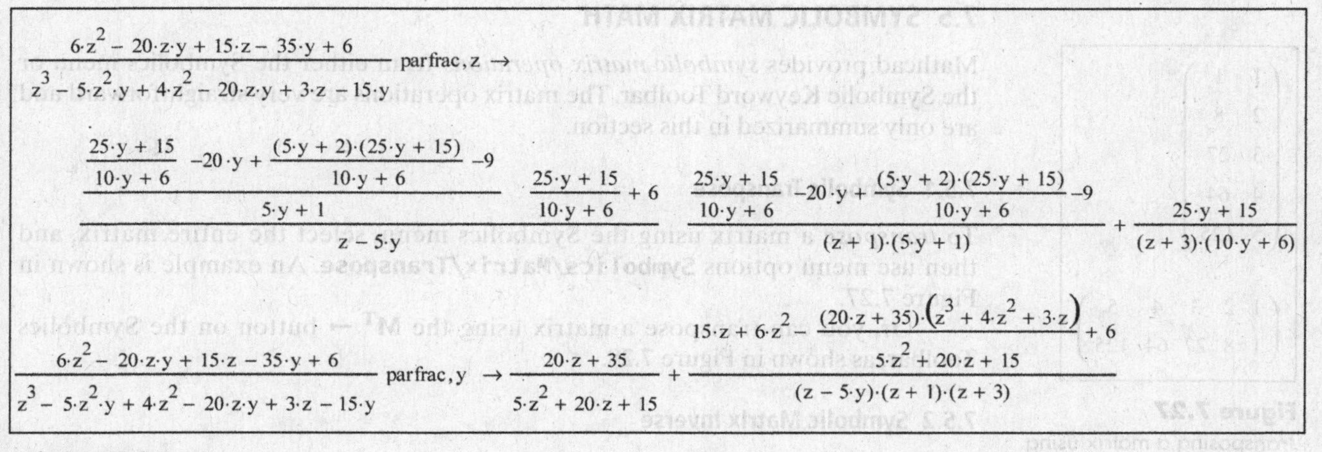

Figure 7.24
The result of a partial-fraction expansion of a multivariable expression can depend on the variable used in the expansion.

A partial-fraction expansion on **z** requires that each denominator be linear or quadratic in **z** and precludes any function of **z** from every numerator. The result is shown in the upper portion of Figure 7.24.

On the other hand, a partial-fraction expansion on **y** precludes functions of **y** in the numerators, but allows functions of **z**. The result of a partial-fraction expansion on **y** is shown in the lower portion of Figure 7.24.

7.4 POLYNOMIAL COEFFICIENTS

If you have an expression that can be written as a polynomial, Mathcad will return a vector containing the coefficients of the polynomial. This operation is available from the Symbolics menu using **Symbolics/*Polynomial Coefficients***, or through the **coeffs** keyword on the Symbolics Toolbar.

The coefficients of the polynomial $x^2 - 7x + 12$ are pretty obvious. If you select any **x** in the expression, and then ask for the vector of polynomial coefficients using menu options **Symbolics/Polynomial Coefficients**, the coefficients are returned as a vector, as shown in Figure 7.25.

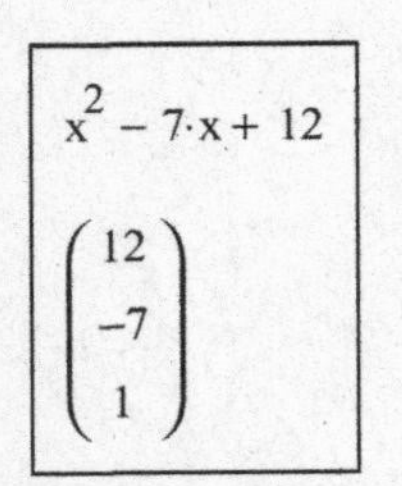

Figure 7.25
The polynomial coefficients of the expression $x^2 - 7x + 12$ have been returned as a vector.

Note: While Mathcad's symbolic math functions typically display the higher powers first (i.e., the x^2 before the $7x$), the first polynomial coefficient in the returned vector is the constant, 12.

When your expression is written as a polynomial, the coefficients are apparent. However, Mathcad's symbolic processor can return the polynomial coefficients of expressions that are not displayed in standard polynomial form. For example, we know that $x^2 - 7x + 12$ is algebraically equivalent to **(x − 3)·(x − 4)**. Using the **coeffs** keyword on **(x − 3)·(x − 4)** returns the same polynomial coefficients as in the standard-form case, as shown in Figure 7.26.

$$(x-3)(x-4) \text{ coeffs} \rightarrow \begin{pmatrix} 12 \\ -7 \\ 1 \end{pmatrix}$$

Figure 7.26
Finding the polynomial coefficients using the **coeffs** keyword on the Symbolics Toolbar.

7.5 SYMBOLIC MATRIX MATH

Mathcad provides *symbolic matrix operations* from either the Symbolics menu or the Symbolic Keyword Toolbar. The matrix operations are very straightforward and are only summarized in this section.

7.5.1 Symbolic Transpose

To *transpose* a matrix using the Symbolics menu, select the entire matrix, and then use menu options **Symbolics/Matrix/Transpose**. An example is shown in Figure 7.27.

Or, you can transpose a matrix using the $\mathbf{M^T} \rightarrow$ button on the Symbolics Toolbar, as shown in Figure 7.28.

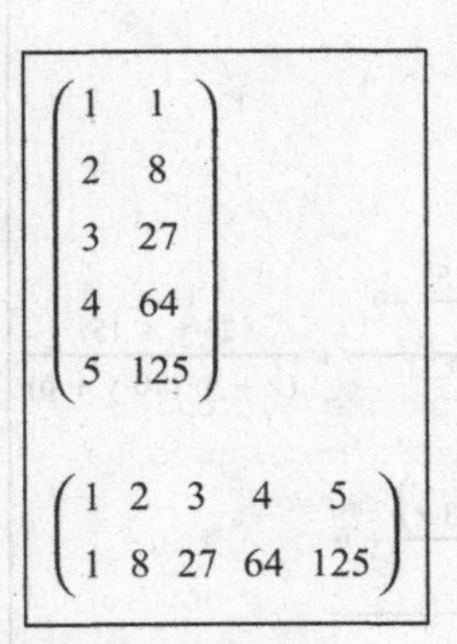

Figure 7.27
Transposing a matrix using menu options **Symbolics/Matrix/Transpose**.

7.5.2 Symbolic Matrix Inverse

To *invert* a matrix using symbolic math, select the entire matrix, and then use menu option **Symbolics/Matrix/Invert**. An example is shown in Figure 7.29.

$$\begin{pmatrix} 1 & 1 \\ 2 & 8 \\ 3 & 27 \\ 4 & 64 \\ 5 & 125 \end{pmatrix}^T \rightarrow \begin{pmatrix} 1 & 2 & 3 & 4 & 5 \\ 1 & 8 & 27 & 64 & 125 \end{pmatrix}$$

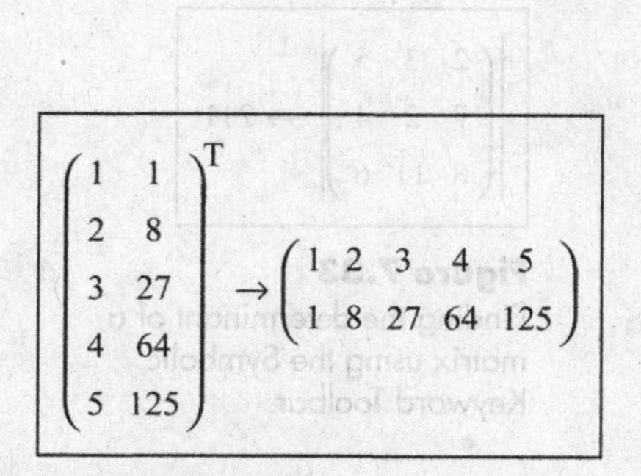

Figure 7.28
Transposing a matrix using the Symbolics toolbar.

$$\begin{pmatrix} 2 & 3 & 5 \\ 7 & 2 & 4 \\ 8 & 11 & 6 \end{pmatrix}$$

$$\begin{pmatrix} -\frac{32}{211} & \frac{37}{211} & \frac{2}{211} \\ \frac{10}{211} & -\frac{28}{211} & \frac{27}{211} \\ \frac{61}{211} & \frac{2}{211} & -\frac{17}{211} \end{pmatrix}$$

Figure 7.29
Inverting a matrix using menu options **Symbolics/Matrix/Invert**.

The $\mathbf{M^{-1}} \rightarrow$ button on the Symbolic Keyword Toolbar can also be used to invert a matrix. This is illustrated in Figure 7.30.

$$\begin{pmatrix} 2 & 3 & 5 \\ 7 & 2 & 4 \\ 8 & 11 & 6 \end{pmatrix}^{-1} \rightarrow \begin{pmatrix} -\frac{32}{211} & \frac{37}{211} & \frac{2}{211} \\ \frac{10}{211} & -\frac{28}{211} & \frac{27}{211} \\ \frac{61}{211} & \frac{2}{211} & -\frac{17}{211} \end{pmatrix}$$

Figure 7.30
Inverting a matrix using the Symbolic Keyword Toolbar.

If you want to see the inverted matrix represented as values rather than fractions, then select the entire matrix result and

- use menu options **Symbolics/Evaluate/Floating Point ...** and enter the number of digits to display, or
- use the **float** keyword on the Symbolic Toolbar and enter the number of digits to display in the placeholder. This result is shown in Figure 7.31.

$$\begin{pmatrix} 2 & 3 & 5 \\ 7 & 2 & 4 \\ 8 & 11 & 6 \end{pmatrix}^{-1} \rightarrow \begin{pmatrix} -\frac{32}{211} & \frac{37}{211} & \frac{2}{211} \\ \frac{10}{211} & -\frac{28}{211} & \frac{27}{211} \\ \frac{61}{211} & \frac{2}{211} & -\frac{17}{211} \end{pmatrix} \text{float}, 4 \rightarrow \begin{pmatrix} -0.1517 & 0.1754 & 0.009479 \\ 0.04739 & -0.1327 & 0.128 \\ 0.2891 & 0.009479 & -0.08057 \end{pmatrix}$$

Figure 7.31
Displaying the result as decimal values using the **float** keyword.

7.5.3 Symbolic Matrix Determinant

You can use either the Symbolics menu or the Symbolic Keyword Toolbar to calculate the *determinant* of a matrix. The use of Symbolics menu options **Symbolics/Matrix/Determinant** is illustrated in Figure 7.32, while the use of the **|M|** button on the Symbolic Keyword Toolbar is shown in Figure 7.33.

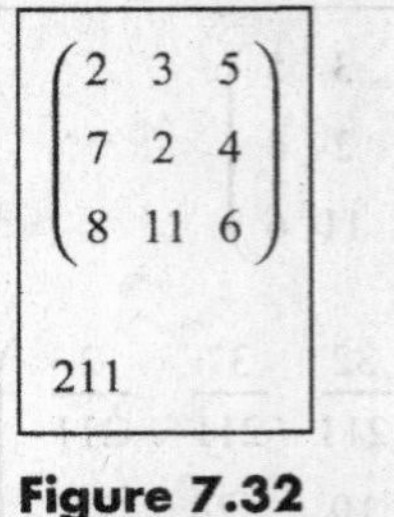

Figure 7.32
Finding the determinant of a matrix using menu options `Symbolics/Matrix/Determinant.`

$$\left|\begin{pmatrix} 2 & 3 & 5 \\ 7 & 2 & 4 \\ 8 & 11 & 6 \end{pmatrix}\right| \rightarrow 211$$

Figure 7.33
Finding the determinant of a matrix using the Symbolic Keyword Toolbar.

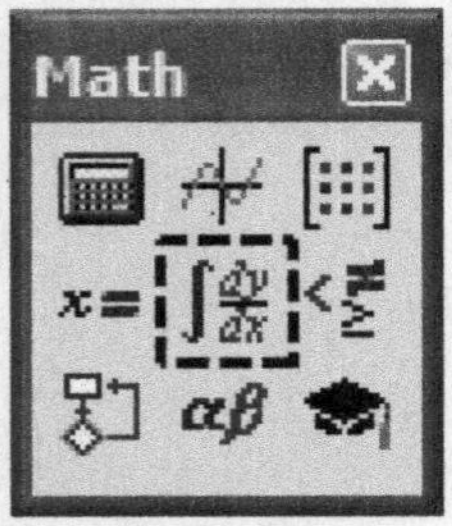

Figure 7.34
The Calculus Toolbar is accessed via the Math Toolbar.

7.6 SYMBOLIC INTEGRATION

Mathcad provides a number of ways to integrate and differentiate functions. In this chapter, we focus on *symbolic integration*. The integration and differentiation operators are found on the Calculus Toolbar (see Figure 7.35), which is available from the Math Toolbar (see Figure 7.34).

In this section, we use a simple polynomial as a demonstration function and the same polynomial throughout to allow us to compare the various integration and differentiation methods, except where multiple variables are required to demonstrate multivariate integration. The example polynomial is $12 + 3x - 4x^2$.

7.6.1 Indefinite Integrals

Symbolic Integration Using the Indefinite Integral Operator

The most straightforward of Mathcad's integration methods uses the ***indefinite integral operator*** from the Calculus Toolbar. When you click on the button for the indefinite integral operator (see Figure 7.37), the operator is placed on your worksheet with two empty placeholders, the first for the function to be integrated and the second for the integration variable:

$$\int \blacksquare \, d\blacksquare$$

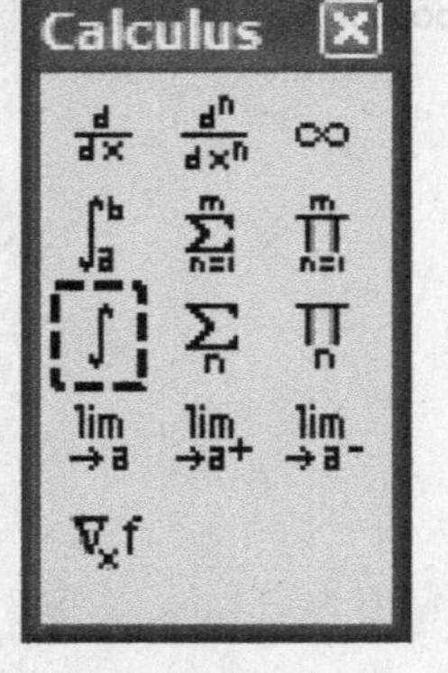

Figure 7.35
The Calculus Toolbar with Indefinite Integral operator indicated.

To integrate the sample polynomial, enter the polynomial in the first placeholder and an x in the second placeholder:

$$\int (12 + 3\cdot x - 4\cdot x^2)\, dx$$

Complete the integration using the "evaluate symbolically" symbol, the →. This symbol is available on the Symbolic Keyword Toolbar, or may be entered by pressing [Ctrl-.] (Hold down the Control key while pressing the period key.) The result is shown in Figure 7.36.

$$\int \left(12 + 3\cdot x - 4\cdot x^2\right) dx \rightarrow \frac{3\cdot x^2}{2} - \frac{4\cdot x^3}{3} + 12\cdot x$$

Figure 7.36
Symbolic integration of the example polynomial.

Notice that Mathcad does not add the constant of integration you might expect when integrating without limits. Mathcad shows the functional form of the integrated expression, but you have to add your own integration constant if you want to evaluate the result. Copy the result to an editable equation region then add the constant.

Symbolic Integration Using the Symbolics Menu

Symbolic integration using the Symbolics menu is an alternative method when integrating with respect to a single variable. Symbolic integration does not use the indefinite integral operator at all. Instead, you enter your function, for instance,

$$12 + 3 \cdot x - 4 \cdot x^2$$

and then select the variable of integration, one of the x's in this example. To perform the integration, use Symbolics menu option **Symbolics/Variable/Integrate**, as shown in Figure 7.37. The result is placed below the original expression.

Figure 7.37
Using the Symbolics menu to integrate.

The result (again, without the constant of integration) is placed below the original function as shown in Figure 7.38. This approach is very handy, but it is useful only for indefinite integrals with a single integration variable.

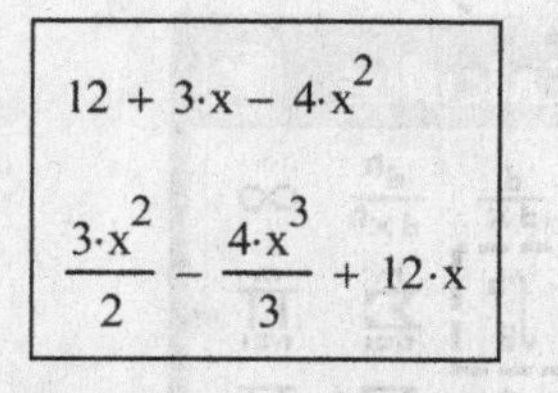

Figure 7.38
The result of Symbolic integration of the example polynomial.

Symbolic Integration with Multiple Variables

You can use the indefinite integral operator from the Calculus Toolbar (see Figure 7.35) to integrate over multiple variables; simply use one indefinite integral operator for each integration variable.

For example, consider integrating the following expression to determine the volume of a cylinder:

$$\int\int\int r \, dr \, d\theta \, dl$$

This triple integral will integrate over three variables: **r**, **θ**, and **l** (the letter el, not the number one).

To enter this expression in Mathcad, simply use one indefinite integral operator for each integration variable. The second integration operator is placed in the

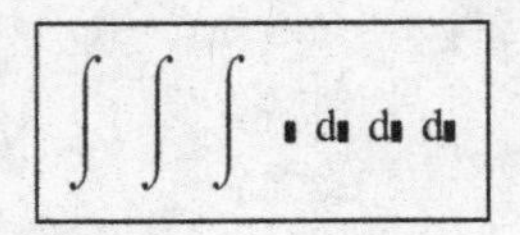

Figure 7.39
After placing three integration operators on the worksheet.

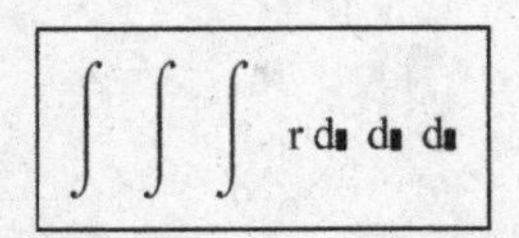

Figure 7.40
After filling the function placeholder.

first operator's function placeholder, and so on. To integrate over **r**, **θ**, and **l** we use three integration operators, as shown in Figure 7.39. (These can be entered simply by clicking the indefinite integral operator button on the Calculus Toolbar three times.)

Note that there is now only one function placeholder (before the first **d** in Figure 7.39), but there are three integration variable placeholders (after each **d**). In order to compute the volume of a cylinder, the function placeholder will simply hold an **r**, as illustrated in Figure 7.40.

The remaining placeholders are for the three integration variables: **r**, **θ**, and **l**. Figure 7.41 shows the completed integration, after using the "evaluate symbolically" operator →.

You may have expected $\pi \cdot r^2 \cdot l$ to be the formula for the volume of a cylinder, but this formula in Figure 7.41 was integrated as an indefinite integral (no limits). The π in the equation for the volume of a cylinder comes from the integration limits on **θ**. This will become apparent later, when we use the ***definite integral*** operator to integrate this function with limits.

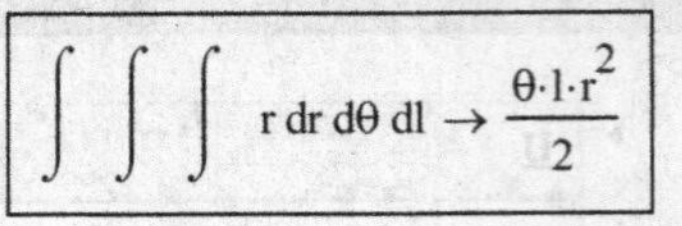

Figure 7.41
The completed integration.

Practice!

Use Mathcad's indefinite integral operator to evaluate the following integrals (starting with the obvious):

a. $\int x\, dx.$

b. $\int (a + bx + cx^2)\, dx.$

c. $\int\int 2x^2 y\, dx\, dy.$

d. $\int \frac{1}{x}\, dx.$

e. $\int e^{-ax}\, dx.$

f. $\int \cos(2x)\, dx.$

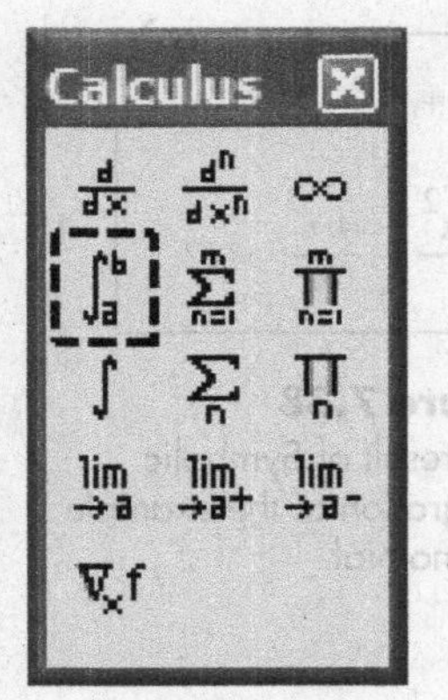

Figure 7.42
The Calculus Toolbar with the definite integral operator indicated.

7.6.2 Definite Integrals

Mathcad evaluates definite integrals using the *definite integral operator* from the Calculus Toolbar. (See Figure 7.42.) Definite integrals can be evaluated in a number of ways, with symbolic or numeric ***limits of integration***:

- symbolic evaluation with variable limits
- symbolic evaluation with numeric limits
- symbolic evaluation with mixed (variable and numeric) limits
- numerical evaluation (requires numeric limits)

Definite Integrals: Symbolic Evaluation with Variable Limits

The definite integral operator comes with four placeholders—the function and integration variable placeholders and two limit placeholders:

$$\int_{\blacksquare}^{\blacksquare} \blacksquare \, d\blacksquare$$

To evaluate the polynomial integrated from $x = A$ to $x = B$, simply include the limits in the appropriate placeholders:

$$\int_A^B (12 + 3 \cdot x - 4 \cdot x^2)\, dx.$$

You instruct Mathcad to evaluate the integral by using the symbolic evaluation symbol, →. The result is shown in Figure 7.43.

$$\int_A^B \left(12 + 3 \cdot x - 4 \cdot x^2\right) dx \to \frac{4 \cdot A^3}{3} - \frac{3 \cdot A^2}{2} - 12 \cdot A - \frac{4 \cdot B^3}{3} + \frac{3 \cdot B^2}{2} + 12 \cdot B$$

Figure 7.43
Symbolic integration with variable limits.

Definite Integrals: Symbolic Evaluation with Numeric Limits

To evaluate an integral with numeric limits, you can place the numbers in the limit placeholders, as shown in Figure 7.44.

$$\int_{-1}^{2} \left(12 + 3 \cdot x - 4 \cdot x^2\right) dx \to \frac{57}{2}$$

Figure 7.44
Symbolic integration with numeric limits.

Or, you can assign values to variables before the integration and use the variables as limits, as shown in Figure 7.45.

$$A := -1 \qquad B := 2$$

$$\int_A^B \left(12 + 3 \cdot x - 4 \cdot x^2\right) dx \to \frac{57}{2}$$

Figure 7.45
Symbolic integration with numeric limits; limit values assigned to variables.

Definite Integrals: Symbolic Evaluation with Mixed Limits

It is fairly common to have a numeric value for one limit and to want to integrate from that known value to an arbitrary (variable) limit. Mathcad handles this type of integration as well, as illustrated in Figure 7.46.

$$\int_{-1}^{C} \left(12 + 3 \cdot x - 4 \cdot x^2\right) dx \to \frac{(C + 1) \cdot \left(17 \cdot C - 8 \cdot C^2 + 55\right)}{6} \text{ collect } \to \frac{3 \cdot C^2}{2} - \frac{4 \cdot C^3}{3} + 12 \cdot C + \frac{55}{6}$$

Figure 7.46
Symbolic integration with mixed limits.

Note that Mathcad's symbolic math processor does not automatically collect variables after integration, but this can be done as a second step, as illustrated in Figure 7.46.

In the preceding example, the known limit was evaluated and generated the 55/6 in the result. The unknown limit was evaluated in terms of the variable *C*. The result can then be evaluated for any value of *C*.

Note: In evaluating an integral from a known limit to a variable limit, a common notation style is to use the integration variable as the variable limit as well. Thus, for the preceding example, we would have

$$\int_{-1}^{x} (12 + 3 \cdot x - 4 \cdot x^2)\, dx.$$

Mathcad, however, will not evaluate this expression. For Mathcad, the integration variable (the **x** in **dx**) is a dummy variable, but the limits are not. In Mathcad, you cannot use the same symbol to represent both a dummy variable and a limit variable in a single equation.

Mixed Limits with Multiple Integration Variables

To obtain a formula for the volume of a cylinder, we would integrate the cylinder function over the variable **r** from **r** = 0 to some arbitrary radius **R**, over from $\theta = 0$ to 2π (for a completely round cylinder), and over l from l = 0 to an arbitrary length **L**. The result is the common expression for the volume of a cylinder, shown in Figure 7.47.

$$\int_0^L \int_0^{2\cdot\pi} \int_0^R r\, dr\, d\theta\, dl \rightarrow \pi \cdot L \cdot R^2$$

Figure 7.47
Symbolic integration with multiple integration variables.

Note the order of the integration symbols and the integration variables. Mathcad uses the limits on the inside integration symbol (*0* to **R**) with the inside integration variable (**dr**), the limits on the middle integration operator (*0* to 2π) with the middle integration variable **dθ**, and so forth.

Definite Integrals: Numerical Evaluation

Numerical evaluation of an integral does not really fit in this chapter on symbolic math, but it is the final way that Mathcad can evaluate an integral. To request a numerical evaluation, use the equal sign instead of the → symbol, as illustrated in Figure 7.48.

$$\int_{-1}^{2} \left(12 + 3 \cdot x - 4 \cdot x^2\right) dx = 28.5$$

Figure 7.48
Integrating using numerical (not symbolic) evaluation.

Note: Units can be used with numerical evaluation (and with the limits on symbolic integration as well).

Practice!

Use Mathcad's definite integral operator to evaluate these expressions:

a. $\int_0^4 x\, dx.$

b. $\int_1^3 (ax + b)\, dx.$

c. $\int_{-3}^{0} \frac{1}{3}\, dx.$

d. $\int_0^{\pi} \int_0^{2\text{ cm}} r\, dr\, d\theta.$

Practice!

Find the area under the following curves in the range from $x = 1$ to $x = 5$. (Part a has been completed as an example):

a. $3 + 1.5x - 0.25x^2$

$$\text{Area} = \int_1^5 3 + 1.5x - 0.25x^2\,dx$$

$$= 19.667$$

b. $0.2 + 1.7x^3$

c. $\sin\left[\frac{x}{2}\right]$

Find the area between the following curves in the range from $x = 1$ to $x = 5$. (Part a has been completed as an example):

a. $3 + 1.5x - 0.25x^2$ and $0 + 1.5x - 0.25x^2$

$$\text{Area} = \int_1^5 (3 + 1.5x - 0.25x^2) - (0 + 1.5x - 0.25x^2)\,dx$$

$$= \int_1^5 3\,dx$$

$$= 12$$

b. $3 + 1.5x - 0.25x^2$ and $0.1x^2$

APPLICATION

Energy Required to Warm a Gas

Warming up a gas is a pretty common thing to do, partly because we like to live in warm buildings and partly because we tend to burn things to warm those buildings. It is a common need to know how much energy is required to warm a gas.

The amount of energy needed to warm a gas depends on the amount, heat capacity, and temperature change of the gas. Also, heat capacities of gases are strong functions of temperature, so the relationship between heat capacity and temperature must be taken into account. All of this is included in the equation

$$\Delta H = n \int_{T_1}^{T_2} C_p\,dT$$

where

ΔH is the change in enthalpy of the gas, which is equal to the amount of energy required to warm the gas if all of the energy added to the gas is used to warm it (i.e., if the energy is not used to make the gas move faster, etc.),

n is the number of moles of gas present (say, 3 moles),
C_p is the heat capacity of the gas at constant pressure,
T_1 is the initial temperature of the gas (say, 25°C), and
T_2 is the final temperature of the gas (say, 400°C).

Since heat capacities change with temperature, the relationship between heat capacity and temperature is often given as an equation. For example, for CO_2 the heat capacity is related to the temperature by the expression[1]

$$C_p = 36.11 + 4.233 \cdot 10^{-2} T - 2.887 \cdot 10^{-5} T^2 + 7.464 \cdot 10^{-9} T^3 (\text{J/mole} \cdot {}^\circ\text{C}),$$

and the equation is valid for temperatures between 0 and 1500°C.

We can use Mathcad to integrate this expression and determine the amount of energy required to warm 3 moles of gas from 25°C to 400°C. The Mathcad solution is shown in Figure 7.49.

$n := 3$ << moles

$C_p(T) := 36.11 + 4.233 \cdot 10^{-2} \cdot T - 2.887 \cdot 10^{-5} \cdot T^2 + 7.464 \cdot 10^{-9} \cdot T^3$ << Joules/mole K

$\Delta H := n \cdot \int_{25}^{400} C_p(T)\, dT$ << Joules

$\Delta H = 49 \times 10^3$ << Joules

Figure 7.49
Integrating a heat capacity polynomial.

The energy required is 49 kJ.

[1]From *Elementary Priniciples of Chemical Processes* by R.M. Felder and R.W. Rousseau, 2nd ed., Wiley, New York (1986).

7.7 SYMBOLIC DIFFERENTIATION

You can evaluate ***derivatives*** with respect to one or more variables using the **Derivative** or **Nth Derivative** buttons on the Calculus Toolbar. For a first derivative with respect to a single variable, you can also use **Variable/Derivative** from the Symbolics menu.

7.7.1 First Derivative with Respect to One Variable

When you click on the derivative button, [d/dx], on the Calculus Toolbar, the derivative operator is placed on the worksheet:

$$\frac{d}{d\blacksquare}\blacksquare$$

The operator contains two placeholders: one for the function, and the other for the differentiation variable. To take the derivative of the sample polynomial with respect to the variable **x**, the polynomial and the variable are inserted into their respective placeholders:

$$\frac{d}{dx}(12 + 3 \cdot x - 4 \cdot x^2)$$

You tell Mathcad to evaluate the derivative by using the → symbol, as shown in Figure 7.50.

Note that Mathcad does not have a "derivative evaluated at" operator. To evaluate the result at a particular value of x, simply give x a value before performing the differentiation, as shown in Figure 7.51.

$$\frac{d}{dx}\left(12 + 3\cdot x - 4\cdot x^2\right) \to 3 - 8\cdot x$$

Figure 7.50
Using the derivative operator from the Calculus Toolbar.

$$x := -1$$

$$\frac{d}{dx}\left(12 + 3\cdot x - 4\cdot x^2\right) \to 11$$

Figure 7.51
Evaluating the derivative at a specific value of x.

As an alternative to using the derivative operator from the Calculus Toolbar, you can enter your function and select one variable (i.e., one of the x's in the polynomial example) as shown in Figure 7.52.

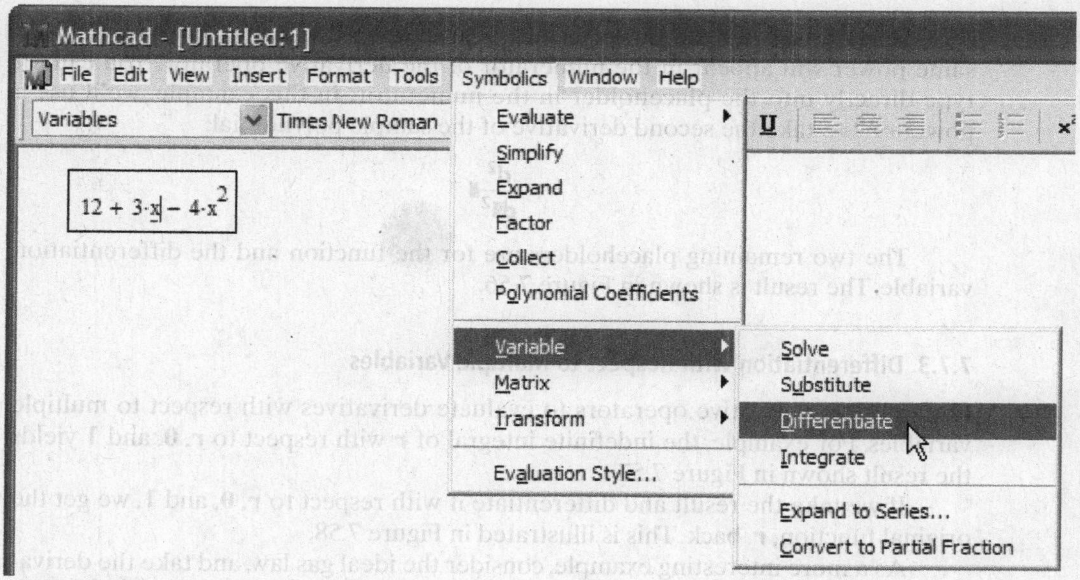

Figure 7.52
Using the Symbolics menu to take a derivative.

Then you differentiate the function with respect to the selected variable using **`Symbolics/Variable/Differentiate`** from the Symbolics menu. The result is shown in Figure 7.53.

This method works only for evaluating a first derivative with respect to a single variable.

$$12 + 3\cdot x - 4\cdot x^2$$

$$3 - 8\cdot x$$

Figure 7.53
Evaluating a derivative using menu options **`Symbolics/Variable/Differentiate`**.

7.7.2 Higher Derivatives with Respect to a Single Variable

For higher derivatives, use the Nth derivative operator from the Calculus Toolbar (Figure 7.54). This operator (see Figure 7.55) comes with four placeholders, but you can use only three.

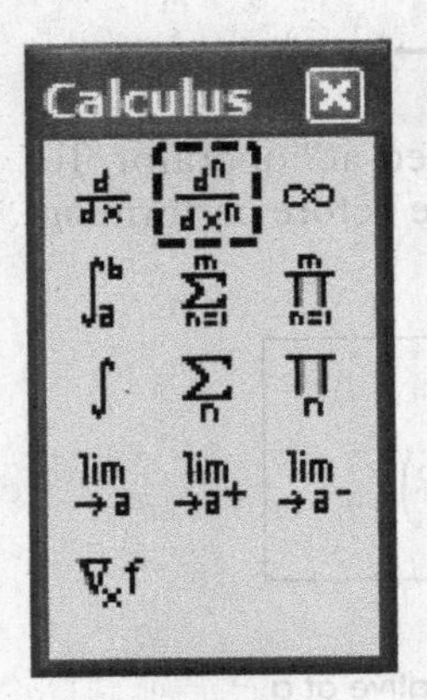

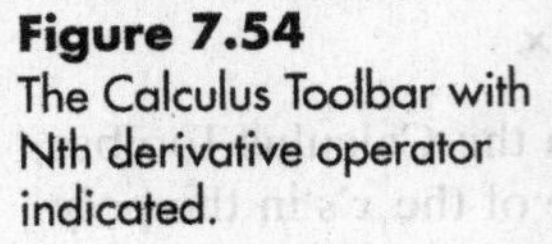

Figure 7.54
The Calculus Toolbar with Nth derivative operator indicated.

Figure 7.55
Nth derivative operator placeholders.

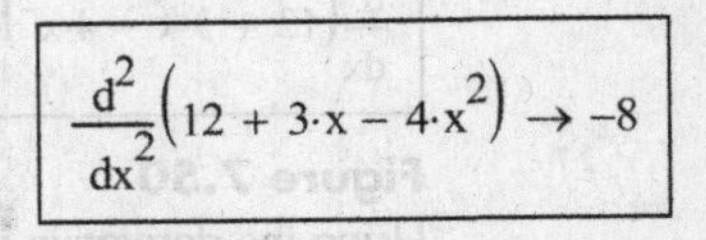

Figure 7.56
Taking the second derivative of a function.

When you enter the power in the right placeholder in the denominator, the same power will appear in the numerator of the derivative operator. You cannot type directly into the placeholder in the numerator. In this example, we'll use a power of 2 to take the second derivative of the sample polynomial:

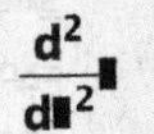

The two remaining placeholders are for the function and the differentiation variable. The result is shown in Figure 7.56.

7.7.3 Differentiation with Respect to Multiple Variables

Use multiple derivative operators to evaluate derivatives with respect to multiple variables. For example, the indefinite integral of **r** with respect to **r**, **θ**, and **l** yields the result shown in Figure 7.57.

If we take the result and differentiate it with respect to **r**, **θ**, and **l**, we get the original function, **r**, back. This is illustrated in Figure 7.58.

As a more interesting example, consider the ideal gas law, and take the derivative of pressure with respect to temperature and volume, as shown in Figure 7.59.

Figure 7.57
Integrating *r* with respect to **r**, **θ**, and **l**.

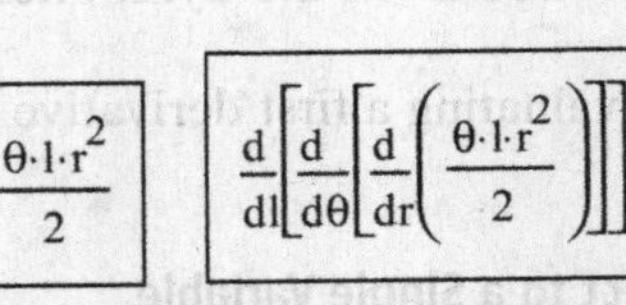

Figure 7.58
Differentiating the result from Figure 7.57 with respect to **r**, **θ**, and **l**.

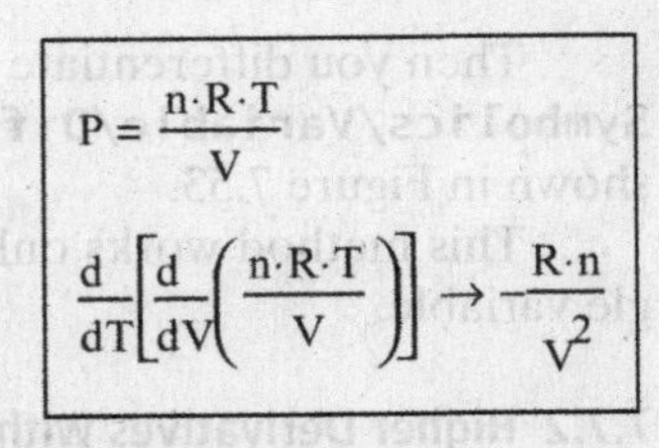

Figure 7.59
Derivative of pressure with respect to temperature and volume.

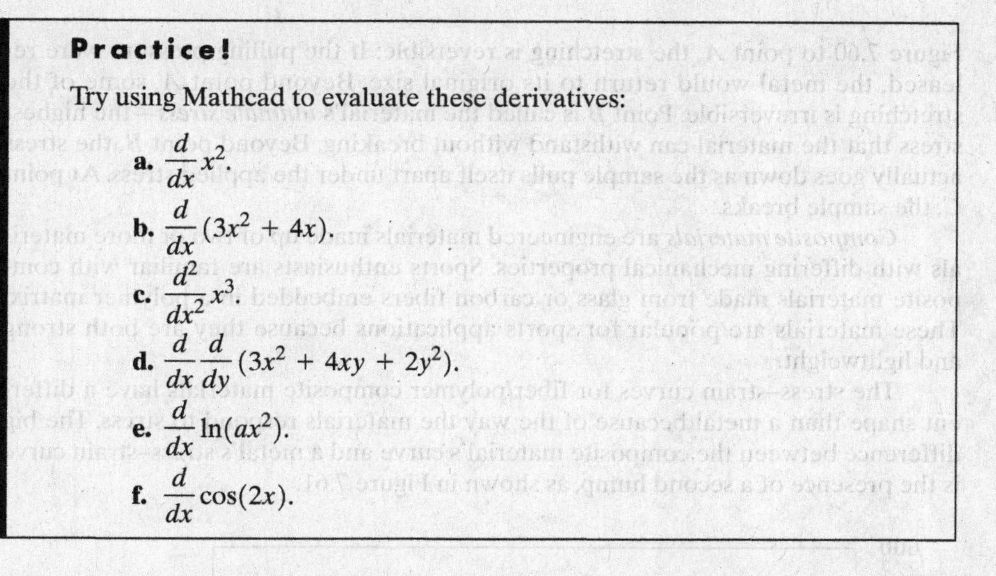

Practice!

Try using Mathcad to evaluate these derivatives:

a. $\frac{d}{dx}x^2$.

b. $\frac{d}{dx}(3x^2 + 4x)$.

c. $\frac{d^2}{dx^2}x^3$.

d. $\frac{d}{dx}\frac{d}{dy}(3x^2 + 4xy + 2y^2)$.

e. $\frac{d}{dx}\ln(ax^2)$.

f. $\frac{d}{dx}\cos(2x)$.

APPLICATION

Analyzing Stress–Strain Diagrams

A fairly standard test for new materials is the tensile test. In essence, a sample of the material is very carefully prepared and then slowly pulled apart. The stress on the sample and the elongation of the sample are recorded throughout the test.

Consider a *tensile test* on a metal sample. The stress–strain diagram obtained from testing is shown in Figure 7.60.

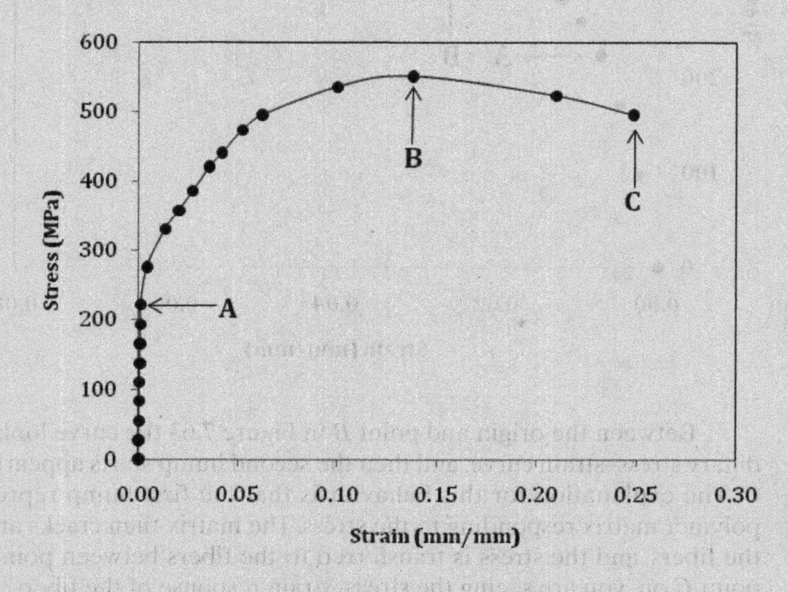

Figure 7.60
Metal stress–strain diagram.

Note the units on strain in Figure 7.60 are mm/mm. This is millimeters of elongation divided by the original length of the sample. The test begins with no stress and no strain. As the pulling begins, the metal starts to stretch. From the origin of

Figure 7.60 to point *A*, the stretching is reversible: If the pulling pressure were released, the metal would return to its original size. Beyond point *A*, some of the stretching is irreversible. Point *B* is called the material's *ultimate stress*—the highest stress that the material can withstand without breaking. Beyond point *B*, the stress actually goes down as the sample pulls itself apart under the applied stress. At point *C*, the sample breaks.

Composite materials are engineered materials made up of two or more materials with differing mechanical properties. Sports enthusiasts are familiar with composite materials made from glass or carbon fibers embedded in a polymer matrix. These materials are popular for sports applications because they are both strong and lightweight.

The stress–strain curves for fiber/polymer composite materials have a different shape than a metal because of the way the materials respond to stress. The big difference between the composite material's curve and a metal's stress–strain curve is the presence of a second hump, as shown in Figure 7.61.

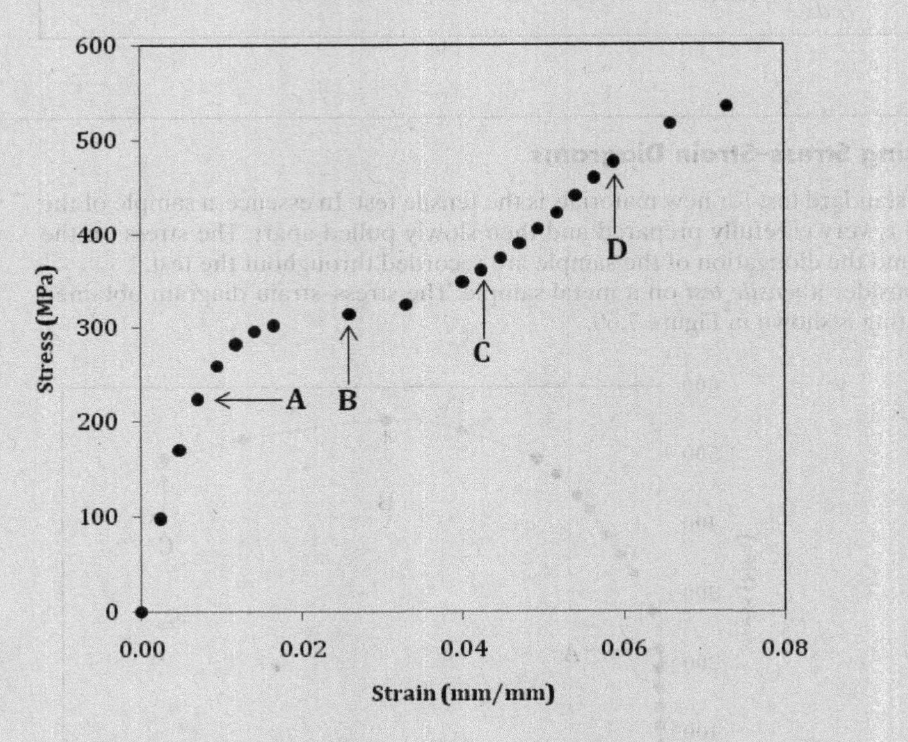

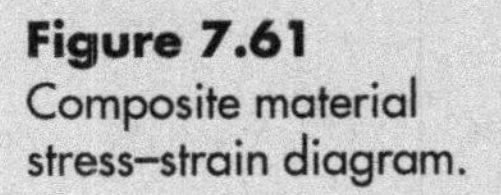

Figure 7.61
Composite material stress–strain diagram.

Between the origin and point *B* in Figure 7.63 the curve looks a lot like an ordinary stress–strain curve, and then the second hump starts appearing at about point *C*. The explanation for this behavior is that the first hump represents mostly the polymer matrix responding to the stress. The matrix then cracks and separates from the fibers, and the stress is transferred to the fibers between points *B* and *C*. From point *C* on, you are seeing the stress–strain response of the fibers.

There are a couple of analyses we can perform on these data:

a. Estimate Young's modulus for the matrix and the fibers.
b. Calculate the work done on the sample during the test.

YOUNG'S MODULUS

Young's modulus is the proportionality factor relating stress and strain in the linear sections of the graph between the origin and point A and (sometimes visible) between points C and D. Because the material is a composite, neither of the values we will calculate truly represents Young's modulus for the pure materials, but they will help quantify how this composite material behaves under stress.

The linear region near the origin of Figure 7.61 includes approximately the first four or five data points. Young's modulus for the matrix can be calculated from the change in stress and the measured change in strain using the first four data points, as shown in Figure 7.62.

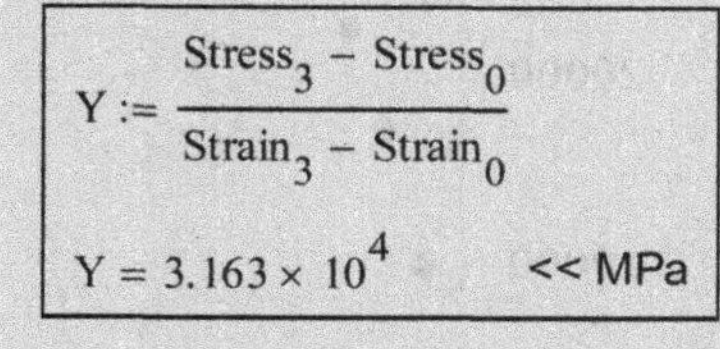

$Y := \frac{Stress_3 - Stress_0}{Strain_3 - Strain_0}$
$Y = 3.163 \times 10^4$ << MPa

Figure 7.62
Finding Young's modulus for the polymer matrix.

Similarly, Young's modulus for the fibers (the region between C and D) can be determined using data points 12 to 19, as shown in Figure 7.63.

$$Y_{fiber} := \frac{Stress_{19} - Stress_{12}}{Strain_{19} - Strain_{12}}$$

$Y_{fiber} = 6.686 \times 10^3$ << MPa

Figure 7.63
Determining Young's modulus for the fibers using points between C and D.

WORK

A little rearranging of variables can turn a stress–strain diagram into a force–displacement diagram. The area under a force–displacement diagram is the work done on the sample.

Stress is the force per unit cross-sectional area of the sample. If the sample tested is $L = 10$ mm by $W = 10$ mm, the area and force on the sample can be computed as

$$A = L \cdot W$$
$$F = Stress \cdot A.$$

To obtain a displacement, **x**, we need to multiply the strain by the original sample length (or height; the samples are usually vertical when tested). If $H = 10$ mm as well, then

$$x = Strain \cdot H.$$

We can now replot the stress–strain diagram as a force–displacement graph. The result will look like Figure 7.64.

The area under the graph is the work, but in order to use Mathcad's integration operator, we need a function relating force to displacement, not data points. We need to fit a polynomial to the data. This is accomplished in the Mathcad example shown in Figure 7.65.

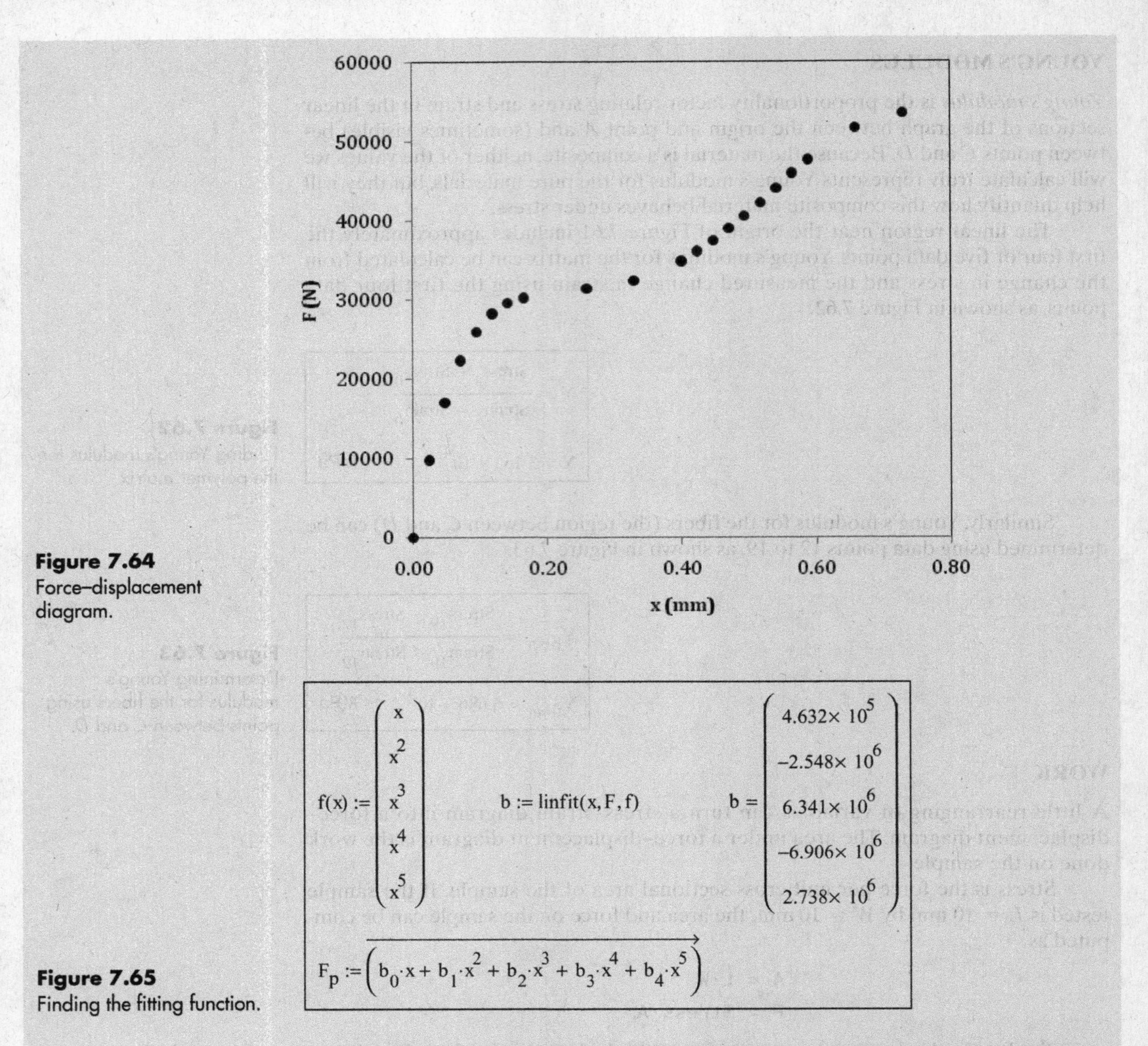

Figure 7.64
Force–displacement diagram.

Figure 7.65
Finding the fitting function.

From the graph in Figure 7.66, it looks as though the fifth-order polynomial fits the data nicely. But F_P is a vector of values; we still need a *function*. This function is shown in Figure 7.67.

This function can be integrated using the definite integration operator from the Calculus Toolbar. As the graph shows, the upper limit on **x** is almost 0.8. The actual value can be found by using the **max** function on the *x*-vector, as shown in Figure 7.68.

The work has a value of 2.592×10^4, but what are the units? *F* was determined from stress (MPa) and area (mm^2) and *x* came from strain (mm/mm) and length (mm). This work has units of MPa·mm^3. We can convert the units as shown in Figure 7.69.

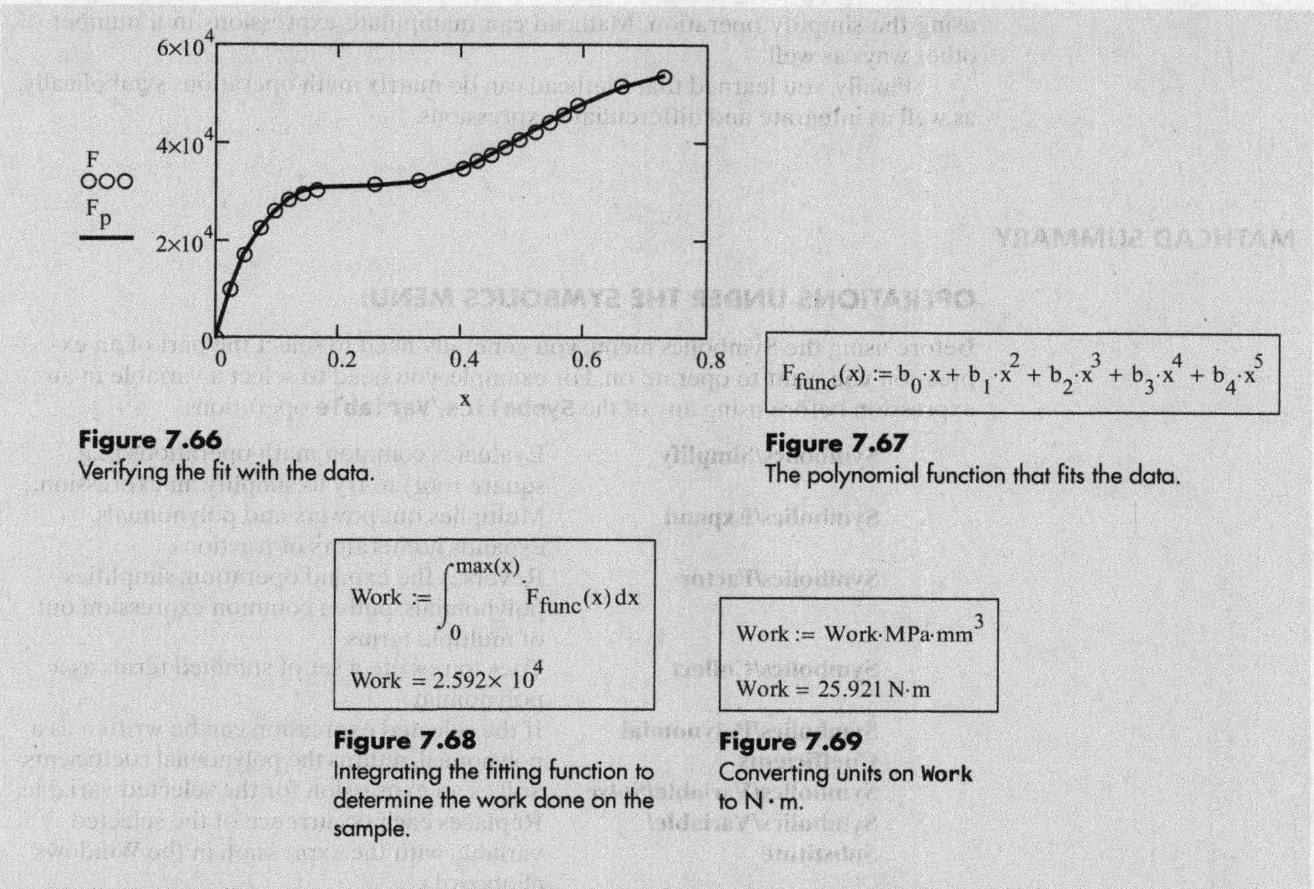

6×10^4
4×10^4
2×10^4
0
F
ooo
F_p
0
0.2
0.4
0.6
0.8
x
$F_{func}(x) := b_0 \cdot x + b_1 \cdot x^2 + b_2 \cdot x^3 + b_3 \cdot x^4 + b_4 \cdot x^5$
$Work := \int_0^{max(x)} F_{func}(x)\,dx$
$Work = 2.592 \times 10^4$
$Work := Work \cdot MPa \cdot mm^3$
$Work = 25.921\ N \cdot m$

Figure 7.66
Verifying the fit with the data.

Figure 7.67
The polynomial function that fits the data.

Figure 7.68
Integrating the fitting function to determine the work done on the sample.

Figure 7.69
Converting units on **Work** to N · m.

Note: The integration operator can handle units, but the **linfit** function does not. That's why this problem was worked without units.

SUMMARY

In this chapter, we looked at Mathcad's symbolic math capabilities—its ability to work directly with mathematical expressions, rather than numerical results. We saw that Mathcad's symbolic math features are housed in two areas and are used in slightly different ways. For example, you can solve for a variable in an equation from either the Symbolics menu or the Symbolic Keyword Toolbar. With the latter, there is a "live operator," so that if you make changes to the worksheet, the **solve** operation will be automatically recalculated, and you specify the variable you want to solve for as part of the operation. In order to solve for a variable using the Symbolics menu, you first select the variable of interest and then use **Symbolics/Variable/Solve** from the menu. The result is placed on the worksheet, but it is not a live operator, so the result will not be automatically recalculated if the worksheet changes. Both approaches are useful.

You also saw that Mathcad can replace (substitute) every occurrence of a variable with another mathematical expression and can factor common terms out of complex expressions. You can use symbolic math to find a common denominator using the simplify operation. Mathcad can manipulate expressions in a number of other ways as well.

Finally, you learned that Mathcad can do matrix math operations symbolically, as well as integrate and differentiate expressions.

MATHCAD SUMMARY

OPERATIONS UNDER THE SYMBOLICS MENU:

Before using the Symbolics menu, you generally need to select the part of an expression you want to operate on. For example, you need to select a variable in an expression before using any of the `Symbolics/Variable` operations.

Symbolics/Simplify	Evaluates common math operations (e.g., square root) to try to simplify an expression.
Symbolics/Expand	Multiplies out powers and polynomials, expands numerators of fractions.
Symbolics/Factor	Reverses the expand operation: simplifies polynomials, pulls a common expression out of multiple terms.
Symbolics/Collect	Tries to rewrite a set of summed terms as a polynomial.
Symbolics/Polynomial Coefficients	If the selected expression can be written as a polynomial, returns the polynomial coefficients.
Symbolics/Variable/Solve	Solves an expression for the selected variable.
Symbolics/Variable/ Substitute	Replaces each occurrence of the selected variable with the expression in the Windows clipboard.
Symbolics/Variable/ Differentiate	Returns the first derivative of the expression with respect to the selected variable.
Symbolics/Variable/ Integrate	Integrates the expression with respect to the selected variable—does not add a constant of integration.
Symbolics/Variable/ Convert to Partial Fractions	Expands an expression into a sum of expressions with denominators containing only linear and quadratic terms and with no functions of the selected variable in the numerator.
Symbolics/Matrix/ Transpose	Interchanges rows and columns in the matrix.
Symbolics/Matrix/Invert	Inverts the matrix using symbolic math operations, rather than decimal numbers—this preserves accuracy by avoiding round-off errors, but the resulting fractions can get unwieldy.
Symbolics/Matrix/ Determinant	Calculates the determinant of a matrix using symbolic math operations.
Symbolics/Evaluation Style...	Opens a dialog box that allows you to change the way the results of the calculations are presented.

OPERATIONS USING THE SYMBOLIC KEYWORD TOOLBAR:

With the Symbolic Keyword Toolbar, the symbolic evaluation operator → is used and any variables or expressions that must be specified in order to perform the operation are entered into placeholders as needed. You do not need to select a variable or a portion of the expression before performing the symbolic operation.

solve	Solves the expression for the specified variable.
simplify	Evaluates common math operations to try to simplify an expression.
substitute	Replaces each occurrence of the specified variable with the specified expression.
factor	Pulls the specified variable out of all terms in an expression.
expand	Evaluates powers and polynomials involving the specified variable; expands numerators of fractions involving the variable.
coeffs	Returns the polynomial coefficients, if there are any, for the specified variable.
collect	Tries to rewrite summed terms as a polynomial in the specified variable.
parfrac	Expands an expression into a sum of expressions, with denominators containing only linear and quadratic terms and with no functions of the selected variable in the numerator.
$\mathbf{M^T} \rightarrow$	Transposes the matrix (interchanges rows and columns in the matrix).
$\mathbf{M^{-1}} \rightarrow$	Inverts the matrix using symbolic math operations.
|M|	Calculates the determinant of a matrix using symbolic math operations.

Operations Using the Calculus Palette:

Indefinite integral	Integrates the expression in the function placeholder with respect to the variable in the integration variable placeholder. If the symbolic evaluation operator, →, is used after the integral, the integral is evaluated symbolically. Mathcad does not add a constant of integration. If an equal sign is used after the integral, the integral is evaluated numerically.
Definite integral	Integrates the expression in the function placeholder with respect to the variable in the integration variable placeholder, using the specified limits. (The dummy integration variable cannot be used in a limit.) If the symbolic evaluation operator, →, is used after the integral, the integral is evaluated symbolically. If an equal sign is used after the integral, the integral is evaluated numerically.
Derivative	Takes the derivative of the expression in the function placeholder with respect to the variable in the differentiation variable placeholder. If the symbolic evaluation operator, →, is used, the derivative is evaluated symbolically. If an equal sign is used, the

	derivative is evaluated numerically. To evaluate the derivative at a particular value, assign the value to the differentiation variable before taking the derivative.
Nth derivative	Takes second- and higher-order derivatives.

KEY TERMS

collect	factor	polynomial coefficients
definite integral	indefinite integral	simplify
derivative	limits of integration	substitute
differentiation	live symbolic operator, →	symbolic math
expand	partial-fraction expansion	

PROBLEMS

7.1 Solving for a Variable in an Expression

The equation

$$A = 2 \cdot \pi \cdot r^2 + 2 \cdot \pi \cdot r \cdot L$$

is used to calculate the surface area of a cylinder. The equation is written using the symbolic equality, [CTRL- =].

(a) Use the **solve** operation on the Symbolic Keyword Toolbar to solve for the L in this equation.

(b) Use the expression returned by the **solve** operation to determine the cylinder length required to get 1 m^2 of surface area on a cylinder with a radius of 12 cm.

7.2 Polynomial Equations: Coefficients and Roots

The following equation clearly has three solutions (roots), at $x = 3, x = 1$, and $x = -2$:

$$(x - 3) \cdot (x - 1) \cdot (x + 2) = 0$$

(a) Create a QuickPlot of $(x - 3) \cdot (x - 1) \cdot (x + 2)$ vs. x in the range from -3 to 3. Verify that the curve does cross the x-axis at $x = 3, x = 1$, and $x = -2$.

(b) Use the **solve** operation on **x** to have Mathcad find the roots of:

$$(x - 3) \cdot (x + 1) \cdot (x + 2) \;\; \textbf{solve}, \;\; x \rightarrow$$

(c) Expand the function in **x** to see the expression written as a polynomial:

$$(x - 3) \cdot (x - 1) \cdot (x - 2) \;\; \textbf{expand}, \;\; x \rightarrow$$

(d) Obtain the polynomial coefficients as a vector using the **coeffs** operator from the Symbolic Math Toolbar:

$$(x - 3) \cdot (x - 1) \cdot (x - 2) \;\; \textbf{coeffs}, \;\; x \rightarrow$$

7.3 Symbolic Integration

Evaluate the following indefinite integrals symbolically:

(a) $\int \sin(x)\,dx$.

(b) $\int \ln(x)\,dx$.

(c) $\int [\sin(x)^2 + \cos(x)]\,dx$.

(d) $\int_{-3}^{0} \frac{x}{x+b}\,dx$.

Note: In part (c), notice how Mathcad indicates the sine-squared term: $\sin(x)^2$, not $\sin^2(x)$.

7.4 Definite Integrals

Evaluate the following definite integrals, using either symbolic or numeric evaluations:

(a) $\int_0^{\pi} \sin(\theta)\,d\theta$.

(b) $\int_0^{2\pi} \sin(\theta)\,d\theta$.

(c) $\int_{-3}^{0} \frac{x}{x-3}\,dx$.

(d) $\int_0^{\infty} e^{\frac{-t}{4}}\,dt$.

Note: The numerical integrator cannot handle the infinite limit in part (d); you must use symbolic integration. The infinity symbol is available on the Calculus Palette.

7.5 Integrating for the Area Under a Curve

Integrate $\int y\,dx$ to find the area under the curves represented by these functions:

(a) $y = -x^2 + 16$ between $-2 \le x \le 4$.

(b) $y = e^{\frac{-t}{4}}$ between $0 \le t \le \infty$.

(c) $y = e^{\frac{-t}{4}}$ between $0 \le t \le 4$.

7.6 Integrating for the Area between Curves

Find the area between the curves represented by the functions

$$x = -x^2 + 16$$

and

$$y = -x^2 + 9$$

over the range $-3 \le x \le 3$.

7.7 Area of a Sector

The technical term for a pie-slice-shaped piece of a circle is a *sector*. The area of a sector can be found by integrating over r and θ:

$$A = \int_{\theta=0}^{\alpha} \int_{r=0}^{R} r\,dr\,d\theta.$$

(a) Check the preceding equation by symbolically integrating over the entire circle ($\alpha = 2\pi$). Do you get the expected result, $A = \pi R^2$?

(b) Find the area of a 37° sector using Mathcad's numerical integration capability (i.e., use an equal sign rather than a symbolic evaluation operator, →, to evaluate the integral.)

7.8 Designing an Irrigation System

An essential element of center-pivot irrigation systems is their ability to distribute water fairly evenly. Because the end of the pipe covers a lot more ground than the pipe near the center, water must be applied at a faster rate at the outside of the circle than near the center in order to apply the same number of gallons per square foot of ground.

(a) Use Mathcad's symbolic math capabilities to integrate the equation

$$\text{Area}_{\text{ring}} = \int_{0}^{2\pi} \int_{R_i}^{R_o} r\,dr\,d\theta$$

to obtain a formula for the area of a ring with inside radius R_i and outside radius R_o.

(b) At what rate must water be applied to provide one inch (depth) of water

1. to the innermost ring: $R_i = 0$, $R_o = 20$ feet.
2. to the outermost ring: $R_i = 1300$ feet, $R_o = 1320$ feet.

(c) Write a Mathcad function that accepts the inside and outside radii and the desired water depth as inputs and returns the flow rate required to provide one inch of water. Use your function to determine the water flow rate for $R_i = 1000$ feet, $R_o = 1020$ feet, and depth $= 2$ inches.

7.9 Finding the Equation for the Area of an Ellipse

The equation of an ellipse (illustrated in Figure 7.70) centered at the origin is

$$\frac{x^2}{a^2} + \frac{y^2}{b^2} = 1.$$

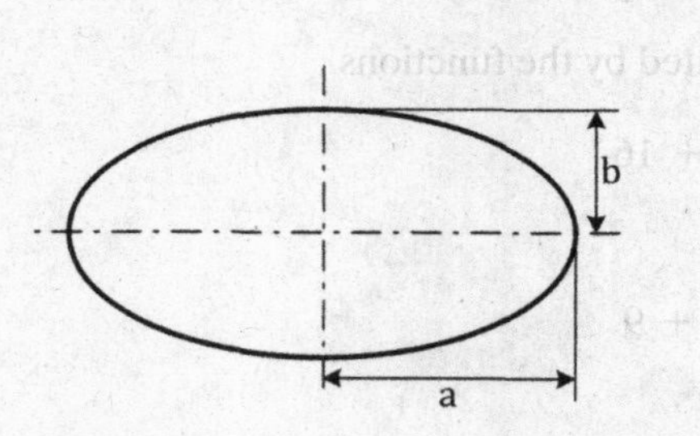

Figure 7.70
Area of an ellipse.

The area of the upper half of the ellipse can be determined by finding the area between the ellipse and the x-axis ($y = 0$) as illustrated in Figure 7.71.

Figure 7.71
Upper half of ellipse.

The total area of the ellipse is twice the area of the upper half.

(a) Solve the equation of the ellipse for y. (You will obtain two solutions, since there are two y values on the ellipse at every x value.) Verify that the positive y values are returned by

$$y = \frac{b}{a} \cdot \sqrt{-(x^2) + a^2}.$$

(b) Substitute the expression for y from part a in the integral for the area of an ellipse (twice the area of the upper half of the ellipse), and symbolically solve the following equation for the area of an ellipse:

$$2 \cdot \int_{-a}^{a} y\, dx \rightarrow$$

7.10 Work Required to Stretch a Spring

Hooke's law says that the force exerted against a spring as the spring is being stretched is proportional to the extended length. That is,

$$F = k\,x.$$

Work is defined as

$$W = \int F\, dx.$$

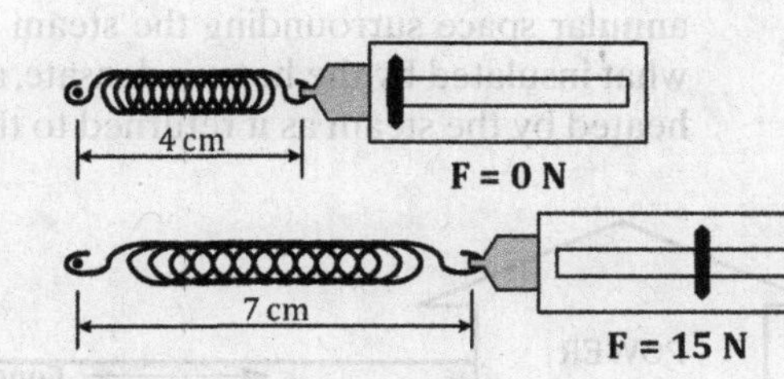

Figure 7.72
Work required to stretch a spring.

From Figure 7.72, it can be seen that the length of the spring with no applied force is 4 cm. An applied force of 15 N extends the spring 3 cm to a total length of 7 cm. This information can be used to find the spring constant k. We write

$$15\text{ N} = k \times 3\text{ cm},$$

$$k = 5\frac{N}{\text{cm}}.$$

Calculate the work done on the spring as it was stretched from 4 to 7 cm (i.e., as the extended length went from 0 to 3 cm).

7.11 Work Required to Compress an Ideal Gas at Constant Temperature

One equation for work is

$$W = \int P\,dV.$$

For an ideal gas, pressure and volume are related through the ideal gas law:

$$PV = nRT.$$

(a) Use Mathcad's ability to solve an expression for a variable to solve the ideal gas law for pressure P.

(b) Substitute the result for P from part (a) into the work integral.

(c) Determine the work required to compress 10 moles of an ideal gas at 400 K from a volume of 300 liters to a volume of 30 liters. Express your result in kJ. (Assume that cooling is provided to maintain the temperature at 400 K throughout the compression.)

(d) Does the calculated work represent work done on the system (the 10 moles of gas) by the surroundings (the outside world) or by the system on the surroundings?

7.12 Energy Required to Warm a Gas

At what rate must energy be added to a stream of methane to warm it from 20°C to 240°C? Data are as follows:

Methane flow rate: 20,000 mole/min (these are gram moles);

Heat capacity equation:[1] $C_p = 34.31 + 5.469 \cdot 10^{-2}T + 0.3361 \cdot 10^{-5}T^2 - 11.00 \cdot 109T^3\ J/\text{mole} \cdot °C.$

7.13 Annular Piping Systems

One pipe may be placed inside of another pipe, to carry two fluids, creating an *annulus*. One fluid flows in the center pipe and the other in the annular space between the two pipes. Annular flow arrangements are often used to allow heat transfer between the two fluids. For example, hot steam from a power plant might flow in the center pipe out to a distant laboratory to provide heat (illustrated in Figure 7.73). In the building's heating system, the steam is condensed to keep the building warm. The liquid condensate is returned to the boiler in the annular space surrounding the steam pipe. In this way, the hot steam is somewhat insulated by the hot condensate, and the condensate would be partially reheated by the steam as it returned to the power plant.

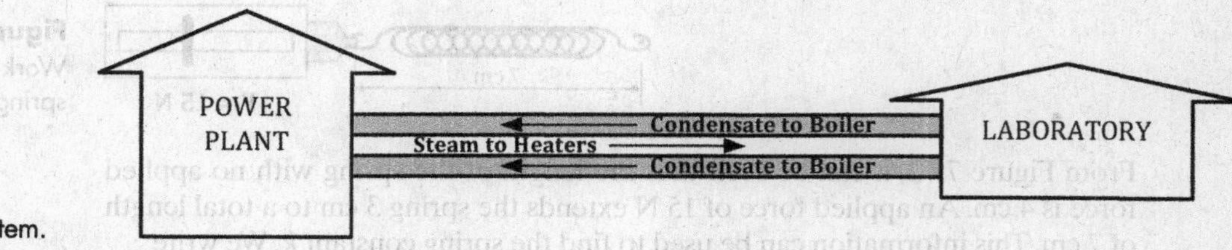

Figure 7.73
Steam Distribution System.

The annulus is the space between the two concentric pipes (see Figure 7.74). The cross-sectional area of the annulus is an important design parameter used

[1]From R. M. Felder and R. W. Rousseau, *Elementary Principles of Chemical Processes,* 2nd ed. (New York: Wiley, 1986).

in sizing this type of piping system. The area of the annulus depends on the outside radius of the small pipe, r_o, and the inside radius of the large pipe, R_I.

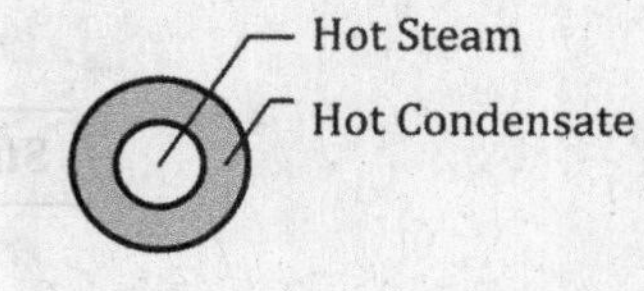

Figure 7.74
Annular piping.

The area can be calculated in two ways:

- by subtracting the area of the small circle from the area of the large circle as illustrated in Figure 7.75; that is,

$$A = \pi R_I^2 - \pi r_o^2.$$

Figure 7.75
Determining the area of an annulus.

- by integrating the expression from r_O to R_I

$$A = \int_{\theta=0}^{2\pi} \int_{r=r_o}^{R_I} r \, dr \, \theta.$$

Using symbolic integration, demonstrate that the two methods give equivalent results.

7.14 Young's Modulus

In the Application analyzing stress–strain diagrams, values of Young's modulus for the matrix and the fiber were determined using algebra. An alternative approach would be to take the derivative of a function fit to the stress–strain data and evaluate the derivative in the linear regions of the stress–strain curve.

Let

$$f(x) := \begin{bmatrix} x \\ x^2 \\ x^3 \\ x^4 \\ x^5 \end{bmatrix}$$

```
b := linfit(Strain, Stress, f)
```

$$b = \begin{bmatrix} 5.951 \cdot 10^4 \\ -4.006 \cdot 10^6 \\ 1.195 \cdot 10^8 \\ -1.542 \cdot 10^9 \\ 7.184 \cdot 10^9 \end{bmatrix}$$

be a fifth-order polynomial fit to the stress–strain curves. Use Mathcad's derivative operator to differentiate the polynomial and then evaluate the derivative at strain values of 0.001 and 0.048 to find the values of Young's modulus for the matrix and the fibers, respectively.

7.15 Calculating Work

The stress–strain data for the metal sample mentioned in the Application box at the end of this chapter have been abridged and reproduced as follows:

Strain (mm/mm)	Stress (MPa)
0.00000	0
0.00028	55
0.00055	110
0.00083	165
0.00110	221
0.00414	276
0.01324	331
0.02703	386
0.04193	441
0.06207	496
0.13793	552
0.20966	524
0.24828	496

(a) Convert the stress–strain diagram shown in Figure 7.76 to a force–displacement graph. (The sample size is 10 mm × 10 mm × 10 mm.)

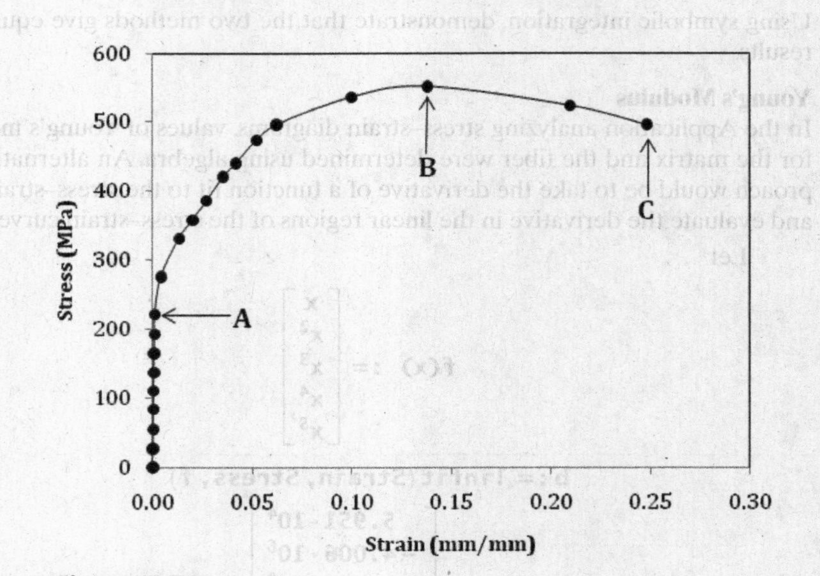

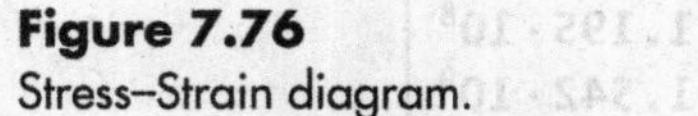

Figure 7.76
Stress–Strain diagram.

(b) Fit a polynomial to the force–displacement data. (A fifth-order polynomial with no intercept works well.)

(c) Integrate the polynomial to determine the work done on the sample during the tensile test.

SOLUTIONS TO PRACTICE! PROBLEMS

Section 7.2 Practice!

a. $(x-1)\cdot(x+1)$ solve $\rightarrow \begin{pmatrix} -1 \\ 1 \end{pmatrix}$

b. $x^2 - 1$ solve $\rightarrow \begin{pmatrix} -1 \\ 1 \end{pmatrix}$

c. $(y-2)\cdot(y+3)$ solve $\rightarrow \begin{pmatrix} 2 \\ -3 \end{pmatrix}$

d. $3\cdot z^2 - 2\cdot z + 8$ solve $\rightarrow \begin{bmatrix} \frac{1}{3} - \left(\frac{\sqrt{23}}{3}\right)\cdot i \\ \frac{1}{3} + \frac{1}{3}\cdot\sqrt{23}\cdot i \end{bmatrix}$

e. $\dfrac{x+4}{x^2 + 6\cdot x - 2}$ solve $\rightarrow -4$

Section 7.3.2. Practice!

a. $6\cdot x^2 + 4\cdot x$ << select entire expression, then **Symbolics/Factor**

$2\cdot x\cdot(3\cdot x + 2)$

b. $3\cdot x\cdot y + 4\cdot x - 2\cdot y$ << select first two terms, then **Symbolics/Factor**

$x\cdot(3\cdot y + 4) - 2\cdot y$

Section 7.3.4 Practice!

Using Symbolics Menu

a. $x^3 - 11\cdot x^2 + 31\cdot x - 21$ $\quad (x-1)\cdot(x-3)\cdot(x-7)$

b. $x^2 + \sqrt[3]{216}\cdot x - \sqrt{50 + 2^3 - 3^2}$ $\quad (x-1)\cdot(x+7)$

c. $\dfrac{x^2}{x-1} - \dfrac{10\cdot x}{x-1} + \dfrac{21}{x-1}$ $\quad \dfrac{(x-3)\cdot(x-7)}{x-1}$

Using Symbolic Toolbar

a. $x^3 - 11\cdot x^2 + 31\cdot x - 21$ simplify $\rightarrow (x-1)\cdot(x-3)\cdot(x-7)$

b. $x^2 + \sqrt[3]{216}\cdot x - \sqrt{50 + 2^3 - 3^2}$ simplify $\rightarrow (x-1)\cdot(x+7)$

c. $\dfrac{x^2}{x-1} - \dfrac{10\cdot x}{x-1} + \dfrac{21}{x-1}$ simplify $\rightarrow \dfrac{(x-3)\cdot(x-7)}{x-1}$

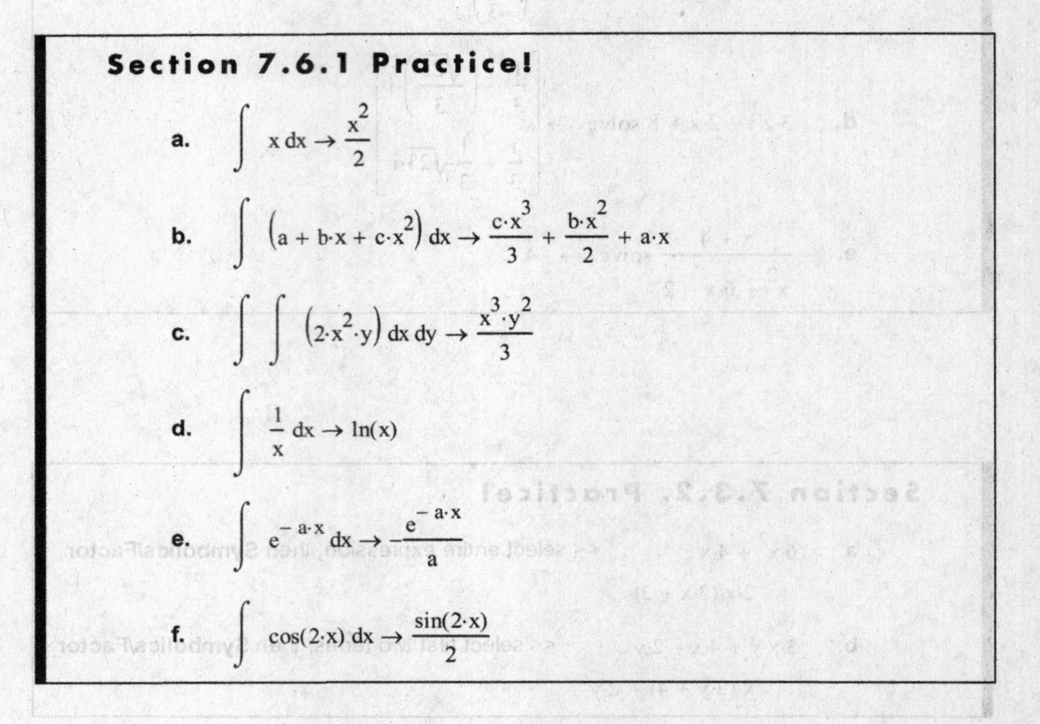

Section 7.6.1 Practice!
a. $\int x\,dx \rightarrow \frac{x^2}{2}$
b. $\int \left(a + b\cdot x + c\cdot x^2\right) dx \rightarrow \frac{c\cdot x^3}{3} + \frac{b\cdot x^2}{2} + a\cdot x$
c. $\int\int \left(2\cdot x^2\cdot y\right) dx\,dy \rightarrow \frac{x^3\cdot y^2}{3}$
d. $\int \frac{1}{x}\,dx \rightarrow \ln(x)$
e. $\int e^{-a\cdot x}\,dx \rightarrow -\frac{e^{-a\cdot x}}{a}$
f. $\int \cos(2\cdot x)\,dx \rightarrow \frac{\sin(2\cdot x)}{2}$

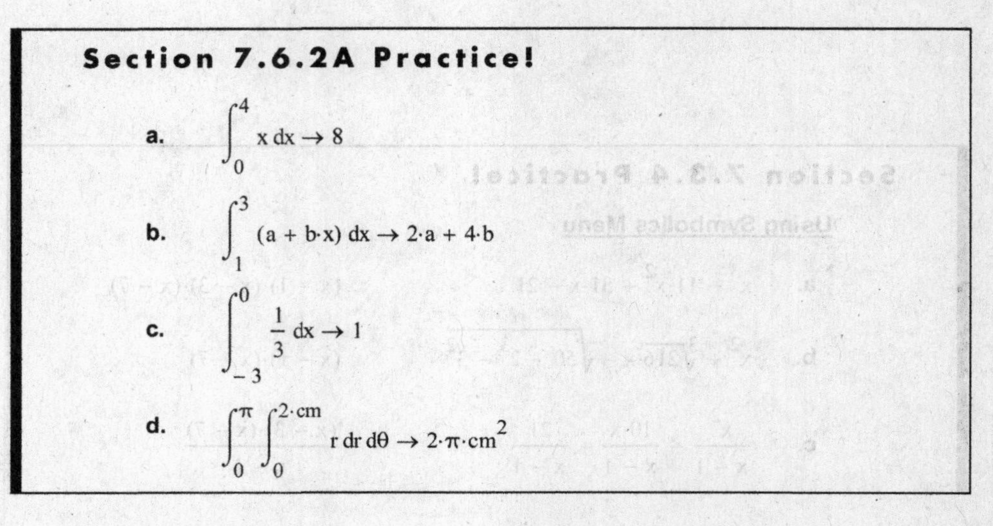

Section 7.6.2A Practice!
a. $\int_0^4 x\,dx \rightarrow 8$
b. $\int_1^3 (a + b\cdot x)\,dx \rightarrow 2\cdot a + 4\cdot b$
c. $\int_{-3}^0 \frac{1}{3}\,dx \rightarrow 1$
d. $\int_0^{\pi}\int_0^{2\cdot cm} r\,dr\,d\theta \rightarrow 2\cdot\pi\cdot cm^2$

Section 7.6.2B Practice!

Area Under Curves

a. $\int_1^5 \left(3 + 1.5\cdot x - 0.25\cdot x^2\right) dx \rightarrow 19.666666666666666667$ << using Symbolic processor

$\int_1^5 \left(3 + 1.5\cdot x - 0.25\cdot x^2\right) dx = 19.667$ << using Numeric processor

b. $\int_1^5 \left(0.2 + 1.7\cdot x^3\right) dx = 266$

c. $\int_1^5 \sin\left(\frac{x}{2}\right) dx = 3.357$

Area Between Curves

a. $\int_1^5 \left(3 + 1.5\cdot x - 0.25\cdot x^2\right) - \left(0 + 1.5\cdot x - 0.25\cdot x^2\right) dx = 12$

$\int_1^5 \left(3 + 1.5\cdot x - 0.25\cdot x^2\right) - \left(1.5\cdot x - 0.25\cdot x^2\right) dx = 12$

b. $\int_1^5 \left(3 + 1.5\cdot x - 0.25\cdot x^2\right) - 0.1\cdot x^2\, dx = 15.533$

Section 7.7 Practice!

a. $\frac{d}{dx}x^2 \rightarrow 2\cdot x$

b. $\frac{d}{dx}\left(3\cdot x^2 + 4\cdot x\right) \rightarrow 6\cdot x + 4$

c. $\frac{d^2}{dx^2}x^3 \rightarrow 6\cdot x$

d. $\frac{d}{dx}\left[\frac{d}{dy}\left(3\cdot x^2 + 4\cdot x\cdot y + 2\cdot y^2\right)\right] \rightarrow 4$

e. $\frac{d}{dx}\left(\ln\left(a\cdot x^2\right)\right) \rightarrow \frac{2}{x}$

f. $\frac{d}{dx}\cos(2\cdot x) \rightarrow -2\cdot\sin(2\cdot x)$

ANSWERS TO SELECTED PROBLEMS

Problem 1	b. 1.2 m
Problem 3	b. $\int \ln(x)\, dx \rightarrow x \cdot \ln(x) - x$
Problem 4	c. 0.921
Problem 5	a. 72
Problem 8	b. outer ring, 43 gal/min
Problem 11	c. −76.6 kJ
Problem 14	$Y_{matrix} = 5.2 \times 10^4 \text{MPa}$

CHAPTER

Numerical Techniques

Objectives

After reading this chapter, you will

- know how to use Mathcad's powerful iterative solver
- be able to integrate by fitting data with a model equation, then integrating the equation
- be able to perform integration on a data set using numerical methods directly on the data values
- be able to differentiate by fitting data with a model equation, then differentiating the equation
- know how to use finite differences to calculate derivatives from data values

8.1 ITERATIVE SOLUTIONS

Equations (or systems of equations) that cannot be solved directly occur often in engineering. For example, to determine the flow rate in a pipe with a given pressure drop, the friction loss must be known. But the friction loss depends on the flow rate. Thus, in order to calculate the friction loss, you must know the flow rate. But to calculate the flow rate, you must know the friction loss. To solve this very common problem, it is necessary to guess either the flow rate or the friction loss. If you guess the friction loss, you would calculate the flow rate and then calculate the friction loss at that flow rate. If the calculated friction loss equals the guessed friction loss, the problem is solved. If not, guess again. This repeated calculate and check process is termed ***iteration***.

Mathcad provides a better way to solve problems like this: the ***iterative solver***. As a simple example, consider finding the value of x that satisfies the equation

$$x^3 + 12x - 21 = 0.$$

The x^3 suggests that there will be three solutions, but there could be duplicate roots or imaginary roots. In this example, two of the roots are imaginary. We'll try to find the one real root.

Using the Worksheet for Trial-and-Error Calculations

You can simply do trial-and-error calculations in the Mathcad workspace. For example, if we try **x = 0**, the expression **x³ + 12x − 21** yields **−21**. Since **−21 ≠ 0**, the guessed value of x is incorrect. We might then try $x = 1$, as illustrated in Figure 8.1.

This result is closer to zero than the first attempt, so we're moving in the right direction, but the guessed **x** is still too small. Try **x = 2**, as shown in Figure 8.2.

The equation is still not satisfied, but this guess was too big. Try `x = 1.5`. (The result is shown in Figure 8.3.)

We're getting close, but `x = 1.5` is still a little high. With a few more tries, you'll find that a value of `x = 1.48` comes very close to satisfying this equation.

$x := 1$

$x^3 + 12 \cdot x - 21 = -8$

Figure 8.1
Trial and error, first attempt.

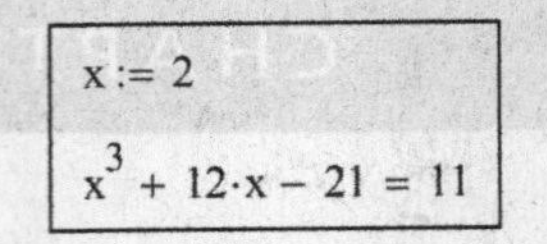

Figure 8.2
Trial and error, second attempt.

$x := 1.5$

$x^3 + 12 \cdot x - 21 = 0.375$

Figure 8.3
Trial and error, third attempt.

Note: Iterative solutions always attempt to find values that "nearly" solve the equation. The process of choosing ever-closer guessed values could go on forever. To decide when the solution is "close enough" to stop the process, iterative methods test the calculated result against a preset ***tolerance***. For Mathcad's iterative solver, the two sides of the equation are evaluated by using the computed value, and the difference between the two values is called the *error*. When the error falls below the preset tolerance, Mathcad stops iterating and presents the result. You can change the value of the tolerance from the default value of 0.001 to a larger number for less accurate solutions or (more likely) a smaller value for more accurate solutions. Very small tolerance values, such as 10^{-15} may make it hard for Mathcad to find a solution because of computer round-off error.

Automating the Iterative Solution Process

Mathcad provides a better iterative solver than manual trial and error. Mathcad's approach uses an ***iterative solve block***, which will allow you to solve multiple equations simultaneously. The ***solve block*** is bounded by two keywords: **`given`** and **`find`**. The equations between these two words will be included in the iterative search for a solution. Before the **`given`**, you must provide an ***initial guess*** for each variable in the solve block. The iterative solve block for the preceding cubic example, with an initial guess of $x = 0$, would look like the example shown in Figure 8.4.

$x := 0$	<< guess value
given	<< "given" starts the solve block
$x^3 + 12 \cdot x - 21 = 0$	<< use a symbolic equality inside the solve block
$x := \text{find}(x)$	<< "find" ends the solve block
$x = 1.4799$	<< the iteration result

Figure 8.4
Using a **given – find** solve block.

The equation between the **`given`** and **`find`** is inside the solve block and has been indented for readability. The line

***x* := `find(x)`**

terminates the solve block and assigns the solution (returned by the **`find`** function) to the variable *x*. Here the *x* variable was used to hold both the initial guess and the computed solution. This is common but not necessary; you could assign the solution to any variable.

You can check the solution by using the computed value in the equation, as illustrated in Figure 8.5.

$x := 1.4799$

$x^3 + 12 \cdot x - 21 = -6.508 \times 10^{-5}$

Figure 8.5
Testing the solution found by the iterative solver.

The result is not quite zero, but is very close (within the preset tolerance).

You can adjust the displayed precision of any result by double-clicking on the displayed value and changing the number of displayed digits, but Mathcad's default is to drop trailing zeros. A zero on the right side of the equation is actually zero to at least three decimal places. (By default, Mathcad shows three decimal places.) The computed solution definitely satisfies the equation.

Keep in mind the following comments on using iterative solve blocks in Mathcad:

- Finding "a" solution does not imply that you have found "the" solution or all solutions. Any time you anticipate that there are multiple solutions, you should try different starting values to search for the other solutions (called *roots*). If you want to see if there are complex solutions, use an imaginary (or complex) value as the guess. This is illustrated in Figure 8.6.
- Solve blocks do support units, but if you are solving for more than one variable, each iterated variable must have the same units. If the variables in your problem do not all have the same units (flow rate and friction loss, for example), you must solve the set of equations without any units on any variable.

```
x := 3i

Given

  x^3 + 12·x − 21 = 0

x := Find(x)

x = −0.73995 + 3.69359i
```

Figure 8.6
Using an imaginary guess to search for complex solutions.

The root Function: An Alternative to Using Solve Blocks

Mathcad provides a **root** function that can be used to find a single solution to a single equation. The **root** function does not require the use of a **given find** solve block. The **root** function is an iterative solver, so an initial guess is still required. The procedure is illustrated in Figure 8.7.

```
f(x) := x^3 + 12·x − 21      << define the function

x := 0                       << provide an initial guess

soln := root(f(x), x)        << solve for the solution (or root)

soln = 1.48                  << show the solution
```

Figure 8.7
Using the **root** function to find a solution.

The **root** function will find only one solution. For functions that have multiple solutions, you can provide other guesses (starting values) to search for other roots. To search for an imaginary root, use an imaginary guess value.

Finding All Roots of Polynomials

For polynomial functions, Mathcad provides a function that will find all of the roots at one time: the **polyroots** function. To use the **polyroots** function, the ***polynomial coefficients*** must be written as a column vector, starting with the constant. The polynomial we have been using as an example is

$$x^3 + 0x^2 + 12x - 21 = 0.$$

The $0x^2$ term was included in the polynomial as a reminder that the zero must be included in the coefficient vector **v**.

The coefficients of this polynomial would be written as the column vector

$$\mathbf{v} := \begin{pmatrix} -21 \\ 12 \\ 0 \\ 1 \end{pmatrix}$$

The **polyroots** function could then be used to find all solutions of this polynomial, as shown in Figure 8.8.

$$x^3 + 12\cdot x - 21 \text{ coeffs} \rightarrow \begin{pmatrix} -21 \\ 12 \\ 0 \\ 1 \end{pmatrix}$$ << optional step: let Mathcad show the coefficients

$$v := \begin{pmatrix} -21 \\ 12 \\ 0 \\ 1 \end{pmatrix}$$ << define the vector of polynomial coefficients

Soln := polyroots (v) << solve for all roots of the polynomial

$$\text{Soln} = \begin{pmatrix} -0.74 + 3.694i \\ -0.74 - 3.694i \\ 1.48 \end{pmatrix}$$ << display the roots

Figure 8.8
Using the **polyroots** function to find all roots of a polynomial.

The **polyroots** function is a quick way to find all solutions, but only for polynomials.

Note: In Figure 8.8 the **coeffs** keyword from the Symbolic Toolbar was used to determine the polynomial coefficient values.

Professional Success

Use a QuickPlot to find good initial guesses.

1. Create an X-Y Plot (from the Graphics Toolbar or by pressing [Shift-2].)
2. Enter your function on the *y*-axis, as shown in Figure 8.9.

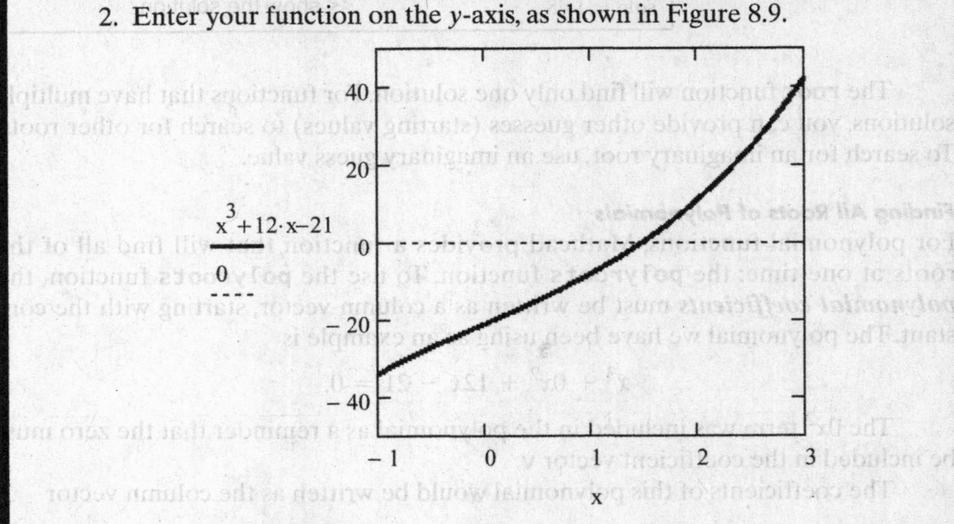

Figure 8.9
Plotting the function to find approximate roots.

3. Add a second curve to the plot (select your entire function and then press [comma]), and enter a zero in the *y*-axis placeholder for the second

plot. This will draw a horizontal line across the graph at $y = 0$. [Or, double-click on the graph and add a *marker* (at $y = 0$) to the y axis.]

4. Enter the variable used in your function in the placeholder on the x-axis.
5. Adjust the limits on the x-axis as needed to see where your function crosses the $y = 0$ line. The locations where your function crosses the $y = 0$ line are the roots, or solutions, of the function.

This function clearly has a root between 1 and 2, but Mathcad provides a way to get a more accurate value off the graph. This is described in the following steps.

6. Click on the graph to select it.
7. Bring up the X-Y Trace dialog box (shown in Figure 8.10) from the Format menu: **Format/Graph/Trace...**

Figure 8.10
The X-Y Trace dialog box is used to read values from graphs.

8. Position the X-Y Trace dialog box so that the entire graph is visible.
9. Click on the spot where your function and the $y = 0$ curves intersect (illustrated in Figure 8.11).

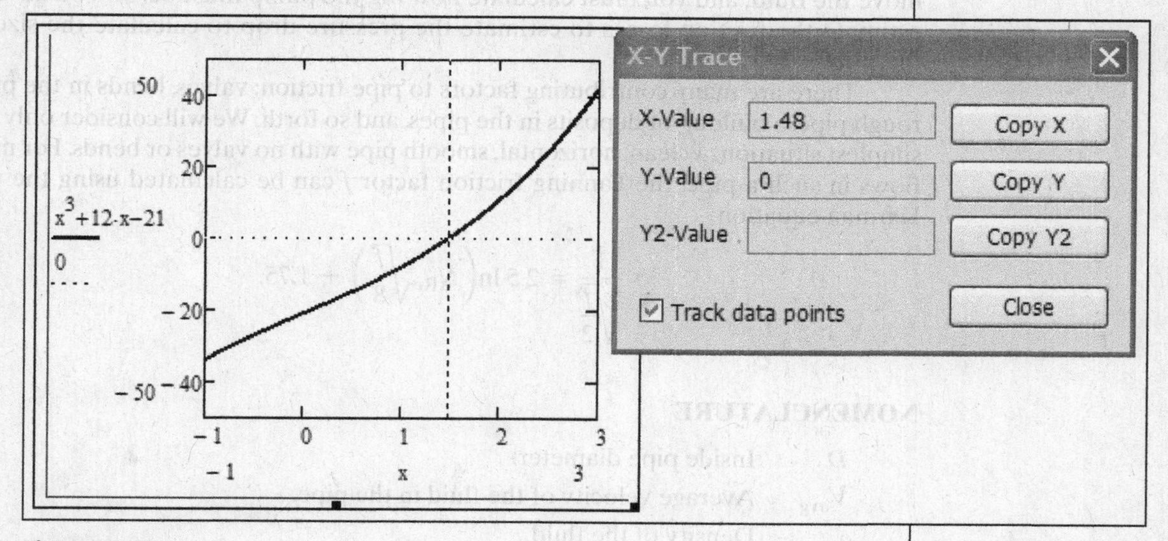

Figure 8.11
Using the X-Y Trace dialog box.

10. Read the x-value at that location from the X-Y Trace dialog box.

A guessed value of 1.48 (shown in the X-Y Trace dialog box in Figure 8.11) should be very close to the actual root. Using this value as the initial guess should allow the iterative solver to converge quickly.

Practice!

Use Mathcad's iterative solver to find solutions for each of the expressions that follow. For each expression, how many solutions should there be? Use different initial guesses to search for multiple solutions.

a. $(x - 3) \cdot (x - 4) = 0.$

b. $x^2 - 1 = 0.$

c. $x^3 - 2x^2 + 4x = 3.$

d. $e^{3x} - 4 = 0.$

e. $\sqrt{x^3} + 7x = 10.$

APPLICATION

Friction Losses and Pressure Drop in Pipe Flows

Here's a typical pump-sizing problem:

> *What size pump (HP) is required to move water at an average velocity of 3.0 ft/s through a 5,000-foot-long pipe with a 1-inch inside diameter? Assume that the viscosity of water is 0.01 poise at room temperature and the pump has an efficiency of 0.70.*

Anytime you are designing a piping system, you will need to estimate the friction in the system, since friction can be responsible for much of the pressure drop from one end of the pipe to another. Because of this friction, you need a pump to move the fluid, and you must calculate how big the pump must be. So you have to estimate the friction losses to estimate the pressure drop to calculate the size of the pump.

There are many contributing factors to pipe friction: valves, bends in the pipe, rough pipe, a buildup of deposits in the pipes, and so forth. We will consider only the simplest situation: a clean, horizontal, smooth pipe with no valves or bends. For most flows in such a pipe, the Fanning friction factor f can be calculated using the von Karman equation,

$$\frac{1}{\sqrt{\frac{f}{2}}} = 2.5 \ln\left(N_{\mathrm{Re}}\sqrt{\frac{f}{8}}\right) + 1.75.$$

NOMENCLATURE

D	Inside pipe diameter.
V_{avg}	Average velocity of the fluid in the pipe.
ρ	Density of the fluid.
μ	Viscosity of the fluid.
L	Length of the pipe.
η	Efficiency of the pump (no units).
P_P	Pump's power rating (HP or kW).
$\dot{m}$	Mass flow rate of fluid in the pipe.

The von Karman equation is valid for smooth pipes (e.g., PVC pipe, not steel pipe) with Reynolds numbers greater than 6,000. The Reynolds number is defined as

$$N_{Re} = \frac{D V_{avg} \rho}{\mu}.$$

This equation is used in Figure 8.12 to find the value of the Reynolds number for this example.

Information from the Problem Statement

$V_{avg} := 3 \frac{ft}{sec}$ $L := 5000. ft$ $D_i := 1in$

$\mu := 0.01 \cdot poise$ $\eta := 0.70$

Readily Available Data

$\rho := 1000. \frac{kg}{m^3}$ **water density**

Reynolds Number Calculation

$Re := \frac{D_i \cdot V_{avg} \cdot \rho}{\mu}$ $Re = 23226$

Since Re > 6000, von Karman equation is ok

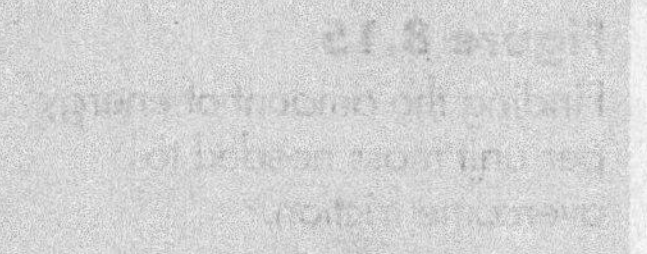

Figure 8.12
Finding the Reynolds number.

Since the Reynolds number was found to be greater than 6,000, the von Karman equation can be used to find the friction factor, as shown in Figure 8.13.

Solve for Friction Factor

guess: $f := 0.001$

Given

$$\frac{1}{\sqrt{\frac{f}{2}}} = 2.5 \cdot \ln\left(Re \cdot \sqrt{\frac{f}{8}}\right) + 1.75$$

$f := Find(f)$

$f = 0.0062$ ***calculated friction factor***

Figure 8.13
Solving for friction factor.

Once you know the friction factor, you can calculate the pressure drop in a horizontal pipe (see Figure 8.14) as

$$\Delta P = 4 f \frac{L}{D} \frac{\rho (V_{avg})^2}{2}.$$

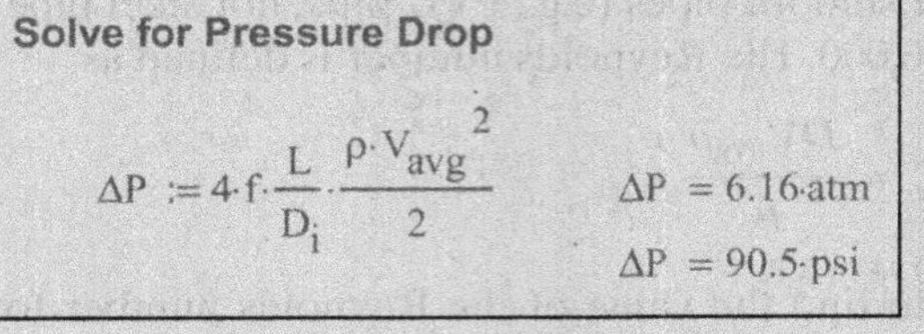

Solve for Pressure Drop

$$\Delta P := 4 \cdot f \cdot \frac{L}{D_i} \cdot \frac{\rho \cdot V_{avg}^2}{2} \qquad \Delta P = 6.16 \cdot atm$$

$$\Delta P = 90.5 \cdot psi$$

Figure 8.14
Determining the pressure drop.

Once you have the pressure drop, you can determine the energy per unit mass required to overcome friction, as shown in Figure 8.15:

$$h_f = \frac{\Delta P}{\rho}.$$

Solve for Energy/Mass to Overcome Friction

$$h_f := \frac{\Delta P}{\rho} \qquad h_f = 624.0 \cdot \frac{N \cdot m}{kg}$$

$$h_f = 208.8 \cdot \frac{ft \cdot lbf}{lb}$$

Figure 8.15
Finding the amount of energy per unit mass needed to overcome friction.

Finally, you can determine the mass flow rate and pump power required, as shown in Figure 8.16.

$$\eta P_P = h_f \dot{m}$$

Solve for Mass Flow Rate

$$A_{flow} := \pi \cdot \left(\frac{D_i}{2}\right)^2 \qquad A_{flow} = 507 \cdot mm^2$$

$$A_{flow} = 0.785 \cdot in^2$$

$$m_{dot} := V_{avg} \cdot A_{flow} \cdot \rho \qquad m_{dot} = 1668 \cdot \frac{kg}{hr}$$

$$m_{dot} = 3677 \cdot \frac{lb}{hr}$$

Solve for Required Pump Power

$$P_p := \frac{h_f \cdot m_{dot}}{\eta} \qquad P_p = 0.41 \cdot kW$$

$$P_p = 0.55 \cdot hp$$

Figure 8.16
Finding the required pump power.

Surprised by the result? It doesn't take much of a pump just to overcome friction in a well-designed pipeline. (It takes a lot more energy to lift the water up a hill, but that wasn't considered here.) If the pipeline is not designed correctly, the friction losses can change dramatically. That's the subject of Problem 8.7.

8.2 NUMERICAL INTEGRATION

Numerical integration and ***numerical differentiation*** of functions are very straightforward in Mathcad. But many times the relationship between the dependent and independent variables is known only through a set of data points. For example, we will soon present a data set representing a relationship between temperature and time. If you want to integrate the temperature data over time, you have two choices:

- Fit the data with an equation, and then integrate the equation.
- Use a numerical integration method on the data set itself.

Both approaches are common, and both will be described in this section.

8.2.1 Integrating Functions Numerically

If you have a function, such as the polynomial relating temperature and time calculated in the example shown in Figure 8.17, then integrating temperature over time from 0 to 9 minutes is easily performed with Mathcad's definite integral operator.

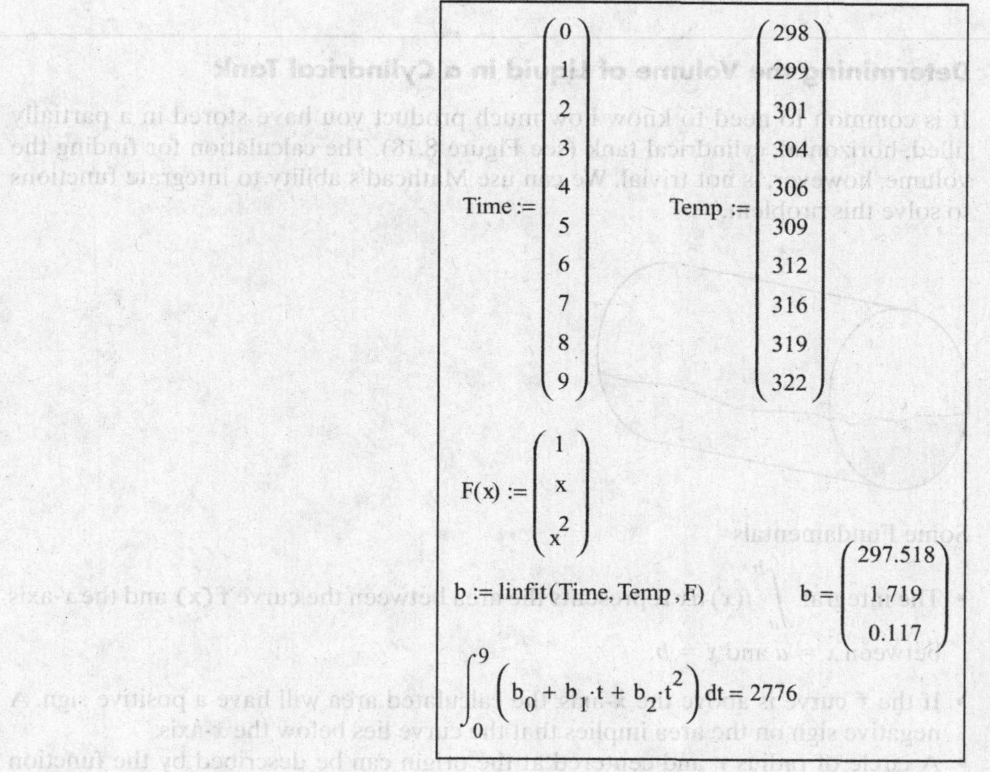

Time := (0 1 2 3 4 5 6 7 8 9)
Temp := (298 299 301 304 306 309 312 316 319 322)
F(x) := (1 x x^2)
b := linfit(Time, Temp, F)
b = (297.518 1.719 0.117)
∫0^9 (b0 + b1·t + b2·t^2) dt = 2776

Figure 8.17
Integration using the definite integral operator from the Calculus Toolbar.

The symbol t was used instead of **`Time`** in the integration. The choice is irrelevant, because the integration variable is a dummy variable. Units were not used here. The integration operator does allow units on the limit values, but since the **`linfit`** function does not support units, the problem is more easily solved without them.

Practice!

Evaluate the following integrals, and check your results using the computational formulas shown to the right of each integral:

- Area under the curve $y = 1.5x$ from $x = 1$ to $x = 3$ (a trapezoidal region):

$$\int_1^3 1.5x\,dx \quad A_{trap} = \frac{1}{2}(y_{left} + y_{right})\cdot(x_{right} - x_{left}).$$

- Volume of a sphere of radius 2 cm:

$$\int_0^{2\text{ cm}} 4\pi r^2\,dr \quad V_{sphere} = \frac{4}{3}\pi R^3.$$

- Volume of a spherical shell with inside radius $R_i = 1$ cm and outside radius $R_o = 2$ cm.

$$\int_{1\text{ cm}}^{2\text{ cm}} 4\pi r^2\,dr \quad V_{shell} = \frac{4}{3}\pi(R^3 - R_i^3)$$

APPLICATION

Determining the Volume of Liquid in a Cylindrical Tank

It is common to need to know how much product you have stored in a partially filled, horizontal, cylindrical tank (see Figure 8.18). The calculation for finding the volume, however, is not trivial. We can use Mathcad's ability to integrate functions to solve this problem.

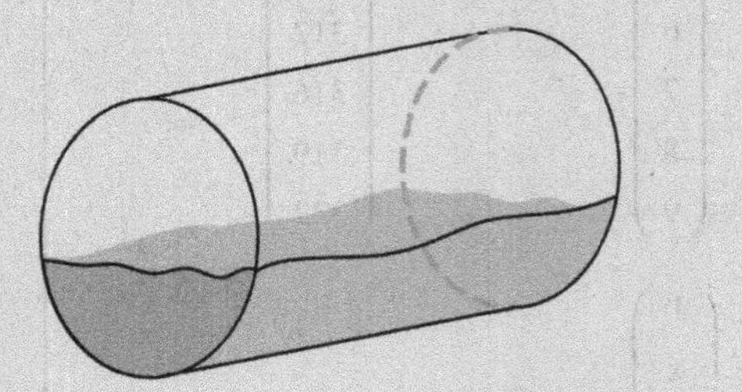

Figure 8.18
A partially filled cylindrical tank.

Some Fundamentals

- The integral $\int_a^b f(x)\,dx$ represents the area between the curve **f(x)** and the x-axis between $x = a$ and $x = b$.
- If the **f** curve is above the **x**-axis, the calculated area will have a positive sign. A negative sign on the area implies that the curve lies below the x-axis.
- A circle of radius **r** and centered at the origin can be described by the function $x^2 + y^2 = r^2$.
- A horizontal line (representing the level of liquid in the tank) is described by the function **Y** = *constant*. We will relate the constant to the depth of liquid later in this example.

CASE 1: THE TANK IS LESS THAN HALF FULL

When the tank is less than half full, we can compute the volume by multiplying the cross-sectional area of the fluid (shown shaded in the circle at the left in the figure that follows) by the length of the tank, L. The trick is determining the cross-sectional area of the fluid. One method is illustrated in Figure 8.19.

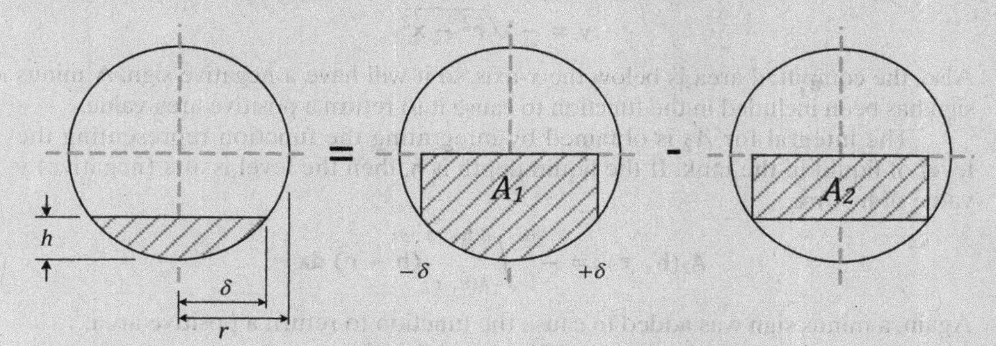

Figure 8.19
A method for finding the cross-sectional area of the fluid in the tank.

The shaded area in the center circle (marked A_1) represents the area between the circle and the x-axis. Note that when the tank is less than half full, there is fluid between $-\delta$ and $+\delta$. We can integrate the formula for the circle $-\delta$ and $+\delta$ to compute the shaded area in the center circle.

The shaded area in the rightmost circle (marked A_2) represents the area between the function describing the level of the liquid and the x-axis. The integration limits are again $-\delta$ and $+\delta$. Subtracting A_2 from A_1 gives the desired cross-sectional area of the fluid in the tank. All that remains is to carry out these integrations, but first we need to know how δ depends on the level in the tank, **h**, and the tank radius **r**.

There is a right triangle involving the level of the liquid and the origin of the axis, shown in Figure 8.20.

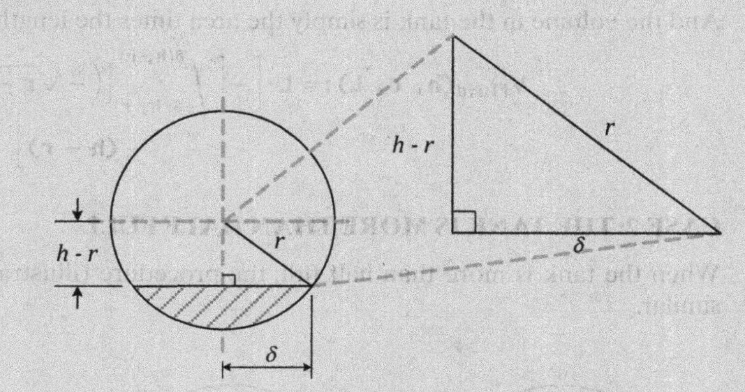

Figure 8.20
There is a right triangle that relates **h**, **r**, and δ.

Using the Pythagorean theorem, we can relate the lengths of the sides of the triangle.

$$(h - r)^2 + \delta^2 = r^2$$

Solving for δ as a function of **h** and **r** yields the first required Mathcad function:

$$\delta(\mathbf{h}, \mathbf{r}) := \sqrt{\mathbf{r}^2 - (\mathbf{r} - \mathbf{h})^2}$$

The integral for the area A_1 is then written as a function of **r** and **h** as well:

$$A_1(\mathbf{h}, \mathbf{r}) = -\left[\int_{-\delta(\mathbf{h},\ \mathbf{r})}^{\delta(\mathbf{h},\ \mathbf{r})} \left(-\sqrt{\mathbf{r}^2 - \mathbf{x}^2}\right) d\mathbf{x}\right]$$

Here, the function representing the circle was solved for **y**, and a negative sign was introduced to calculate the **y** values below the x-axis. (Since Mathcad's square root operator always returns the positive root, we need to change the sign to obtain the negative **y** values.) Hence, we have

$$\mathbf{y} = -\sqrt{\mathbf{r}^2 - \mathbf{x}^2}$$

Also, the computed area is below the *x*-axis, so it will have a negative sign. A minus sign has been included in the function to cause it to return a positive area value.

The integral for A_2 is obtained by integrating the function representing the level of liquid in the tank. If the liquid depth is **h**, then the level is at a (negative) *y* value of **h − r**:

$$A_2(h,\ r) := -\left[\int_{-\delta(h,\ r)}^{\delta(h,\ r)} (h - r)\,dx\right]$$

Again, a minus sign was added to cause the function to return a positive area.

The cross-sectional area of fluid in the tank is then

$$A_{fluid}(h,\ r) := -\left[\int_{-\delta(h,\ r)}^{\delta(h,\ r)} \left(-\sqrt{r^2 - x^2}\right) dx - \int_{-\delta(h,\ r)}^{\delta(h,\ r)} (h - r)\,dx\right]$$

or, simplifying slightly,

$$A_{fluid}(h,\ r) := -\int_{-\delta(h,\ r)}^{\delta(h,\ r)} \left[\left(-\sqrt{r^2 - x^2}\right) - (h - r)\right] dx$$

And the volume in the tank is simply the area times the length:

$$V_{fluid}(h,\ r,\ L) := L \cdot \left[-\left[\int_{-\delta(h,\ r)}^{\delta(h,\ r)} \left[\left(-\sqrt{r - x^2}\right) - (h - r)\right] dx\right]\right]$$

CASE 2: THE TANK IS MORE THAN HALF FULL

When the tank is more than half full, the procedure (illustrated in Figure 8.21) is similar.

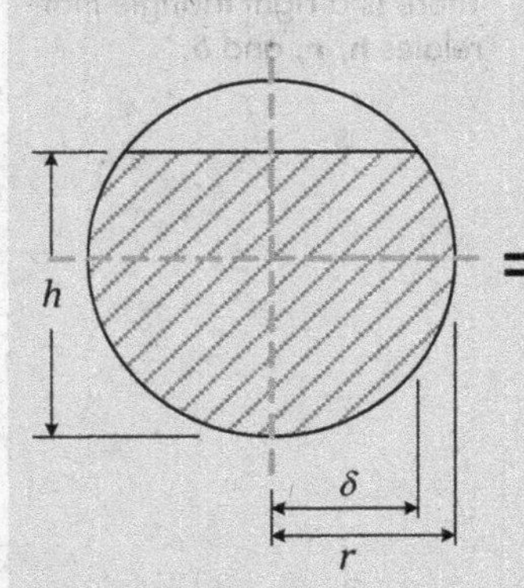

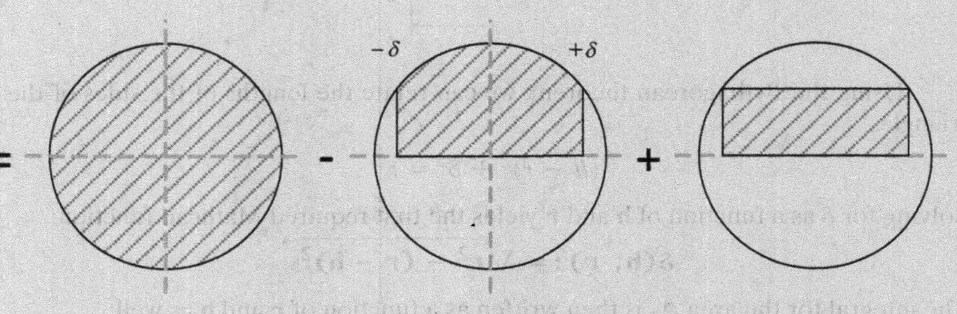

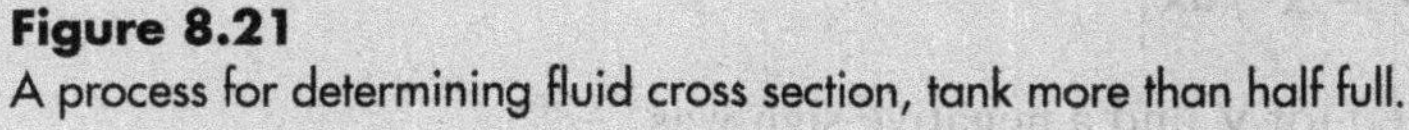

Figure 8.21
A process for determining fluid cross section, tank more than half full.

$$A_{fluid} = A_{total} - \int_{-\delta}^{\delta} f_{circle}(x)\,dx + \int_{-\delta}^{\delta} f_{level}(x)\,dx$$

The function relating δ to **r** and **h** is unchanged, and since the areas are above the x-axis, we don't have to worry about changing the signs. The result is the following function for the volume of the liquid in the tank:

$$V_{fluid}(h, r, L) = L \cdot \left[\pi \cdot r^2 - \int_{-\delta(h,r)}^{\delta(h,r)} \left(-\sqrt{r^2 - x^2}\right) - (h - r)\right] dx$$

CREATING A GENERAL FUNCTION FOR EITHER CASE

Mathcad's **if** function can be used to automatically select the appropriate formula:

$$V_{tank}(h, r, L) := L \cdot \text{if}\left[(h < r), \left[-\int_{-\delta(h,r)}^{\delta(h,r)} \left[\left(-\sqrt{r^2 - x^2}\right) - (h - r)\right] dx\right], \left[\pi \cdot r^2 - \int_{-\delta(h,r)}^{\delta(h,r)} \left[\left(\sqrt{r^2 - x^2}\right) - (h - r)\right] dx\right]\right]$$

Here, if $h < r$ then the first formula is used; otherwise, the second formula is used. Since both functions work for $h = r$ (a half-full tank), deciding which formula to use at that point is arbitrary.

8.2.2 Integrating Data Sets

Integration via Curve Fitting

One approach to integrating a data set is to first fit an equation to the data and then to integrate the equation. This was demonstrated in a previous example: A polynomial was fit to the temperature–time data (see Figure 8.10), and then the polynomial was integrated using Mathcad's definite integral operator.

Integrating without Curve Fitting

An alternative to curve fitting is simply to use the data values themselves to compute the integral. This is fairly straightforward if you recall that the integral represents the area between the curve and the x-axis when the data points are plotted, as shown in Figure 8.22.

Any method that computes the area under the curve[1] computes the value of the integral. A number of methods are commonly used. One of the simplest divides the area under the curve into a series of trapezoids. The area of each trapezoid is calculated from the data values, and the sum of the areas represents the result of the integration. Since there are 10 data points, there will be nine trapezoids to cover the entire time range. We will keep track of these nine regions by defining a range variable

$$i := 0..8$$

[1]Be sure to calculate the area all the way to the x-axis ($y = 0$) not just to $y = 290°$ shown in this figure.

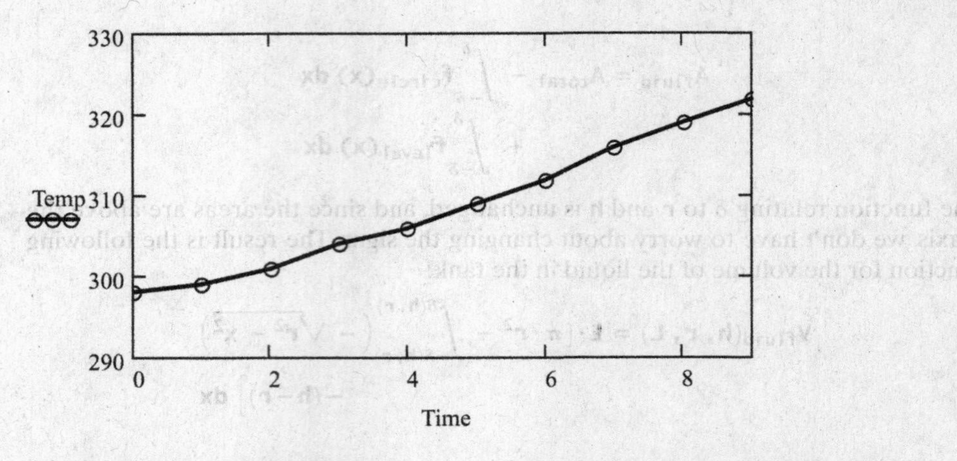

Figure 8.22
Plotting temperature and time.

Or, alternatively, we could define the range variable in more general terms as

```
i := 0..(last(Time)-1)
```

We then need a function that computes the area of the leftmost trapezoid:

$$A_0 := \frac{1}{2} \cdot (Temp_0 + Temp_1) \cdot (Time_1 - Time_0) \qquad A_0 = 298.7$$

We can now generalize this equation to obtain a function capable of calculating the area of any of the nine trapezoids, as shown in Figure 8.23.

Finally, we sum the areas of all of these trapezoids to estimate the total area under the curve. This summation is carried out by using Mathcad's *range variable summation operator* from the Calculus Toolbar (see Figure 8.24). The result is shown in Figure 8.25.

$$i := 0..8$$

$$A_i := \frac{1}{2} \cdot \left(Temp_i + Temp_{i+1}\right)\left(Time_{i+1} - Time_i\right)$$

$$A = \begin{pmatrix} 298.5 \\ 300 \\ 302.5 \\ 305 \\ 307.5 \\ 310.5 \\ 314 \\ 317.5 \\ 320.5 \end{pmatrix}$$

Figure 8.23
Integrating with trapezoids.

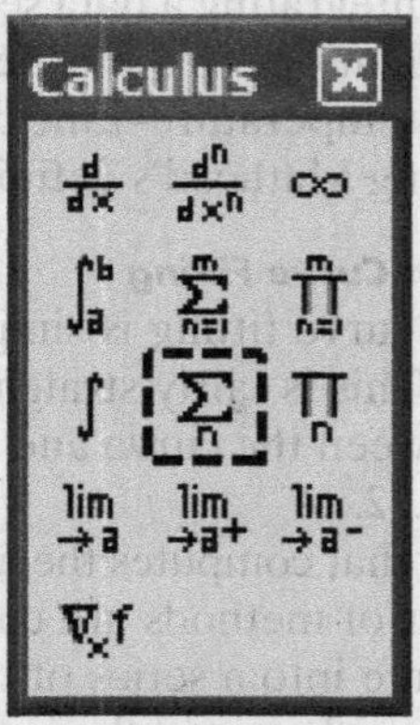

Calculus

Figure 8.24
The range variable summation operator on the Calculus Toolbar.

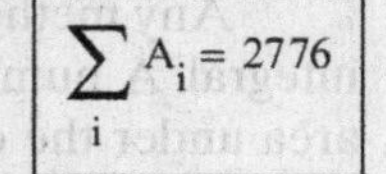

Figure 8.25
Summing the trapezoid regions to get the area under the curve.

The result compares well with that computed by using numerical integration of the polynomial in the earlier example (see Figure 8.17). Again, this calculation was performed without using Mathcad's units capability, but units could have been used.

8.2.2.1 A Trapezoidal Rule Function We can push this process one step further and write a function that will perform *trapezoidal-rule integration* on any data set. This is illustrated in Figure 8.26.

$$\text{trap}(x, y) := \sum_{i=0}^{\text{length}(x)-2} \left[\left(\frac{y_i + y_{i+1}}{2} \right) \cdot \left(x_{i+1} - x_i \right) \right]$$

$$A_{total} := \text{trap}(\text{Time}, \text{Temp}) \qquad A_{total} = 2776$$

Figure 8.26
The user-written **trap** function.

In the user-written **trap** function, the range variable is replaced by defined limits on the summation (Mathcad's standard *summation operator*, shown in Figure 8.27, is used), and the `length` function is used to determine the size of the array, from which the number of trapezoids can be computed. The expressions

```
length (x) - 2  used in the trap function and
last (x) - 1    used in the previous example
```

are equivalent as long as the array origin is set at zero, which is assumed in the **trap** function, since the summation index, **i**, starts at zero.

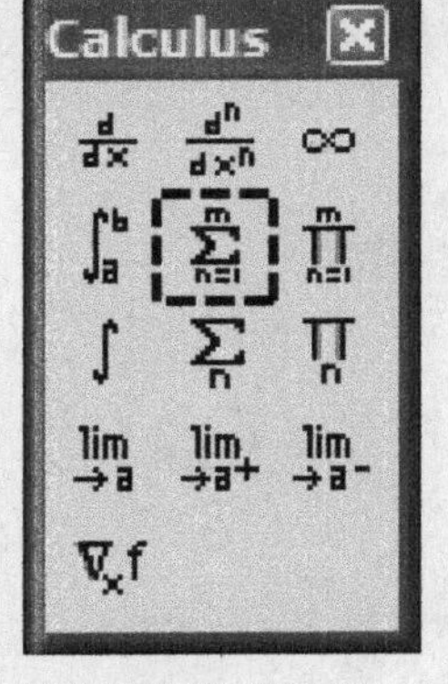

Figure 8.27
The summation operator on the Calculus Toolbar.

Practice!

Create a test data set, and then use the **trap** function to integrate $y = \cos(x)$ from $x = 0$ to $x = \pi/2$. Vary the number of points in the data set to see how the size of the trapezoids (over the same *x* range) affects the accuracy of the result. Then check your result using Mathcad's symbolic integrator. The following are the details of the procedure.

- Create the test data set:

$$N_{pts} := 20$$
$$i := 0 .. (N_{pts} - 1)$$
$$x_i := \frac{\pi}{2} \cdot \frac{i}{N_{pts} - 1}$$
$$y_i := \cos(x_i)$$

- Integrate by using the **trap** function, and vary the number of points in the data set.
- Use symbolic integration to evaluate $\int_0^{\pi/2} \cos(x)\,dx$.

APPLICATION

Controlled Release of Drugs

When a patient takes a pill, there is a rapid rise in the concentration of the drug in the patient's bloodstream, which then decreases with time as the drug is removed from the bloodstream, often by the kidneys or the liver. Then the patient takes another pill. The result is a time-varying concentration of drug in the blood.

In some situations, there may be therapeutic benefits to maintaining a more constant (perhaps lower) drug concentration for prolonged periods of time. For example, a chemotherapy drug might be active only at concentrations greater than 2 mg/L. The pills for this drug might be designed to raise the concentration in the blood to 15 mg/L to try to keep the concentration above 2 mg/L for as long as possible. If you could keep the concentration of a cancer-fighting drug from falling below 2 mg/L for a month or more, it might do a better job of killing the cancer cells. If you could also reduce the maximum concentration from 15 mg/L to perhaps 10 mg/L, the side effects of the drug might be reduced.

In this example, we will consider an implanted "drug reservoir" for chemotherapy. This drug reservoir is little more than a plastic bag containing a solution of the drug. The shape, materials of construction, and volume of the bag, as well as the concentration of the dissolved drug, can all be varied to change the drug release characteristics. This example considers only the volume and drug concentration in the reservoir.

For preliminary testing of the release characteristics, human subjects would not be used. Instead, the computer model used to generate the accompanying graph assumes that the drug is being released into a body simulator (a 50-liter tank) with a slow (1 mL/min) feed of fresh water and removal of drug solution at the same rate. Two tests were simulated, with the drug reservoir volume and concentration adjusted to give a maximum drug concentration of 10 mg/L in the simulator. The results are shown in Figure 8.28 where concentrations on the y-axis have units of mg/L, and the units on time are hours).

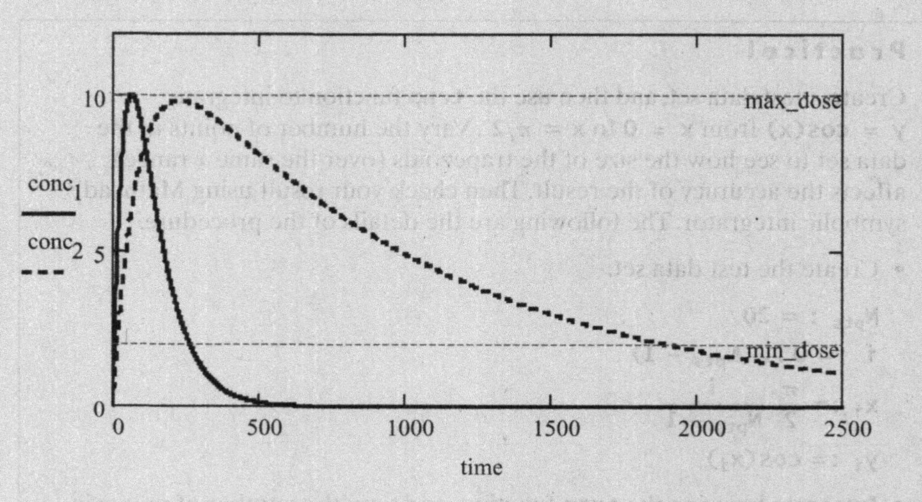

Figure 8.28
Two different drug delivery options.

The $\mathbf{conc_1}$ curve was produced using a small reservoir containing a high drug concentration, and $\mathbf{conc_2}$ used a larger reservoir with a much lower drug concentration. The curves show that, by varying the reservoir volume and concentration, the active period (concentration above 2 mg/liter) of the drug can be adjusted from approximately 250 hours (about 10 days) to almost 1,800 hours (75 days), without ever causing blood concentrations to exceed 10 mg/L.

While this simple drug delivery system is a long way from delivering a good, constant concentration of the chemotherapy drug, it does demonstrate that it is quite possible to change the way drugs are administered. By designing better drug delivery systems, we may be able to improve the performance of some drugs and the quality of life of patients.

TOTAL DRUG RELEASE

With the *concentration* vs. *time* data, and knowing the flow rate of fluid through the simulator, we can determine the total amount of drug that is released from the reservoir. Multiplying the time by the volumetric flow rate gives the volume that has passed through the simulator. The concentrations can then be plotted against *volume* (liters). This is shown in Figure 8.29, where

$$Q := 60\,\frac{mL}{hr}$$

$$vol := time \cdot Q$$

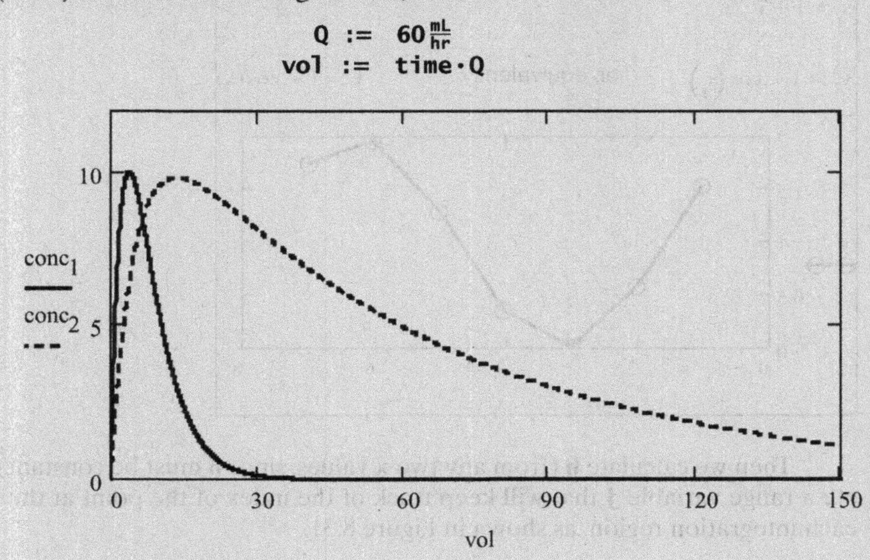

Figure 8.29
Drug concentration plotted against volume. Area under curve represents amount of drug delivered.

The area under each curve represents the amount of drug released from the associated reservoir in 2,500 hours. The **trap** function can be used to perform the integrations:

$$D_1 := trap(vol, conc_1) \qquad D_1 = 108\,mg$$

$$D_2 := trap(vol, conc_2) \qquad D_2 = 674\,mg$$

8.2.2.2 Simpson's-Rule Integration *Simpson's rule* is a popular numerical integration technique that takes three data points, fits a curve through the points, and computes the area of the region below the curve. This operation is repeated for each set of three points in the data set. The common formula for Simpson's rule looks something like

$$A_{total} = \frac{h}{3} \sum_{\substack{all \\ regions}} (y_{i-1} + 4y_i + y_{i+1})$$

where the unusual summation over "all regions" is necessary because an integration region using Simpson's method requires three data points. So, using Mathcad's default array indexing, we see that points 0, 1, and 2 make up the first integration region, points 2, 3, and 4 make up the second region, and so on. The number of integration regions is approximately half the number of data points. The distance between two adjacent points is h, a constant. The use of Simpson's rule comes with two restrictions:

- You must have an odd number of data points.
- The independent values (usually called x) must be uniformly spaced, with $h = \Delta x$.

We will look at a way to get around these restrictions later, but first we try an example of applying Simpson's rule when the conditions are met. In Figure 8.30, we create a data set containing seven values and strong curvature.

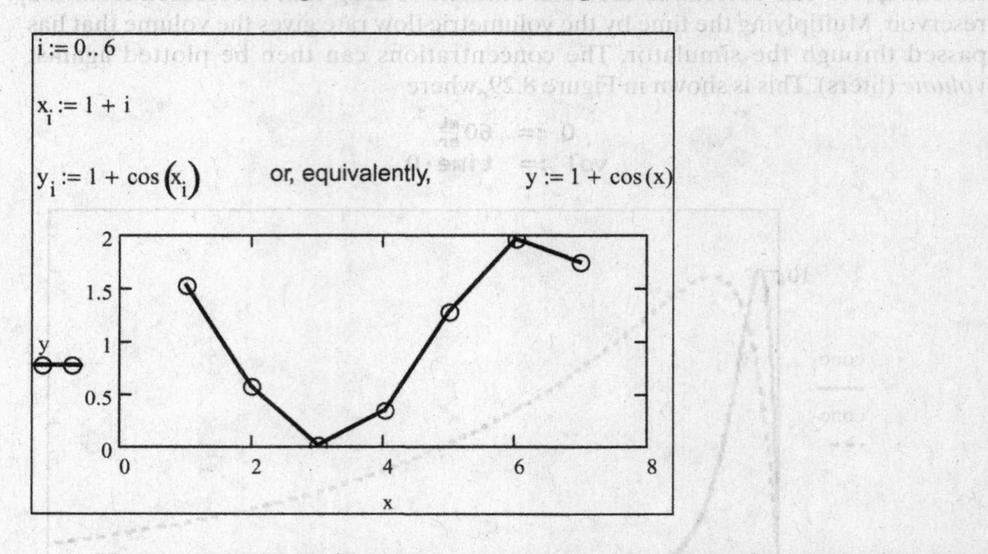

i := 0..6
$x_i := 1 + i$
$y_i := 1 + \cos(x_i)$
or, equivalently,
$y := 1 + \cos(x)$
y
x
0
0.5
1
1.5
2
2
4
6
8

Figure 8.30
Creating a data set for testing Simpson's-rule integration.

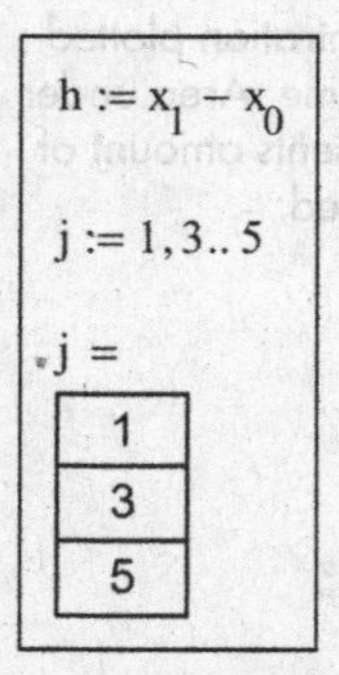

$h := x_1 - x_0$
j := 1, 3..5
j =
1
3
5

Figure 8.31
Defining integration step, **h**, and range variable, **j**, to track integration regions.

Then we calculate **h** (from any two **x** values, since **h** must be constant) and create a range variable **j** that will keep track of the index of the point at the center of each integration region, as shown in Figure 8.31.

Integration Region, Point Numbers	Central Point
0, 1, 2	1
2, 3, 4	3
4, 5, 6	5

Next, we apply Simpson's rule to determine the area under the curve (see Figure 8.32), again by using the range variable summation operator.

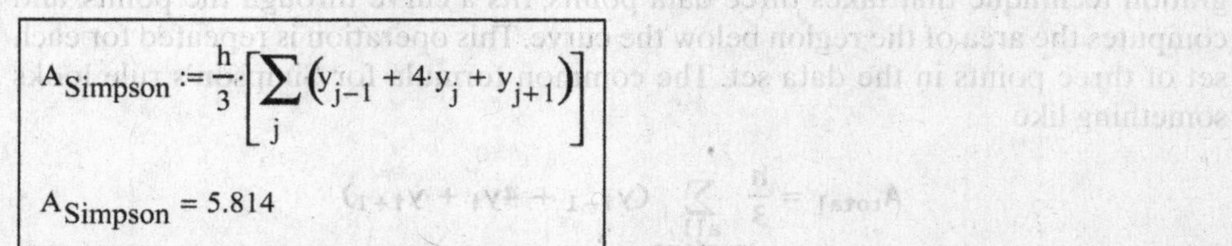

$A_{Simpson} := \frac{h}{3} \cdot \left[\sum_j \left(y_{j-1} + 4 \cdot y_j + y_{j+1} \right) \right]$
$A_{Simpson} = 5.814$

Figure 8.32
Simpson's-rule integration.

In Figure 8.33 we compare the results with trapezoidal-rule integration and exact integration of the cosine function.

$$A_{Simpson} := \frac{h}{3} \cdot \left[\sum_j \left(y_{j-1} + 4 \cdot y_j + y_{j+1} \right) \right] \qquad A_{Simpson} = 5.814$$

$$A_{trap} := \text{trap}(x, y) \qquad A_{trap} = 5.831$$

$$A_{exact} := \int_1^7 (1 + \cos(x))\, dx \qquad A_{exact} = 5.816$$

Figure 8.33
Comparing integration results.

Simpson's method came a lot closer to the exact result than trapezoidal integration did—it usually does. Because Simpson's rule connects data points with smooth curves rather than straight lines, it typically fits data better than the ***trapezoidal rule*** does. But Simpson's rule has those two restrictions that make it useless in many situations. Is there a way to get around these restrictions? Yes, there is: You can use a *cubic spline* to fit any data set with a smooth curve and then use *cubic spline interpolation* to compute a set of values that covers the same range as the original data but has an odd number of data points and uniform point spacing. After that, you can use Simpson's rule on the interpolated data.

To test this approach, we will use the temperature–time data employed in earlier examples. These data consist of 10 uniformly spaced values. The uniform spacing is good, but 10 values won't work with Simpson's rule.[2] A cubic spline fit to the data is shown in Figure 8.34.

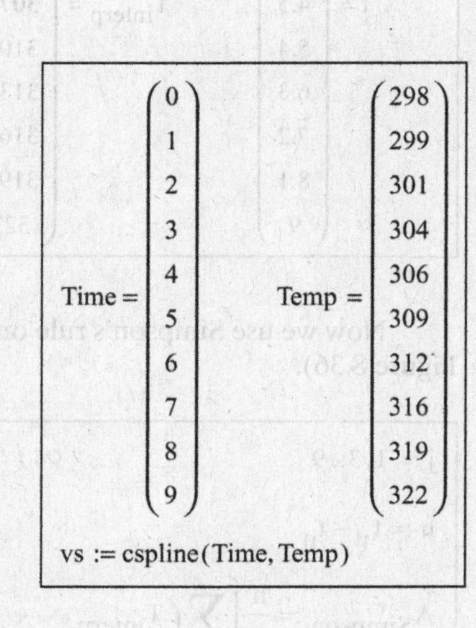

Time = (0 1 2 3 4 5 6 7 8 9)
Temp = (298 299 301 304 306 309 312 316 319 322)
vs := cspline(Time, Temp)

Figure 8.34
Fitting the temperature and time data with a cubic spline.

[2]Another common way to get around the odd-number-of-data-points restriction is to use Simpson's rule as far as possible and then, if there is an even number of points, finish the integration by using a trapezoid for the last two points.

Then we can use the **interp** function to find temperature values at 11 points over the same time interval, 0 to 9 minutes, as shown in Figure 8.35.

$i := 0..10$ << define range variable containing 11 values

$t_i := \frac{9}{10} \cdot i$ << define new t vector with 11 values

Or, more generally:

$N_{pts} := 11$

$i := 0..(N_{pts} - 1)$

$t_i := \frac{\max(Time) - \min(Time)}{N_{pts} - 1} \cdot i$

Use the **interp** function to interpolate new T_{interp} values at all 11 values of t.

$T_{interp} := interp(vs, Time, Temp, t)$

Results...

$$t = \begin{pmatrix} 0 \\ 0.9 \\ 1.8 \\ 2.7 \\ 3.6 \\ 4.5 \\ 5.4 \\ 6.3 \\ 7.2 \\ 8.1 \\ 9 \end{pmatrix} \qquad T_{interp} = \begin{pmatrix} 298 \\ 298.9 \\ 300.5 \\ 303.2 \\ 305.2 \\ 307.4 \\ 310.1 \\ 313.2 \\ 316,7 \\ 319.3 \\ 322 \end{pmatrix}$$

Figure 8.35 Interpolating the **Time** and **Temp** data to create a data set that meets the requirements for Simpson's-rule integration (odd number of points, evenly spaced).

Now we use Simpson's rule on **t** and $\mathbf{Temp_{interp}}$ instead of **Time** and **Temp** (see Figure 8.36).

$j := 1, 3..9$ *or, j := 1,3 .. (N pts-2)*

$h := t_1 - t_0$

$$A_{Simpson} := \frac{h}{3} \cdot \left[\sum_j \left(T_{interp_{j-1}} + 4 \cdot T_{interp_j} + T_{interp_{j+1}} \right) \right]$$

$A_{Simpson} = 2776$

Figure 8.36 Applying Simpson's-rule integration to the interpolated data.

The result compares well with that obtained from the trapezoidal integration (see Figure 8.25), which is not surprising for this data set, since it does not show a lot of curvature. (There is not a lot of difference between connecting the points with lines or curves for the data set, so there is little difference between the results computed by the two methods.)

8.3 NUMERICAL DIFFERENTIATION

8.3.1 Evaluating Derivatives of Functions Numerically

Mathcad can take the derivative of a function such as

$$y = b_0 e^{b_1 t}$$

If you use the live symbolic operator, →, to evaluate an expression symbolically, Mathcad will use its symbolic processor and give you another function, as shown in Figure 8.24.

$$\frac{d}{dt}\left(b_0 \cdot e^{b_1 \cdot t}\right) \rightarrow e^{t \cdot b_1} \cdot b_0 \cdot b_1$$

Figure 8.37
Using Mathcad's symbolic math processor to evaluate a derivative.

Note: This derivative was evaluated in a new Mathcad worksheet (**t** must be undefined to obtain a symbolic result).

If the values of $\mathbf{b_0}$, $\mathbf{b_1}$, and **t** are specified before evaluating the derivative, Mathcad's symbolic processor will calculate a numeric result, as shown in Figure 8.38.

$b_0 := 3.4$

$b_1 := 0.12$

$t := 16$

$$\frac{d}{dt}\left(b_0 \cdot e^{b_1 \cdot t}\right) \rightarrow 2.7829510554706259327$$

Figure 8.38
Using Mathcad's symbolic math processor to evaluate a derivative and return a numeric result.

The extreme number of significant figures is a reminder that Mathcad solved for the value using the symbolic processor. The default number of digits displayed is 20.

On the other hand, if you use the "numerical evaluation" symbol (a plain equal sign, =), then Mathcad will use its numeric processor to calculate the value of the derivative, as shown in Figure 8.39.

If you use the numeric processor, you must specify the values of $\mathbf{b_0}$, $\mathbf{b_1}$, and **t** before Mathcad evaluates the derivative.

$b_0 := 3.4$

$b_1 := 0.12$

$t := 16$

$$\frac{d}{dt}\left(b_0 \cdot e^{b_1 \cdot t}\right) = 2.783$$

Figure 8.39
Using Mathcad's numeric math processor to evaluate a derivative.

8.3.2 Derivatives from Experimental Data

If the relationship between your variables is represented by a set of values (a data set), rather than a mathematical expression, you have two choices for trying to determine the derivative at some specified point:

- Fit the data with a mathematical expression (i.e., ***curve fitting***) and then differentiate the expression at the specified value.
- Use numerical approximations for derivatives on the data points themselves. (This approach is practical only when you have "clean" data.)

8.3.2.1 Using a Fitting Function The data that we will use to demonstrate the curve-fitting approach are shown in Figure 8.40.

The data are expected to fit the exponential model

$$y = b_0 e^{b_1 t},$$

which can be rewritten in linear form as

$$\ln(y) = \ln(b_0) + b_1 t.$$

We can then use linear regression and a bit of math to determine $\mathbf{b_0}$ and $\mathbf{b_1}$ (see Figure 8.41).

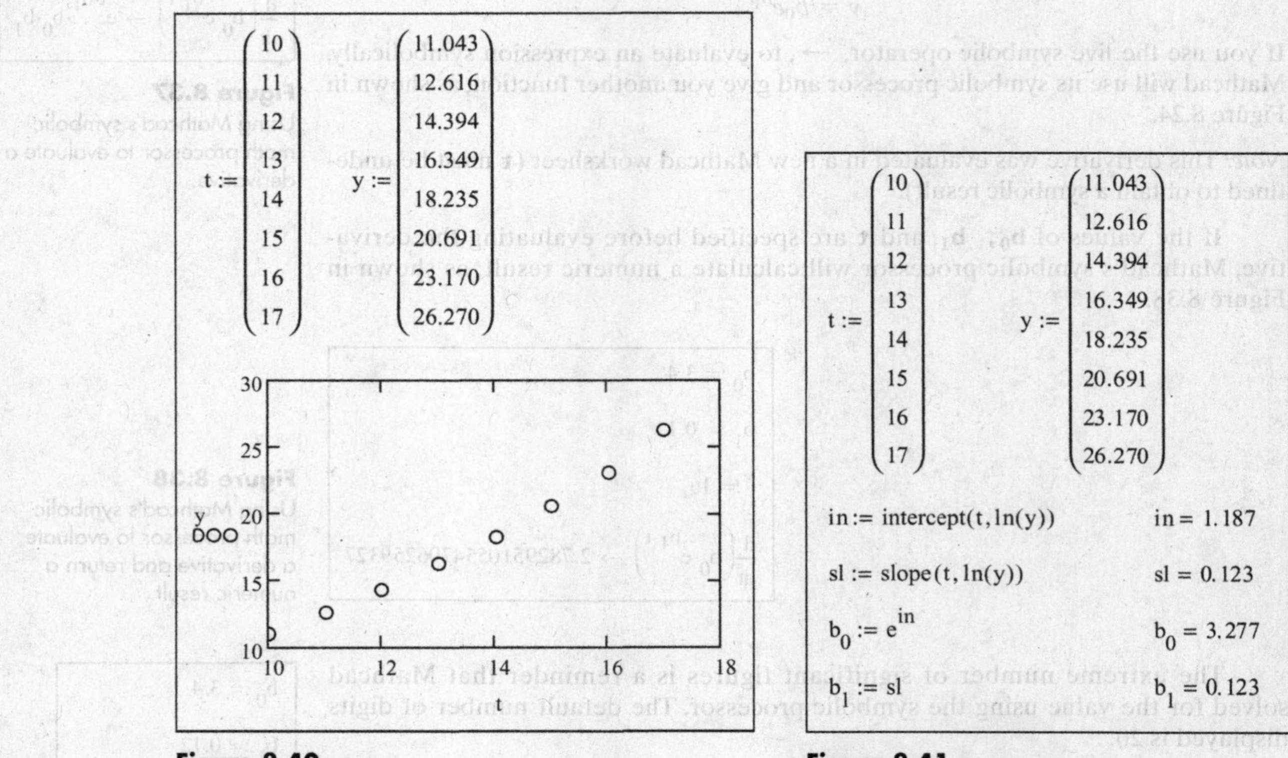

t := (10, 11, 12, 13, 14, 15, 16, 17)
y := (11.043, 12.616, 14.394, 16.349, 18.235, 20.691, 23.170, 26.270)
y
t
30
25
20
15
10
10
12
14
16
18
t := (10, 11, 12, 13, 14, 15, 16, 17)
y := (11.043, 12.616, 14.394, 16.349, 18.235, 20.691, 23.170, 26.270)
in := intercept(t, ln(y))
in = 1.187
sl := slope(t, ln(y))
sl = 0.123
b0 := e^in
b0 = 3.277
b1 := sl
b1 = 0.123

Figure 8.40
Data for the fitting function example.

Figure 8.41
Calculating regression coefficients.

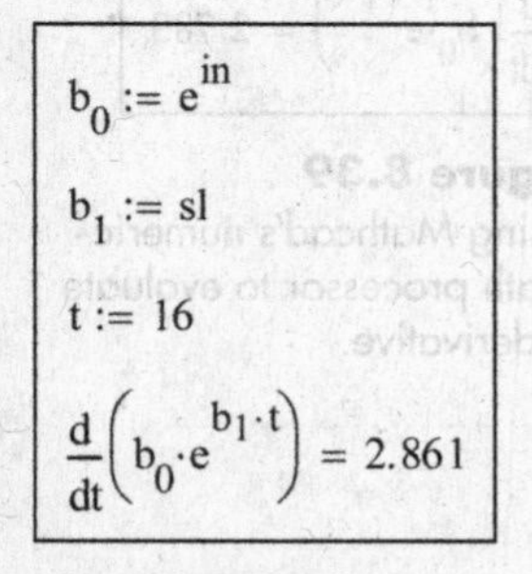

Figure 8.42
Evaluating the derivative.

Once we have a mathematical expression, we can evaluate the derivative at **t = 16** by using either the symbolic or numeric processor, as described earlier. The results of using the numeric processor are shown in Figure 8.42.

8.3.2.2 Using Numerical Approximations for Derivatives The derivative physically represents the slope of a plot of the data at a specified point. You can approximate the derivative at any point by estimating the slope of the graph at that point. Just as there are several ways to estimate the slope, there are several ways to compute numerical approximations for derivatives. Since array index values will be used in these calculations, the example data (again), along with the array index values, are shown in Figure 8.43.

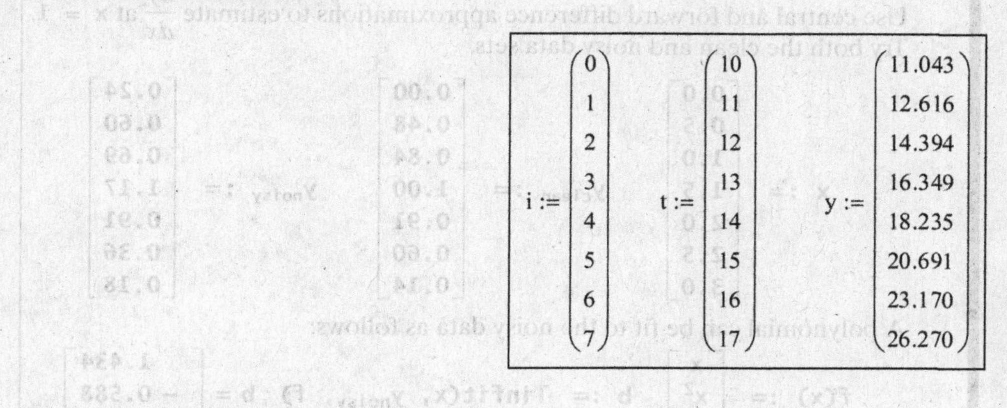

Figure 8.43
Sample data with array index `i`.

One way to estimate the slope at point `i = 6` (where `t = 16`) would be

$$\left.\frac{dy}{dt}\right|_{i=6} \approx \frac{y_7 - y_6}{t_7 - t_6}$$

If that equation is valid, then the following equation is equally valid. (Both are approximations.)

$$\left.\frac{dy}{dt}\right|_{i=6} \approx \frac{y_6 - y_5}{t_6 - t_5}$$

The first expression uses the point at `t = 16` and the point to the right (`i = 7` or `t = 17`) to estimate the slope and is called a *forward finite-difference approximation for the first derivative* at `i = 6`. The term ***finite difference*** is used because there is a finite distance between the `t` values used in the calculation. Finite-difference approximations are truly equal to the derivatives only in the limit as Δt approaches zero.

The second expression uses the point at `t = 16` and the point to the left (`t = 15`) to estimate the slope and is called a *backward finite-difference approximation for the first derivative* at `i = 6`. You can also write a *central finite-difference approximation for the first derivative* at `i = 6`:

$$\left.\frac{dy}{dt}\right|_{i=6} \approx \frac{y_7 - y_5}{t_7 - t_5}.$$

Central differences tend to give better estimates of the slope and are the most commonly used. Applying the preceding equation to calculate the derivative at point `i = 6 (t = 16)` in the data set (see Figure 8.44), we find that it has a value of 2.79.

$$\text{slope}_{16} := \frac{y_7 - y_5}{t_7 - t_5}$$

$$\text{slope}_{16} = 2.79$$

Figure 8.44
Central difference approximation of the slope at `t = 16`.

There are also finite-difference approximations for higher order derivatives. For example, a *central finite-difference approximation for a second derivative* at `i = 6` can be written (assuming uniform point spacing, i.e., Δt is constant) as

$$\left.\frac{d^2y}{dt^2}\right|_{i=6} \approx \frac{y_7 - 2y_6 + y_5}{(\Delta t)^2}.$$

Practice!

Use central and forward difference approximations to estimate $\frac{dy}{dx}$ at x = 1. Try both the clean and noisy data sets.

$$x := \begin{bmatrix} 0.0 \\ 0.5 \\ 1.0 \\ 1.5 \\ 2.0 \\ 2.5 \\ 3.0 \end{bmatrix} \quad y_{clean} := \begin{bmatrix} 0.00 \\ 0.48 \\ 0.84 \\ 1.00 \\ 0.91 \\ 0.60 \\ 0.14 \end{bmatrix} \quad y_{noisy} := \begin{bmatrix} 0.24 \\ 0.60 \\ 0.69 \\ 1.17 \\ 0.91 \\ 0.36 \\ 0.18 \end{bmatrix}$$

A polynomial can be fit to the noisy data as follows:

$$f(x) := \begin{bmatrix} x \\ x^2 \\ x^3 \end{bmatrix} \quad b := \text{linfit}(x, y_{noisy}, f) \quad b = \begin{bmatrix} 1.434 \\ -0.588 \\ 0.041 \end{bmatrix}$$

Try using the differentiation operator on the Calculus Toolbox to evaluate

$$\frac{d}{dx}(1.434x - 0.588x^2 + 0.041x^3)$$

at x = 1.

Note: The derivative values you calculate in this "Practice!" box will vary widely. Calculating derivatives from noisy data is highly prone to errors, and you should try to avoid doing that if possible. If you must take derivatives from experimental data, try to get good, clean data sets, or use regression to find the best-fit curve through the data and find the derivative from the regression result.

APPLICATION

Biomedical Instrumentation

When someone takes his or her pulse, the person is feeling pressure fluctuations in an artery due to the beating of the heart. If you want to design an instrument to measure a pulse, a likely design would include a pressure-sensing device. The data would be some sort of time-varying pressure signal, such as that shown in Figure 8.45. The arterial pulses superimposed on a time-varying mean pressure is fairly typical in data collected on humans. The slow change in mean pressure could be the result of the instrument (or the artery) moving slightly as the subject breathes, for example.

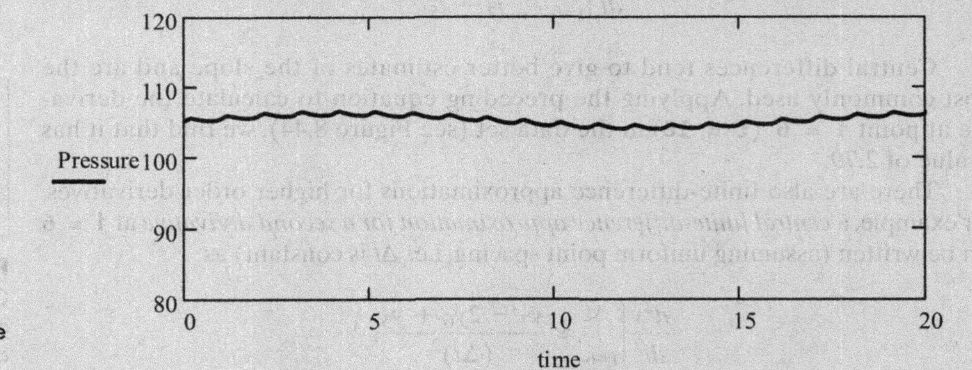

Figure 8.45
Pressure signal from a pulse detection system.

To determine the pulse, you need to detect and count the number of pressure ridges over a period of time, typically one minute. Because the mean value of pressure changes during the testing period, it can be hard to design an instrument that can identify the pulses, which are of smaller amplitude than the changes in mean pressure. While it is possible to devise a technique that could identify the pressure ridges in Figure 8.45, it might be easier to detect the pulses if you work with the time derivative of the pressure signal, shown in Figure 8.46.

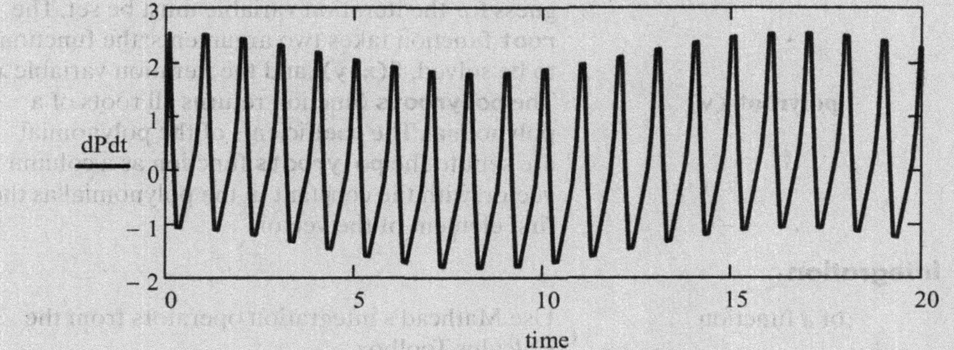

Figure 8.46 Time derivative of the pressure signal.

In the derivative signal, the amplitude of the pulses is much larger relative to the mean, making it easy to detect the pulses and count them.

If the time and pressure vectors are available in a Mathcad worksheet, the derivative vector (dPdt) can be calculated as shown in Figure 8.47.

$$i := 1 .. \text{last}(\text{time}) - 1$$

$$dPdt_i := \frac{\text{Pressure}_{i+1} - \text{Pressure}_{i-1}}{\text{time}_{i+1} - \text{time}_{i-1}}$$

Figure 8.47 Calculating the derivative vector.

SUMMARY

In this chapter, you learned to use several standard numerical methods in Mathcad, including an iterative solve block and methods for numerical integration and differentiation. The numerical integration techniques involved using Mathcad's integration operators (from the Calculus Toolbox) when you were working with a function, and numerical techniques like trapezoidal or Simpson's-rule integration when you were working with a data set. Similarly, Mathcad's differentiation operators apply when you need to take the derivative of a function, and finite difference methods were presented for evaluating derivatives when you have a data set.

MATHCAD SUMMARY

Iterative Solutions

given The keyword that begins an iterative solve block. Remember to specify an initial guess value before using this keyword.

`find`	The keyword that closes an iterative solve block, and the function that performs the iteration and returns the solution. Remember to assign the value returned by **find** to a variable.
`root(f(x, y), x)`	The **root** function uses an iterative solver to find a single root of a function of one or more variables. Before using the **root** function, the function to be solved must be defined and an initial guess for the iteration variable must be set. The **root** function takes two arguments: the function to be solved, **f(x,y)**, and the iteration variable *x*.
`polyroots(v)`	The **polyroots** function returns all roots of a polynomial. The coefficients of the polynomial are sent to the **polyroots** function as a column vector, with the constant in the polynomial as the first element of the vector.

Integration

of a function	Use Mathcad's integration operators from the Calculus Toolbox.
of a data set	Trapezoidal and Simpson's rules can be used to approximate the integral, or you can fit a function to the data and then integrate the function.

Differentiation

of a function	Use Mathcad's differentiation operators from the Calculus Toolbar.
of a data set	Finite differences can be used to estimate the values of the derivative, or you can fit a function to the data and then differentiate the function. (The latter approach is preferred if you have "noisy" data.)

KEY TERMS

curve fitting	iterative solver	solve block
finite difference	numerical differentiation	tolerance
initial guess	numerical integration	trapezoidal rule
iteration	polynomial coefficients	
iterative solve block	Simpson's rule	

PROBLEMS

8.1 Finding Solutions with the root( ) Function
Use the **root** function to find the solution(s) of the following equations:

(a) $x - 3 = 0$ (admittedly, this one is kind of obvious)
(b) $x^2 - 3 = 0$
(c) $x^2 - 4x + 3 = 0$.

8.2 Finding Intersections
Use the **root** function to find the intersection(s) of the listed curves (Part **a** has been completed as an example):

(a) $y = x^2 - 3$ and $y = x + 4$.

The intersections are at the locations where the y values are equal, so start by eliminating y from the equations:

$$x^2 - 3 = x + 4.$$

Next, get all terms on one side of the equation, so that the equation is set equal to zero:

$$x^2 - x - 7 = 0.$$

Create a QuickPlot to find good initial guesses for x (as shown in Figure 8.48), then use the **root** function to find more precise x values at each intersection.

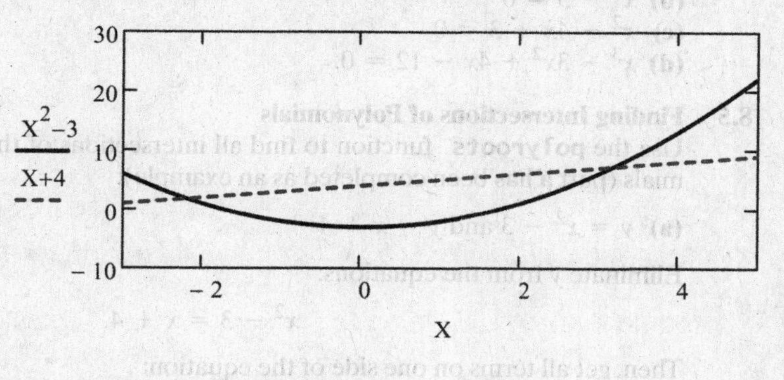

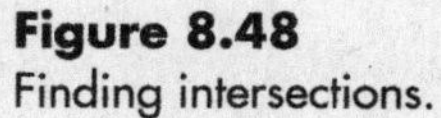

Figure 8.48
Finding intersections.

```
guess:  x :=  - 2
root(x^2 - x - 7, x) = -2.193
x := 3
root(x^2 - x - 7, x) = -3.193
```

(b) $y = x^2 - 3$ and $y = -(x^2) + 4$

(c) $y = x^2 - 3$ and $y = e^{-x}$.

8.3 Finding the Zeroes of Bessel Functions

Bessel functions are commonly used when solving differential equations in cylindrical coordinates. Two Bessel functions, J_0 and J_1 are shown in Figure 8.49.[3]

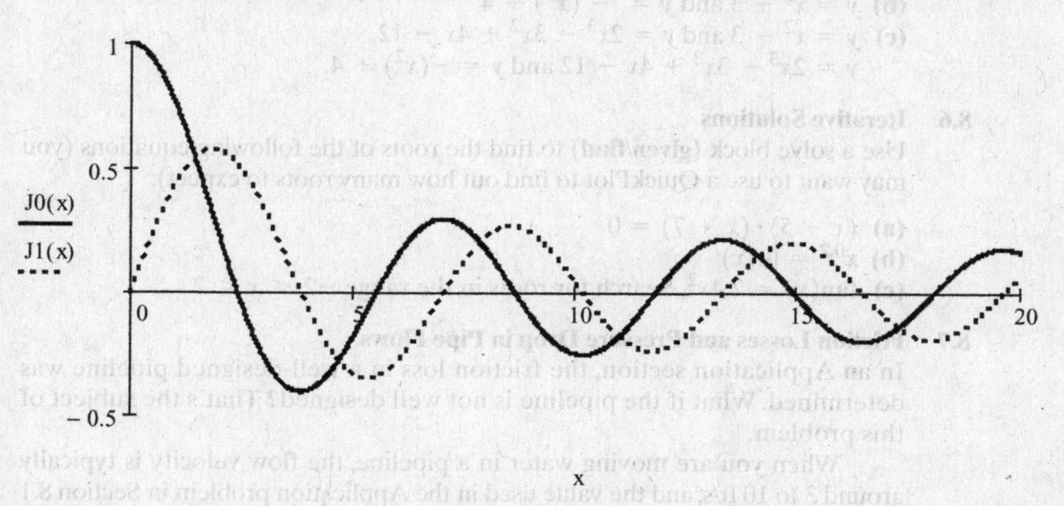

Figure 8.49
Bessel functions.

[3]More precisely, these are Bessel functions of the first kind, of order zero and one, respectively.

These Bessel functions are available as built-in functions in Mathcad and are called **J0** and **J1**. Use the **root** function with different initial guesses to find the roots of J0(x) and J1(x) in the range $0 < x < 20$.

Note: The **J0** function has a zero in its name, not the letter "o."

8.4 Finding Polynomial Roots

Use the **polyroots** function to find all roots for the following polynomial expressions:

(a) $x - 3 = 0$ (admittedly, not much of a polynomial)
(b) $x^2 - 3 = 0$
(c) $x^2 - 4x + 3 = 0$
(d) $x^3 - 3x^2 + 4x - 12 = 0$.

8.5 Finding Intersections of Polynomials

Use the **polyroots** function to find all intersections of the following polynomials (part **a** has been completed as an example):

(a) $y = x^2 - 3$ and $y = x + 4$

Eliminate y from the equations:

$$x^2 - 3 = x + 4.$$

Then, get all terms on one side of the equation:

$$x^2 - x - 7 = 0.$$

Create a column vector containing the coefficients of the polynomial (with the constant on top). Then use the **polyroots** function to find each intersection:

$$v := \begin{pmatrix} -7 \\ -1 \\ 1 \end{pmatrix}$$

$$\text{polyroots}(v) = \begin{pmatrix} -2.193 \\ 3.193 \end{pmatrix}$$

(b) $y = x^2 - 3$ and $y = -(x^2) + 4$
(c) $y = x^2 - 3$ and $y = 2x^3 - 3x^2 + 4x - 12$
$y = 2x^3 - 3x^2 + 4x - 12$ and $y = -(x^2) + 4$.

8.6 Iterative Solutions

Use a solve block (**given/find**) to find the roots of the following equations (you may want to use a QuickPlot to find out how many roots to expect):

(a) $(x - 5) \cdot (x + 7) = 0$
(b) $x^{0.2} = \ln(x)$
(c) $\tan(x) = 2.4x^2$. Search for roots in the range $-2 < x < 2$.

8.7 Friction Losses and Pressure Drop in Pipe Flows

In an Application section, the friction loss in a well-designed pipeline was determined. What if the pipeline is not well designed? That's the subject of this problem.

When you are moving water in a pipeline, the flow velocity is typically around 3 to 10 ft/s, and the value used in the Application problem in Section 8.1 was in this range. This is a commonly velocity range that will allow the pipe to carry a reasonable amount of water quickly and without too much friction loss. What happens to the friction loss and the required pump power when you try to move water through the same pipeline at a velocity of 15 ft/s?

8.8 Real Gas Volumes

The Soave–Redlich–Kwong (SRK) equation of state is a commonly used equation that relates the temperature, pressure, and volume of a gas under conditions when the behavior of the gas cannot be considered ideal (e.g., moderate temperature and high pressure). The equation is

$$P = \frac{R \cdot T}{\hat{V} - b} - \frac{\alpha a}{\hat{V}(\hat{V} + b)},$$

where α, a, and b are parameters specific to the gas, R is the ideal gas constant, P is the absolute pressure, T is absolute temperature, and

$$\hat{V} = \frac{V}{n}$$

is the molar volume (volume per mole of gas).

For common gases, the parameters α, a, and b can be readily determined from available data. Then, if you know the molar volume of the gas, the SRK equation is easy to solve for either pressure or temperature. However, if you know the temperature and pressure and need to find the molar volume, an iterative solution is required.

Determine the molar volume of ammonia at 300°C and 1200 kPa by

(a) using the ideal gas equation.
(b) using the SRK equation and an iterative solve block.

For ammonia at these conditions,

$$\alpha = 0.7007$$
$$a = 430.9 \text{ kPa} \cdot \text{L}^2/\text{mole}^2$$
$$b = 0.0259 \text{ L/mole}.$$

8.9 Required Size for a Water Retention Basin

In urban areas, as fields are turned into streets and parking lots, water runoff from sudden storms can become a serious problem. To prevent flooding, retention basins are often built to hold excess water temporarily during storms. In arid regions, these basins are dry most of the time, so designers sometimes try to incorporate alternative uses into their design. A proposed design is a half-pipe for skateboarders that can hold 100,000 cubic feet of water during a storm. The radius of the half-cylinder needs to be scaled to fit the skateboarders, and a radius of 8 feet is proposed.

(a) What is the required length of the basin to hold the 100,000 ft^3 of storm water? (There may be several short sections of half-pipes to provide the required total volume.)
(b) If the basin is filled with water to a depth of 4 feet after a storm, what volume of water is held in the basin?

8.10 Work Required to Stretch a Spring

The device shown in Figure 8.50 can be used to determine the work required to extend a spring.

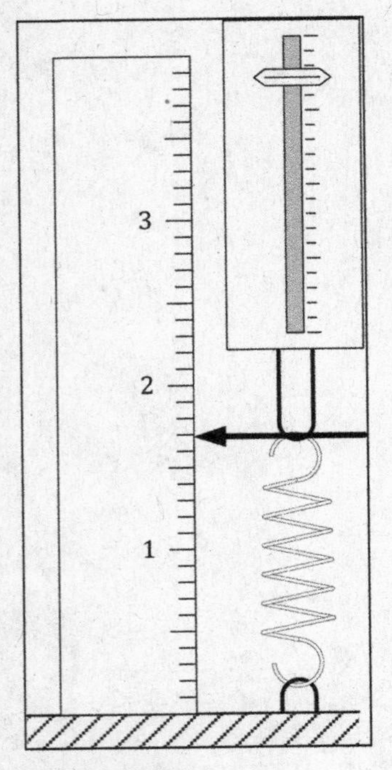

Figure 8.50
Work to stretch a spring.

This device consists of a spring, a spring balance, and a ruler. Before stretching the spring, its length is measured and found to be 1.3 cm. The spring is then stretched 0.4 cm at a time, and the force indicated on the spring balance is recorded. The resulting data set is shown in the following table:

Measurement (cm)	Unextended Length (cm)	Extended Length (cm)	Force (n)
1.3	1.3	0.0	0.00
1.7	1.3	0.4	0.88
2.1	1.3	0.8	1.76
2.5	1.3	1.2	2.64
2.9	1.3	1.6	3.52
3.3	1.3	2.0	4.40
3.7	1.3	2.4	5.28
4.1	1.3	2.8	6.16
4.5	1.3	3.2	7.04
4.9	1.3	3.6	7.92

Work can be computed as

$$W = \int F\,dx,$$

where x is the extended length of the spring.

(a) Calculate the work required to stretch the spring from an extended length of 0 cm to 3.6 cm. Watch the signs on the forces in this problem. Force is a vector quantity, so there is an associated direction.

(b) Is the force being used to stretch the spring in the same direction as the movement of the spring or in the opposite direction?

8.11 Work Required to Stretch a Nonlinear Spring

Springs come in a variety of sizes and shapes. The shape can sometimes affect the performance of the spring.

The spring shown in Figure 8.51 pulls easily at first, but gets stiffer the more it is stretched. (See the accompanying table.) It is a nonlinear spring. Calculate the work required to stretch this spring

(a) from an extended length of 0 cm to 1.6 cm.

(b) from an extended length of 2.0 cm to 3.6 cm.

Extended Length (cm)	Force (n)
0.0	0.00
0.4	0.15
0.8	0.41
1.2	0.72
1.6	1.07
2.0	1.46
2.4	1.89
2.8	2.35
3.2	2.83
3.6	3.33

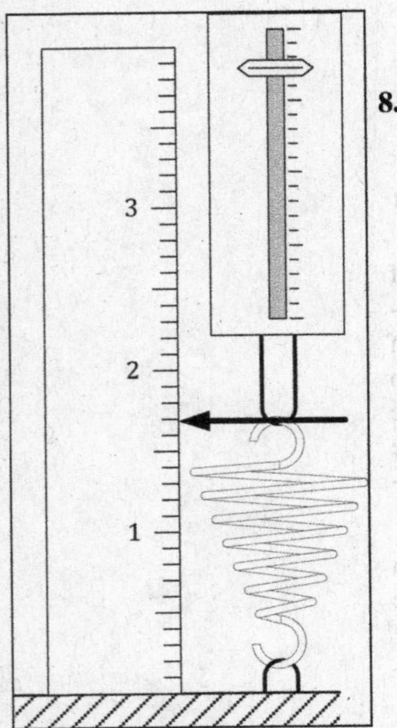

Figure 8.51
Work to stretch a non-linear spring.

8.12 Work Required to Expand a Gas

The work required to expand a gas is calculated as

$$W = \int F\,dx.$$

Because force divided by area equals pressure, and length times area equals volume, work can also be found as

$$W = \int P\,dV.$$

This form is handier for dealing with gas systems such as the one shown in Figure 8.52.

As the piston is lifted, the pressure in the sealed chamber will fall. By monitoring the change in position on the ruler and knowing the cross-sectional area of the chamber, we can calculate the chamber volume at each pressure. Care must be taken to allow the system to equilibrate at room temperature before taking the readings. The pressure and volume after equilibrium are listed in the accompanying table.

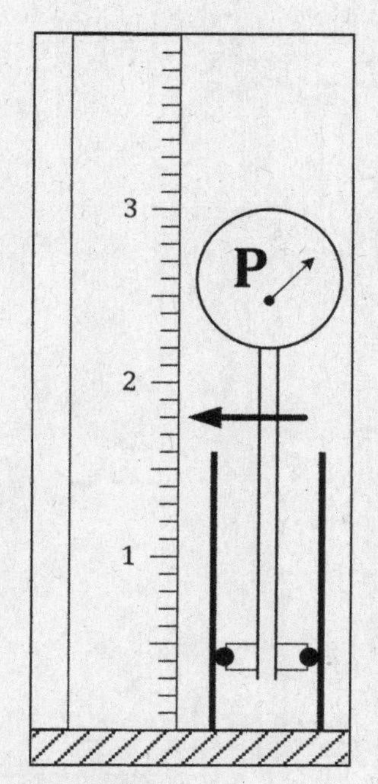

Figure 8.52
Work from an expanding gas.

Volume (mL)	Pressure (atm)
1.3	4.50
1.7	3.44
2.1	2.79
2.5	2.34
2.9	2.02
3.3	1.77
3.7	1.58
4.1	1.43
4.5	1.30
4.9	1.19

(a) Calculate the work required to expand the gas from a volume of 1.3 mL to 4.9 mL.

(b) Is this work being done by the gas on the surroundings or by the surroundings on the gas?

8.13 Calculating Spring Constants

The extension of a linear spring is described by Hooke's law,

$$F = kx,$$

where x is the extended length of the spring (i.e., the total length of the stretched spring minus the length of the spring before stretching) and k is the spring constant. The spring constant quantifies the "stiffness" of the spring. Hooke's law can also be written in differential form as

$$\frac{dF}{dx} = k.$$

For a linear spring, dF/dx is a constant, the *spring constant*. For a nonlinear spring, dF/dx will not be constant, but taking this derivative at various spring extensions can help you see how the spring characteristics change as the spring is pulled.

Use the data from Problems 8.10 and 8.11 to compute

(a) the spring constant for the linear spring (Problem 8.10).
(b) the derivative dF/dx as a function of x for the nonlinear spring. Use a central difference approximation for the derivative whenever possible, and create a plot of dF/dx vs. x.

8.14 Thermal Conductivity

Thermal conductivity is a property related to a material's ability to transfer energy by conduction. Good conductors, such as copper and aluminum, have large thermal conductivity values. Insulating materials should have low thermal conductivities to minimize heat transfer.

Figure 8.53 illustrates a device that could be used to measure thermal conductivity. A rod of the material to be tested is placed between a resistance heater (on the right) and a cooling coil (on the left). Five (numbered) thermocouples have been inserted into the rod at 5-cm intervals. The entire apparatus is placed in a bell jar, and the air around the rod is pumped out to reduce heat losses.

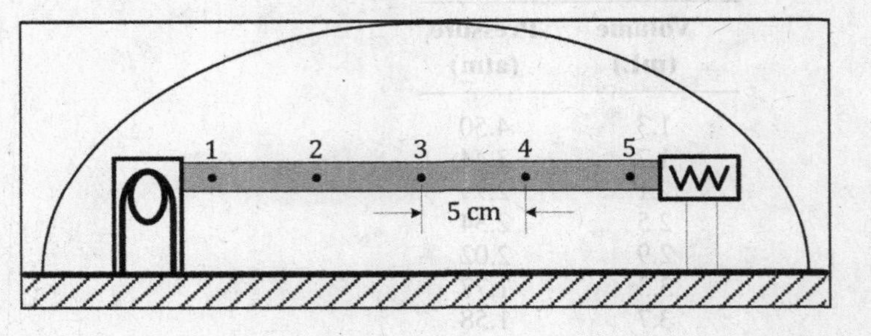

Figure 8.53
Thermal conductivity apparatus.

To run the experiment, a known amount of power is sent to the heater, and the system is allowed to reach steady state. Once the temperatures are steady, the power level and temperatures are recorded. A data sheet might look like the following:

Rod Diameter:	**2 cm**
TC Spacing:	**5 cm**
Power:	**100 watts**

TC #	Temp. (K)
1	348
2	387
3	425
4	464
5	503

The thermal conductivity can be determined from Fourier's law,

$$\frac{q}{A} = -k\frac{dT}{dx}.$$

where q is the power being applied to the heater and A is the cross-sectional area of the rod.

Note: The term q/A is called the *energy flux* and is a vector quantity—that is, it has a direction as well as a magnitude. As drawn, with the energy source on the right, the energy will be flowing in the **–x** direction, so the flux in this problem is negative.

(a) Use finite-difference approximations to estimate dT/dx at several locations along the rod, and then calculate the thermal conductivity of the material.

(b) Thermal conductivity is a function of temperature. Do your dT/dx values indicate that the thermal conductivity of the material changes appreciably between 348 and 503 K?

8.15 Calculating Work

More than one method can be used to calculate work in a force–displacement graph. The method discussed here is numerical integration using a function such as **trap**.

The stress–strain data (shown in Figure 8.54) for a composite material have been converted to force–displacement data, abridged, and listed in the accompanying table.

Calculate the work done on the sample.

F (n)	X (mm)
0	0.000
5500	0.009
8300	0.014
13800	0.023
16500	0.033
22100	0.051
24800	0.066
30300	0.137
33100	0.248
38600	0.380
41400	0.407
46900	0.461
49600	0.488
55200	0.561
57900	0.628
60700	0.749

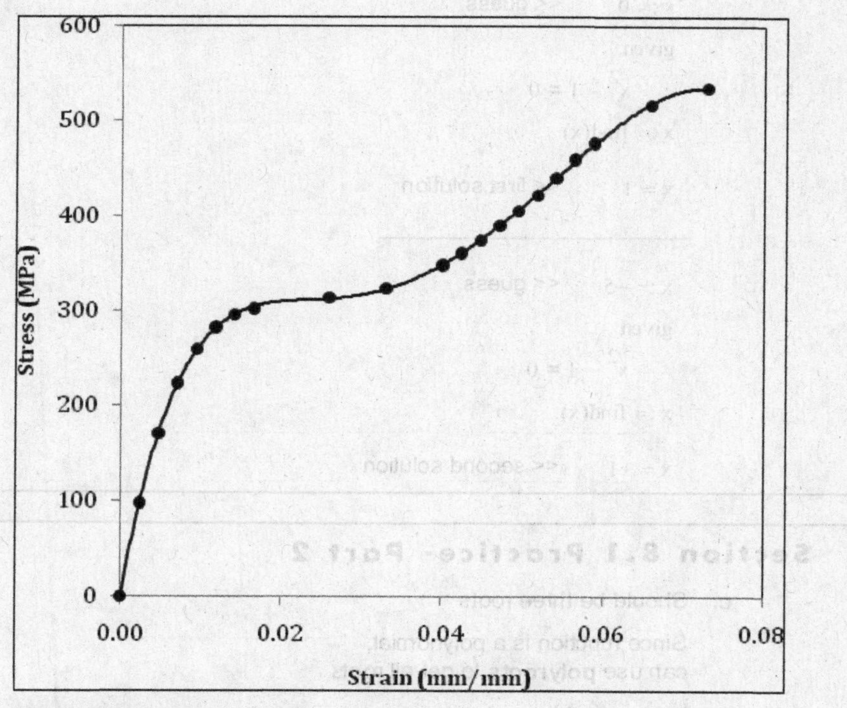

Figure 8.54
Calculating work from stress–strain data.

SOLUTIONS TO PRACTICE! PROBLEMS

Section 8.1 Practice! - Part 1

a. Should be two roots: x = 3, x = 4

$x := 0$ << guess

given

$$(x-3)\cdot(x-4) = 0$$

$x := \mathrm{find}(x)$

$x = 3$ << first solution

$x := 6$ << guess

given

$$(x-3)\cdot(x-4) = 0$$

$x := \mathrm{find}(x)$

$x = 4$ << second solution

b. Should be two roots: x = 1, x = -1

$x := 0$ << guess

given

$$x^2 - 1 = 0$$

$x := \mathrm{find}(x)$

$x = 1$ << first solution

$x := -5$ << guess

given

$$x^2 - 1 = 0$$

$x := \mathrm{find}(x)$

$x = -1$ << second solution

Section 8.1 Practice- Part 2

c. Should be three roots

Since function is a polynomial,
can use **polyroots** to get all roots

$$\mathrm{coef} := \begin{pmatrix} -3 \\ 4 \\ -2 \\ 1 \end{pmatrix} \qquad \mathrm{polyroots}(\mathrm{coef}) = \begin{pmatrix} 0.5 + 1.658i \\ 0.5 - 1.658i \\ 1 \end{pmatrix}$$

Or, can use **given / find** solve blocks

$x := 0$ << guess

given

$$x^3 - 2\cdot x^2 + 4\cdot x = 3$$

$x := \text{find}(x)$

$x = 1$ << first solution

$x := 3i$ << guess

Given

$$x^3 - 2\cdot x^2 + 4\cdot x = 3$$

$x := \text{Find}(x)$

$x = 0.5 + 1.658i$ << second solution

$x := -3i$ << guess

Given

$$x^3 - 2\cdot x^2 + 4\cdot x = 3$$

$x := \text{Find}(x)$

$x = 0.5 - 1.658i$ << third solution

Section 8.1 Practice- Part 3

d. One root? Use a QuickPlot to confirm.

$x := 0.5$ << guess

Given

$$e^{3x} - 4 = 0$$

$x := \text{Find}(x)$

$x = 0.462$ << solution

Note: xx used instead of x in QuickPlot because x-axis variable must be previously undefined. ("x" was used in parts a - c.)

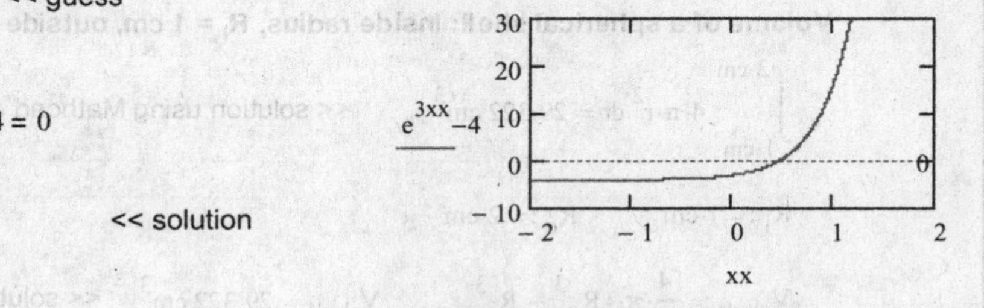

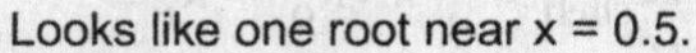

Looks like one root near x = 0.5.

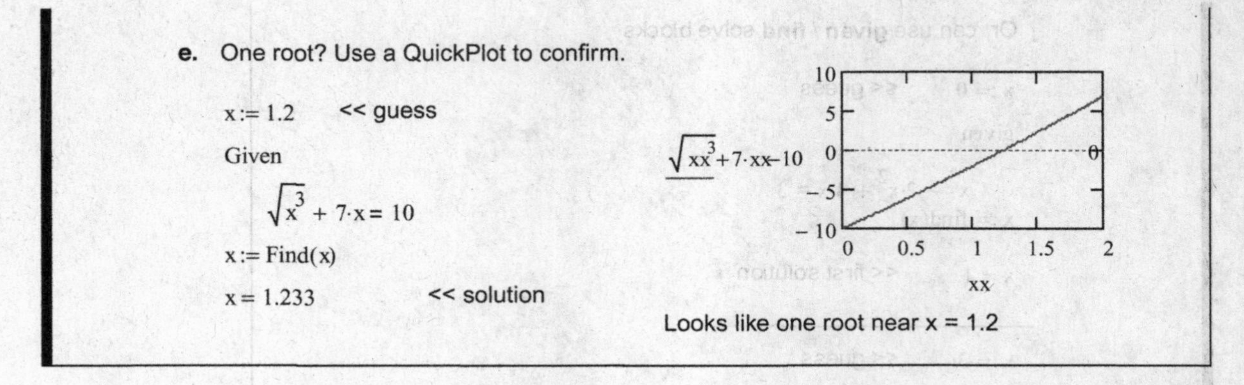

e. One root? Use a QuickPlot to confirm.

$x := 1.2$ << guess

Given

$\sqrt{x^3} + 7\cdot x = 10$

$x := \text{Find}(x)$

$x = 1.233$ << solution

Looks like one root near x = 1.2

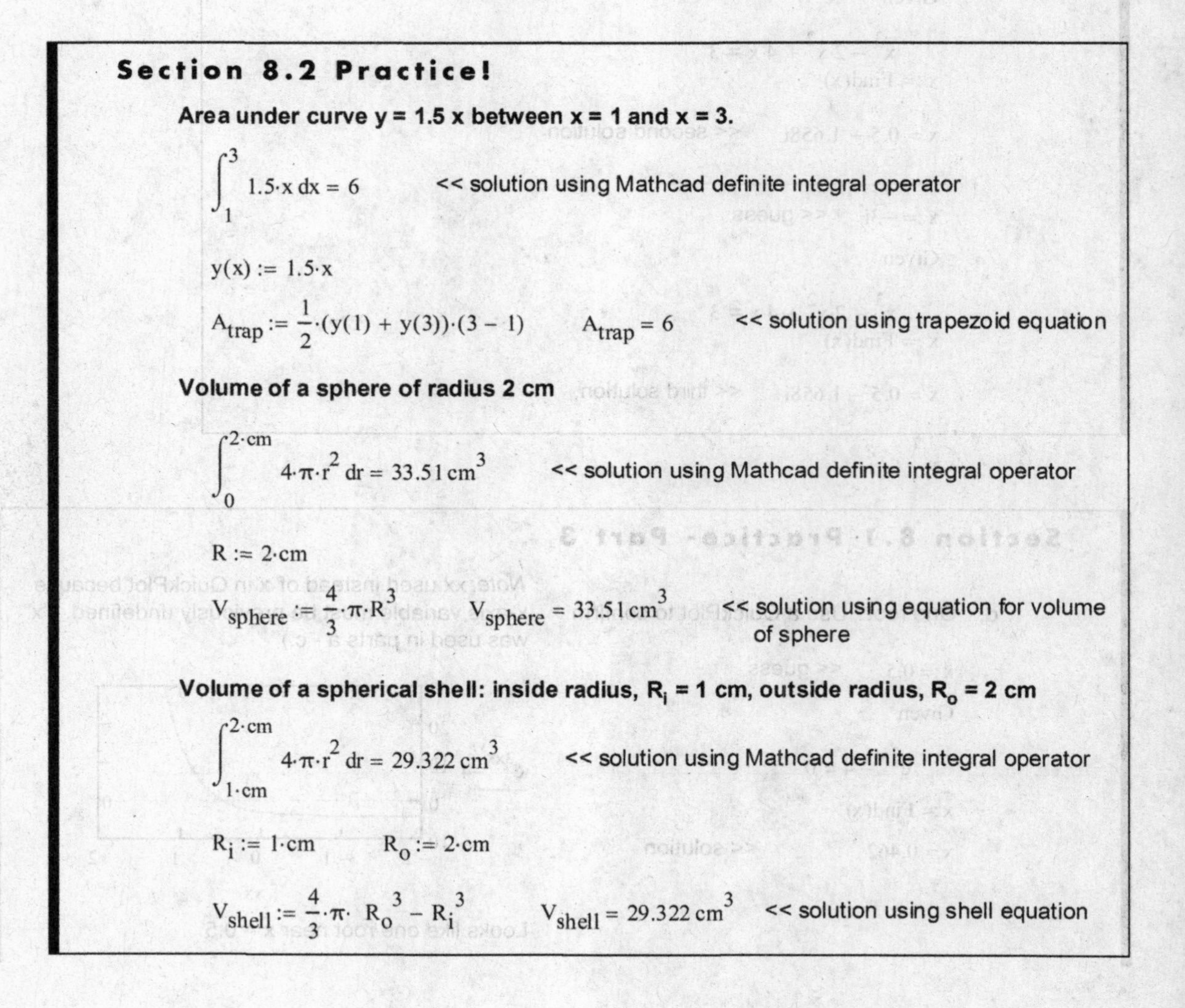

Section 8.2 Practice!

Area under curve y = 1.5 x between x = 1 and x = 3.

$\int_1^3 1.5\cdot x\,dx = 6$ << solution using Mathcad definite integral operator

$y(x) := 1.5\cdot x$

$A_{trap} := \frac{1}{2}\cdot(y(1) + y(3))\cdot(3 - 1)$ $A_{trap} = 6$ << solution using trapezoid equation

Volume of a sphere of radius 2 cm

$\int_0^{2\cdot cm} 4\cdot\pi\cdot r^2\,dr = 33.51\,cm^3$ << solution using Mathcad definite integral operator

$R := 2\cdot cm$

$V_{sphere} := \frac{4}{3}\cdot\pi\cdot R^3$ $V_{sphere} = 33.51\,cm^3$ << solution using equation for volume of sphere

Volume of a spherical shell: inside radius, R_i = 1 cm, outside radius, R_o = 2 cm

$\int_{1\cdot cm}^{2\cdot cm} 4\cdot\pi\cdot r^2\,dr = 29.322\,cm^3$ << solution using Mathcad definite integral operator

$R_i := 1\cdot cm$ $R_o := 2\cdot cm$

$V_{shell} := \frac{4}{3}\cdot\pi\cdot\left(R_o{}^3 - R_i{}^3\right)$ $V_{shell} = 29.322\,cm^3$ << solution using shell equation

Section 8.2.2 Practice!

Create a data set

$N_{pts} := 20$

$i := 0 .. (N_{pts} - 1)$

$x_i := \frac{\pi}{2} \cdot \frac{i}{N_{pts} - 1}$

$y_i := \cos(x_i)$

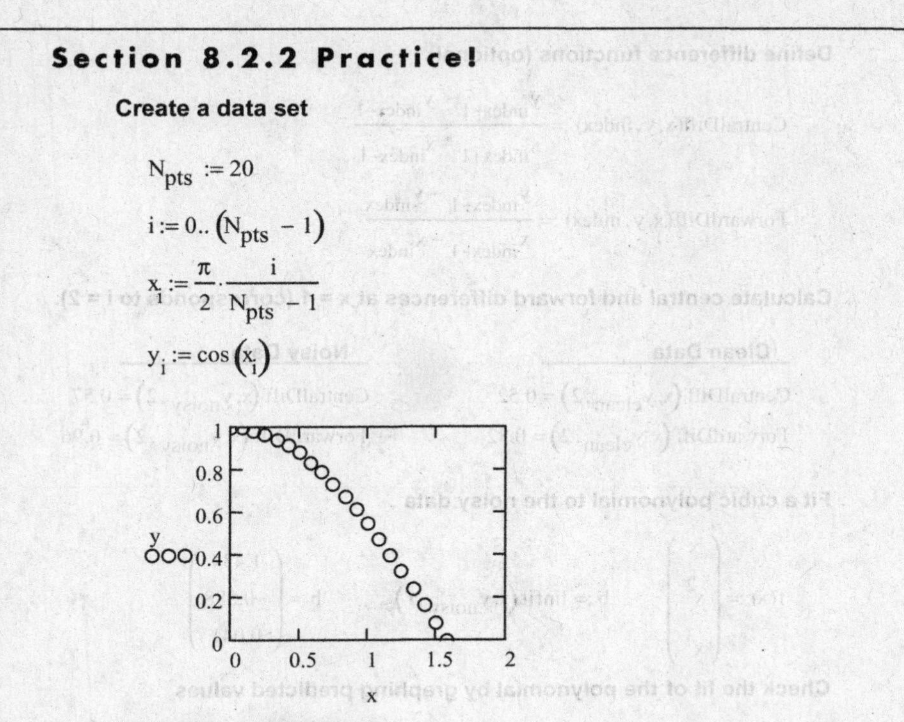

Define the trap function

$$\text{trap}(x, y) := \sum_{i=0}^{\text{length}(x)-2} \left[\frac{y_i + y_{i+1}}{2} \cdot (x_{i+1} - x_i) \right]$$

Use the trap function

$\text{Area} := \text{trap}(x, y) \qquad \text{Area} = 0.99943$

Compare with symbolic integration result

$$\int_0^{\frac{\pi}{2}} \cos(x)\, dx = 1$$

The ***trap*** *function came pretty close.*

Section 8.3.2 Practice!

$$i := \begin{pmatrix} 0 \\ 1 \\ 2 \\ 3 \\ 4 \\ 5 \\ 6 \end{pmatrix} \quad x := \begin{pmatrix} 0.0 \\ 0.5 \\ 1.0 \\ 1.5 \\ 2.0 \\ 2.5 \\ 3.0 \end{pmatrix} \quad y_{clean} := \begin{pmatrix} 0.00 \\ 0.48 \\ 0.84 \\ 1.00 \\ 0.91 \\ 0.60 \\ 0.14 \end{pmatrix} \quad y_{noisy} := \begin{pmatrix} 0.24 \\ 0.60 \\ 0.69 \\ 1.17 \\ 0.91 \\ 0.36 \\ 0.18 \end{pmatrix}$$

Define difference functions (optional)

$$\text{CentralDiff}(x, y, \text{index}) := \frac{y_{\text{index}+1} - y_{\text{index}-1}}{x_{\text{index}+1} - x_{\text{index}-1}}$$

$$\text{ForwardDiff}(x, y, \text{index}) := \frac{y_{\text{index}+1} - y_{\text{index}}}{x_{\text{index}+1} - x_{\text{index}}}$$

Calculate central and forward differences at x = 1 (corresponds to i = 2).

Clean Data	Noisy Data
$\text{CentralDiff}(x, y_{\text{clean}}, 2) = 0.52$	$\text{CentralDiff}(x, y_{\text{noisy}}, 2) = 0.57$
$\text{ForwardDiff}(x, y_{\text{clean}}, 2) = 0.32$	$\text{ForwardDiff}(x, y_{\text{noisy}}, 2) = 0.96$

Fit a cubic polynomial to the noisy data

$$f(x) := \begin{pmatrix} x \\ x^2 \\ x^3 \end{pmatrix} \qquad b := \text{linfit}(x, y_{\text{noisy}}, f) \qquad b = \begin{pmatrix} 1.434 \\ -0.588 \\ 0.041 \end{pmatrix}$$

Check the fit of the polynomial by graphing predicted values

$$y_{\text{pred}} := b_0 \cdot x + b_1 \cdot x^2 + b_2 \cdot x^3$$

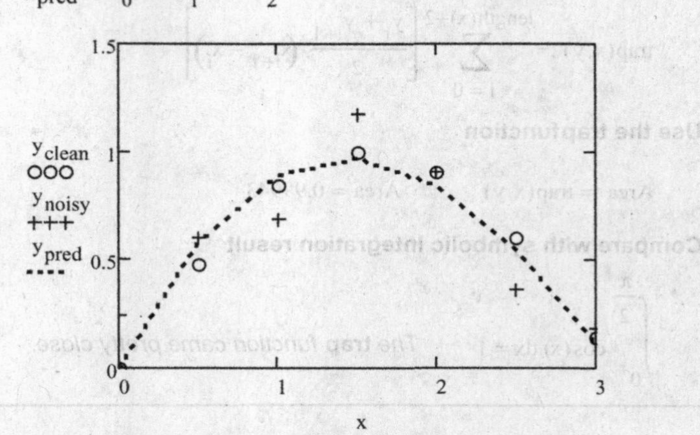

Differentiate the polynomial (derivative evaluated at x = 1)

$$x := 1$$

$$\frac{d}{dx}\left(b_0 \cdot x + b_1 \cdot x^2 + b_2 \cdot x^3\right) = 0.38$$

This is a more accurate result than using the difference formulas for this data because of the large ?x between data points.

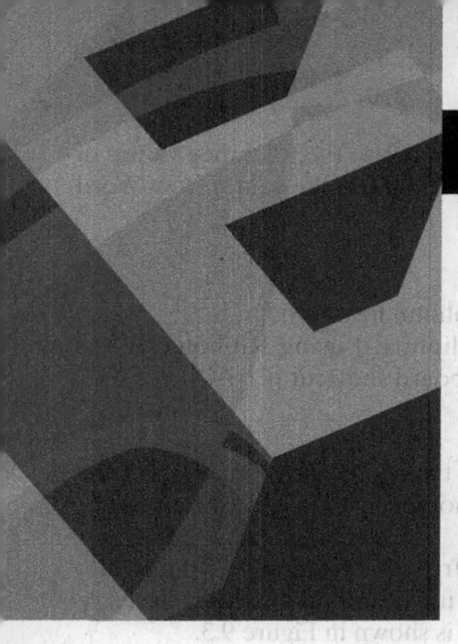

CHAPTER 9

Using Mathcad with Other Programs

Objectives

After reading this chapter you will know:

- how to use data from Word tables in Mathcad
- how to insert Mathcad results into Word documents
- eight ways to share information between Mathcad and Excel
- how to move arrays between Mathcad and Matlab

9.1 INTRODUCTION

One of the powerful features of modern software products is the ability to share information among programs, allowing you to use the best features of each program. In this chapter we will explore Mathcad's ability to work with Microsoft Word, Microsoft Excel,[1] and Matlab.[2]

Note: Mathcad 14 predates the major changes in Excel that were introduced with Microsoft Office 2007. When using Mathcad 14 you will need to save Excel files in Excel 2003 formal (.xls). Mathcad 15 supports the new Excel file format (.xlsx).

9.2 MATHCAD AND MICROSOFT WORD

Microsoft Word is not designed as a mathematical problem solver, but it is commonly used to prepare reports. You might receive a report as a Word document and want to use data from the report in a Mathcad worksheet; it is easy to get information from Word tables into Mathcad. It is even more likely that you will solve problems using Mathcad and want to get the results into a Word document in order to communicate your results to others. Getting information from Mathcad into Word is also a simple, straightforward process.

9.2.1 Using Data from Word Tables in Mathcad

You can copy values from a Word table to the ***Windows clipboard*** and from there directly to a Mathcad array. As an example, consider the Word table displayed in Figure 9.1, which contains product quality data for a particular production run.

[1]Microsoft Word and Microsoft Excel are trademarks of Microsoft Corporation, Inc.
[2]Matlab is a trademark of the MathWorks, Inc.

Sample	*Tolerance* (μm)
1	0.013
2	−0.015
3	0.018
4	0.022
5	−0.010
6	0.014
7	0.016

Figure 9.1
An MS Word table of part tolerances.

If you receive the report electronically, and want to use the tolerance values in a statistical analysis in Mathcad, you can easily ***copy*** and ***paste*** the values from Word into a Mathcad array. Here's the procedure:

In Microsoft Word,

1. Select the values to be copied (without any column heading).
2. Copy the selected values to the Windows clipboard using Ribbon options **Home tab/Clipboard Group/Copy** or the keyboard shortcut [Ctrl-C].

In Mathcad,

3. Begin the definition of a Mathcad array by typing the array name and the colon key [:]. Mathcad will display a placeholder to the right of the assignment operator, as shown in Figure 9.2.
4. Click in the placeholder, and paste the values from the Windows clipboard into the place-holder. To paste in Mathcad, either use menu options **Edit/Paste**, or use the keyboard shortcut [Ctrl-V]. The result is shown in Figure 9.3.

tols := ∎

Figure 9.2
Beginning the definition of a Mathcad array.

$$\text{tols} := \begin{pmatrix} 0.013 \\ -0.015 \\ 0.018 \\ 0.022 \\ -0.010 \\ 0.014 \\ 0.016 \end{pmatrix}$$

mean(tols) = 0.0083
Stdev(tols) = 0.0146

Figure 9.3
The tolerance values have been pasted into Mathcad as the **tols** array, then used in Mathcad functions.

9.2.2 Inserting Mathcad Information into Word

All of the calculations in a Mathcad worksheet are housed in *equation regions*, while *text regions* are used to hold text. Either type of region can be copied and pasted into Word. In order to copy and paste, you must first select each Mathcad region that you want to copy, and then copy the regions to the Windows clipboard.

Selecting Regions in Mathcad

There are two ways to select regions in Mathcad using your computer's mouse:

- ***Drag Select***—First, click outside of all regions, and then drag the mouse over the regions to be selected.
- ***Click Select***—Hold down the [Shift] key while you click on each region you wish to select.

Copying Selected Mathcad Regions to the Windows Clipboard

Copy the selected regions to the Windows clipboard using menu options **Edit/Copy** or the keyboard shortcut [Ctrl-C].

Pasting Mathcad Regions into Word

Once the desired Mathcad regions are on the Windows clipboard, you can paste them into Word, or use ***Paste Special...*** to change the way the regions are inserted into the Word document.

- **Pasting as Rich Text Regions** (equations as images)

  If you simply paste the copied Mathcad regions into Word, using either Ribbon options **Home tab/Clipboard Group/Paste** or the keyboard shortcut [Ctrl-V], the Mathcad regions are pasted into Word as individual rich text regions containing either images of the Mathcad equations or editable text from Mathcad text regions. This is the simplest way to paste Mathcad regions into Word and is often sufficient.

If you use menu options Ribbon options **Home tab/Clipboard Group/Paste Special...** you have several additional options for inserting Mathcad regions into Word:

- **Pasting as an Image**

  If you use Ribbon options **Home tab/Clipboard Group/Paste Special...** you have the option of pasting the Mathcad regions (previously copied to the Windows clipboard) as a graphic image (either Windows Metafile or Enhanced Metafile). When using this option, the regions are inserted into Word as a single image. The advantage here is that the original layout of the images on the Mathcad worksheet (indenting, for example) is preserved. This method creates an image that accurately represents the Mathcad worksheet, but it cannot be edited.

  - ***Windows Metafile***—most accurately preserves large operators and Mathcad graphs.
  - ***Enhanced Metafile***—creates a scalable (resizable) graphic image, but multiline vertical bars drop out, and Mathcad graphs are not displayed correctly.

The problems with Enhanced Metafiles with Mathcad graphics are illustrated in Figure 9.4 and Figure 9.5.

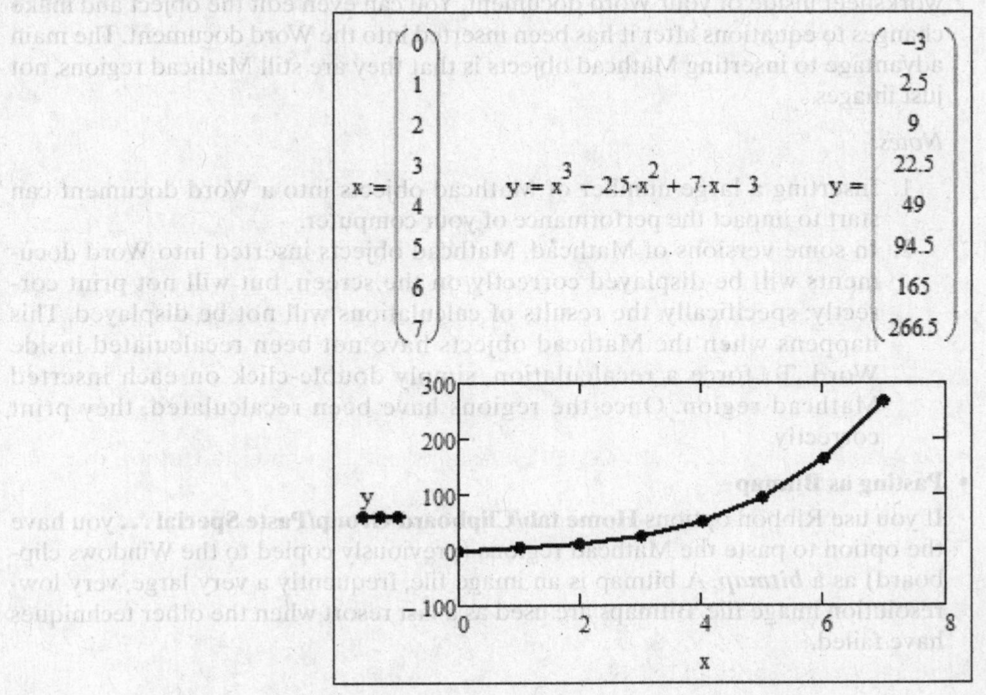

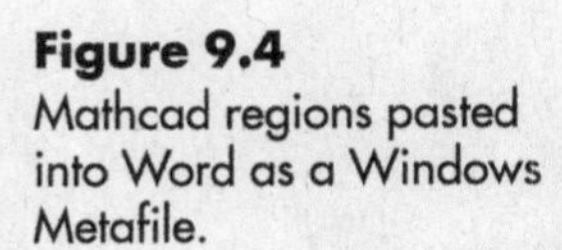
Figure 9.4
Mathcad regions pasted into Word as a Windows Metafile.

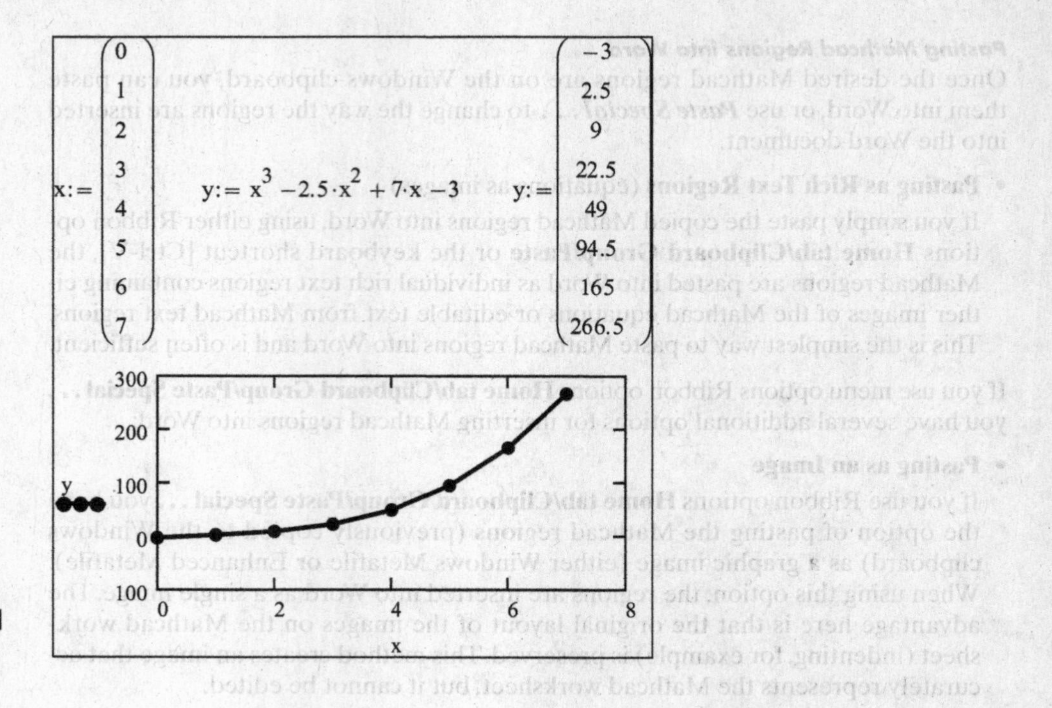

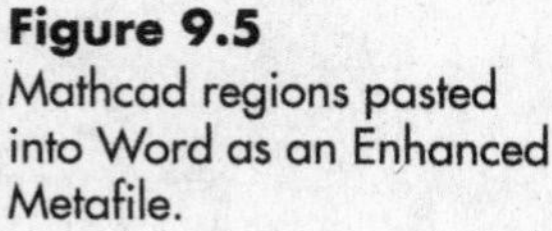

Figure 9.5
Mathcad regions pasted into Word as an Enhanced Metafile.

- **Pasting as Mathcad Objects**
  If you use Ribbon options **Home tab/Clipboard Group/Paste Special...** you have the option of pasting the Mathcad regions (previously copied to the Windows clipboard) as a **Mathcad Document Object.** The regions are inserted into Word as a single ***Mathcad object***. The Mathcad object is like imbedding a small Mathcad worksheet inside of your Word document. You can even edit the object and make changes to equations after it has been inserted into the Word document. The main advantage to inserting Mathcad objects is that they are still Mathcad regions, not just images.

  Notes:

  1. Inserting a large number of Mathcad objects into a Word document can start to impact the performance of your computer.
  2. In some versions of Mathcad, Mathcad objects inserted into Word documents will be displayed correctly on the screen, but will not print correctly; specifically, the results of calculations will not be displayed. This happens when the Mathcad objects have not been recalculated inside Word. To force a recalculation, simply double-click on each inserted Mathcad region. Once the regions have been recalculated, they print correctly.

- **Pasting as Bitmap**
  If you use Ribbon options **Home tab/Clipboard Group/Paste Special . . .** you have the option to paste the Mathcad regions (previously copied to the Windows clipboard) as a ***bitmap***. A bitmap is an image file, frequently a very large, very low-resolution image file. Bitmaps are used as a last resort when the other techniques have failed.

9.3 MATHCAD AND MICROSOFT EXCEL

Mathcad and Excel are both numeric problem solvers, and there are numerous ways they can work together. Table 9.1 summarizes the options.

Table 9.1 Sharing Information between Mathcad and Excel

Method	Info. Flow	Type	Description	Sec.
Cut and Paste	Excel → Mathcad, or Mathcad → Excel	One Time	Values in one program (Excel cell values, or Mathcad array values) can be copied and pasted into the other program.	9.3.1
Data Table	Excel → Mathcad	One Time	Excel values can be imported into Mathcad using a Mathcad Data Table.	9.3.2
File Input Component	Excel → Mathcad	One Time	Excel values can be imported into Mathcad using a Mathcad File Input component.	9.3.3
File Output Component	Mathcad → Excel	One Time	Mathcad array values can be exported into Excel using a Mathcad File Output component.	9.3.4
Mathcad Data Import Wizard	Excel → Mathcad	Reloads	Excel values can be imported into Mathcad using the Mathcad Data Import Wizard.	9.3.5
READFILE Function	Excel → Mathcad	Reloads	Excel values can be imported into Mathcad using the **READFILE** function.	9.3.6
READEXCEL Function	Excel → Mathcad	Reloads	Similar to **READFILE** function, but designed to support the Excel 2007 .xlsx file format.	9.3.6
Excel Component	Mathcad ↔ Excel	Reloads	The Excel component places a small Excel worksheet into the Mathcad worksheet. Information can be moved from Excel to Mathcad and from Mathcad to Excel.	9.3.7
Mathcad Add-In for Excel	Mathcad ↔ Excel	Reloads	The Mathcad Add-In for Excel places a small Mathcad worksheet into the Excel worksheet. Information can be moved from Excel to Mathcad and from Mathcad to Excel.	9.3.8

The differences between these methods are

- the direction that the information is moving (e.g., Mathcad to Excel, or Excel to Mathcad), and
- whether the information is transferred once or reloaded each time Mathcad recalculates the worksheet.

Each of these methods will be presented in the following sections.

Note: Before you attempt to import data from an Excel file into Mathcad, be sure to save the Excel file. This ensures that the imported data are current.

The new file format used with Excel 2007 files (.xlsx file extension) requires updated Mathcad functions for accessing Excel data. That is why Mathcad 14 only works with Excel files filed in the 2003 format (.xls file extension). Mathcad 15 supports the new Excel file format, but there are limitations. All Mathcad 15 functions listed in Table 9.1 can load all data from the first worksheet in the Excel workbook, but if you need to import data from a specific cell range, or data on a specific worksheet, you will need to choose the import method carefully. (This shortcoming will likely be addressed in future releases of Mathcad.) Table 9.2 indicates the capabilities of the various methods for accessing Excel data from Mathcad.

Table 9.2 Capabilities of Mathcad 15 Excel Data Import Methods

Excel 2003	**Entire First Sheet**	**Partial First Sheet**	**Other Sheet**
Data Table	✓	✓	✓
File Input Component	✓	✓	✓
Data Import Wizard	✓	✓	✓
READFILE function	✓	✓	×
Excel 2007	**Entire First Sheet**	**Partial First Sheet**	**Other Sheet**
Data Table	✓	✓*	✓*
File Input Component	✓	✓*	✓*
Data Import Wizard	✓	×	×
READFILE function	✓	✓	×
READEXCEL function	✓	✓	✓

*Sheet name (except first sheet) and cell range must be entered explicitly in Data Range dialog box.

9.3.1 Copying and Pasting Information between Excel and Mathcad

The simplest way to get information back and forth between Excel and Mathcad is copying the values from one program to the Windows clipboard and pasting the values from the clipboard into the other program.

The copy and paste operation is a one-time transfer of information from one program to another. It is not automatically updated; the only way to update the information is to repeat the copy and paste procedure.

9.3.1.1 Turning Excel Cell Values into a Mathcad Array The procedure for copying Excel cell values and pasting them into Mathcad as an array is as follows:

1. In Excel, select the cells containing the values to be copied. (See Figure 9.6.)
2. Use Excel Ribbon options **Home tab/Clipboard Group/Copy** or the keyboard shortcut [Ctrl-C] to copy the cell values to the Windows clipboard.
3. In Mathcad, begin declaring a new array by entering the array name and pressing the colon [:] key to enter the assignment operator. Mathcad will display a placeholder to the right of the assignment operator, as illustrated in Figure 9.7.
4. Click on the placeholder and paste the values from the clipboard. To paste, use either Mathcad menu options **Edit/Paste** or the keyboard shortcut [Ctrl-V]. The result is shown in Figure 9.8, along with the position data from Excel.

	A	B	C	D
1	Excel Datasheet			
2				
3		time (s)	position (m)	
4		0.0	2.45	
5		0.1	2.45	
6		0.2	2.46	
7		0.3	2.48	
8		0.4	2.51	
9		0.5	2.58	
10		0.6	2.67	
11		0.7	2.79	
12		0.8	2.96	
13		0.9	3.18	
14		1.0	3.45	
15				

Figure 9.6
Select the Excel cell values to be copied.

time := ▪

Figure 9.7
Before posting, begin the definition of a new array, called *time* in this example.

$$time := \begin{pmatrix} 0.0 \\ 0.1 \\ 0.2 \\ 0.3 \\ 0.4 \\ 0.5 \\ 0.6 \\ 0.7 \\ 0.8 \\ 0.9 \\ 1.0 \end{pmatrix} \quad position := \begin{pmatrix} 2.45 \\ 2.45 \\ 2.46 \\ 2.48 \\ 2.51 \\ 2.58 \\ 2.67 \\ 2.79 \\ 2.96 \\ 3.18 \\ 3.45 \end{pmatrix}$$

Figure 9.8
The result of pasting the *time* and *position* values from Excel into Mathcad.

9.3.1.2 Turning Mathcad Array Values into Excel Cell Values The procedure for copying Excel cell values and pasting them into Mathcad as an array is as follows:

1. In Mathcad, select the array of values to be copied (just the array, not the entire equation region).

   Note: Mathcad can display arrays either as a matrix or as a table. If you want to select only part of the array, the array must be displayed as a table. To change the display, double-click on the array and select the desired **Matrix Display Style** on the **Display Options** panel of the Result Format dialog box.

2. Use Mathcad menu options **Edit/Copy** or the keyboard shortcut [Ctrl-C] to copy the selected array values to the Windows clipboard.
3. In Excel, click in the cell that should contain the top-left value of the array. Be sure there are enough empty cells to receive the entire array since the pasting operation will overwrite the cell contents.
4. Paste the values from the clipboard using either Excel Ribbon options **Home tab/Clipboard Group/Paste** or the keyboard shortcut [Ctrl-V].

9.3.2 Mathcad Data Table

A ***Data Table*** in Mathcad is a spreadsheet-like interface that can be used to manually enter values into an array. Alternatively, the Data Table can be filled by pasting values copied from a spreadsheet such as Excel, or by importing values from an Excel file.

Remember: Mathcad 14 cannot read Excel 2007 files saved in the .xlsx format (but Mathcad 15 can). If you are using Mathcad 14 and Excel 2007, the Excel files will have to be saved with the .xls extension by saving them as Excel 2003 files.

Loading a Data Table is a one-time transfer of information from Excel to Mathcad. The Data Table is never automatically updated; the only way to update the information is to manually reimport the values into the table.

Inserting a Data Table into a Mathcad Worksheet

You insert a Data Table into a Mathcad worksheet by using menu options **Insert/Data/Table**, as shown in Figure 9.9.

Figure 9.9
Inserting a Data Table.

When the Data Table is inserted, there is a placeholder (above and to the right of the table as shown in Figure 9.10) for the name of the variable that will receive the data in the table.

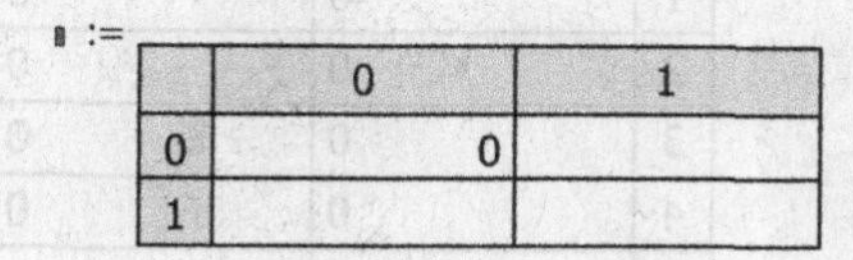

■ :=

	0	1
0	0	
1		

Figure 9.10
The inserted Data Table.

The Data Table is shown as a 2 × 2 grid by default, but you can expand the size by clicking on the Data Table to select it (this causes the ***handles*** (small black squares) on the edges to be displayed). By grabbing and dragging a handle, the table is resized. In Figure 9.11, the Data Table has been expanded (and the variable name, D, has been assigned).

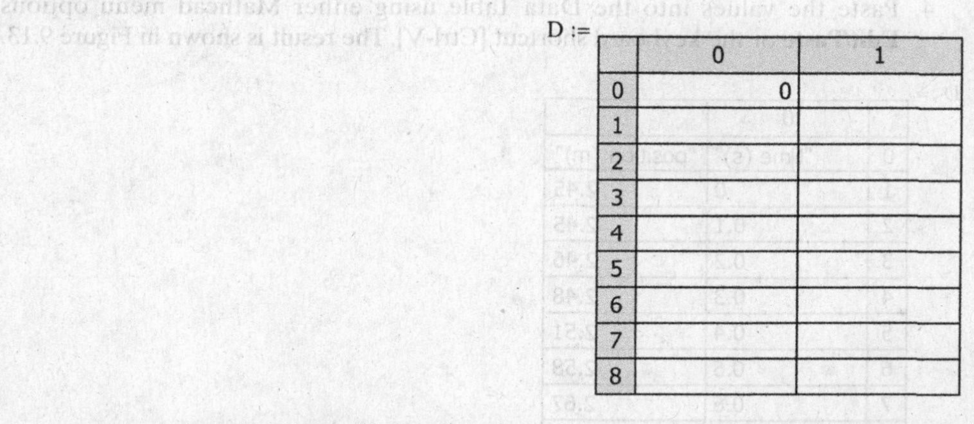

D :=

	0	1
0	0	
1		
2		
3		
4		
5		
6		
7		
8		

Figure 9.11
The expanded Data Table.

The Data Table uses cell numbers based on the value of the array *ORIGIN* variable, which is zero by default. If you want the cell numbers to start at 1,1, change the *ORIGIN* value using menu options **Tools/Worksheet Options** or use the ORIGIN := 1 statement on the worksheet.

Filling the Data Table with Values

There are three ways to fill the Data Table: manually entering values, pasting values from another source, and importing the values from another source.

- **Manually Filling the Data Table**
  To manually enter values, simply click on any cell and enter a value. Pressing [Enter] or the down arrow key after an entry causes the selected cell to move down the grid. Use the arrow keys to move in other directions around the grid.
  If you enter a value into a cell that leaves empty cells between the new value and the home cell (such as the 1 in cell 7,1 shown in Figure 9.12), all empty cells will be filled with zeroes.
- **Pasting Values into the Data Table**
  If data are already available in a spreadsheet, such as the time and position data shown in Figure 9.6, you can copy and paste the values into the Data Table, as follows:

  1. In Excel, select the cells containing the values to be copied. (See Figure 9.6.)
  2. Use Excel Ribbon options **Home tab/Clipboard Group/Copy** or the keyboard shortcut [Ctrl-C] to copy the cell values to the Windows clipboard.
  3. Click on a cell in the Mathcad Data Table to select it.

D :=

	0	1
0	0	0
1	0	0
2	0	0
3	0	0
4	0	0
5	0	0
6	0	0
7	0	1
8		

Figure 9.12
Empty cells are automatically filled with zeroes.

4. Paste the values into the Data Table using either Mathcad menu options **Edit/Paste** or the keyboard shortcut [Ctrl-V]. The result is shown in Figure 9.13.

D :=

	0	1
0	"time (s)"	"position (m)"
1	0	2.45
2	0.1	2.45
3	0.2	2.46
4	0.3	2.48
5	0.4	2.51
6	0.5	2.58
7	0.6	2.67
8	0.7	2.79
9	0.8	2.96
10	0.9	3.18
11	1	3.45

Figure 9.13
The contents of the Excel spreadsheet (from Figure 9.6) pasted into the Mathcad Data Table.

Note: In Figure 9.13, the headings from the Excel spreadsheet were copied and pasted into the Data Table along with the numeric values. This was done to demonstrate that you can use text strings in Data Tables—but you probably don't want to. The text strings in the data table will make array math operations impossible using array D.

- **Importing Values into the Data Table**

If data are already available in a spreadsheet, you can also ask Mathcad to import them directly into the Data Table. To do so, follow these steps:

1. Click on any cell in the Data Table to select it.

   Note: You can start this procedure from any cell in the Data Table, but the imported data will be placed in the top-left corner of the Data Table, starting with cell 0,0.

2. Right-click on the cell to open a pop-up menu, as shown in Figure 9.14.
3. Select **Import . . .** from the pop-up menu. (See Figure 9.14.) The File Options dialog box will open, as shown in Figure 9.15.

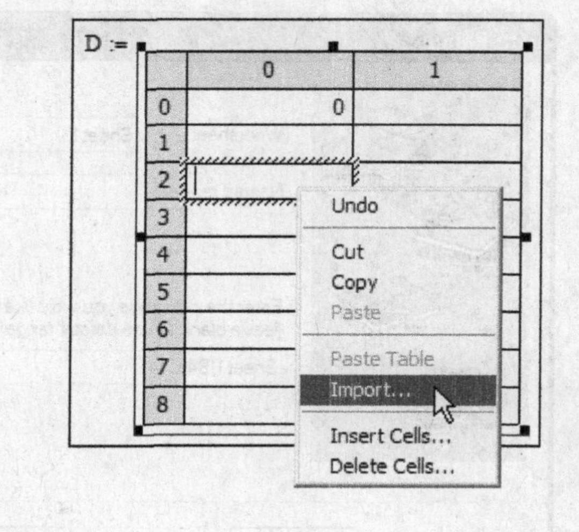

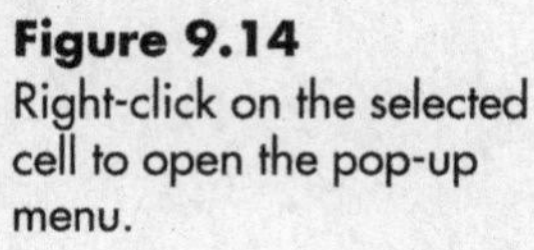

Figure 9.14
Right-click on the selected cell to open the pop-up menu.

Figure 9.15
The File Options dialog box.

4. Use the **File Format:** drop-down list to select **Microsoft Excel** as the type of file to be imported.
5. Use the **Browse...** button (shown in Figure 9.15) to locate the Excel file containing the data to be imported. After the file has been identified, click the **Next >** button to continue. The Data Range dialog box will open, as shown in Figure 9.16.

   Note: With Mathcad 15 and Excel 2007 (.xlsx) files, you cannot select a worksheet or named range using the Data Range dialog box, but you can type in a cell range (including sheet names if the cell range is not on the first worksheet).
6. In the **Worksheet:** field (see Figure 9.16), select the worksheet in the Excel worksheet that contains the data to be imported.
7. If the data set is a named cell range in the Excel workbook, select the name from the **Named range:** drop-down list; if not, enter the desired cell range (e.g., Sheet1!B4:C14) as shown in Figure 9.16. If you do not enter a cell range, Mathcad will import all of the data on the selected worksheet.

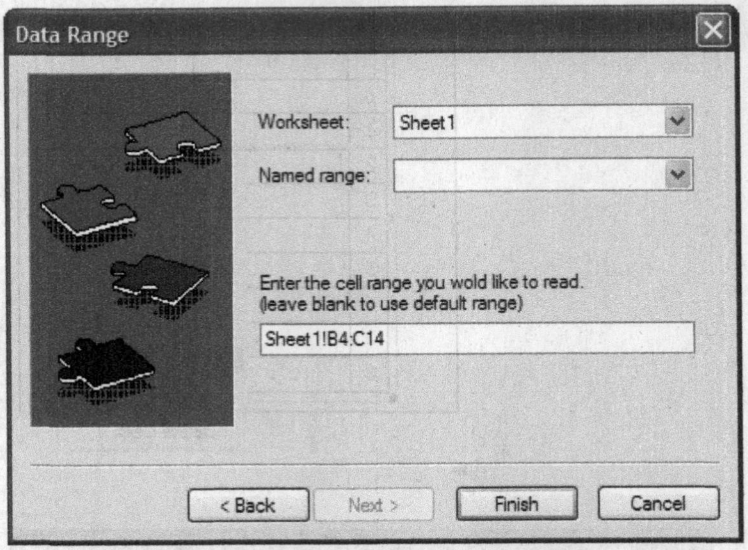

Data Range

Worksheet: Sheet1

Named range:

Enter the cell range you wold like to read.
(leave blank to use default range)

Sheet1!B4:C14

< Back | Next > | Finish | Cancel

Figure 9.16
The Data Range dialog box.

D :=

	0	1
0	0	2.45
1	0.1	2.45
2	0.2	2.46
3	0.3	2.48
4	0.4	2.51
5	0.5	2.58
6	0.6	2.67
7	0.7	2.79
8	0.8	2.96
9	0.9	3.18
10	1	3.45
11		

Figure 9.17
The data table with the imported data.

Once you have indicated the location of the data to be imported, click **Finish**. The imported data are shown in Figure 9.17.

9.3.3 Mathcad File Input Component

Using a Mathcad ***File Input component*** is very similar to importing data into a Data Table; the only difference is that manual input and pasting values are not options when using a File Input component. A File Input component will only read values from files.

Remember: Mathcad 14 cannot read Excel 2007 files saved in the .xlsx format (but Mathcad 15 can). If you are using Mathcad 14 and Excel 2007, the Excel files will have to be saved with the .xls extension by saving them as Excel 2003 files.

Loading a File Input component is a one-time transfer of information from Excel to Mathcad. The File Input component is never automatically updated.

You insert a File Input component into a Mathcad worksheet using menu options **Insert/Data/File Input . . . .** This causes the File Options dialog box to open, as shown in Figure 9.15. After locating the Excel file containing the data to be imported, click the **Next >** button to continue. The Data Range dialog box will open, as shown in Figure 9.16. Once you have identified the sheet (in Excel) and the cell range that contains the values to be imported, click **Finish**. The imported data will not be shown on the Mathcad worksheet, but an icon (picture of a floppy disk) for the File Input component will be placed on the worksheet with a placeholder. The placeholder is for the variable name that will hold the data. In Figure 9.18, the data read by the File Input component from file *C:\ExcelData.xls* has been assigned to variable *E*.

Figure 9.18
The File Input component is indicated by a floppy disk icon.

Once the data have been assigned to variable *E*, you can use array *E* just like any other array. An example of using data imported from Excel is shown in Figure 9.19.

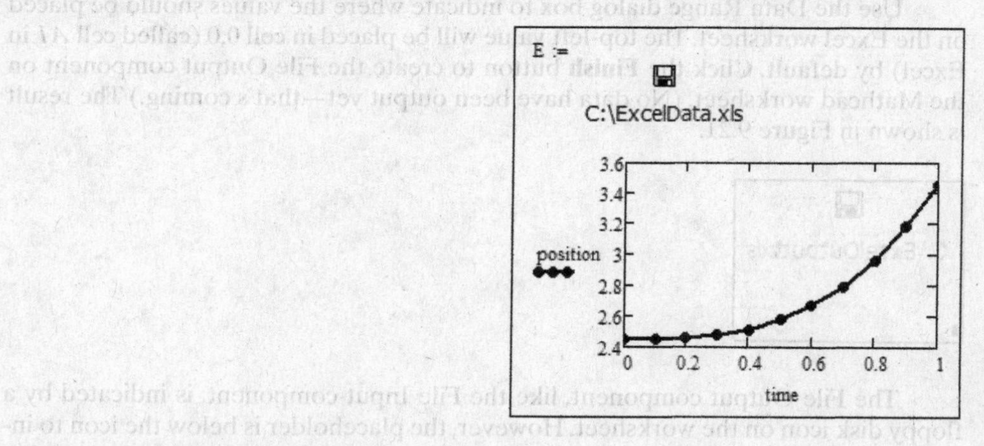

Figure 9.19
Using the imported data in array *E*.

9.3.4 Mathcad File Output Component

The Mathcad ***File Output component*** provides an easy way to send Mathcad results to an Excel file. Outputting data with a File Output component is a one-time transfer of information from Mathcad to Excel at the moment that the component is created. After the File Output component is created, it is never updated.

To insert a File Output component into a Mathcad worksheet, use menu options **Insert/Data/File Output . . . .** This causes the File Options dialog box (shown in Figure 9.15) to open.

Note: The File Output dialog box will create a new file or overwrite an existing file; it does not add data to an existing file.

Remember: Mathcad 14 cannot write to Excel 2007 files saved in the .xlsx format (but Mathcad 15 can). If you are using Mathcad 14 and Excel 2007, the Excel files will have to be saved with the .xls extension by saving them as Excel 2003 files.

Once you have entered a file name (and file path) on the File Options dialog box, click **Next >** to open the Data Range dialog box, as shown in Figure 9.20.

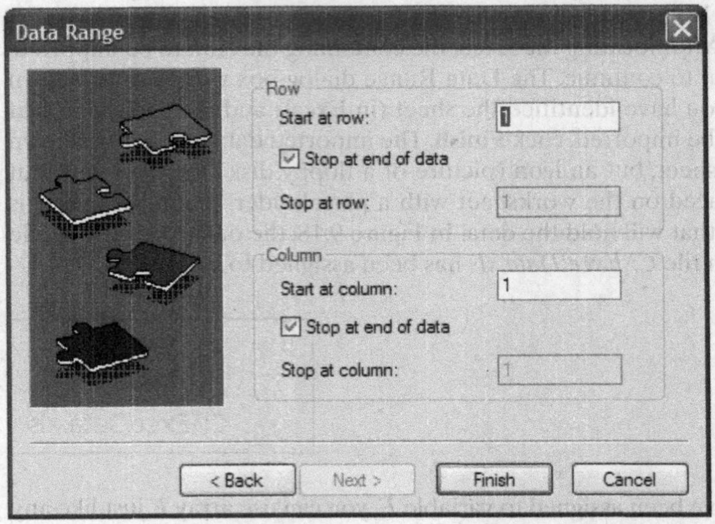

Figure 9.20
The Data Range dialog box for the File Output component.

Use the Data Range dialog box to indicate where the values should be placed on the Excel worksheet. The top-left value will be placed in cell 0,0 (called cell *A1* in Excel) by default. Click the **Finish** button to create the File Output component on the Mathcad worksheet. (No data have been output yet—that's coming.) The result is shown in Figure 9.21.

Figure 9.21
The File Output component inserted into the Mathcad worksheet.

The File Output component, like the File Input component, is indicated by a floppy disk icon on the worksheet. However, the placeholder is below the icon to indicate that the file will receive the data from the variable placed in the placeholder.

Note: By default, the border of the File Output component is not shown. The border has been included here to better indicate the connection between the data in the Excel file and the Mathcad variable that will be entered in the placeholder shown in Figure 9.21.

In Figure 9.22, the variable `position` has been entered into the placeholder (the `position` array was previously defined on the Mathcad worksheet). The moment the placeholder is filled with the `position` variable, Mathcad creates the Excel file with the values in the position vector.

The resulting Excel file is shown in Figure 9.23.

9.3.5 Using the Mathcad Data Import Wizard

Using the ***Data Import Wizard*** is similar to inserting a File Input component, but there are two significant differences between the component placed on the

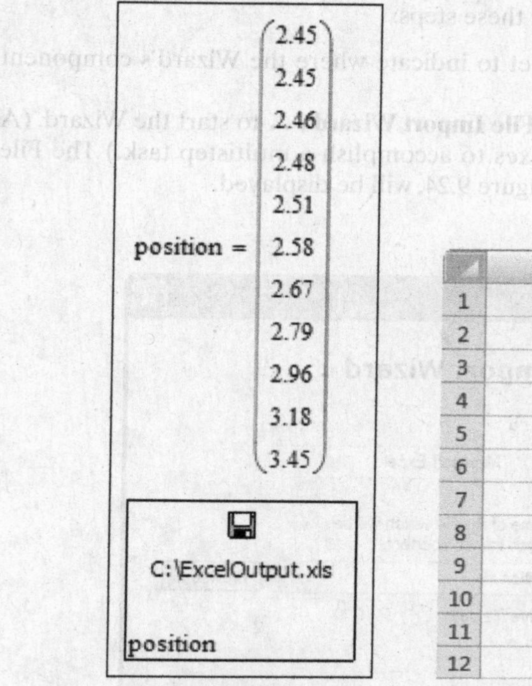

Figure 9.22
Sending the values in the *position* vector to an Excel file using a Mathcad File Output component.

	A	B	C
1	2.45		
2	2.45		
3	2.46		
4	2.48		
5	2.51		
6	2.58		
7	2.67		
8	2.79		
9	2.96		
10	3.18		
11	3.45		
12			

Figure 9.23
The Excel file created by the Mathcad File Output component.

worksheet by the Data Import Wizard and a File Input component. Table 9.2 summarizes these differences.

Table 9.2 Differences between File Import Wizard Component and File Input Component

	File Import Wizard Component	File Input Component
Setting the Excel cell range to be imported	The Excel file contents are displayed and you can select the desired import values with a mouse.	You must type in the desired cell range.
After the component has been created	If the values in the Excel file are updated, the Mathcad worksheet is updated the next time the worksheet recalculates.	Changes to the Excel worksheet are not communicated to Mathcad.

So, the primary reasons you might want to use the File Import Wizard are:

- it allows you to see the Excel file contents when setting the range of cells to be imported, and
- the component it places on the worksheet updates so that the Mathcad worksheet can always use the latest values in the Excel file.

To use the File Import Wizard, follow these steps:

1. Click on the Mathcad worksheet to indicate where the Wizard's component should be placed.
2. Use menu options **Insert/Data/File Import Wizard . . .** to start the Wizard. (A *Wizard* is a series of dialog boxes to accomplish a multistep task.) The File Options dialog box, shown in Figure 9.24, will be displayed.

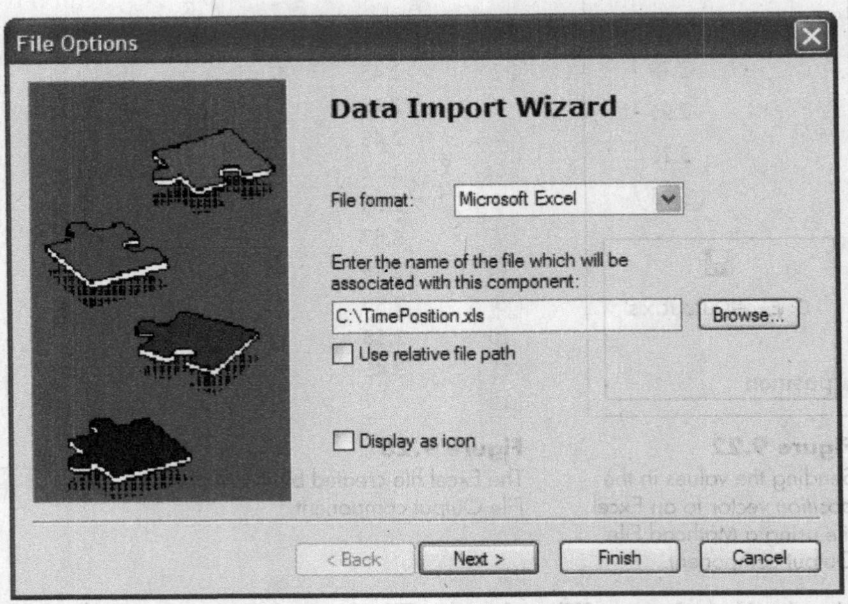

Figure 9.24
The File Options dialog box of the Data Import Wizard.

Note: If you click the Finish button on the File Options dialog box, all of the data on the first Excel worksheet will be imported.

3. Be sure that the **File format:** field indicates **Microsoft Excel**. (This is the default.)
4. Use the **Browse . . .** button to locate the Excel file containing the data to be imported. In this example, a file named *C:\TimePosition.xls* will be used.
5. Check the **Display as icon** box if you do not want the imported values displayed in a data table. (Showing the imported values in a data table is the default.)
6. Once the correct file has been identified, click **Next >** to open the Excel Options dialog box, shown in Figure 9.25.
7. Be sure that the **Worksheet:** field indicates the Excel sheet that contains the data to be imported. The selected worksheet is displayed in the **Preview** panel on the Excel Options dialog box. (See Figure 9.25.)
8. Click on the top-left cell that is to be imported, and then hold down the [Shift] key and use the arrow keys to select the cell range to be imported. The selected cell range is displayed in the **Range:** field. (You can also type into the **Range:** field.)
9. After selecting the cell range to import, click **Finish**. The imported values will be displayed in a data table as shown in Figure 9.26. A placeholder will

Excel Options

Named range: | Blank rows: skip
Worksheet: Sheet1 | Missing value: NaN
Range: Sheet1!B3:C13

Preview

	A	B	C
1	Excel Datas	NaN	NaN
2	NaN	time (s)	position (n
3	NaN	0	2.45
4	NaN	0.1	2.45
5	NaN	0.2	2.46
6	NaN	0.3	2.48

< Back | Next > | Finish | Cancel

Figure 9.25
The Excel Options dialog box of the Data Import Wizard.

■ :=

	0	1
0	0	2.45
1	0.1	2.45
2	0.2	2.46
3	0.3	2.48
4	0.4	2.51
5	0.5	2.58
6	0.6	2.67
7	0.7	2.79
8	0.8	2.96
9	0.9	3.18
10	1	3.45

Figure 9.26
The imported values, with a placeholder for the variable that will receive the values.

be included; the variable that should receive the array values is entered in the placeholder.

10. In Figure 9.27, the array of values has been assigned to variable *F*.

Once the imported values have been assigned to a Mathcad array variable, that variable can be used just like any other variable in Mathcad.

9.3.6 Using the READFILE Function

The **READFILE** function provides the same capabilities that are available with the Data Import Wizard, but without the dialog boxes. It is less user friendly, and is now mostly used as a way to read data from an Excel file into a Mathcad program. The use of the **READFILE** function is summarized here.

Remember: The **READFILE** function in Mathcad 14 cannot read from Excel 2007 files saved in the .xlsx format. If you are using Mathcad 14 and Excel 2007, the Excel

F :=

	0	1
0	0	2.45
1	0.1	2.45
2	0.2	2.46
3	0.3	2.48
4	0.4	2.51
5	0.5	2.58
6	0.6	2.67
7	0.7	2.79
8	0.8	2.96
9	0.9	3.18
10	1	3.45

$$F = \begin{pmatrix} 0 & 2.45 \\ 0.1 & 2.45 \\ 0.2 & 2.46 \\ 0.3 & 2.48 \\ 0.4 & 2.51 \\ 0.5 & 2.58 \\ 0.6 & 2.67 \\ 0.7 & 2.79 \\ 0.8 & 2.96 \\ 0.9 & 3.18 \\ 1 & 3.45 \end{pmatrix}$$

Figure 9.27 Assigning the array values to variable *F*.

files will have to be saved with the .xls extension by saving them as Excel 2003 files. Mathcad 15 provides the **READEXCEL** function to read Excel 2007 files.

The **READFILE** function is executed each time the Mathcad worksheet recalculates, so it will update the Mathcad worksheet each time the Excel file changes.

- Option 1 is to supply just the Excel path name and the file type, such as

  READFILE(*filePath*, "*Excel*")

  When only the path name and file type are supplied, Mathcad will attempt to read all contiguous data from the first sheet of the Excel workbook.
- Option 2 is used if you want Mathcad to read a cell range that does not start in cell A1. Option 2 is the four-parameter version, such as

  READFILE(*filePath*, "*Excel*", *startRow*, *startCol*)

  where *startRow* is the row number of the first row containing data to be read, and *startCol* is the column number of the first column containing the data you want Mathcad to read. Mathcad will read all continuous values starting from cell *startRow*, *startCol*.
- Option 3 is a variation on Option 2 in which you pass the **READFILE** function a two-element vector indicating the starting and stopping rows, and a two-element vector indicating the starting and stopping columns:

  READFILE(*filePath*, "*Excel*", *vRows*, *vCols*)

  As an example, the time and position values in cells B4:C14 of file *C:\TimePosition.xls* could be imported using the **READFILE** function as shown in Figure 9.28. The *ORIGIN* statement is used so that array indexing (including the counting of rows and columns in the Excel worksheet) starts at 1 rather than 0.

The **READEXCEL** function (Mathcad 15 only) provides an easy way to read Excel 2007 (.xlsx) files. The **READEXCEL** function cannot be used with Excel 2003 (.xls) files. There are two options for using the **READEXCEL** function:

- Option 1 is to supply just the Excel path name and the file type, such as

  READEXCEL(*filePath*)

ORIGIN := 1

filePath := "C:\TimePosition.xls"

$vRows := \begin{pmatrix} 4 \\ 14 \end{pmatrix}$ $vCols := \begin{pmatrix} 2 \\ 3 \end{pmatrix}$

G := READFILE(filePath, "Excel", vRows, vCols)

G =

	1	2
1	0	2.45
2	0.1	2.45
3	0.2	2.46
4	0.3	2.48
5	0.4	2.51
6	0.5	2.58
7	0.6	2.67
8	0.7	2.79
9	0.8	2.96
10	0.9	3.18
11	1	3.45

Figure 9.28
Using the READFILE function (Option 3) to import data from an Excel file.

When only the path name is supplied, Mathcad will attempt to read all contiguous data from the first sheet of the Excel workbook.

- Option 2 allows you to specify the cell range (including sheet names, if desired):

```
READFILE(filePath, "B2:B4")
```

As an example, the time and position values in cells B4:C14 on Sheet2 of file *C:\TimePos2007.xlsx* (shown in Figure 9.29) could be imported using the **READEXCEL** function, as shown in Figure 9.30.

9.3.7 Using an Excel Component

Using an ***Excel component*** is like placing a small Excel worksheet right into your Mathcad worksheet. You can send data from Mathcad into the Excel component, and you can send data from the Excel component to the Mathcad worksheet.

The typical reason for using an Excel component is to gain access to Excel's capabilities. In the example that follows, two vectors of data values will be sent from Mathcad to Excel for a statistical analysis to see if the two samples are likely to be from the same population. The result is returned from Excel to the Mathcad worksheet.

To insert an Excel component into a Mathcad worksheet, follow these steps:

1. Click on the Mathcad worksheet to indicate the location of the top-left corner of the Excel component.
2. Use menu options **Insert/Component . . .** to start the Component Wizard, as shown in Figure 9.31.
3. Select **Microsoft Excel** as the type of component to insert (see Figure 9.27), and then click **Next >** to open the Excel Setup Wizard, as seen in Figure 9.32.

TimePos2007.xlsx

	A	B	C	D	E	F	G
1	Excel 2007 Datasheet						
2							
3		time (s)	position (m)				
4		0.0	2.45				
5		0.1	2.45				
6		0.2	2.46				
7		0.3	2.48				
8		0.4	2.51				
9		0.5	2.58				
10		0.6	2.67				
11		0.7	2.79				
12		0.8	2.96				
13		0.9	3.18				
14		1.0	3.45				
15							

Sheet1 Sheet2 Sheet3

Figure 9.29
Excel 2007 data file.

Comments on the Excel Setup Wizard, page 1:

- You can insert an empty Excel worksheet into the Mathcad worksheet, or you can insert an Excel worksheet that looks like an existing Excel file. The existing Excel file (if selected) is not opened for editing through Mathcad; instead, a copy of the existing file is used as a template for the new Excel file that will be displayed on the Mathcad worksheet. An Excel component is not a way to access and modify an existing Excel file, and the existing file will not be changed.
- Any changes you make to the Excel worksheet displayed on the Mathcad worksheet are stored as part of the Mathcad file (although you can save the Excel file as a separate file after the Excel component has been created; this is described later).

data := READEXCEL("C:\TimePos2007.xlsx" , "Sheet2!B4:C14")

data =

	0	1
0	0	2.45
1	0.1	2.45
2	0.2	2.46
3	0.3	2.48
4	0.4	2.51
5	0.5	2.58
6	0.6	2.67
7	0.7	2.79
8	0.8	2.96
9	0.9	3.18
10	1	3.45

Figure 9.30
Reading data from an Excel 2007 (.xlsx) file.

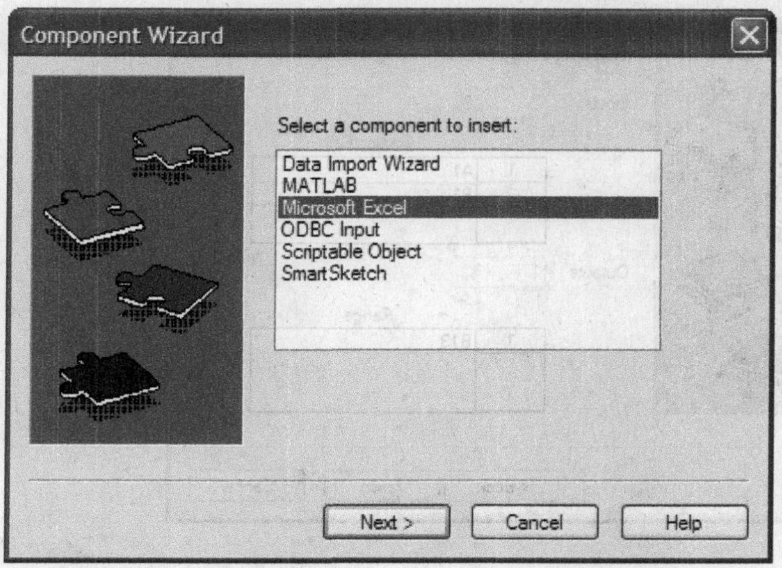

Figure 9.31
The Component Wizard, used to install an Excel component on the Mathcad worksheet.

- You can choose to display the Excel component as an icon (rather than as an Excel worksheet) by checking a box on the Excel Setup Wizard, but you will lose the ability to edit the contents of the Excel worksheet.

4. Once you have indicated whether to create an empty Excel worksheet (as in this example), or create a new Excel worksheet modeled after an existing file, click **Next >** to move to page 2 of the Excel Setup Wizard, shown in Figure 9.33.

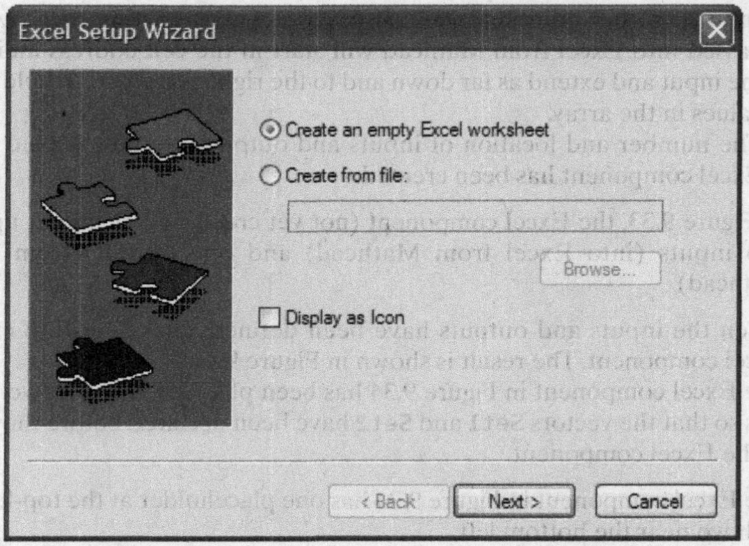

Figure 9.32
The Excel Setup Wizard, page 1.

Figure 9.33
The Excel Setup Wizard, page 2.

Comments on the Excel Setup Wizard, page 2

- The inputs and outputs described on page 2 of the Excel Setup Wizard are used to move information back and forth between Mathcad and Excel. The terms input and output are interpreted from Excel's point of view as
  - input—information into Excel from Mathcad
  - output—information from Excel into Mathcad
- You can have many inputs and/or outputs. If you want to send the contents of 12 Mathcad variables into Excel, you will need 12 inputs. If you want to send 6 cell ranges of values from Excel into 6 Mathcad arrays, you will need 6 outputs.
- Outputs are complete cell ranges, and the contents of those cell ranges are sent to Mathcad as arrays.
- Inputs are not complete cell ranges, but only "starting cells." An array passed into Excel from Mathcad will start in the cell address indicated on the input and extend as far down and to the right as needed to hold all of the values in the array.
- The number and location of inputs and outputs can be changed after the Excel component has been created.

In Figure 9.33, the Excel component (not yet created) is being set up to have two inputs (into Excel from Mathcad) and one output (from Excel to Mathcad).

5. When the inputs and outputs have been defined, click **Finish** to create the Excel component. The result is shown in Figure 9.34.
6. The Excel component in Figure 9.34 has been placed below the two data vectors so that the vectors **Set1** and **Set2** have been declared before they are sent to the Excel component.

   The Excel component in Figure 9.34 has one placeholder at the top-left corner and two near the bottom left.

   - The one placeholder at the top left is for the one output; the variable in this placeholder will receive a value from the Excel component.

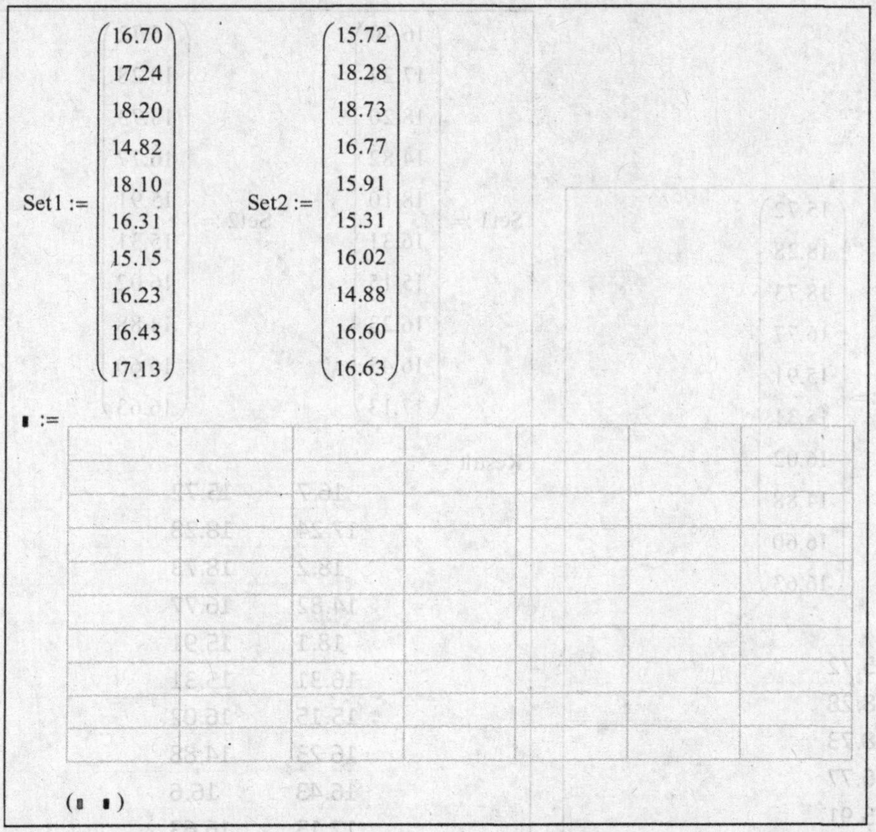

$$\text{Set1} := \begin{pmatrix} 16.70 \\ 17.24 \\ 18.20 \\ 14.82 \\ 18.10 \\ 16.31 \\ 15.15 \\ 16.23 \\ 16.43 \\ 17.13 \end{pmatrix} \qquad \text{Set2} := \begin{pmatrix} 15.72 \\ 18.28 \\ 18.73 \\ 16.77 \\ 15.91 \\ 15.31 \\ 16.02 \\ 14.88 \\ 16.60 \\ 16.63 \end{pmatrix}$$

▪ :=

(▪ ▪)

Figure 9.34
The Excel component inserted into the Mathcad worksheet.

- The placeholders at the bottom left are for the two inputs; the values of the variables in these placeholders will be sent into the Excel component.

After assigning values to some variables, the (unfinished) Mathcad worksheet is shown in Figure 9.35.

7. In Figure 9.35, the placeholders at the bottom left have been filled with variable names **Set1** and **Set2**, and the values in these vectors now appear in the Excel component in cells A1:A10 and B1:B10, respectively.

So far, the Excel component is not doing much, but by double-clicking on the grid, Excel becomes active, and you can use the data with Excel functions. In this example, we will use Excel's **TTEST** function, which has no counterpart in Mathcad. (However, Mathcad does provide the basic statistical functions that would allow you to perform the t-test without using Excel. A Mathcad t-test function will be presented later.)

The **ttest** function requires four parameters, as

```
ttest(Array1, Array2, tails, type)
```

In this example, **Array1** = *A*1:*A*10, **Array2** = *B*1:*B*10, **tails** = 2(a two-tailed test for equality), and *type* = 2 (assumed equal variance). The test types are described in Excel's help files.

The **TTEST** function returns a probability, equal to 0.7827 in this example. That value, in cell B13, is sent out of the Excel component and received by the Mathcad variable **Result**. This is illustrated in Figure 9.36.

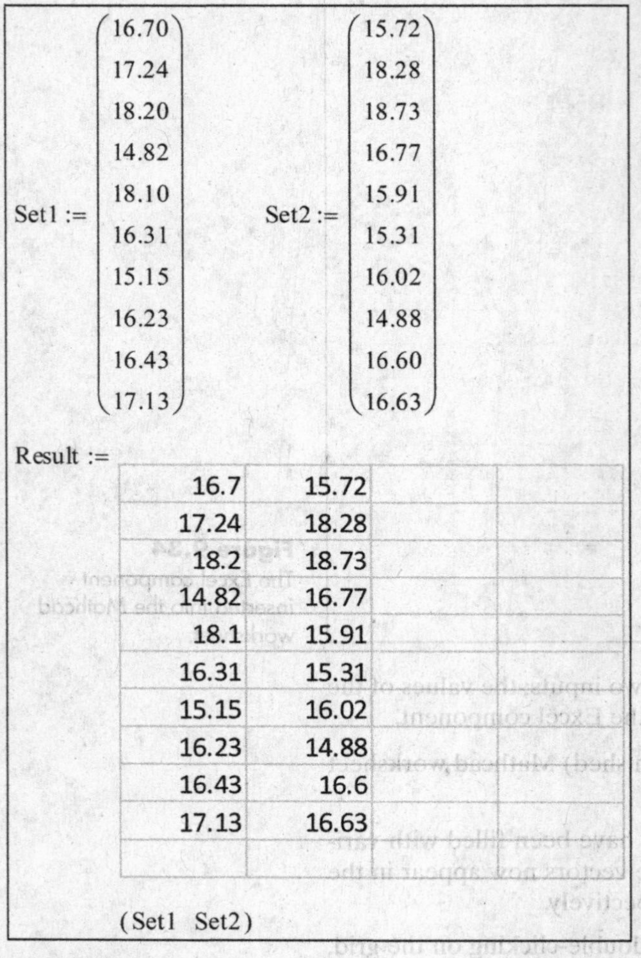

$$\text{Set1} := \begin{pmatrix} 16.70 \\ 17.24 \\ 18.20 \\ 14.82 \\ 18.10 \\ 16.31 \\ 15.15 \\ 16.23 \\ 16.43 \\ 17.13 \end{pmatrix} \qquad \text{Set2} := \begin{pmatrix} 15.72 \\ 18.28 \\ 18.73 \\ 16.77 \\ 15.91 \\ 15.31 \\ 16.02 \\ 14.88 \\ 16.60 \\ 16.63 \end{pmatrix}$$

Result :=

16.7	15.72		
17.24	18.28		
18.2	18.73		
14.82	16.77		
18.1	15.91		
16.31	15.31		
15.15	16.02		
16.23	14.88		
16.43	16.6		
17.13	16.63		

(Set1 Set2)

Figure 9.35
The partially completed Mathcad worksheet.

$$\text{Set1} := \begin{pmatrix} 16.70 \\ 17.24 \\ 18.20 \\ 14.82 \\ 18.10 \\ 16.31 \\ 15.15 \\ 16.23 \\ 16.43 \\ 17.13 \end{pmatrix} \qquad \text{Set2} := \begin{pmatrix} 15.72 \\ 18.28 \\ 18.73 \\ 16.77 \\ 15.91 \\ 15.31 \\ 16.02 \\ 14.88 \\ 16.60 \\ 16.63 \end{pmatrix}$$

Result :=

16.7	15.72		
17.24	18.28		
18.2	18.73		
14.82	16.77		
18.1	15.91		
16.31	15.31		
15.15	16.02		
16.23	14.88		
16.43	16.6		
17.13	16.63		
t-Test			
P:	0.78275		

(Set1 Set2)

Result = 0.78275

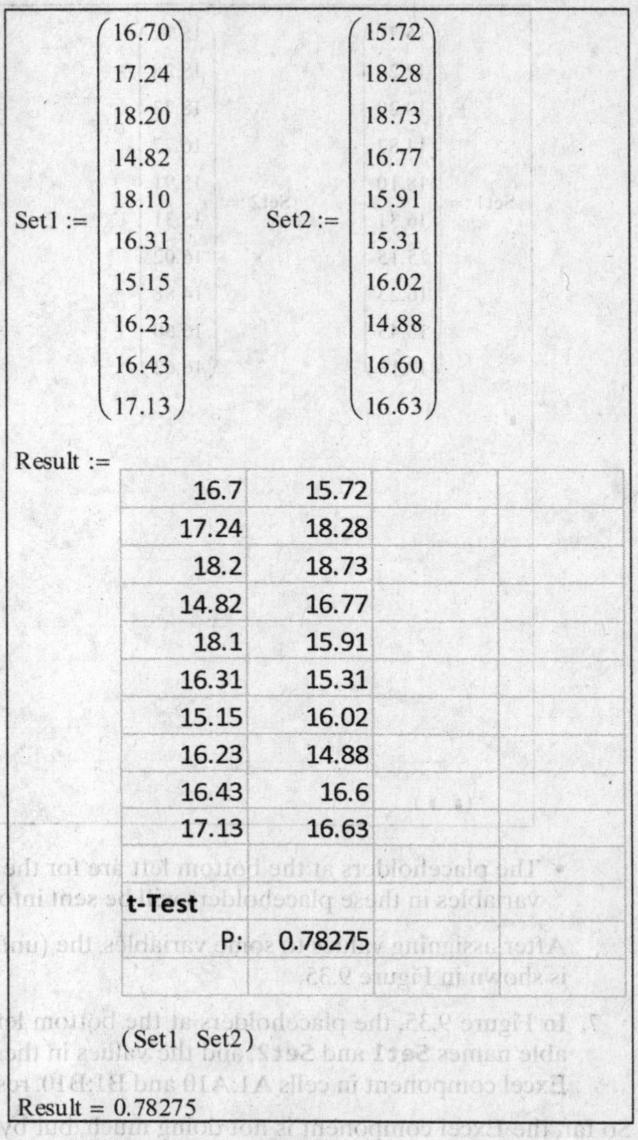

Figure 9.36
The completed Mathcad worksheet.

Note: The Excel calculations performed in an Excel component are, by default, stored with the Mathcad worksheet (not as an Excel file). However, you can right-click on the Excel component and select **Save As . . .** from the component's pop-up menu to save a separate Excel file containing the contents of the Excel component.

Aside: Mathcad provides all of the statistical tools needed to perform the t-test without using Excel, but not in a convenient form. The user-written function shown in Figure 9.37 is a Mathcad version of Excel's **TTEST** function (it assumes unequal variance in the data files).

$$
\text{ttest}(X_1, X_2, \text{tails}) := \left|
\begin{array}{l}
\text{"mimics Excel's TTEST functior} \\
avg_1 \leftarrow \text{mean}(X_1) \\
avg_2 \leftarrow \text{mean}(X_2) \\
s_1 \leftarrow \text{Stdev}(X_1) \\
s_2 \leftarrow \text{Stdev}(X_2) \\
N_1 \leftarrow \text{length}(X_1) \\
N_2 \leftarrow \text{length}(X_2) \\
DoF \leftarrow N_1 + N_2 - 2 \\
t_{stat} \leftarrow \dfrac{avg_1 - avg_2}{\sqrt{\dfrac{s_1^{\,2}}{N_1} + \dfrac{s_2^{\,2}}{N_2}}} \\
p \leftarrow \text{tails} \cdot \displaystyle\int_{t_{stat}}^{\infty} \text{dt}(t, DoF)\, dt
\end{array}
\right.
$$

$$\text{ttest}(\text{Set1}, \text{Set2}, 2) = 0.78275$$

Figure 9.37
A Mathcad function that mimics Excel's TTEST function.

Excel components are an interesting and powerful way to bring the best features of Excel into a Mathcad worksheet. There is a similar component for inserting a small Mathcad worksheet into an Excel worksheet; this is presented in the next section.

9.3.8 Using the Mathcad Add-In for Excel

The ***Mathcad Add-In for Excel*** inserts a small Mathcad worksheet directly into an Excel worksheet. It is typically used when you want to access Mathcad features from within Excel. In the example that follows, Excel data will be sent to Mathcad where it will be fit with a cubic spline. (Excel has no cubic spline fitting function.)

Before you can use the Mathcad Add-In in Excel, the Add-In must be installed and activated.

Has the Mathcad Add-In for Excel Been Activated?

From Excel, look for an **Add-Ins tab** on the Ribbon. If the **Add-Ins tab** is missing, no Add-Ins have been activated in your installation of Excel. If there is an **Add-Ins tab** on the Ribbon, look for a **Mathcad** drop-down menu in the Menu Commands group, or a **New Mathcad Object** in a Custom Toolbars group. If these appear on the Ribbon in Excel, the Mathcad Add-In for Excel has been installed and activated. If not, the Mathcad Add-In is not active. (It may or may not be installed.)

Has the Mathcad Add-In for Excel Been Installed?

In Excel 2007, Add-Ins are managed as part of the Excel Options. To access the Excel Options dialog box (shown in Figure 9.38), use the Office button (top-left corner of the Excel window), as **Office/Excel Options/Add-Ins Panel.**

Excel Options

Popular
Formulas
Proofing
Save
Advanced
Customize
Add-Ins
Trust Center
Resources

View and manage Microsoft Office add-ins.

Add-ins

Name	Location	Type
Active Application Add-ins		
Analysis ToolPak	C:\...ce\Office12\Library\Analysis\ANALYS32.XLL	Excel Add-in
Mathcad	C:\...osoft Office\Office12\Library\Mathcad.xla	Excel Add-in
National Instruments TDM Importer for MS Excel	C:\...nts\Shared\TDM Excel Add-In\ExcelTDM.dll	COM Add-in
OmniPage SE 2.0 COM Addin	C:\...anSoft\OmniPageSE2.0\OfficeAddinSE2.dll	COM Add-in
Solver	C:\... Office\OFFICE11\Library\SOLVER\solver.xla	Excel Add-in
Solver Add-in	C:\...ce\Office12\Library\SOLVER\SOLVER.XLAM	Excel Add-in

Add-in: Analysis ToolPak
Publisher: Microsoft Corporation
Location: C:\Program Files\Microsoft Office\Office12\Library\Analysis\ANALYS32.XLL

Description: Provides data analysis tools for statistical and engineering analysis

Manage: Excel Add-ins Go...

OK Cancel

Figure 9.38
The Excel Options dialog box; Add-Ins panel is used to change the currently active Excel Add-Ins.

On the **Add-Ins** panel, select **Excel Add-Ins** in the **Manage:** field, and click the **Go . . .** button. This opens the Add-Ins dialog box, shown in Figure 9.39.

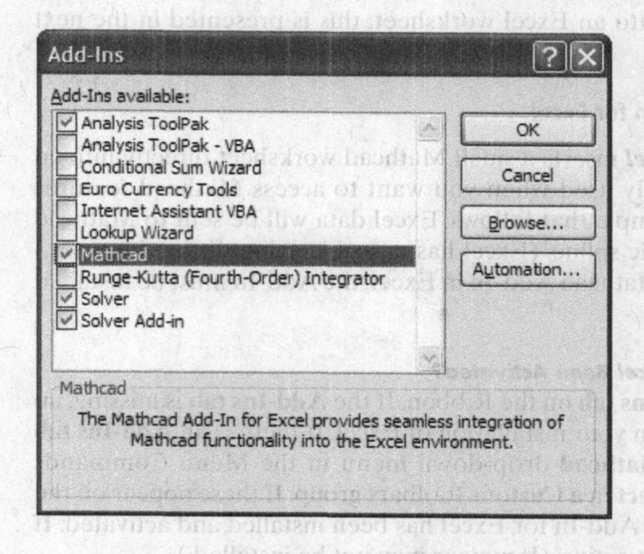

Figure 9.39
Excel Add-Ins dialog box.

If **Mathcad** appears in the list of available Add-Ins, it has been installed. Check the **Mathcad** box (if necessary) to activate the Mathcad Add-In for Excel.

If Mathcad is not in the list, you will need to install the Add-In from the original Mathcad CDs.

Note: Because new versions of Mathcad and Excel are released at different times, you may want to download the most current version of the Mathcad Add-In for Excel from the Mathcad website: http://www.ptc.com/products/mathcad/ (search for "Mathcad Add-In for Excel")

Once the Mathcad Add-In Is Installed and Active

When the Mathcad Add-In is installed and active, Mathcad options appear on Excel's Ribbon, on the **Add-Ins** tab.

To insert a Mathcad object into an Excel workbook, follow these steps:

1. Click in a cell on the Excel worksheet to indicate where the top-left corner of the Mathcad object should be placed.
2. Use Excel Ribbon options **Add-Ins tab/Mathcad/New . . .** to open the New Mathcad Object dialog box (Figure 9.40).

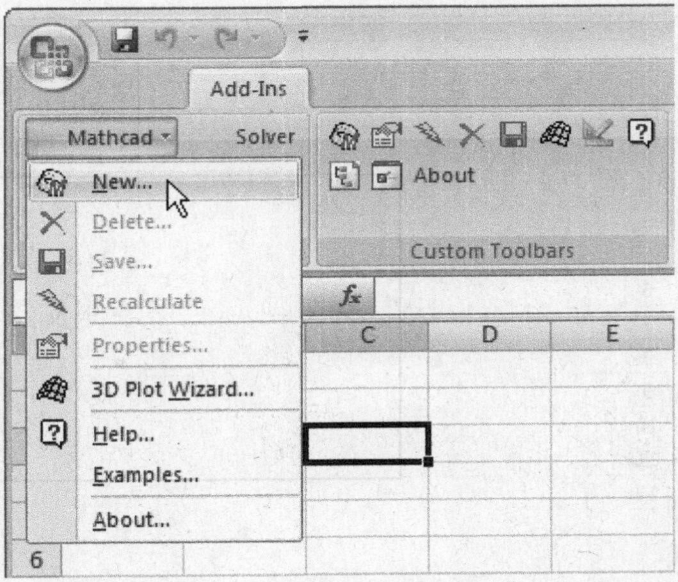

Figure 9.40
The Mathcad menu on the Ribbon's Add-In tab.

In the New Mathcad Object dialog box (shown in Figure 9.41), you have the option of creating an empty (new) Mathcad object, or creating a Mathcad object by copying an existing file. Only a copy of an existing file is used to create the Mathcad object, so no existing file will be overwritten by creating the Mathcad object in Excel.

In this example, we will create a new Mathcad object and click **OK** to insert the Mathcad object into the Excel worksheet (Figure 9.42.)

The new Mathcad object is not much to look at, just a white rectangle, but it is a functioning Mathcad worksheet right in the middle of the Excel worksheet.

3. Before using the Mathcad object, use Excel Ribbon options **Add-Ins/Mathcad/Properties . . .** to open the Set Mathcad Object Properties dialog box, shown in Figure 9.43.

Comments on Mathcad object inputs and outputs.

- The **Inputs** and **Outputs** described on the Set Mathcad Object Properties dialog box are used to move information back and forth between Mathcad

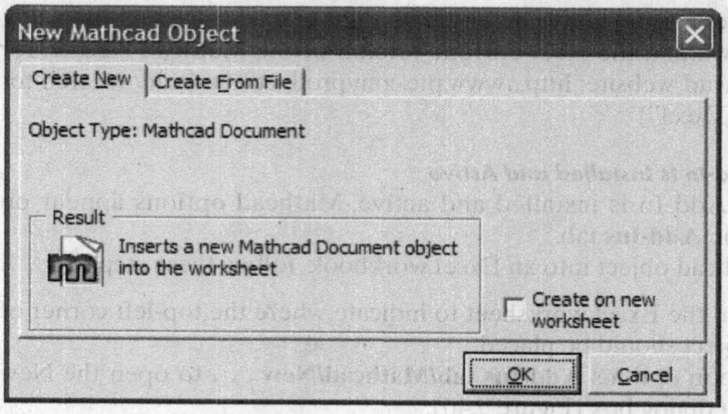

Figure 9.41
The New Mathcad Object dialog box.

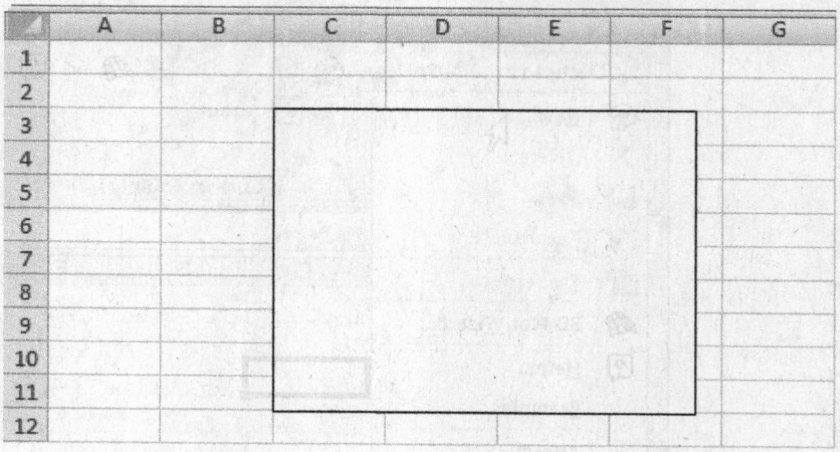

Figure 9.42
The empty Mathcad object in the Excel worksheet.

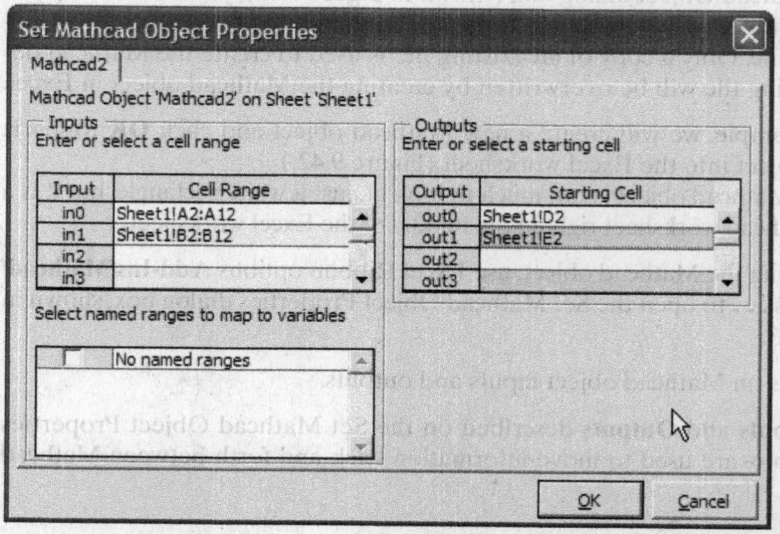

Figure 9.43
The Set Mathcad Object Properties dialog box.

and Excel. The terms input and output are interpreted from Mathcad's point of view as

- o input—information into Mathcad from Excel
- o output—information from Mathcad into Excel
- o You can have up to ten inputs and ten outputs. The names are fixed as **`in0`** to **`in9`**, and **`out0`** to **`out9`**.

- Inputs are complete cell ranges, and the contents of those cell ranges are sent into Mathcad as arrays.
- Outputs are not complete cell ranges, but only "starting cells." An array passed out of Mathcad and into Excel will start in the cell address indicated on the output and extend as far down and to the right as needed to hold all of the values in the array.
- The number and location of inputs and outputs can be changed as needed.
- In this example (Figure 9.36), the Mathcad object is being configured to have two inputs (two Excel arrays will be passed into Mathcad) and two outputs (two Mathcad arrays will be passed back into Excel).

4. Click **OK** to close the Set Mathcad Object Properties dialog box.

At this point the Mathcad object has been created in Excel. Excel data in cell range A2:A12 are available inside the Mathcad object as variable **`in0`**, and data in cell range B2:B12 are available inside the Mathcad object as variable **`in1`**. These have been renamed inside the Mathcad object to **`time`** and **`temp`**, respectively, as shown in Figure 9.44.

	A	B	C	D	E
1	Time(s)	Temp (°C)		fitTime	fitTemp
2	0	23		0.00	23.0
3	1	27		0.25	24.6
4	2	32		0.50	25.7
5	3	45		0.75	26.4
6	4	54		1.00	27.0
7	5	58		1.25	27.7
8	6	63		1.50	28.6
9	7	67		1.75	30.0
10	8	72		2.00	32.0
11	9	74		2.25	34.8
12	10	80		2.50	38.1
13				2.75	41.6
14				3.00	45.0
15				3.25	47.9
16				3.50	50.4
17				3.75	52.4
18				4.00	54.0
19				4.25	55.2
20				4.50	56.2
21				4.75	57.0
22				5.00	58.0
23				5.25	59.2
24				5.50	60.5
25				5.75	61.8
26				6.00	63.0
27				6.25	64.0
28				6.50	65.0
29				6.75	65.9
30				7.00	67.0

Mathcad object (columns G–J):

time := in0 temp := in1

time =	0
0	0
1	1
2	2
3	3
4	4
5	5
6	6
7	7
8	8
9	9
10	10

temp =	0
0	23
1	27
2	32
3	45
4	54
5	58
6	63
7	67
8	72
9	74
10	80

vs := cspline(time, temp)

i := 0.. 40

$plotTime_i := \frac{i}{4}$

plotTemp := interp(vs, time, temp, plotTime)

out0 := plotTime

out1 := plotTemp

Figure 9.44
Using the Mathcad object to fit a spline curve to Excel data.

Once the **time** and **temp** vectors have been filled with data passed in from Excel, a cubic spline is fit to the data, and new time and temperature values for plotting the spline fit are calculated. Finally, output variables **out0** and **out1** are used to return the plotting data to Excel with starting addresses *D2* and *E2*, respectively. (The headings **fitTime** and **fitTemp** were typed directly into cells *D1* and *E1* in Excel.)

Finally, we can use Excel's graphing capability to compare the original data and the spline fit. This result is shown in Figure 9.45.

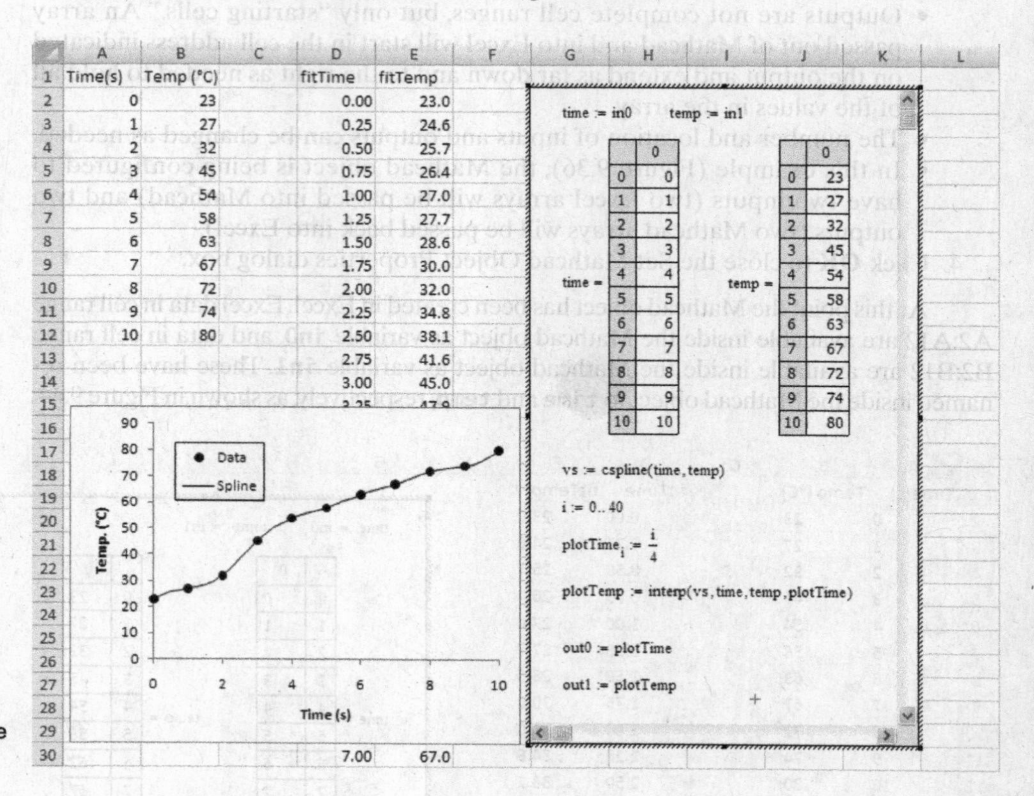

Figure 9.45
Plotting the spline fit with the original data in Excel.

9.4 MATHCAD AND MATLAB

Matlab is a commonly used problem solver in the scientific and engineering communities, and Mathcad can exchange information with Matlab. Here, we simply illustrate how to copy and paste array values between the two programs.

Copying a Mathcad Array to Create a Matlab Array

Matlab shows all defined variables in a current session in a Workspace window, and you can inspect the contents of any variable using Matlab's Array Editor. Values displayed in the Array Editor can be copied to the Windows clipboard, and values on the Windows clipboard can be pasted into Matlab's Array Editor.

To create an array in Matlab based on an existing array in Mathcad, such as array **mathX** shown in Figure 9.46, follow these steps:

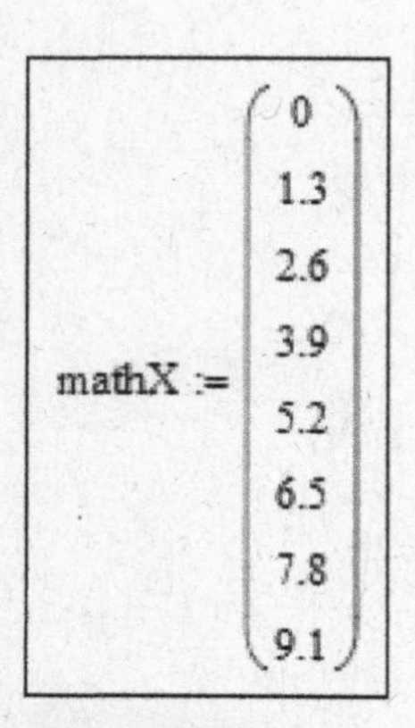

Figure 9.46
Mathcad array *mathX* will be copied and pasted into Matlab.

1. In Mathcad, select the array of values to be copied.

   Note: Mathcad can display arrays either as a matrix or as a table. If you want to select only part of the array, the array must be displayed as a table. To change the display, double-click on the array and select the desired matrix display style on the Display Options panel of the Result Format dialog box.

2. Use Mathcad menu options **Edit/Copy**, or the keyboard shortcut [Ctrl-C] to copy the selected array values to the Windows clipboard.
3. In Matlab's Array Editor, use menu options **File/New** to create a new array; it will initially be called *unnamed*. Enter a name for the new variable. In Figure 9.47, the new variable has been named **`matX`**.

Figure 9.47
New variable, *matX*, created in Matlab's Workspace window.

4. Double-click on the grid icon to the left of the variable name (**`matX`** in this example) to open Matlab's Array Editor. (See Figure 9.48.)

Figure 9.48
Matlab's Array Editor.

5. Click in cell 1,1 and then paste the contents of the Windows clipboard into Matlab's Array Editor, using Matlab menu options **Edit/Paste** or the keyboard shortcut [Ctrl-V]. The result is shown in Figure 9.49.

Array **`matX`** is now a part of the Matlab environment and can be used like any other Matlab array.

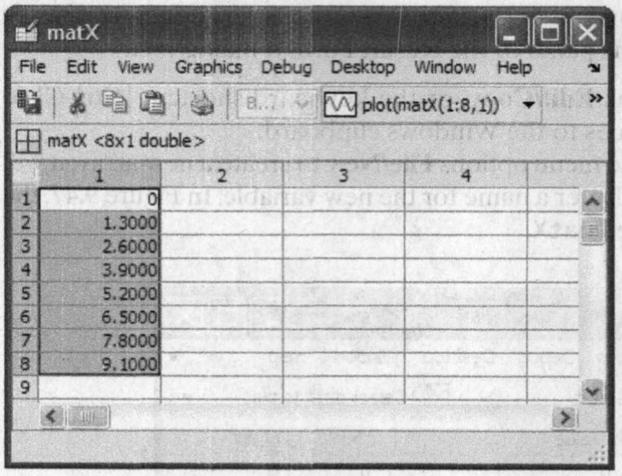

Figure 9.49
Matlab array *matX* was filled by pasting values from the Windows clipboard.

Copying a Matlab Array to Create a Mathcad Array

To create an array in Mathcad from values in a Matlab array, simply reverse the process:

1. In Matlab, locate the array containing the desired values in the workspace window. Double-click on the grid icon next to the variable name to open Matlab's Array Editor.
2. Select the values to be copied.
3. In Mathcad, begin declaring a new array by entering the array name and pressing the colon [:] key to enter the assignment operator, (:=). Mathcad will display a placeholder to the right of the assignment operator.
4. Click on the placeholder and paste the values from the clipboard. To paste, either use Mathcad menu options **Edit/Paste** or the keyboard shortcut [Ctrl-V].

SUMMARY

MATHCAD AND WORD

Using Data from Word tables in Mathcad—Copy and Paste

1. Select data in the Word table
2. Copy to the Windows clipboard (**Home tab/Clipboard group/Copy**, or [Ctrl-C])
3. Create a variable definition in Mathcad
4. Paste the clipboard data into the Mathcad placeholder (**Edit/Paste**, or [Ctrl-V]).

Inserting Mathcad regions into a Word document—Copy and Paste, or Paste Special . . .

1. Select regions in Mathcad.
2. Copy to the Windows clipboard (**Edit/Copy**, or [Ctrl-C]).
3. Choose the location to insert into Word.
4. Paste the clipboard data into Word at the current cursor location (**Home tab/Clipboard group/Paste**, or [Ctrl-V]), or

5. Use **Home tab/Clipboard group/Paste Special . . .** to get more control over how the regions are pasted.
   - paste as **rich text fields** (Mathcad equations as images)
   - paste as **Windows Metafile** (all selected regions paste as one image)
   - paste as **Enhanced Metafile** (all selected regions paste as one image; some Mathcad features may not display correctly)
   - paste as **Mathcad objects** (working Mathcad regions in the Word document)

MATHCAD AND EXCEL

Eight ways to share information between Mathcad and Excel

Method	Info. Flow	Type	Description	Sec.
Cut and Paste	Excel → Mathcad, or Mathcad → Excel	One Time	Values in one program (Excel cell values, or Mathcad array values) can be copied and pasted into the other program.	9.3.1
Data Table	Excel → Mathcad	One Time	Excel values can be imported into Mathcad using a Mathcad Data Table.	9.3.2
File Input Component	Excel → Mathcad	One Time	Excel values can be imported into Mathcad using a Mathcad File Input component.	9.3.3
File Output Component	Mathcad → Excel	One Time	Mathcad array values can be exported into Excel using a Mathcad File Output component.	9.3.4
Mathcad Data Import Wizard	Excel → Mathcad	Reloads	Excel values can be imported into Mathcad using the Mathcad Data Import Wizard.	9.3.5
READFILE Function	Excel → Mathcad	Reloads	Excel values can be imported into Mathcad using the **READFILE** function.	9.3.6
Excel Component	Mathcad ↔ Excel	Reloads	The Excel component places a small Excel worksheet into the Mathcad worksheet. Information can be moved from Excel to Mathcad and from Mathcad to Excel.	9.3.7
Mathcad Add-In for Excel	Mathcad ↔ Excel	Reloads	The Mathcad Add-In for Excel places a small Mathcad worksheet into the Excel worksheet. Information can be moved from Excel to Mathcad and from Mathcad to Excel.	9.3.8

MATHCAD AND MATLAB

Creating a Matlab array from a Mathcad array

1. In Mathcad, select the array values to be copied.
2. Copy to the Windows clipboard (**Edit/Copy**, or [Ctrl-C]).
3. In Matlab, use the Workspace window to create a new variable. Double-click on the grid icon to open Matlab's Array Editor.
4. Click on cell 1,1 in the Array Editor.
5. Paste the clipboard data into the Array Editor (**Edit/Paste**, or [Ctrl-V]).

Creating a Mathcad array from a Matlab array

1. In Matlab's Workspace window, double-click on the desired array's grid icon to open Matlab's Array Editor.
2. Select the array values to be copied.

3. Copy to the Windows clipboard (**Edit/Copy**, or [Ctrl-C]).
4. In Mathcad, define a new variable.
5. Click on the placeholder created as part of the variable definition.
6. Paste the clipboard data into the placeholder (**Edit/Paste**, or [Ctrl-V]).

KEY TERMS

bitmap	Enhanced Metafile	paste
click select	Excel component	Paste Special . . .
copy	File Input component	READEXCEL
Data Import Wizard	File Output component	READFILE
Data Table	handles	Windows clipboard
drag select	Mathcad Add-In for Excel	Windows Metafile
	Mathcad object	

PROBLEMS

9.1 Using Excel Data in Mathcad: Statistics, I

The exam scores listed in Table 9.3 are stored in an Excel file. Import the scores into a Mathcad worksheet and compute the average and median scores.

9.2 Using Excel Data in Mathcad: Statistics, II

The data shown in Table 9.4 were created in Excel with the use of a random-number generator. The data are supposed to be normally distributed with a mean of 4 and a standard deviation of 0.2. Check these values by getting the Excel data into Mathcad (copy and paste, or data file input), and then use Mathcad functions to compute the mean and standard deviation of the values in the data set. Finally, plot the data, using Mathcad's `histogram` function to see if the data appear to be normally distributed.

Table 9.3 Exam Scores

73	71	62	68	67	78
65	89	72	69	67	77
92	74	69	57	77	95
52	67	100	69	51	87
71	86	69	65	70	70
69	84	90	95	79	87
82	71	71	66	69	65
79	75	49	65	92	
57	77	99	82	84	
56	84	84	73	55	

These data are available in Excel file *Ch0901.xls* at www.chbe.montana.edu/mathcad.

Table 9.4 Normally Distributed Data

3.98	4.07	4.04	3.68
3.72	4.23	4.16	3.74
4.01	4.16	3.98	4.40
4.11	3.81	4.07	3.92
3.97	3.98	3.68	4.08
3.88	3.95	4.21	4.05
4.08	4.36	4.21	3.96
4.09	3.83	3.98	4.18
3.87	3.86	4.29	4.13
4.05	4.27	4.20	4.23
4.11	4.37	3.68	4.10
3.95	4.01	4.00	3.87
4.06	3.91	4.02	3.28
4.08	3.78	3.80	4.09
3.70	4.03	4.40	4.33

These data are available in Excel file *Ch0902.xls* at www.chbe.montana.edu/mathcad.

9.3 Using Excel Data in Mathcad: Interpolation
Calibration data for a mass flowmeter (mass flowmeters are rare, but they do exist) are shown in Table 9.5. Use the data with Mathcad's `linterp` function to determine the flow rate corresponding to a meter reading of 48.6.

Table 9.5 Calibration Data

Meter Reading	Mass Flow (kg/hr)
0.0	0.0
10.0	72.6
20.0	193.2
30.0	361.8
40.0	578.4
50.0	843.0
60.0	1156
70.0	1516
80.0	1925
90.0	2381
100.0	2886

These data are available in Excel file *Ch0903.xls* at www.chbe.montana.edu/mathcad.

9.4 Using Excel Data in Mathcad: Regression
Use Mathcad to graph the data in Table 9.5, and then use Mathcad's regression functions (e.g., the `linfit` function) to find a calibration equation that fits the data.

9.5 Using Excel Data in Mathcad: Curve Fitting
Someone spilled a cup of coffee on a lab notebook, and there is now a big coffee stain where some valuable data used to be. The missing values are shown in Table 9.6. Use Mathcad's spline fitting functions (e.g., `cspline`) to estimate the missing values.

Table 9.6 Missing Data

x	y
0.0	5.0
0.5	7.4
1.0	9.1
1.5	10.2
2.0	10.9
2.5	11.1
3.0	10.9
3.5	10.4
4.0	9.7
4.5	8.8
5.0	7.9
5.5	Missing
6.0	Missing
6.5	Missing
7.0	Missing
7.5	4.1
8.0	4.1
8.5	4.4
9.0	5.2
9.5	6.6
10.0	8.5

These data are available in Excel file *Ch0905.xls* at www.chbe.montana.edu/mathcad.

9.6 Using Mathcad Results in Microsoft Word, I
After solving Problem 9.5, write a memo in Microsoft Word describing how you went about estimating values for the missing data. Your memo should include the following items:

- A memorandum heading, such as
  Date: <Current Date>
  To: <Boss' Name, Title>
  Fr: <Your Name>
  Re: Estimating Values for the Missing Data
- A short paragraph describing the reason (the spilled coffee) it is necessary to estimate these values.
- A paragraph describing how you went about finding estimates for the missing y values.
- The Mathcad equations that you used to fit the data and estimate the missing y values.
- Your results (numerical values for y at $x = 5.5, 6.0, 6.5$, and 7.0).
- A Mathcad graph showing that the estimated values appear to fill in the gap in the data.

9.7 Using Mathcad Results in Microsoft Word, II

The goal at a breakfast cereal packaging plant is to put 1.0 kg of cereal in each box, but in practice some get a little more and some less. Every 10th box is weighed before and after filling so that the weight of the contents can be determined. Data for the last 20 weighed boxes are shown in Table 9.7.

- If the average mass is less than 0.98 kg, the packaging equipment must be immediately shut down and recalibrated.
- If the average mass is greater than 1.06 kg or the standard deviation of the sample exceeds 0.05 kg, the packaging equipment should be recalibrated after the next work shift.

Table 9.7 Cereal Box Data (mass in grams)

Before Filling	After Filling
14.1	1087.2
13.8	1014.1
14.0	1028.8
14.1	1194.8
14.0	996.2
14.0	1137.2
14.0	889.6
14.4	1084.7
14.2	877.7
14.2	1102.7
13.8	992.8
13.8	1047.0
14.3	971.6
14.1	1039.3
14.0	1085.4
13.9	1028.4
14.2	978.9
14.0	999.0
14.0	920.8
14.2	968.2

These data are available in Excel file *Ch0907.xls* at www.chbe.montana.edu/mathcad.

Use the data in Table 9.7 to determine whether the packaging equipment is still working correctly or whether recalibration is needed. If recalibration is necessary, decide whether it can wait until the end of the work shift or whether the equipment must be shut down immediately.

Write a memo in Microsoft Word documenting your results. Your memo should include the following items:

- A memorandum heading, such as
  Date: <current date>
  To: <boss's name, title>
  Fr: <your name>
  Re: Packaging Equipment Calibration Status
- A brief paragraph stating your results ("Leave the equipment running," "Shut it down immediately," or "Shut it down at the end of the work shift") and your reason (e.g., average is too high or low, or standard deviation is too large).
- The data used in your analysis.
- The Mathcad equations that you used to make your decision.

Index

F